GREEN BIORENEWABLE
BIOCOMPOSITES

From Knowledge to Industrial Applications

GREEN BIORENEWABLE BIOCOMPOSITES

BIOCOMPOSITES

From Knowledge to Industrial Applications

Edited by
Vijay Kumar Thakur, PhD, and
Michael R. Kessler, PhD

AAP | APPLE
ACADEMIC
PRESS

Apple Academic Press Inc.	Apple Academic Press Inc.
3333 Mistwell Crescent	9 Spinnaker Way
Oakville, ON L6L 0A2	Waretown, NJ 08758
Canada	USA

©2015 by Apple Academic Press, Inc.

First issued in paperback 2021

Exclusive worldwide distribution by CRC Press, a member of Taylor & Francis Group
No claim to original U.S. Government works

ISBN 13: 978-1-77463-347-2 (pbk)
ISBN 13: 978-1-77188-032-9 (hbk)

Library and Archives Canada Cataloguing in Publication

Green biorenewable biocomposites : from knowledge to industrial applications/edited by Vijay Kumar Thakur, PhD, and Michael R. Kessler, PhD.

Includes bibliographical references and index.
ISBN 978-1-77188-032-9 (bound)
1. Composite materials--Industrial applications. 2. Biopolymers--Industrial applications.
I. Thakur, Vijay Kumar, 1981-, editor
II. Kessler, Michael R., editor

TA418.9.C6 G74 2014   620.1'18   C2014-907726-2

Library of Congress Cataloging-in-Publication Data

Green biorenewable biocomposites : from knowledge to industrial applications / [editors], Vijay Kumar Thakur, PhD, and Michael R. Kessler, PhD.

pages cm
Includes bibliographical references and index.
ISBN 978-1-77188-032-9 (hardback: acid-free paper)
1. Polymeric composites. 2. Biopolymers. 3. Biodegradable plastics. I. Thakur, Vijay Kumar. II. Kessler, Michael R.

TA418.9.C6G7435 2015   620.1'180286--dc23   2014044127

Apple Academic Press also publishes its books in a variety of electronic formats. Some content that appears in print may not be available in electronic format. For information about Apple Academic Press products, visit our website at **www.appleacademicpress.com** and the CRC Press website at **www.crcpress.com**

ABOUT THE EDITORS

Vijay Kumar Thakur, PhD
Staff Scientist School of Mechanical and Materials Engineering, Washington State University, USA

Vijay Kumar Thakur, PhD, PDF, is a Staff Scientist in the School of Mechanical and Materials Engineering at Washington State University, Pullman, Washington, USA. He is an editorial board member of several international journals including *Advanced Chemistry Letters, Lignocelluloses, Drug Inventions Today International Journal of Energy Engineering* and *Journal of Textile Science and Engineering (USA)*, to name a few, and also member of scientific bodies around the world. His former appointments include Research Scientist in Temasek Laboratories at the Nanyang Technological University, Singapore; Visiting Research Fellow in the Department of Chemical and Materials Engineering at Languha University, Taiwan, and Post Doctorate in the Department of Materials Science and Engineering at Iowa State University, USA. In his academic career, he has published more than 100 research articles, patents and conference proceedings in the field of polymers and materials science. He has published 10 books and 25 books on the advanced state of the art of polymers/materials science with numerous publishers. He has extensive expertise in the synthesis of polymers (natural/synthetic), nano materials, nanocomposites, biocomposites, graft copolymers, high performance capacitors and electrochromic materials.

Michael R. Kessler, PhD, PE

Professor and Berry Family Director
School of Mechanical and Materials Engineering, Washington State University, Pullman, USA

Professor Kessler is an expert in the mechanics, processing, and characterization of polymer matrix composites and nanocomposites. His research thrusts include the development of multifunctional materials (including the development of self-healing structural composites), polymer matrix composites for extreme environments, bio-renewable polymers and composites, and the evaluation of these materials using experimental mechanics and thermal analysis. These broad-based topics span the fields of organic chemistry, applied mechanics, and processing science. He has extensive experience in processing and characterizing thermosets including those created through ring-opening metathesis polymerization (ROMP), such as poly dicyclopentadiene, and the cyclotrimerization of cyanate ester resins. In addition to his responsibilities as professor of Mechanical and Materials Engineering at Washington State University, he serves as the Director of the school. He has developed an active research group with external funding of over 10 million dollars—including funding from the National Science Foundation, ACS Petroleum Research Fund, Strategic Environmental Research and Development Program (SERDP), Department of Defense, Department of Agriculture, and NASA. His honors include the Army Research Office Young Investigator Award, the Air Force Office of Scientific Research Young Investigator Award, the NSF CAREER Award, and the Elsevier Young Composites Researcher Award from the American Society for Composites. He has published more than 110 journal papers and 3700 citations, holds 6 patents, edited a book on the characterization of composite materials, presented more than 200 talks at national and international meetings, and serves as a frequent reviewer and referee in his field.

CONTENTS

LIST OF CONTRIBUTORS

Pratheep K. Annamalai
Australian Institute for Bioengineering and Nanotechnology, The University of Queensland, Old Cooper Road, Brisbane 4072, Australia. E-mail: p.annamalai@uq.edu.au

Maurizio Avella
Institute for Chemistry and Technology of Polymer ICTP-CNR, Napoli-Italy

Ansou Malang Badji
Université Gaston Berger de Saint-Louis, Senegal

Csaba Balázsi
Institute for Materials Science and Technology, Bay Zoltán Nonprofit Ltd. for Applied Research, Fehérváristr. 130, 1116 Budapest, Hungary

Katalin Balázsi
Institute for Technical Physics and Materials Science, Research Centre for Natural Sciences, Hungarian Academy of Sciences, Konkoly Thege M. str. 29–33, 1121 Budapest, Hungary

Saurabh Chaitanya
Department of Mechanical and Industrial Engineering, I.I.T. Roorkee, Roorkee, India

Tyler Chuang
University of the Pacific, Department of Biology, USA

Dilip Depan
Center for Structural and Functional Materials, University of Louisiana at Lafayette, P.O. Box 44130, Lafayette, LA 70504-4130, USA. E-mail: ddepan@gmail.com

Adina Maria Dobos
"Petru Poni" Institute of Macromolecular Chemistry, 41A Grigore Ghica Voda Alley, Iasi 700487, Romania

Maria E. Errico
Institute for Chemistry and Technology of Polymer ICTP-CNR, Napoli-Italy

S. Fatima
Department of Mechanical Engineering, Indian Institute of Technology Kharagpur, 721302, India

Gennaro Gentile
Institute for Chemistry and Technology of Polymer ICTP-CNR, Napoli-Italy

David Grewell
Polymer Composites Research Group: Agricultural and Biosystems Engineering, College of Agriculture and Life Science, Iowa State University, Ames, IA, 50011, USA

Anita Grozdanov
Faculty of Technology and Metallurgy, Rugjer Boskovic 16, Skopje, R. Macedonia

Mamadou Gueye
Université Cheikh Anta Diop de Dakar, Senegal

Murshid Iman
Department of Chemical Sciences, Tezpur University, Assam - 784028, India

Silvia Ioan
"Petru Poni" Institute of Macromolecular Chemistry, 41A Grigore Ghica Voda Alley, Iasi 700487, Romania

Igor Jordanov
Faculty of Technology and Metallurgy, Rugjer Boskovic 16, Skopje, R. Macedonia

Albert Lin
University of the Pacific, Department of Biology, USA

Lucian A. Lucia
North Carolina State University, Departments of Forest Biomaterials, Chemistry, Campus Box 8005, 8204, Raleigh, North Carolina 27695

Tarun K. Maji
Department of Chemical Sciences, Tezpur University, Assam – 784028, India, Tel.: +91 3712 275053; Fax: +91 3712 267005. E-mail: tkm@tezu.ernet.in (T. K. Maji)

A. R. Mohanty
Department of Mechanical Engineering, Indian Institute of Technology Kharagpur, 721302, India, Mobile: +91-94340-16966; Email: amohanty@mech.iitkgp.ernet.in

Virginia Hernandez Montoya
División de Estudios de Posgrado e Investigación del Instituto Tecnológico de Aguascalientes. Ave. López Mateos # 1801 Ote. Fracc. Bona Gens, CP 20256, Aguascalientes, México

Tri-Dung Ngo
Research Officer Boucherville, Quebec, Canada, J4B 6Y4, Diène NDIAYE Université Gaston Berger de Saint-Louis, Senegal

Rogers Harry O'kuru
Bio-oils Research Unit, National Center for Agricultural Utilization Research, Agricultural Research Service, United States Department of Agriculture, Peoria, IL 61604, USA

Mihaela-Dorina Onofrei
"Petru Poni" Institute of Macromolecular Chemistry, 41A Grigore Ghica Voda Alley, Iasi 700487, Romania

Pierre Ouagne
Laboratoire PRISME University of Orleans, 8 rue Leonard de Vinci, 45072 Orleans, France, E-mail: pierre.ouagne@univ-orleans.fr

Adrian Bonilla Petriciolet
División de Estudios de Posgrado e Investigación del Instituto Tecnológico de Aguascalientes. Ave. López Mateos # 1801 Ote. Fracc. Bona Gens, CP 20256, Aguascalientes, México

Nancy B. Powell
College of Textiles, North Carolina State University, Raleigh, NC 27695, USA

Louis Reifschneider
Department of Technology, College of Applied Science and Technology, Illinois State University, Normal IL 61790, USA

J. Sahari
School of Science and Technology, Universiti Malaysia Sabah,Jalan UMS, 88400 Kota Kinabalu, Sabah, Malaysia

L. Sanyang
Materials Synthesis and Characterization Laboratory, Institute of Advanced Technology

S. M. Sapuan
Department of Mechanical and Manufacturing Engineering, Universiti Putra Malaysia, 43400 UPM Serdang, Selangor, Malaysia

Mitsuhiro Shibata
Department of Life and Environmental Sciences, Faculty of Engineering, Chiba Institute of Technology, 2-17-1, Tsudanuma, Narashino, Chiba 275-0016, Japan

Deusanilde Silva
Universidade Federal de Viçosa, Departamento do Quimica, 36570-000 Viçosa, Brazil

Teresa Cristina Fonseca Silva
North Carolina State University, Department of Forest Biomaterials, Campus Box 8005, Raleigh, North Carolina 27695-8005

Inderdeep Singh
Department of Mechanical and Industrial Engineering, I.I.T. Roorkee, Roorkee, India

Damien Soulat
GEMTEX, ENSAIT Roubaix 2 allée Louise et Victor Champier, 59056 Roubaix, France

Gowrishanker Srinivasan
Polymer Composites Research Group: Agricultural and Biosystems Engineering, College of Agriculture and Life Science, Iowa State University, Ames, IA, 50011 USA

Minh-Tan Ton-That
 Senior Research Officer National Research Council Canada, 75 De Mortagne Blvd, Canada

Coumba Thiandoume
Université Cheikh Anta Diop de Dakar, Senegal

Adams Tidjani
Université Cheikh Anta Diop de Dakar, Senegal

Brent Tisserat
Functional Foods Research Unit, National Center for Agricultural Utilization Research, Agricultural Research Service, United States Department of Agriculture, 1815 N. University St., Peoria IL 61604, USA; E-mail: Brent.Tisserat@ars.usda.gov

Norma-Aurea Rangel-Vazquez
División de Estudios de Posgrado e Investigación del Instituto Tecnológico de Aguascalientes. Ave. López Mateos # 1801 Ote. Fracc. Bona Gens, CP 20256, Aguascalientes, México

Craig Vierra
University of the Pacific, Department of Biology, USA

Nazire Deniz Yılmaz
Department of Textile Engineering, Pamukkale University, Denizli 20020, Turkey

LIST OF ABBREVIATIONS

AAT	Accelerated Aging Temperature
AATD	Accelerated Aging Time Duration
AES	Auger Electron Spectroscopy
ATR	Attenuated Total Reflectance
BMC	Bone Mineral Content
BMD	Bone Mineral Density
CA	Cellulose Acetate
CAGR	Compound Annual Growth Rate
CAN	Acrylonitrile
CAP	Cellulose Acetate Phthalate
CD	Cardanol
CFL	Critical Fiber Length
CH	Chitosan
CNC	Cellulose Nanocrystals
D	Diameter
DDGS	Distillers Dried Grains With Soluble
DMA	Dynamic Mechanical Analysis
DS	Substitution Degree
DSC	Differential Scanning Calorimeter
DTA	Differential Thermal Analyzer
EDM	Electric Discharge Machining
EM	Electron Microscopy
ERGS	Exploratory Research Grant Scheme
ESO	Epoxidized Soybean Oil
FM	Flexural Modulus
GA	Gly-Ala Couplets
GPE	Polyglycidyl Ether
GTMAC	Glycidyltrimethylammonium Chloride
HAP	Hydroxyapatite
HIPS	High Impact Polystyrene
HMW	High Molecular Weight
LC	Lignocelluloses
LMW	Low-Molecular-Weight
LOI	Limiting Oxygen Index
MA	Major Ampullate
MAPE	Maleated Polyethylene

MCC	Microcrystalline Cellulose
MFC	Microfibrillated Cellulose
MI	Minor Ampullate
MMT	Montmorillonite
NC	Noise Criterion
NCC	Nanocrystalline Cellulose
NF	Natural Fibers
NFC	Nanofibrillated Cellulose
NFCs	Natural Fiber Composites
NVH	Noise, Vibration and Harshness
NVNVN	Asn-Val-Asn-Val-Asn
PA	Polyacetal
PCL	Polycaprolactone
PE	Polyethylene
PEA	Polyether Amine
PET	Polyethylene Terepthalate
PF	Plant Fiber
PG	Pyrogallol
PGPE	Polyglycerol Polyglycidyl Ether
PHB	Polyhydroxybutyrate
PMCs	Polymer Matrix Composites
PMMA	Polymethylmethacrylate
PS	Polystyrene
PSIL	Preferred Speech Interference Level
PTFE	Polytetrafluoroethylene
PU	Polyurethane
PVC	Polyvinyl Chloride
PW	Paulownia Wood
QC	Quercetin
QPGSG	Gln-Pro-Gly-Ser-Gly
RH	Rice Husk
ROP	Ring Opening Polymerization
SAED	Selected Area Electron Diffraction
SEM	Scanning Electron Microscopy
SF	Soy Flour
SPC	Soy Protein Concentrate
SPE	Sorbitol Polyglycidyl Ether
SPF	Sugar Palm Fiber
SPI	Soy Protein Isolate
SPS	Sugar Palm Starch
STC	Sound Transmission Class
TA	Tannic Acid

TGA	Thermogravimetric Analysis
TIC	Titanium Carbide
TL	Transmission Loss
TMC	Tissue Mineral Content
TMD	Tissue Mineral Density
TO	Tung Oil
TPS	Thermoplastic Starch
TS	Tensile Strength
UPE	Unsaturated Polyester
UTM	Universal Testing Machine
VN	Vanillin
WAXD	Wide Angle X-Ray Diffraction
WF	Wood Flour
WPC	Wood Plastic Composite

LIST OF SYMBOLS

A	area under the light absorption
a	fiber radius
A_i	normal surface area
b	width of the specimen bars
c	speed of sound
c*	critical concentration
c_0	sound propagation velocity in air
d	thickness of the specimen bars
E	modulus of elasticity of the material
E'	temperature dependency
f	frequency
f_c	coincident frequency
G"	function of frequency
h	thickness of the panel
I_i	sound intensity
K	consistency index
k	wave number
L	length of fiber
l	thickness of the material
m	slope of the tangent
m	surface density
m_o	oven-dried weight
m_t	weight after soak time
n	flow behavior
P	maximum applied load
p	pressure
R	flow resistance
r	resistance
r_0	flow resistivity
$r_{0,g}$	flow resistivity values of glassfiber
$r_{0,w}$	flow resistivity values of wool
S	area
T	total area of the curve
t_0	flow time
T_1	initial thickness

T_2	final thickness
T_f	thickness at end of soak time
T_o	initial thickness of the specimen
TS	thickness swelling
u	average volume flow rate
u	volumetric velocity
V	velocity
W_1	original weight
W_2	final weight
z	acoustic impedance
ρ'	density of the jute fiber
ρ	density of the jute sample
$[\eta]$	intrinsic viscosities
ν	poison's ratio
$\Re$	reflection coefficient
ΔP	static pressure difference
$\wedge'$	thermal characteristic length
$2\pi r l$	total perimeter of fiber
DH_{exp}	experimental heat of fusion
μ	dynamic viscosity of air
μCT	micro-computerized tomograms
ρ	density of the fluid
ρ_0	density of air
ρc	impedance of the fluid
χ	reactance
Ψ	ratio of fiber volume
ω	angular frequency
r_f	density of the fiber
r_w	density of fibrous material

PREFACE

During the last few decades, the environmental problems primarily caused by the excessive use of conventional petrochemical based materials such as plastics have become one of the major public concerns all around the globe. Many countries are making enormous efforts to overcome these problems by making and adopting various policies as well as management programs. With the growing concerns towards the environment, industries and researchers are also looking forward to the use of environmentally friendly materials for product development in a number of applications. Among various environmentally friendly materials, renewable resources-derived polymers (e.g., bio-based polymers) and composites (e.g., natural fibers reinforced composites) are attracting a great deal of attention because of the inherent advantages of these polymers such as conservation of limited petroleum resources, biodegradability, low toxicity, easy availability, economy and the control of carbon dioxide emissions that lead to global warming.

From the view of sustainable development, the new materials associated with biorenewable sources are enormously being explored. Indeed the concerns over new materials from renewable resources, especially in the automotive and biomedical industries, have recently increased because of the economic consequences of depleting petroleum resources, the demands from industrialists and customers for high performance low-cost materials and environmental regulations. From biorenewable natural resources, ecofriendly materials can be obtained as native biopolymers, raw materials for monomers and bio-engineered biopolymers to name a few. Biopolymer based materials such as cellulose, starch, chitosan/chitin, poly (lactic acid), and poly (hydroxyalkanoate), etc. are among the most abundantly available biopolymers on the planet Earth. They are replacing the materials for many industrial applications where synthetic polymers have been materials of choice, traditionally. One of the important aspects of biobased materials is that they can be designed and tailored to meet different desires. As the native biopolymers are not conventionally processable, research efforts have been focused on the processing and meeting the requirements of particular applications. The biobased materials are most frequently being used in the form of biocomposites. These biocomposites materials contain at least one component form the biorenewable resources that may either be the polymer matrix/ reinforcement or may contain both.

Using the recently developed techniques and technologies, biocomposites with better mechanical properties and thermal stability can be efficiently developed depending upon the applications. In fact the use of biocomposites may provide us a healthier environment owing to their multifaceted advantages over conventional polymers.

Keeping in mind the advantages of bio-based materials, this book focuses on the potential efficacy of different biocomposites procured from diverse natural resources and preparation and processing of the biocomposites to be used for a variety of applications. The book consists of 17 chapters, and each chapter gives an overview of a particular biocomposite material, its processing, and successful utilization for selected applications. The chapters summarize recently developed research concerning spider silk biocomposites; biogenic hydroxyapatite based implant biocomposites; liquid crystals and cellulose derivatives biocomposites; bio-based epoxy resins, bio-based polyphenols and lignocellulosic fibers; wood based biocomposites; flame retardant biocomposites; biocomposites for industrial noise control, cellulose based bionanocomposites, etc. Each individual chapter also focuses on the knowledge and understanding of the interfaces manifested in these biocomposites systems and optimization of different parameters for novel properties. In addition to this, the book also summarizes the recent developments made in the area of injection molding of biocomposites; chemical functionalization of natural fibers, processing of biocomposites and their applications in automotive and biomedical. A number of critical issues and suggestions for future work are discussed in a number of chapters, underscoring the roles of researchers for the efficient development of biocomposite materials through value addition to enhance their use.

As the editors of *Green Biorenewable Biocomposites: From Knowledge to Industrial Applications*, we have attempted to compile, unify, and present the emerging research trends in biopolymers based biocomposites. We hope that this book will contribute to the advancement of both science and technology in this exciting area.

—**Vijay Kumar Thakur, PhD**
Washington State University – U.S.A.

—**Michael R. Kessler, PhD, P.E.**
Washington State University – U.S.A.

CHAPTER 1

SPIDER SILK BIOCOMPOSITES: FROM RECOMBINANT PROTEIN TO FIBERS

TYLER CHUANG, ALBERT LIN, and CRAIG VIERRA

ABSTRACT

Spider silk has extraordinary mechanical properties, containing a unique balance of high-tensile strength and extensibility. Spider silks outperform several well-known manmade materials including high-tensile steel, Kevlar (body armor), and nylon. Modern spiders spin at least six to seven different fiber types that have distinct mechanical properties. These fibers have been shown to be biocompatible, heat stable and environmentally green materials. Because of their unique mechanical features, scientists are pursuing spider silks as next generation biomaterials that can be used for a broad range of applications. Given the cannibalistic nature of spiders, combined with their venomous nature, farming spiders for silk becomes a dangerous and impractical method for obtaining large amounts of silk for industrial applications. This has prompted scientists to develop DNA methodologies and heterologous protein expression systems that produce vast quantities of recombinant silk proteins. Some potential applications of silks spun from recombinant proteins include body armor, ropes and cords, sutures, tissue scaffolds, tires, and shoes. Here we cover the following topics: the diverse protein nature of spider silks and the compositions of threads, the method and steps for synthetic silk production, and the translation of fiber production from a small scale to a large-scale spinning platform. By drawing upon advances in spinning methodologies, we will describe the development of spider silk fiber composites using full-length silk proteins retrieved from the major ampullate silk-producing gland blended with other recombinant proteins expressed and purified from microorganisms.

1.1 DIVERSITY OF SPIDER SILK

1.1.1 DIFFERENT SILK-PRODUCING GLANDS

There are at least seven different fiber types that can be spun from black widow spiders. Specialized abdominal silk-producing glands display distinct morphological features and are responsible for manufacturing the different silk types.[1] Each gland has evolved the ability to express different fibroins that can be spun into distinct fibers that perform specific biological tasks, including threads used for locomotion, prey wrapping, web construction, and protection of eggs (females), as well as reproduction and adhesion of threads (Table 1.1).[1] These glands, which are normally found in pairs in a single individual, include the major and minor ampullate, tubuliform, flagelliform, aggregate, aciniform and pyriform. Because most laboratories across the world are focusing on elucidating the properties and characteristics of dragline silk, this review, in particular will emphasize the latest information on this fiber type and current progress made with synthetic silk fiber production using recombinant dragline silk proteins.

TABLE 1.1   Different Spider Silk-Producing Glands, their Function and Fibroins Spun into the Fibers

The seven different silk types, uses and proteins		
Silk gland	Use	Proteins
Major ampullate	Frame, web radii, dragline	MaSp1, MaSp2
Minor ampullate	Dragline reinforcement	MiSp1
Tubuliform	Egg sac	TuSp1
Pyriform	Attachment disc	PySp1
Flagelliform	Capture spiral; coatings*	Flag, SCP-1, SCP-2
Aciniform	Wrapping	AcSp1
Aggregate	Sticky glues; connection joints	AgSF1, AgSF2

1.1.2   MAJOR AMPULLATE GLAND

The major ampullate (MA) gland, which is named after its morphological features, represents the most extensively studied silk-producing gland in the spider silk community (Fig. 1.1). From an anatomical perspective, it has been divided into three regions: the tail region, the ampulla and the spinning duct. The tail region is responsible for the synthesis of large quantities of silk protein, the ampulla for silk protein storage, and the spinning duct functions to facilitate the conversion of the liquid dope into a solid. Recent studies have pinpointed three different types of epithelial cells that span the tail and ampulla region; these regions are denoted zones A, B and C.[2] The MA gland produces silk for the web frame and radii and serves as a lifeline for spiders to move and escape predators (Table 1.1). MA silk evolved long before its use in orb-webs, being used as far back as 350 mya.[3] One primary function of dragline silk includes locomotion; however, it is becoming apparent that MA silk performs a variety of other tasks. In particular, biochemical studies have shown that cob-weavers use MA silk to form web scaffolding and gumfooted lines, two important fiber types that facilitate prey capture.[4] In some instances, it has been reported to be present in prey wrapping silk.[5] There are essentially three reasons that MA silk has been the most extensively studied: 1) The MA gland represents the largest and easiest structure to identify during microdissection of the different silk-producing glands; 2) MA silk represents the easiest silk type to forcibly remove or collect from a spider during a controlled descent; 3) MA silk has extraordinary mechanical properties, which include its high tensile strength and toughness.

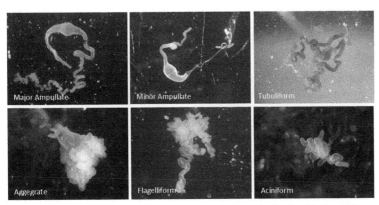

FIGURE 1.1 Different silk-producing glands isolated from the black widow spider, *Latrodectus hesperus.*

1.1.2.1 CHEMICAL COMPOSITION

The ultrastructure for dragline silk has been intensely investigated and different numbers of layers have been reported in the literature, ranging from two to five.[2,6] For the purpose of this discussion, we will assume that dragline silk consists of a core, skin, thin glycoprotein layer, and a lipid outer coating. Chemical analysis of MA silk has revealed this material contains at least two distinct proteins. To date, two different structural polypeptides referred to as fibroins or spidroins (spidroin is a contraction of the words spider and fibroin) have been identified as major constituents (Table 1.1).[7] Full-length DNA sequences have been reported for both constituents, which have been named Major Ampullate Spidroin 1 (MaSp1) and Major Ampullate Spidroin 2 (MaSp2).[8] This nomenclature reflects their restricted pattern of mRNA and protein expression and the order of discovery of the proteins.[9] Manual inspection of the full-length translated sequence for both MaSp1 and MaSp2 predicts large molecular weight proteins that are >250–kDa.[8] Analysis of these protein architectures reveals nonrepetitive N- and C-termini along with internal repetitive block modules. These internal modules are approximately 40–50 amino acids in length and can be further broken into submodules. Amino acid composition studies of the MA luminal contents and the natural fibers using acid hydrolysis or NMR have generated experimental data that are similar to the predicted amino acid composition profiles from full-length MaSp1 and MaSp2 sequences, both showing compositions that are A, G and Glx-rich.[10] However, although the experimental and theoretical amino acid composition profiles are similar, they are not identical and the differences support the assertion that other proteins are spun into MA fibers. MS/MS analyzes of peptides generated from solubilized MA silk digested with trypsin are consistent with this hypothesis; these peptide sequences do not match regions from the full-length translated MaSp1 or MaSp2 sequences (unpublished data).

1.1.2.2   MOLECULAR ARCHITECTURE

Comparison of the protein architectures of the spidroin family members reveals common structural themes. Spidroins contain iterations of internal block segments flanked by nonrepetitive N- and C-termini. The N-terminal domain consists of approximately 155 amino acids that include a secretion signal. X-ray diffraction studies of the N-terminal domain of MaSp1 reveal a dimeric structure with each subunit forming an antiparallel five-helix bundle.[11] In-vitro studies using purified recombinant proteins demonstrate that alterations in pH can affect spidroin aggregation, suggesting that this domain functions as a pH-sensitive relay to control protein aggregation and solubilization.[12] This experimental observation is consistent with reports that have demonstrated that the acidity increases moving down the spinning duct from a pH 6.9 to 6.3, a process apparently regulated by proton pumps positioned right before exiting via the spigots.[13] Specifically, as the pH approaches 6.0, the N-terminal domain has been proposed to change conformations, accelerating the self-assembly process of the fibroins. Consistent with the fact that hydrogen ion concentrations serve an important trigger for fiber formation is the observation that homodimer stability and assembly is favored under acidic pH.[12a] Although the N-terminal domain appears to be more conserved relative to the nonrepetitive C-terminus, the C-terminal domain is also one of the most highly conserved regions within the spidroin family members. From a length or size perspective, the C-terminal domain is approximately 100 amino acids and its sequence shares no similarity relative to the N-terminal domain. Much attention has also been focused on elucidating the function of the C-terminal domain of MaSp1. The C-terminal domain has been implicated in the control of solubility and fiber assembly.[14] Analysis of the amino acid residues within the C-terminal domain of dragline silk fibroins has revealed the presence of a conserved Cys residue. This Cys residue appears to direct fiber formation, an observation supported by the expression of recombinant fibroins lacking the C-terminal domain that assemble improperly when manufactured inside insect cells.[15] Additionally, experimental evidence supports its direct participation in disulfide bonds and association in large molecular weight complexes in the glandular contents of the MA gland.[16] Interestingly, mutation of the conserved Cys to

Ser leads to inefficient dimerization of recombinant proteins purified from bacteria, suggesting this residue controls the aggregation process (unpublished data).

The internal block segments of the MA fibroins, as previously stated, are 40–50 amino acids in length and are repeated approximately 60 times. Each block module can be further broken into submotifs that are classified into four categories: 1) poly A (A_n) blocks or runs of Gly Ala couplets [GA]; 2) Gly X repeats [GGX]; 3) Gly Pro-Gly X-X motifs [GPGXX]; and 4) spacer regions. Different combinations of these submodules control the secondary structure of the protein and ultimately are responsible for a large part of the mechanical properties of the fiber. Solid-state NMR, Raman spectroscopy and X-ray diffraction studies support that the poly A stretches and GA couplets form the crystalline regions of the fibers, contributing to the high tensile strength.[17] The GGX motifs have been hypothesized to be present within the amorphous region of the fibers and biophysical studies support these sub-modules form alpha-helical structures.[18] It also has been proposed that the GPGXX modules, which are present only within MaSp2 and not MaSp1, form a helical shape that contributes extensibility to the fibers. In part, the mechanical properties of the fibers can be controlled by the ratios of MaSp1 and MaSp2; these ratios can vary from spider species to species and can be influenced by diet.[19]

1.1.2.3   NATURAL FIBER EXTRUSION

Our understanding of the extrusion process of spider silks from the MA gland has advanced over the past decade, but much still remains a mystery. A large empha-sis has been focused on the ampulla and the spinning duct. The ampulla acts as a protein repository, whereas the spinning duct mediates much of the chemical and physical processes that result in fiber assembly. Based upon histological studies, the spinning duct has been divided into three limbs. The duct is comprised of a thin cuticle that functions as a dialysis membrane responsible for selective movement of water, surfactants and lubricants and the ions sodium, potassium, chloride and phosphate. A careful analysis of the elemental constituents in the MA gland has revealed that sodium and chloride ion concentrations are higher in the ampulla but decrease as the dope moves down the spinning duct.[13] It has been hypothesized that the elevated levels of the sodium and chloride ions in the ampulla facilitate fibroin storage, resulting in a highly concentrated liquid spinning dope. Quite impressively, the concentration of the protein mixture, often referred to as the spinning dope, has been reported to represent 30–50% (w/v) or 300–500 g/L.[20] During storage in the ampulla, the secondary structure of the fibroin mixture represents a random coil and alpha helical state.[21] This high storage concentration results in the formation of a lyotropic liquid crystalline phase in the duct, which represents a unique phase that flows as a liquid but maintains the molecular orientational order that is characteristic of a crystal.[22] Levels of phosphate and potassium ions have also been reported to in-crease moving down the spinning duct. Based upon the chemical properties of these elements, it has been suggested that these ions promote precipitation of the fibroins. Intriguingly, elemental sulfur levels have been shown to increase down the spinning duct; however, no theories have been proposed to explain the role of elevated levels of sulfur down the spinning duct during the assembly process.[16] One possibility is that changes in the redox state, in some manner, help trigger fiber formation. Consistent with this assertion is the decrease in mass of protein complexes stored in the MA gland after treatment with reducing agents.[16] It has also been suggested that dehydration occurs during the late stages of the extrusion process, a procedure that uses specialized epithelial cells that recover water prior to extrusion near the third limb of the spinning duct. For air-spinning spiders, which represent the predominant focus in this chapter, the process of evaporation also helps remove water after extrusion. In addition to these chemical changes, physical forces influence the conversion of the liquid-crystalline phase to a solid transition. Microscopic analysis of the MA gland reveals the spinning duct narrows as the tubing approaches the spigot or exit point on the spinneret. This geometry increases the flow rate and shear forces as the silk is extruded through the duct. Collectively, the chemical and physical forces facilitate the alignment of the fibroin molecules with the direction of the flow, drive beta-sheet structure formation in the poly A blocks, and dehydrates the fiber to produce a solid material. Further enhancement of the mechanical properties of the extruded threads is accomplished by pulling or tugging on the threads, a process the spider executes with its legs, generating what is referred to as a postspin draw.

1.1.2.4   MECHANICAL PROPERTIES OF NATURAL DRAGLINE SILK

MA silk is well known for its outstanding mechanical properties. It has been intensively studied by a large number of laboratories across the globe. The majority of these studies have focused on the stress-strain behavior of MA fibers because these values can be normalized to the dimensions of the tested fiber, which allows for comparison across different lengths and thicknesses of materials. Relative to high tensile steel, the breaking stress for dragline silk is comparable but it is considerably more extensible, leading to material that is 30 times tougher than steel.[23] Dragline silk also outperforms Kevlar, a synthetic fiber used for ballistic armor, and is 3 times tougher. Analysis of the literature reveals much variability in the published mechanical properties for MA silk. In part, some of the differences can be attributed to the fact that different spider species have been used to collect MA fibers for the tests.[24] However, it must be emphasized that the methodology for fiber collection can have a significant impact on the mechanical properties.[25] In particular, it has been clearly demonstrated that altering the reeling speed or force during the fiber collection can influence the mechanical data.[26] Humidity, the age of the spider, ultraviolet light, and the duration of storage of the fibers before analysis have also been reported to have an impact on the mechanical properties of the materials.[27]

1.1.2.5. *PHYSICAL PROPERTIES OF NATURAL DRAGLINE SILK*

Natural dragline silk can withstand extreme environmental conditions, such as temperature and pressure. MA silks submerged in liquid nitrogen, which at atmospheric pressure corresponds to −196°C, have exhibited increased breaking stress.[28] Under these conditions, spider silk has been reported to have 64% higher tensile strength after cryogenic treatment.[28] Also, cooling the fibers from room temperature to −60°C has been reported to produce increases in breaking stress and strain, leading to fibers that are tougher.[27b] Natural dragline silk is also thermally stable to approximately 230°C.[29] Other factors that have been shown to influence the mechanical properties of spider silk include humidity, acidity, alcohol treatment (methanol, ethanol and butanol), and special chaotrophic solvents.[27a,30] In general, treatment of dragline silks with solvents that have increased polarity, the Young's modulus and breaking strength of dragline silk has been reported to decrease gradually.[27a] Humidity or incubation of the fibers in water or 8 M urea has been shown to result in shrinking or supercontraction of the fibers. For example, exposure to water can cause some dragline silk threads to shrink by 50% of their original length.[31] For the medical industry there are also benefits, for example, spider silk films have been shown to be biocompatible, elicit poor immune responses and have shown benefits in the wound healing process and promoter of coagulation.[32] Natural dragline silk has also been demonstrated to provide a suitable matrix for 3-dimensional skin cell culturing.[33]

1.1.3 *TUBULIFORM GLAND*

Female spiders use the tubuliform gland to spin fibers that are constituents of egg sacs. The tubuliform gland has a cylindrical shape and extrudes fibers referred to as tubuliform silks (Fig. 1.1). The main component found in tubuliform silks represents the spidroin family member Tubuliform Spidroin 1 (TuSp1). Northern blot and quantitative real-time PCR analyzes have demonstrated that TuSp1 transcripts are highly expressed in the tubuliform gland.[34] MS/MS analyzes of tryptic peptides generated from enzymatic digestion of dissolved egg sacs have demonstrated the presence of TuSp1, along with two other fibroins, Egg Case Protein 1 (ECP–1) and Egg Case Protein 2 (ECP-2) in the cob-weaver *L. hesperus*.[35] Immunoblot analysis has also confirmed that TuSp1 is specifically expressed in the tubuliform gland.[36] Full-length cDNA sequences for CySp1 and CySp1 (equivalents of TuSp1 and TuSp2, respectively) from the orb-weaver, *Argiope bruennichi*, have been described.[37] The basic architecture of TuSps resemble other spidroin family members, consisting of nonrepetitive N- and C-terminal domains and internal block repeats that are larger in size relative to repeats within MaSp1 and MaSp2. Repetitive regions from TuSp1 consist of approximately 180–200 amino acid residues. Relative to other spidroin family members, the TuSp1 C-terminal domain shows little similarity to other spidroin family members, and it has been proposed to represent a silk protein that is spun into an evolutionary ancient silk.[38] Inspection of the internal block repeats of TuSp1 reveals little, if any, representation of the spider silk motifs commonly discussed earlier, such as the GPGXX, GGX, poly A and/or GA couplets, and the spacers regions. Instead, different amino acid motifs are used, which include S_n, (SA)$_n$, (SQ)$_n$, and GX (X represents A, V, I, N, Q, Y, P or D).[34b] Interestingly, the ECP–1 and ECP-2 protein sequences lack recognizable internal block repeats as well as the nonrepetitive conserved N- and C-termini. These proteins also have predicted molecular masses that are approximately 80-kDa, which is considerably smaller relative to the spidroin family members. Protein alignments between ECP–1 and ECP-2 reveal a 52% identity at the amino acid level. Despite having well defined internal block repeats, the ECPs contain poly A/(GA) modules that are similar to sequences reported from dragline silk fibroins.[35a] Analysis of protein sequences of ECPs show cysteine-rich N-terminal domains, suggesting these molecules function as intermolecular cross linkers that interact with the TuSps to provide structural roles in tubuliform silks.

Mechanical studies performed using tubuliform silk collected from *L. hesperus* reveal that these fibers contain lower breaking stress than dragline silk, but higher breaking strain[11b]. When considering both properties, tubuliform silks are shown to be tougher relative to dragline silks. Synthetic fibers have been wet spun from truncated, purified TuSp1 and ECP-2 recombinant proteins.[39] Artificial fibers spun from TuSp1 molecules that contain the C-terminal domain showed slightly lower breaking stress relative to truncated ECP-2-spun fibers, having values of 95.1 and 121.9 MPa, respectively (Table 1.2). Interestingly, reconstituted egg case silk fibers, which contain full-length fibroins, display lower tensile strength relative to natural tubuliform silks.[11b,40] This implies that the lower mechanical properties for the synthetic silks, is impart, due to imperfections in the artificial spinning process.

TABLE 1.2 Synthetic Spider Silk Fibers and their Reported Mechanical Properties Reveal Variation

Mechanical properties of synthetic fibers consisting of recombinant proteins					
Recombinant Protein	Species	Diameter (mm)	Strength (MPa) Breaking Stress	Extensibility (%) Breaking Strain	Reference
ECP-2C Stretched	*L. hesperus*	47	121.9	18	40
TuSp1 1xC	*L. hesperus*	30.5	95.1	25	40
TuSp1 1xC/ ECP-2C Stretched	*L. hesperus*	39.1	82.8	16	40
MaSp1 (96-mer)	*Nephila clavipes*	60	508	15	41
MaSp1 (24-mer) As spun	*N. clavipes*	40.90	35.65	3.13	42

TABLE 1.2   *(Continued)*

Mechanical properties of synthetic fibers consisting of recombinant proteins					
Recombinant Protein	Species	Diameter (mm)	Strength (MPa) Break- ing Stress	Extensibility (%) Break- ing Strain	Reference
MaSp1 (24-mer) Stretched	*N. clavipes*	17.44	132.53	22.78	42
Ma rcSp1	*N. inaurata*	44	320	30	43
Ma rcSp2	*N. inaurata*	36	330	35	43
Ma rcSp1/Sp2 70:30	*N. inaurata*	46	350	51	43

1.1.4   MINOR AMPULLATE GLAND

The minor ampullate (MI) gland is morphologically similar to the MA gland, but it is smaller in size (Fig. 1.1). MI silks are often present with MA silks, which form four fibers in dragline silks; two large diameter MA silk fibers and two smaller MI silk threads (Fig. 1.2). MI silk has been proposed to consist of two spidroin family members, Minor Ampullate Spidroin 1 (MiSp1) and Minor Ampullate Spidroin 2 (MiSp2).[44] Recently, the complete cDNA sequence for MiSp from the orb-weaver spider, *Araneus ventricosus,* was deposited in the GenBank database.[45] Analysis of the MiSp1 cDNA sequence predicts a 1766 amino acid residue protein organized into conserved nonrepetitive N- and C-terminal domains and internal repetitive regions composed of four submodules that are iterated in a nonregular manner. Similar to MA fibroins, the predicted sequence is dominated by poly A and GGX, as well as Gly Gly Gly X (GGGX), Gly X (GX) motifs and spacer segments.

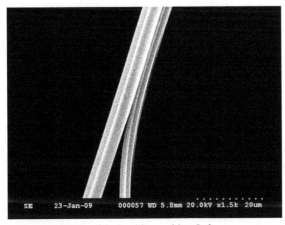

FIGURE 1.2   Dragline silk from a black widow spider, *L. hesperus.*

1.1.5 FLAGELLIFORM GLAND

The flagelliform gland appears to have different biological functions in orb-weavers relative to cob-weavers. Orb-weavers spin flagelliform silk (Flag silk) into two-dimensional webs and these fibers are known as spiral capture silk. The main protein constituent of Flag silk has been shown to have several distinctive features, including iterations of GPGGX motifs, GGX, highly conserved spacer regions with charged and hydrophilic residues, and a nonrepetitive C-terminal domain that is divergent from the other spidroin family members.[46] Flag silk represents the most extensible silk type produced by orb weaver spiders, being able to stretch 200% of its own length before fiber failure. The GPGGX module, which is also present within the protein sequence of MaSp2, is responsible for the elasticity and extensibility of flagelliform silk; iterations of these motifs have been hypothesized to form type II beta-turns that assemble into beta-turn nano-spring structures.[47] Recently, analyzes of flagelliform silk using Raman spectroscopy from three different orb-weaver species have revealed correlations between increased tensile strength and higher amounts of beta-sheet structure, which can be attributed to greater number of spacer regions.[48] This is consistent with the observation that recombinant fibers spun with only spacer regions are stronger relative to synthetic silks spun from recombinant proteins that contain spacer regions, GGX and GPGGX modules.[49] In cob-weavers, which spin three-dimensional webs that lack spiral capture, the flagelliform gland has not been reported to extrude fibroins or fibroin-like proteins (Fig. 1.1); however, two small peptides dubbed Spider Coating Peptide 1 (SCP-1) and Spider Coating Peptide 2 (SCP-2) have been detected.[50] MS/MS studies have demonstrated that the SCPs are present on egg cases, scaffolding threads, gumfooted lines and attachment discs. Recombinant expressed SCP-1 has been shown to bind to a nickel resin, suggesting it has intrinsic metal binding activity and potential antimicrobial effects. Growth studies with the gram-negative bacterium E. coli support this assertion, as addition of either SCP-1 or SCP-2 to rapidly dividing bacterial cells is able to slow cellular division (unpublished data).

1.1.6 ACINIFORM GLAND

Aciniform silks are manufactured by the aciniform gland, a structure that can described in the cob-weaver as somewhat "finger-like" shaped (Fig. 1.1). These silks are used for swathing prey, building sperm webs, web decorations, and egg sacs (Fig. 1.3).[1] Partial cDNA sequences coding the major protein constituent, Aciniform Spidroin 1 (AcSp1), were initially reported, and more recently, the entire genetic blueprint for AcSp1 was completed.[51] Aciniform silks have also been shown to be constituents of egg sacs and MS/MS analysis has demonstrated the presence of AcSp1 in egg sacs and prey wrapping silk.[51b] Inspection of the genomic DNA reveals the AcSp1 gene consists of a single exon that exceeds the largest exons reported from humans, chimpanzees, mouse, and zebrafish, predicting a protein size of 630 kDa. Analyzes of the 16 block repeats of *L. hesperus* AcSp1, which each comprise 375 amino acids, show extreme conservation relative to block repeats reported from other spidroin family members.[51c] This represents approximately twice the size of the 200 amino acid block repeats reported for *Argiope trifasciata* AcSp1.[51a] Although the mechanical data available for aciniform silks is relatively scarce, stress-strain curves collected from a handful of orb-weaver spiders support that this fiber represents one of the toughest spider silk threads.[51a] Purified recombinant AcSp1 proteins from *A. trifasciata* carrying 2, 3, or 4 block repeats (denoted W_2, W_3, and W_4) can be induced to form fiber-like threads by shear forces in physiological buffer.[52] Mechanical analysis of hand-pulled, synthetic fibers produced from purified recombinant protein W_4 have reported breaking stress and strain values of 116 ± 24 MPa and 0.37 ± 0.11, respectively.

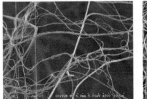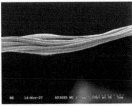

FIGURE 1.3 Aciniform silks are small diameter fibers that are present in wrapping silk and egg sacs. Left to right: prey wrapping silk (smaller fibers) and larger diameter dragline silks, egg case silk, and a bundle of aciniform silks from prey wrapping silk.

1.1.7 PYRIFORM GLAND

Pyriform glands secrete fibers that are important constituents of attachment discs. Attachment discs function to fasten dragline silk fibers to solid supports, including wood, concrete, glass, and other surfaces. This attachment is central to locomotion and web construction. The exact mechanism of the adhesion to the support has yet to be fully elucidated. Biochemical studies have revealed that the attachment discs fibers are embedded in a liquid matrix that rapidly dries. The major constituent of attachment disc silks from *L. hesperus* represents Pyriform Spidroin 1 (PySp1).[53] Real-time quantitative PCR analysis support high levels of PySp1 transcripts in the pyriform gland relative to other silk-producing glands. From a protein architecture standpoint, PySp1 contains the conserved C-terminal domain characteristic of spidroin family members, as well as internal block repeats ranging from 238–300 residues that are rich in A, Q and E, along with a 78 amino acid spacer region that is extremely hydrophilic in nature. Traditional GGX, GPGGX, and poly A stretches are absent. Block modules contain submotifs with the sequence AAARAQAQAEARA-KAE and AAARAQAQAE, which have been shown to form beta-sheet structures when synthetic peptides containing iterations of these sequences were investigated by circular dichroism (unpublished data). Analysis of the amino acid composition profile reveals the protein sequence of PySp1 contains the most hydrophilic residues relative to other spidroin family members, a likely feature that is linked the observation that it is spun into a liquid matrix that readily dries.[53] Because these fibers are difficult to collect from spiders, mechanical data have been difficult to obtain from traditional stress-strain analyzes. In orb-weavers, the equivalent spidroin has been reported and dubbed PySp2 or PiSp1.[54] MS/MS analysis of both the attachment discs and luminal contents from the pyriform gland of orb-weavers, similar to PySp1, has confirmed the presence of PySp2 as one of the major constituents.[54a] Comparable to PySp1, PySp2 contains the highest degree of polarity among the spidroin family members, conservation within its C-terminal domain, but differences in protein sequence within its internal block repeats relative to the cob-weaver PySp1. PySp2 contains block repeats that are approximately 200 amino acids, flanked by spacer regions that are 44 residues; these spacer regions are Pro-rich and have iterations of the submotif PAPRPXPAPX, with X representing a subset of amino acids that mostly contains hydrophobic R groups. The block motifs have repetitive motifs that are Gln-Gln-Ser-Ser-Val-Ala (QQSSVA). Synthetic fibers spun by wet-spinning methodology with purified proteins containing the C-terminal domain, block repeat, and spacer region have been reported.[54a] The high degree of polarity within the protein sequences of the PySp fibroins, along with their ability to form fibers, suggests strategies to allow for fabrication of fibers in liquid environments.(Fig. 1.4.)

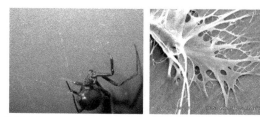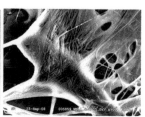

FIGURE 1.4   Attachment discs from a black widow spider, *L. hesperus*. Left to right: digital photograph of a black widow spider with attachment discs holding down dragline silk, followed by two scanning electron microscope (SEM) images of an attachment disc at 450x and 900x, respectively.[53]

1.1.8   AGGREGATE GLAND

Spiders coat their webs with sticky, adhesive substances that have been reported to facilitate prey capture. In orb weavers, the aggregate gland has been characterized as the structure that extrudes aqueous glue that coats the spiral capture threads. Chemical analysis of the aqueous glue solution has revealed high concentrations of organic compounds related to neurotransmitters, small peptides, free amino acids, low concentrations of inorganic salts and glycoproteins.[55] Two cDNAs encoding glycoproteins have been reported in the literature and proposed to be major constituents of the glue droplets found on spiral capture threads; these products are named Aggregate Spider Glue 1 (ASG1) and Aggregate Spider Glue (ASG2).[56] In cob-weaver spiders, data are emerging to suggest a different functional role for the aggregate gland (Fig. 1.1). In part, this could be somewhat anticipated and hypothesized because cob-weavers lack spiral capture silk. In cob-weavers the aggregate gland has been demonstrated to secrete two distinct proteins that are constituents of connection joints in three-dimensional webs, which are structures that glue scaffolding fibers together.[57] These products, named Aggregate Gland Silk Factor 1 (AgSF1) and Aggregate Gland Silk Factor 2 (AgSF2), have markedly divergent protein architectures as well as are highly distinctive relative to protein sequences from traditional fibroins. AgSF2, a 40-kDa nonglycosylated protein, has novel internal amino acid block repeats with the consensus sequence Asn-Val-Asn-Val-Asn (NVNVN), whereas AgSF1 contains pentameric Gln-Pro-Gly Ser-Gly (QPGSG) iterations that are similar to modular elements with mammalian elastin. AgSF1 has the potential to self-assemble into fibers and X-ray diffraction of synthetic threads reveals the presence of noncrystalline domains that resemble classical rubber networks.[57]

1.2 METHODOLOGIES FOR SYNTHETIC SPIDER SILK PRODUCTION

1.2.1 STRATEGIES FOR SPIDER SILK PROTEIN PRODUCTION

Several challenges remain to be resolved before spider silks can be truly manufactured on an industrial scale with mechanical properties that rival or exceed natural fibers. Firstly, scientists will need to develop cost-efficient, rapid methods to produce large amounts of recombinantly expressed spider silk protein that can be easily purified. Secondly, and equally as important, chemical methodologies to spin silks from aqueous solvents needs to undergo substantial advancements. Although the tools of molecular biology and biochemistry have allowed many foreign proteins to be expressed and purified at high levels in bacteria and yeast, the high molecular masses of the silk proteins, combined with their biasness toward specific amino acids within their protein sequence, has presented technical challenges for the production of vast amounts of full-length, and even truncated recombinant dragline silk proteins for synthetic fiber production.

1.2.1.1 EXPRESSION SYSTEMS

Although the Chinese have farmed cocoon silk from *Bombyx mori* for over 5,000 years, the ability to farm spider silk is impractical for several reasons. Firstly, the cannibalistic, territorial, and venomous nature of spiders is less than desirable when considering managing a farm of spiders. Secondly, the "milking" of spiders is labor intensive and yields relatively small quantities of silks. For example, it took 70 individuals over 4 years to collect silk from more than 1 million golden orb-weaver spiders to produce enough material to assemble an 11' × 4' textile. The estimated cost for the single tapestry was approximately $500,000, which from an economical perspective, is staggering. Therefore, attempts to obtain large quantities of spider silk proteins have turned to the implementation of molecular biological approaches.

A variety of different expression systems have been explored to determine whether silk proteins can be expressed at high levels. Both eukaryotic and prokaryotic expressions systems offer advantages and disadvantages. Eukaryotic expression systems offer the benefit of posttranslational modifications, which include glycosylation and phosphorylation. However, the role and importance of these modifications to silk proteins and their necessity for fiber production remains unclear. Thus far, no posttranslational modifications have been shown to be necessary for spidroin function. Prokaryotic systems offer the advantage of faster growth rates, easier genetic manipulation, and custom optimization for expression. Most existing expression systems have focused on producing truncated fibroins that only contain reduced numbers of the internal block repeats. In some cases, the C-terminus has been included in the protein constructs. Synthesizing full-length silk spidroins has been challenging due to their large molecular mass and their bias towards specific amino acids. Native proteins can exceed 3000 amino acids in length and, in the case for MaSp1, have amino acid content that surpasses more than 60% alanine and glycine.[8a] In fact, both DNA replication and expression of spider silk cDNAs have been hampered by their long, repetitive and guanine and cytosine rich nature. Attempts to circumvent the translational challenges via codon optimization have been used with limited success. In the literature, there have been several successful reports of partial or truncated silk fibroins being expressed in prokaryotes. Most of these studies have used *Escherichia coli* as the host. Miniature recombinant proteins that contain C-termini and internal block repeats have been demonstrated to be expressed in prokaryotic organisms as well.[58]

The expression of full-length spidroin cDNAs has not been reported from either prokaryotic or eukaryotic systems. However, a metabolically engineered strain of *E. coli* has been shown to synthesize a large molecular weight recombinant fibroin that approaches the predicted size of native MaSp1.[41] Mammalian and insect cell culture systems have shown promising results involving the expression of large molecular weight spidroins using transformed cells lines from bovine mammary epithelial alveolar cells and *Spodoptera frugiperda* cells, respectively.[59] Transgenic goats that express and secrete fibroins into the milk have also been generated; however, purification of the proteins from the milk has been challenging and expensive to maintain the transgenic goats. In addition, both transgenic mice and silkworms that express spider silk proteins have also been reported.[60] Transgenic silkworms have shown promising possibilities, but some current barriers need to be resolved, such as improving the quantities of the spider silk protein deposited in the final spun product, which is still approximately 94% (w/w) *B. mori* natural product.[60b] The methylotrophic yeast *Pichia pastoris* is a robust eukaryotic expression system that allows for recombinant proteins to be secreted into liquid growth media, and it would appear to offer many advantages. These include the ability to grow the cells at high densities, harvest the protein without breaking or lysing the cells, and cheaper methods for protein purification, making yeast one of the more readily adaptable systems for producing spider silk proteins on an industrial scale.[61]

1.2.2 WET-SPINNING TECHNOLOGIES

A critical step that remains for the commercial production of synthetic silk fibers is successful spinning of recombinant proteins into materials that resemble natural silks, specifically in their mechanical and thermal properties. One technique involves spinning fibers into a liquid coagulation bath, which is referred to as wet spinning. In 1993, it was first described by the DuPont® group with fibroins spun from silkworms[62] and then later modified by Nexia Biotechnologies in 2002.[63] Successful reports that integrate arthropod biomimicry to produce synthetic spider silks also have been reported in the scientific literature.[64] These procedures have relied on purifying recombinant spidroins, followed by spidroin concentration via lyophilization, and then solubilization of spidroins with chaotropic solvents to produce a highly concentrated spinning dope (Fig. 1.5).[64b] This solution is then pushed through a syringe equipped with a needle into an alcohol bath, allowing a slower solidification (Fig. 1.5). The flow of the liquid is best controlled by a syringe infusion pump that can move the liquid through the syringe at a constant rate. Isopropanol is often used as coagulation medium during extrusion and provides a dehydration step that removes water from the fiber (Fig. 1.5). The resulting products are referred to as "as-spun threads," which can be subsequently subject to postspin draw, a process that dramatically improves the breaking stress, toughness, and Young's modulus.[40,63] This methodology has led to fibers (some biocomposites) with mechanical properties that are lower quality relative to natural spider silks. For example, truncated recombinant fibroins spun into fibers have reported breaking stress values that range from 35–350 MPa (Table 1.2), which are lower than the typical 1000 MPa values reported for natural dragline silk. The strongest fibers produced via this technique have been reported from a fiber spun from a recombinant 96-mer MaSp1 protein construct expressed in *E. coli*, which lead to fibers that exceeded a breaking stress of 500 MPa, a value about 50% lower relative to natural dragline silk fibers (Table 1.2).[41] This recombinant protein was approximately 285 kDa and represents the largest molecular weight recombinant protein used for synthetic fiber production.

Although these results are good indicators of progress, several caveats are worth noting for synthetic fiber production. Firstly, the reproducibility of the quality of the synthetic fibers has been a major issue. There is still much variability between the mechanical properties of the fibers generated from different regions of the same spun fiber, which highlights several technical issues that remain to be resolved during the manufacturing process. Much variability is due to the introduction of human errors during the processing steps and can be circumvented by automation of the process. Secondly, the solvent choice for the majority of the laboratories has been a challenge. Hexafluoro-2-propanol (HFIP) and formic acid are two solvents that are excellent at dissolving the silk proteins to achieve the necessary spinning dope concentrations, but are volatile and toxic compounds that are less than ideal for manufacturing conditions. Additionally, maintaining protein solubility during the spinning process can be difficult. Too high of protein concentration can lead to precipitation or gelation, making the spinning process unmanageable. During fiber curation, HFIP has been reported to evaporate, leaving voids within the interior of the material that impact the mechanical properties of the fibers (Fig. 1.6).[40] One laboratory has produced recombinant silk protein constructs that have been demonstrated to form fibers from an aqueous solution, potentially offering an advantage to fiber production without the use of HFIP and formic acid.[65]

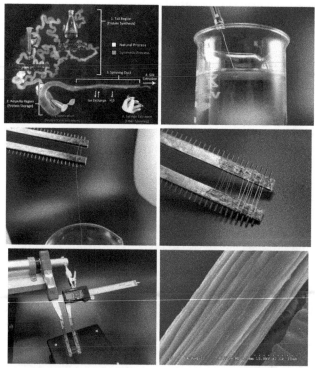

FIGURE 1.5 Biomimicry of the natural spinning process in spiders. Left to right, top to bottom: Biomimicry of the spinning process and use of bacterial for expression, spinning synthetic fibers into an isopropanol bath, spooling, postspin draw using a linear actuator, and SEM of reconstituted egg case silk.

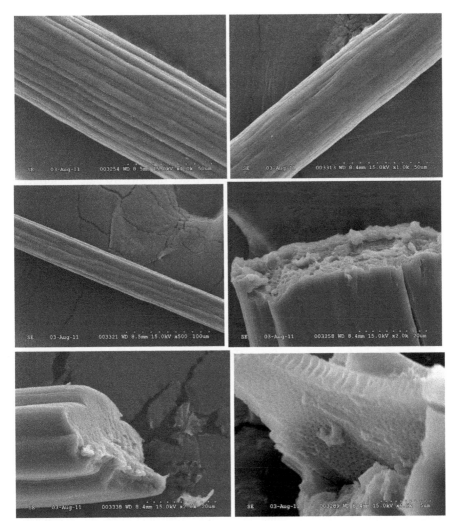

FIGURE 1.6 SEM images of synthetic spider silks spun from purified recombinant proteins using wet-spinning methodology. Left to right, top to bottom: truncated TuSp1 with C-terminus, truncated C-terminus of ECP-2, biocomposite of truncated TuSp1 and ECP-2, fractured TuSp1 fiber, fractured TuSp1/ECP-2 blended fiber, and a fractured TuSp1 thread that has internals voids caused by HFIP evaporation.[39]

1.2.3 ELECTROSPINNING

The process of electrospinning was first investigated by Zeleny in 1914 and was found to be a technique for spinning small diameter fibers.[66] Electrospinning is an adaptable technique, enabling the development of nanofiber-biomaterial scaffolds. These scaffolds can be used for tissue engineering and regenerative medicine because they can mimic the fibrous properties of the natural extracellular matrix in tissues. Electrospinning has also been implemented to produce spider silk fibers. In the process, the spinning dope is placed into a syringe fixed to a needle and a high electric potential is applied to a droplet of the solution at the tip of the needle. When the applied electrical force becomes greater than the surface tension of the silk droplet, it results in a charged jet of the silk solution that is ejected in the direction of the applied field. The jet undergoes dehydration as the solvent is evaporated in the air, resulting in dried fibers that can be collected on a receiving conducting mesh. Many more studies have been reported using silkworm fibroins for electrospinning relative to spider fibroins. To date, nanofibers have been successfully spun from reconstituted natural silk dragline silk as well as recombinantly expressed proteins dissolved in HFIP.[67] The diameters of the fibers were several orders of magnitude smaller relative to natural fibers, ranging from 8 to 200 nm, which is similar to other electrospun polymers.[68] For the reconstituted spun nanofibers, wide angle X-ray diffraction (WAXD) studies demonstrated the fibers contained orientational and crystalline order comparable to that of natural spider silks.[67a] Electrospinning has also been used to prepare spider silk fibroins/poly(D,L-lactide) [PDLLA] composite fibrous nonwoven mats.[69] Addition of the spidroins to PDLLA led to improvements in the hydrophilic and mechanical properties of the composite fiber and its biocompatibility. Collectively, the electrospun nonwoven fabrics have promise for scaffolds for tissue engineering, wound dressing materials, and carriers for drug delivery because of their high hydrophilicity and porous structure.

1.2.4 HAND PULLING APPROACHES

Hand pulling spider fibers has been reported in the literature.[42,52] Applying this methodology, large amounts of variability have been observed in the material properties of the fibers spun from the same spinning dope.[42] Furthermore, the feasibility for large-scale production using this approach is unrealistic. Differences between fibers spun from the same spinning dope can be attributed to several parameters that are difficult to precisely control, which include the draw rate and actual draw ratio. Therefore, it seems intuitive that inherent inconsistencies due to hand drawing would be introduced into the production of fibers and alternatives, such as the integration of automation into the processing of the fibers, which includes postdraw, is more pragmatic.

1.3   TRANSLATION OF FIBER PRODUCTION TO A LARGE-SCALE FORMAT

1.3.1   OTHER FIBROUS MATERIALS PRODUCED ON AN INDUSTRIAL SCALE

DuPont® was one of the first companies to attempt to produce silk fibroins on an industrial scale[70], but perfecting the large-scale production of spider silks has proved arduous and challenging relative to some other materials. DuPont® manufactures approximately 2 million tons of Kevlar each year, requiring 15,800,000 to 18,750,000 pounds of sulfuric acid.[71] The process also requires petroleum products, substantial pressure and temperatures that approach 1,400 degrees Fahrenheit. The U.S. Army uses about 10,000 pounds of Kevlar for composite materials. Manufacturing costs could be lowered by reducing the amount of hazardous waste material generated during the production of Kevlar. Additionally, over the past 70 years, DuPont® has also been manufacturing nylon. This material initially nicknamed the "miracle fiber." One of first applications for nylon was socks, but other uses have expanded into clothing, carpeting, ropes, and the automobile industry. Despite the benefits of nylon's use for a wide range of different applications, its production has a history of environmental concerns that include the reliance of large quantities of crude oil, adipic acid, and production of nitrous oxide, a greenhouse gas.

1.3.2.   SPIDER SILK PRODUCTION IN LARGE QUANTITIES FOR APPLICATIONS

In order to manufacture vast quantities of purified spider silk protein for wet spinning methodologies, the expression system for protein production needs to be robust, cost-effective and offer rapid biochemical strategies to purify the recombinant fibroins for the spinning process. At least five other natural fibroin cDNA sequences have been used for recombinant expression, which include MaSp2, the tubuliform spidroins TuSp1 and ECP-2, PySp2, and AcSp1 (Fig. 1.6).[39,52,54a] However, despite the availability of the seven different spidroin genetic blueprint sequences, the majority of the recombinant expression studies have focused on dragline silk fibroins, specifically MaSp1. Currently, it is unclear how many companies are deeply invested in the large-scale production of spider silk. A few companies with spider silk interest have surfaced in the news, including a San Francisco Bay Area company, Refactored Materials, Inc., as well as a Japanese startup company, named Spiber®. Both have been pursuing the process for large-scale production of synthetic spider silk fibers in either yeast or bacteria, respectively. Spiber® has reported that it can produce several hundreds of grams of recombinant spider silk protein per day.

   If tons of spider silk materials are to be manufactured for global distribution, scaling the procedure to synthesize sufficient quantities of fibroins from transgenic cells or organisms must be optimized. Therefore, identifying the most efficient expression system for recombinant spider protein production has been intensely investigated and still remains as a barrier. So far, it would seem that bacteria or yeast are emerging as the likely candidates. Based upon expression studies, for example, in the methylotrophic yeast *P. pastoris*, it is reasonable to assume that expression of some recombinant fibroins can achieve 1 g/L of culture.[72] Typically, one gram of purified recombinant spidroin could produce about 29,527 feet of silk. To produce one ton of spider silk, which is the equivalent of approximately 998,412 grams, it would require about 100,000 L of culture. This is well within the range of some large industrial sized fermenters that have volume capacities that can exceed 25,000 L. One of the chief advantages for using *P. pastoris* includes the secretion of the recombinant proteins into the extracellular medium, which allows for rapid purification of proteins due to the fact that little, if any, other proteins are secreted into the liquid media. Furthermore, it eliminates the need to analyze the cells and restart cultures for expression, a process that can be tedious, increase production times, and result in higher production costs and manufacturing prices. Additionally, *P. pastoris* can be grown in bioreactors to high densities on mineral salt media and has been shown to be effective for expression of very large and complex proteins, such as collagens, which also require the coexpression of collagen prolyl 4-hydroxylase for the thermal stability of collagens.[73] As more is revealed about the proteins involved in the spider silk assembly pathway, these components can be integrated into the expression system and presumably lead to better products.

1.4 CONCLUSION

Over the past several years, the spider silk community has advanced their understanding regarding the protein compositions of the different fiber types. With these advances, many partial cDNAs and several complete genetic blueprints coding for spider silk proteins are available for recombinant protein expression. Additionally, wet-spinning methodologies that use purified, recombinant spider protein as spinning dopes are becoming more commonly reported in the literature. Still, the need for further progress to increase the quantities of recombinant proteins manufactured by host cells, along with improvements and new strategies for mechanical spinning will need to be developed if spider silk synthesis is to successfully reach large-scale production with material properties that rival natural silks. Undoubtedly, the different applications for spider silk proteins as biocomposites seem endless.

1.5 ACKNOWLEGMENT

We thank Drs. Joan and Geoff Lin-Cereghino with expression studies in *P. pastoris*. In addition, we thank Yang Hsia, Eric Gnesa, Felicia Jeffrey, Thanh Phanm, Connie Liu, Christine Ho, Lisa Pham and Ryan Pacheco for their valuable contributions. We also are indebted to Dr. Mark Brunel at the University of the Pacific, Department of Biology, for his assistance with the scanning electron microscope.

KEYWORDS

- **Biocomposites**
- **Mechanical Properties**
- **Protein**
- **Spider Silk**

REFERENCES

1. Foelix, R. (1996). *Biology of Spiders*. Oxford University Press: New York.
2. Andersson, M., Holm, L., Ridderstrale, Y., Johansson, J., & Rising, A, (2013). Morphology and composition of the spider major ampullate gland and dragline silk. *Biomacromolecules*, 14(8), 2945–2952.
3. Garb, J. E., Ayoub, N. A., & Hayashi, C. Y. (2010). Untangling spider silk evolution with spidroin terminal domains. *BMC Evol Biol*, 10, 243.
4. Blackledge, T. A., Summers, A. P., & Hayashi, C. Y. (2005). Gumfooted lines in black widow cobwebs and the mechanical properties of spider capture silk. *Zoology (Jena)*, 108(1), 41–46.
5. La Mattina, C., Reza, R., Hu, X., Falick, A. M., Vasanthavada, K., McNary, S., Yee, R., & Vierra, C. A. (2008). Spider minor ampullate silk proteins are constituents of prey wrapping silk in the cob weaver *Latrodectus hesperus*. *Biochemistry*, 47(16), 4692–4700.
6. Sponner, A., Vater, W., Monajembashi, S., Unger, E., Grosse, F., & Weisshart, K. (2007). Composition and hierarchical organization of a spider silk. *Plos One*, 2(10), e998.
7. (a) Xu, M., & Lewis, R. V. (1992). Structure of a protein superfiber: spider dragline silk. *Proc. Natl. Acad. Sci. USA* 1990, 87(18), 7120–7124.
   (b) Michael, B. & Hinman, R. V. L. Isolation of a Clone Encoding a Second Dragline Silk Fibroin. *J Biol Chem*, 267(27), 19320–19324.
8. (a) Ayoub, N. A., Garb, J. E., Tinghitella, R. M., Collin, M. A., & Hayashi, C. Y. (2007). Blueprint for a high-performance biomaterial: full-length spider dragline silk genes. *Plos One*, 2, e514.
   (b) Zhang, Y., Zhao, A. C., Sima, Y. H., Lu, C., Xiang, Z. H., & Nakagaki, M. (2013). The molecular structures of major ampullate silk proteins of the wasp spider, *Argiope bruennichi*: a second blueprint for synthesizing de novo silk. *Comp Biochem Physiol B Biochem Mol Biol*, 164(3), 151–158.
9. (a) John Gatesey, C. H., Dagmara Motriuk, Justin Woods, & Randolph Lewis (2001). Extreme Diversity, Conservation, and Convergence of Spider Silk Fibroin Sequences. *Science*, 291, 2603–2605.
   (b) Xu, M., & Lewis, R. V. (1990). Structure of a protein superfiber: Spider Dragline Silk. *Proc. Natl. Acad. Sci.*, 87, 7120–7124.
   (c) Hinman, M. B., & Lewis, R. V. (1992). Isolation of a clone encoding a second dragline silk fibroin. Nephila clavipes dragline silk is a two-protein fiber. *J. Biol. Chem.*, 267(27), 19320–19324.

(d) Guerette, P. A., Ginzinger, D. G., Weber, B. H., & Gosline, J. M. (1996). Silk properties determined by gland-specific expression of a spider fibroin gene family. *Science*, 272(5258), 112–115.

10. (a) Shi, X., Holland, G. P., & Yarger, J. L. (2013). Amino acid analysis of spider dragline silk using (1)H NMR. *Anal Biochem*, 440(2), 150–157.

    (b) Casem, M. L., Turner, D., & Houchin, K. (1999). Protein and amino acid composition of silks from the cob weaver, *Latrodectus hesperus* (black widow). *Int J Biol Macromol*, 24(2–3), 103–108.

    (c) Anderson, S. O. (1970). Amino acid composition of spider silks. *Comp. Biochem. Physiol*, 35, 705–711.

11. (a) Motriuk-Smith, D., Smith, A., Hayashi, C. Y., & Lewis, R. V. (2005). Analysis of the conserved N-terminal domains in major ampullate spider silk proteins. *Biomacromolecules*, 6(6), 3152–3159.

    (b) Hu, X., Vasanthavada, K., Kohler, K., McNary, S., Moore, A. M., & Vierra, C. A. (2006). Molecular mechanisms of spider silk. *Cell Mol Life Sci*, 63(17), 1986–1999.

12. (a) Gaines, W. A., Sehorn, M. G., Marcotte, W. R., & Jr., Spidroin. (2010). N-terminal domain promotes a pH-dependent association of silk proteins during self-assembly. *J Biol Chem*, 285(52), 40745–40753.

    (b) Askarieh, G., Hedhammar, M., Nordling, K., Saenz, A., Casals, C., Rising, A., Johansson, J., & Knight, S. D. (2010). Self-assembly of spider silk proteins is controlled by a pH-sensitive relay. *Nature*, 465(7295), 236–238.

13. Knight, D. P., & Vollrath, F. (2001). Changes in element composition along the spinning duct in a *Nephila* spider. *Naturwissenschaften*, 88(4), 179–182.

14. Ittah, S., Cohen, S., Garty, S., Cohn, D., & Gat, U. (2006). An essential role for the C-terminal domain of a dragline spider silk protein in directing fiber formation. *Biomacromolecules*, 7(6), 1790–1795.

15. Ittah, S., Michaeli, A., Goldblum, A., & Gat, U. (2007). A model for the structure of the C-terminal domain of dragline spider silk and the role of its conserved cysteine. *Biomacromolecules*, 8(9), 2768–2773.

16. Sponner, A., Unger, E., Grosse, F., & Weisshart, K. (2004). Conserved C-termini of Spidroins are secreted by the major ampullate glands and retained in the silk thread. *Biomacromolecules*, 5(3), 840–845.

17. (a) Alexandra Simmons, E. R., & Lynn W. Jelinski. (1994). Solid-State 13C NMR of *Nephila clavipes* Dragline Silk Establishes Structure and Identity of Crystalline Regions. *Macromolecules*, 27 5235–5237.

    (b) Shao, Z., Vollrath, F., Sirichaisit, J., & Young, R. J. (1999). Analysis of spider silk in native and supercontracted states using Raman spectroscopy. *Polymer*, 40(10), 2493–2500.

18. (a) Van Beek, J. D., Hess, S., Vollrath, F., & Meier, B. H. (2002). The molecular structure of spider dragline silk: folding and orientation of the protein backbone. *Proc Natl Acad Sci U S A*, 99(16), 10266–10271.

    (b) Dong, Z., Lewis, R. V., & Middaugh, C. R. (1991). Molecular mechanism of spider silk elasticity. *Arch Biochem Biophys*, 284(1), 53–57.

19. Liu, Y., Sponner, A., Porter, D., & Vollrath, F. (2008). Proline and processing of spider silks. *Biomacromolecules*, 9(1), 116–121.

20. Vollrath, F., & Knight, D. P. (2001). Liquid crystalline spinning of spider silk. *Nature* 410(6828), 541–548.

21. Hijirida, D. H., Do, K. G., Michal, C., Wong, S., Zax, D., & Jelinski, L. W. (1996). 13C NMR of *Nephila clavipes* major ampullate silk gland. *Biophys. J.* 71(6), 3442–3447.

22. Chen, X., Knight, D. P., & Vollrath, F. (2002). Rheological characterization of *Nephila* spidroin solution. *Biomacromolecules*, 3(4), 644–648.

23. Gosline, J. M., Guerrete, P. A., Ortlepp, C. S., & Savage, K. N. (1999). The mechanical design of spider silks: from fibroin sequence to mechanical function. *J. Exp. Biol.* 202, 3295–3303.

24. (a) Swanson, B., Blackledge, T., Beltran, J., & Hayashi, C. (2006). Variation in the material properties of spider dragline silk across species. *Appl Phys A*, 82, 213–218.
(b) Madsen, B., Shao, Z. Z., & Vollrath, F. (1999). Variability in the mechanical properties of spider silks on three levels: interspecific, intraspecific and intraindividual. *Int. J. Biol. Macromol.*, 24(2–3), 301–306.

25. Reed, E. J., Bianchini, L. L., & Viney, C. (2012). Sample selection, preparation methods, and the apparent tensile properties of silkworm (B. mori) cocoon silk. *Biopolymers*, 97(6), 397–407.

26. (a) Vollrath, F. (2000). Strength and structure of spiders' silks. *J. Biotechnol.*, 74(2), 67–83.
(b) Vollrath, F., Madsen, B., & Shao, Z. (2001). The effect of spinning conditions on the mechanics of a spider's dragline silk. *Proc. R. Soc. Lond. B. Biol. Sci.* 268(1483), 2339–2346.

27. (a) Shao, Z. Z., & Vollrath, F. (1999). The effect of solvents on the contraction and mechanical properties of spider silk. *Polymer*, 40(7), 1799–1806.
(b) Yang, Y., Chen, X., Shao, Z. Z., Zhou, P., Porter, D., Knight, D. P., & Vollrath, F. (2005). Toughness of spider silk at high and low temperatures. *Adv Mater*, 17(1), 84–+.
(c) Kitagawa, M., & Kitayama, T. (1997). Mechanical properties of dragline and capture thread for the spider *Nephila clavata*. *J Mater Sci*, 32(8), 2005–2012.

28. Pogozelski, E. M., Becker, W. L., See, B. D., & Kieffer, C. M. (2011). Mechanical testing of spider silk at cryogenic temperatures. *Int J Biol Macromol*, 48(1), 27–31.

29. (a) Magoshi, J., Magoshi, Y., & Nakamura, S. (1985). Crystallization, liquid crystal, and fiber formation of silk fibroins. *J. Appl. Polym. Sci.: Appl. Polym. Symp.*, 41, 187–204.
(b) Cunniff, P. M., Fossey, S. A., Auerback, M. A., Song, J. W., Kaplan, D. L., Adams, W. W., Eby, R. K., Mahoney, D., & Vezie, D. L. (1994). Mechanical and thermal properties of dragline silk from the spider *Nephila clavipes*. *Polymers fro Advanced Technologies*, 5(8), 401–410.

30. Shao, Z., Young, R. J., & Vollrath, F. (1999). The effect of solvents on spider silk studied by mechanical testing and single-fiber Raman spectroscopy. *Int J Biol Macromol*, 24(2–3), 295–300.

31. (a) Boutry, C., & Blackledge, T. A. (2010). Evolution of supercontraction in spider silk: structure-function relationship from tarantulas to orb-weavers. *J Exp Biol*, 213(Pt 20), 3505–3514.
(b) Work, R. W. (1981). A comparative study of the supercontraction of major ampullate silk fibers of orb web-building spiders (Araneae). *Journal of Arachnology*, 9, 299–308.

32. (a) Panilaitis, B., Altman, G. H., Chen, J., Jin, H. J., Karageorgiou, V., & Kaplan, D. L. (2003). Macrophage responses to silk. *Biomaterials*, 24(18), 3079–3085.
(b) Meinel, L., Hofmann, S., Karageorgiou, V., Kirker-Head, C., McCool, J., Gronowicz, G., Zichner, L., Langer, R., Vunjak-Novakovic, G., & Kaplan, D. L. (2005). The inflammatory responses to silk films in-vitro and in-vivo. *Biomaterials*, 26(2), 147–155.

33. Wendt, H., Hillmer, A., Reimers, K., Kuhbier, J. W., Schafer-Nolte, F., Allmeling, C., Kasper, C., & Vogt, P. M. (2011). Artificial Skin Culturing of Different Skin Cell Lines for Generating an Artificial Skin Substitute on Cross-Weaved Spider Silk Fibers. *Plos One*, 6(7).

34. (a) Tian, M., & Lewis, R. V. (2005). Molecular characterization and evolutionary study of spider tubuliform (eggcase) silk protein. *Biochemistry*, 44(22), 8006–8012.
(b) Hu, X., Lawrence, B., Kohler, K., Falick, A. M., Moore, A. M., McMullen, E., Jones, P. R., & Vierra, C. (2005). Araneoid egg case silk: a fibroin with novel ensemble repeat units from the black widow spider, *Latrodectus hesperus*. *Biochemistry*, 44(30), 10020–10027.

35. (a) Hu, X., Kohler, K., Falick, A. M., Moore, A. M., Jones, P. R., Sparkman, O. D., Vierra, C. (2005). Egg case protein-1. A new class of silk proteins with fibroin-like properties from the spider *Latrodectus hesperus*. *J. Biol. Chem*, 280(22), 21220–21230.

(b) Hu, X., Kohler, K., Falick, A. M., Moore, A. M., Jones, P. R., Vierra, C. (2006). Spider egg case core fibers: trimeric complexes assembled from TuSp1, ECP-1, and ECP-2. *Biochemistry*, 45(11), 3506–3516.

36. Huang, W., Lin, Z., Sin, Y. M., Li, D., Gong, Z., & Yang, D. (2006). Characterization and expression of a cDNA encoding a tubuliform silk protein of the golden web spider *Nephila antipodiana*. *Biochimie*, 88(7), 849–858.

37. Zhao, A. C., Zhao, T. F., Nakagaki, K., Zhang, Y. S., Sima, Y. H., Miao, Y. G., Shiomi, K., Kajiura, Z., Nagata, Y., Takadera, M., & Nakagaki, M. (2006). Novel molecular and mechanical properties of egg case silk from wasp spider, *Argiope bruennichi*. *Biochemistry*, 45(10), 3348–3356.

38. Garb, J. E., & Hayashi, C. Y. (2005). Modular evolution of egg case silk genes across orb-weaving spider superfamilies. *Proc. Natl. Acad. Sci. USA*, 102(32), 11379–11384.

39. Gnesa, E., Hsia, Y., Yarger, J. L., Weber, W., Lin-Cereghino, J., Lin-Cereghino, G., Tang, S., Agari, K., & Vierra, C. (2012). Conserved C-terminal domain of spider Tubuliform Spidroin 1 contributes to extensibility in synthetic fibers. *Biomacromolecules*, 13(2), 304–312.

40. Gnesa, E., Hsia, Y., Yarger, J. L., Weber, W., Lin-Cereghino, J., Lin-Cereghino, G., Tang, S., Agari, K., & Vierra, C. (2011). Conserved C-Terminal Domain of Spider Tubuliform Spidroin 1 Contributes to Extensibility in Synthetic Fibers. *Biomacromolecules*.

41. Xia, X. X., Qian, Z. G., Ki, C. S., Park, Y. H., Kaplan, D. L., & Lee, S. Y. (2010). Native-sized recombinant spider silk protein produced in metabolically engineered *Escherichia coli* results in a strong fiber. *Proc Natl Acad Sci U S A*, 107(32), 14059-14063.

42. An, B., Hinman, M. B., Holland, G. P., Yarger, J. L., & Lewis, R. V. (2011). Inducing beta-sheets formation in synthetic spider silk fibers by aqueous postspin stretching. *Biomacromolecules*, 12(6), 2375–2381.

43. Elices, M., Guinea, G. V., Plaza, G. R., Karatzas, C., Riekel, C., Agullo-Rueda, F., Daza, R., & Perez-Rigueiro, J. (2011). Bioinspired Fibers Follow the Track of Natural Spider Silk. *Macromolecules*, 44(5), 1166–1176.

44. Colgin, M. A., & Lewis, R. V. (1998). Spider minor ampullate silk proteins contain new repetitive sequences and highly conserved nonsilk-like "spacer regions." *Protein Sci.*, 7(3), 667–672.

45. Chen, G., Liu, X., Zhang, Y., Lin, S., Yang, Z., Johansson, J., Rising, A., & Meng, Q. (2012). Full-length minor ampullate spidroin gene sequence. *Plos One*, 7(12), e52293.

46. Hayashi, C., & Lewis, R. V. (1998). Evidence from flagelliform silk cDNA for the structural basis of elasticity and modular nature of spider silks. *J. Mol. Biol.*, 275(5), 773–784.

47. (a) Zhou, Y., Wu, S., & Conticello, V. P. (2001). Genetically directed synthesis and spectroscopic analysis of a protein polymer derived from a flagelliform silk sequence. *Biomacromolecules*, 2(1), 111–125.
(b) Van Dijk, A. A., De Boef, E., Bekkers, A., Van Wijk, L. L., Van Swieten, E., Hamer, R. J., & Robillard, G. T. (1997). Structure characterization of the central repetitive domain of high molecular weight gluten proteins. II. Characterization in solution and in the dry state. *Protein Sci*, 6(3), 649–656.

48. Lefevre, T., & Pezolet, M. (2012). Unexpected beta-sheets and molecular orientation in flagelliform spider silk as revealed by Raman spectromicroscopy. *Soft Matter*, 8(23), 6350–6357.

49. Adrianos, S. L., Teule, F., Hinman, M. B., Jones, J. A., Weber, W. S., Yarger, J. L., & Lewis, R. V. (2013). *Nephila clavipes* Flagelliform silk-like GGX motifs contribute to extensibility and spacer motifs contribute to strength in synthetic spider silk fibers. *Biomacromolecules*, 14(6), 1751–1760.

50. Hu, X., Yuan, J., Wang, X., Vasanthavada, K., Falick, A. M., Jones, P. R., La Mattina, C., Vierra, C. A. (2007). Analysis of aqueous glue coating proteins on the silk fibers of the cob weaver, *Latrodectus hesperus*. *Biochemistry*, 46(11), 3294–3303.

51. (a) Hayashi, C. Y., Blackledge, T. A., & Lewis, R. V. (2004). Molecular and mechanical characterization of aciniform silk: uniformity of iterated sequence modules in a novel member of the spider silk fibroin gene family. *Mol. Biol. Evol.*, *21*(10), 1950–9.

(b) Vasanthavada, K., Hu, X., Falick, A. M., La Mattina, C., Moore, A. M., Jones, P. R., Yee, R., Reza, R., Tuton, T., & Vierra, C. (2007). Aciniform spidroin, a constituent of egg case sacs and wrapping silk fibers from the black widow spider *Latrodectus hesperus*. *J Biol Chem*, *282*(48), 35088–35097.

(c) Ayoub, N. A., Garb, J. E., Kuelbs, A., & Hayashi, C. Y. (2013). Ancient properties of spider silks revealed by the complete gene sequence of the prey-wrapping silk protein (AcSp1). *Mol Biol Evol*, 30(3), 589–601.

52. Xu, L., Rainey, J. K., Meng, Q., & Liu, X. Q. (2012). Recombinant minimalist spider wrapping silk proteins capable of native-like fiber formation. *Plos One*, 7(11), e50227.

53. Blasingame, E., Tuton-Blasingame, T., Larkin, L., Falick, A. M., Zhao, L., Fong, J., Vaidyanathan, V., Visperas, A., Geurts, P., Hu, X., La Mattina, C., Vierra, C. (2009). Pyriform spidroin 1, a novel member of the silk gene family that anchors dragline silk fibers in attachment discs of the black widow spider, *Latrodectus hesperus*. *J Biol Chem*, *284*(42), 29097–108.

54. (a) Geurts, P., Zhao, L., Hsia, Y., Gnesa, E., Tang, S., Jeffery, F., Mattina, C. L., Franz, A., Larkin, L., & Vierra, C. (2010). Synthetic spider silk fibers spun from pyriform spidroin 2, a glue silk protein discovered in orb-weaving spider attachment discs (dagger). *Biomacromolecules*, 11(12), 3495–3503.

(b) Perry, D. J., Bittencourt, D., Siltberg-Liberles, J., Rech, E. L., & Lewis, R. V. (2010). Piriform Spider Silk Sequences Reveal Unique Repetitive Elements. *Biomacromolecules*.

55. (a) Vollrath, F., Fairbrother, W. J., Williams, R. J. P., Tillinghast, E. K., Bernstein, D. T., Gallagher, K. S., & Townley, M. A. (1990). Compounds in the droplets of the orb spider's viscid spiral. *Nature*, 345, 526–528.

(b) Vollrath, F., & Tillinghast, E. K. (1991). Glycoprotein Glue Beneath a Spider Web's Aqueous Coat. *Naturwissenschaften*, 78, 557–559.

56. Choresh, O., Bayarmagnai, B., & Lewis, R. V. (2009). Spider web glue: two proteins expressed from opposite strands of the same DNA sequence. *Biomacromolecules*, 10(10), 2852–2856.

57. Vasanthavada, K., Hu, X., Tuton-Blasingame, T., Hsia, Y., Sampath, S., Pacheco, R., Freeark, J., Falick, A. M., Tang, S., Fong, J., Kohler, K., La Mattina-Hawkins, C., & Vierra, C. (2012). Spider glue proteins have distinct architectures compared with traditional spidroin family members. *J Biol Chem*, 287(43), 35986–35999.

58. Stark, M., Grip, S., Rising, A., Hedhammar, M., Engstrom, W., Hjalm, G., & Johansson, J. (2007). Macroscopic fibers self-assembled from recombinant miniature spider silk proteins. *Biomacromolecules*, 8(5), 1695–1701.

59. (a) Anthoula Lazaris, S. A., Yue, Huang., Jiang-Feng Zhou, Francois Duguay, Nathalie Chretien, Elizabeth, A. Welsh, Jason, W. Soares., Costas, N. Karatzas. (2002). Spider Silk Fibers Spun from Soluble Recombinant Silk Produced in Mammalian Cells. *Science*, 295, 472–476.

(b) Huemmerich, D., Scheibel, T., Vollrath, F., Cohen, S., Gat, U., & Ittah, S. (2004). Novel assembly properties of recombinant spider dragline silk proteins. *Curr. Biol.* 14(22), 2070–2074.

60. (a) Xu, H. T., Fan, B. L., Yu, S. Y., Huang, Y. H., Zhao, Z. H., Lian, Z. X., Dai, Y. P., Wang, L. L., Liu, Z. L. Fei, J., & Li, N. (2007). Construct synthetic gene encoding artificial spider dragline silk protein and its expression in milk of transgenic mice. *Anim Biotechnol*, 18(1), 1–12.

(b) Teule, F., Miao, Y. G., Sohn, B. H., Kim, Y. S., Hull, J. J., Fraser, M. J., Jr., Lewis, R. V., Jarvis, D. L. (2012). Silkworms transformed with chimeric silkworm/spider silk genes spin composite silk fibers with improved mechanical properties. *Proc Natl Acad Sci U S A*, 109(3), 923–928.

61. Lin-Cereghino, G. P., Leung, W., & Lin-Cereghino, J. (2007). Expression of protein in *Pichia pastoris*. In *Expression Systems: Methods Express*, Dyson, M., Durocher, Y., (Eds). Scion Publishing Limited: Oxfordshire, 123–145.

62. Lock, R. L. (1993). Process for making silk fibroin fibers.

63. Lazaris, A., Arcidiacono, S., Huang, Y., Zhou, J. F., Duguay, F., Chretien, N., Welsh, E. A., Soares, J. W., Karatzas, C. N. (2002). Spider silk fibers spun from soluble recombinant silk produced in mammalian cells. *Science*, 295(5554), 472–476.

64. (a) Teule, F., Cooper, A. R., Furin, W. A., Bittencourt, D., Rech, E. L., Brooks, A., & Lewis, R. V. (2009). A protocol for the production of recombinant spider silk-like proteins for artificial fiber spinning. *Nat Protoc*, 4(3), 341–355.
   (b) Hsia, Y., Gnesa, E., Pacheco, R., Kohler, K., Jeffery, F., & Vierra, C. (2012). Synthetic spider silk production on a laboratory scale. *J Vis Exp*, (65).

65. Hedhammar, M., Rising, A. Grip, S., Martinez, A. S., Nordling, K., Casals, C., Stark, M., & Johansson, J.,(2008). Structural properties of recombinant nonrepetitive and repetitive parts of major ampullate spidroin 1 from *Euprosthenops australis*: implications for fiber formation. *Biochemistry*, 47(11), 3407–3417.

66. Zeleny, J. (1914). The electrical discharge from liquid points and ahydrostatic method of measuring the electric intensity *J Phy Rev*, 3, 69–91.

67. (a) Zarkoob, S., Eby, R. K., Reneker, D. H., Hudson, S. D., Ertley, D., & Adams, W. W. (2004). Structure and morphology of electrospun silk nanofibers. *Polymer*, 45(11), 3973–3977.
   (b) Stephens, J. S., Fahnestock, S. R., Farmer, R. S., Kiick, K. L., Chase, D. B., & Rabolt, J. F. (2005). Effects of electrospinning and solution casting protocols on the secondary structure of a genetically engineered dragline spider silk analog investigated via Fourier transform Raman spectroscopy. *Biomacromolecules*, 6(3) 1405–1413.

68. Reneker, D. H., Yarin, A. L., Fong, H., & Koombhongse, S. (2000). Bending instability of electrically charged liquid jets of polymer solutions in electrospinning. *J Appl Phys*, 87(9), 4531–4547.

69. Zhou, S., Peng, H., Yu, X., Zheng, X., Cui, W., Zhang, Z., Li, X., Wang, J., Weng, J., Jia, W., & Li, F. (2008). Preparation and characterization of a novel electrospun spider silk fibroin/poly(D,L-lactide) composite fiber. *J Phys Chem B*, 112(36), 11209–11216.

70. Hardy, J. G., Romer, L. M., & Scheibel, T. R. (2008). Polymeric materials based on silk proteins. *Polymer*, 49(20), 4309–4327.

71. Arcidiacono, S. (2003). *Spinning of fibers from aqueous solution*.

72. Fahnestock, S. R., & Bedzyk, L. A. (1997). Production of synthetic spider dragline silk protein in *Pichia pastoris. Appl. Microbiol. Biotechnol.* 47(1), 33–39.

73. (a) Vuorela, A., Myllyharju, J., Nissi, R., Pihlajaniemi, T., & Kivirikko, K. I. (1997). Assembly of human prolyl 4-hydroxylase and type III collagen in the yeast *Pichia pastoris*: formation of a stable enzyme tetramer requires coexpression with collagen and assembly of a stable collagen requires coexpression with prolyl 4-hydroxylase. *Embo J*, 16(22), 6702–6712.
   (b) Myllyharju, J., Nokelainen, M., Vuorela, A., & Kivirikko, K. I. (2000). Expression of recombinant human type I-III collagens in the yeast pichia pastoris. *Biochem Soc Trans*, 28(4), 353–357.

CHAPTER 2

BIOGENIC HYDROXYAPATITE BASED IMPLANT MATERIALS

KATALIN BALÁZSI and CSABA BALÁZSI

ABSTRACT

Biomaterials used for implant should posses some important properties in order to long-term usage in the body without rejection. One of the most important property is biocompatibility. These materials are used in different parts of the human body as artificial valves in the heart, stensts in blood vessels, replacement implant in shoulders, knees, hips and orodental structures. Materials used as different biomaterials should be made with certain properties as excellent biocompatibility, superior corrosion resistance in body environment, excellent combination of high strength and low modulus, high ductility and be without toxicity.

The creation of nanocomposites of ceramic materials with particle size few ten nanometers can significantly improve the bioactivity of the implant and enhance the osteoblast adhesion. One of the most used biomaterial is hydroxyapatite. The major inorganic constituent of bones and teeth is calcium phosphate, whose composition is similar to that of synthetic hydroxyapatite (HAp; $Ca_{10}(PO_4)_6OH)_2$. This similarity provides HAp based materials excellent bioactivity like bone bonding capability, osteoconductivity, and biocompatibility.

On the other hand, titanium (Ti) is most commonly used as orthopedic implant materials or bone substitute materials. Ti has good biocompatibility and sufficient mechanical properties for medical applications. One negative property of Ti is a low abrasion resistance and minute Ti abrasion powders may cause inflammatory reactions. Biomaterials must be chemically inert, stable and mechanically strong enough to wish stand the repeated forces a lifetime. From this point of view, TiC is a very stable phase in comparison to pure Ti or Ti alloys. Titanium carbide (TiC) is useful material for biomedical instruments because has a range of desirable properties. In this work, the combination of excellent bioactive hydroxyapatite with very stable and mechanically strong TiC has been studied. The nanostructured hydroxyapatite has been prepared by high efficient milling starting from biogenic eggshells. TiC thin films were deposited by dc magnetron sputtering in argon atmosphere at different deposition temperatures. Spin coating was applied to obtain HAp decorated TiC films. Structural, mechanical and biological properties of HAp, Polymer-HAp and TiC-HAp coatings are being presented in this study.

2.1   INTRODUCTION

Biomaterials used for implant should posses some important properties in order to long-term usage in the body without rejection. One of most important properties is the biocompatibility. The biomaterial is "any substance, synthetic or natural in origin, which can be used for any period of time, as a whole a part of a system which treats, augments or replaces any tissue, organ or function of the body."[1] Biomaterials are used in different parts of the human body as artificial valves in the heart, stents in blood vessels, replacement implant in shoulders, knees, hips and orodental structures.[2,4] Materials used as different biomaterials should be made with certain properties. The materials used for orthopedic in plants should possess excellent biocompatibility, superior corrosion resistance in body environment, excellent combination of high strength and low modulus, high ductility and be without toxicity [5].

The materials currently used for implants include hydroxyapatite, 316L stainless steel, cobalt-chromium alloys and pure titanium or its alloys. Elements such as Ni, Cr and Co are found to be released from the stainless steel and cobalt chromium alloys due to the corrosion in the body environment [6]. The toxic effects of metals, Ni, Co and Cr released from prosthetic implants have been reviewed by Wapner[7]. Skin related diseases such as dermatitis due to Ni toxicity have been reported and numerous animal studies have shown carcinogenicity due to the presence of Co [8].

The success of a biomaterial or an implant is highly dependent on three major factors; (i) the mechanical, tribological and chemical properties of the biomaterial, (ii) biocompatibility of the implant and (iii) the health conditions of the recipient and competency of the surgeon [9].

The biomaterials are grouped according to use in body. The situation is similar in the case of tissue. The tissue is grouped into hard and soft tissues. Tooth or bone are examples of hard tissue. Cartilage and ligament sor skin are the examples of soft tissues. These two types of tissues have the different properties from the structural or mechanical view. Considering the structural or mechanical compatibility with tissues, metal sor ceramics are chosen for hard tissue applications and polymers for soft tissue applications. The different mechanical properties of both types of tissues are shown in Table. 2.1.[10].

TABLE 2.1 Mechanical Properties of Hard and Softtissues[10]

Hard tissue	Modulus (MPa)	Strength (MPa)
Cortical bone – longitudinal direction	17.7	133
Cortical bone – transverse direction	12.8	52
Cancellous bone	0.4	7.4
Enamel	84.3	10
Dentime	11.0	39.3
Soft tissue	**Modulus (MPa)**	**Strength (MPa)**
Articular cartilage	10.5	27.5
Fibrocartilage	159.1	10.4
Ligament	303.0	29.5
Tendon	401.5	46.5
Skin	0.1– 0.2	7.6
Intraocular lens	5.6	2.3

In this chapter, THA hydroxyapatite based biomaterials developed as hard and soft tissue replacement were studied. The structural, mechanical and biological properties of bioinerttic—bioactive hydroxyapatite, biogen hydroxyapatite prepared from eggshells and polymer—hydroxyapatite composites were characterized.

2.2 BIOGENIC HYDROXYAPATITE IMPLANTS PREPARED FROM EGGSHELLS

Hydroxyapatite (HAp) has been widely used as an artificial bone substitute because of the its high biocompatibility and good bioaffinity, as well as osteoconductabil-ity. HAp powders have been produced using bio products like corals[11], cuttlefish shells[12], natural gypsum[13], natural calcite[14], bovine bone[15], eggshell[16,17], etc. Chemi-cal analysis has shown that these products which are otherwise considered as bio-waste are rich sources of calcium in the form of carbonates and oxide.

Several papers reported to produce the materials for implant or prosthesis pur-poses with chemical characteristics similar to HAp[18–21]. Eggshell is a nonexpensive and environmental friendly material for HAp production. The eggshell is consist-ing of a three-layered structure, namely the cuticle, the spongeous layer and the lamellar layer. The cuticle layer represents the outermost surface and it consists of a number of proteins. Spongeous and lamellar layers form a matrix constituted by protein fibers bonded to calcite-calcium carbonate crystals. The eggshell represents the 11% of the total weight of the egg and is composed by calcium carbonate 94%, calcium phosphate 1%, organic matter 4% and magnesium carbonate 1%[22]. Bone replacements are frequently required to substitute damaged tissue due to trauma, disease or surgery. Resorbable porous bioceramics, such as β-tricalcium phosphate (β-$Ca_3(PO_4)_2$, β-TCP) and hydroxyapatite ($Ca_{10}(PO_4)_6(OH)_2$, Fig. 2.1) have been widely used as bone defect filling materials due to their remarkable biocompatibility and close chemical similarity to biological apatite present in bone tissues[23,26].

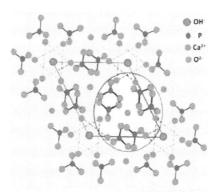

FIGURE 2.1   Schematic view of hydroxyapatite, $Ca_{10}(PO_4)_6(OH)_2$.

2.2.1 PREPARATION OF HYDROXYAPATITE

Eggshells were collected and washed with detergent, then calcined in air at 900 °C for 10 h. During the first 30 min most of the organic materials were burnt out, then the eggshells were converted to calcium oxide. Calcined shells were crushed and milled in a ball mill or an attritor mill. The ball mill (Fritsch GmBH, Fig. 2.2a) was equipped with alumina balls and bowls, the attritor mill (Union Process, Fig. 2.2b) was fitted with zirconia tanks and zirconia balls (Ø2 mm). The crushed eggshells were reacted with phosphoric acid (H_3PO_4) in an exothermic reaction. The mixtures were milled for 5 h at 4000 rpm (attritor milling) or for 10 h at 350 rpm (ball milling), to achieve homogenous mixtures and to prevent agglomeration. In all cases, the used shell: H_3PO_4 ratio was 50 : 50 wt.%[16,17].

After milling, a small amount (approximately 0.5 g) of each type of HAp powder was sintered at 900 °C for 2 h in air. HAp powders usually degrade at high temperatures, the most common problem being CaO formation.

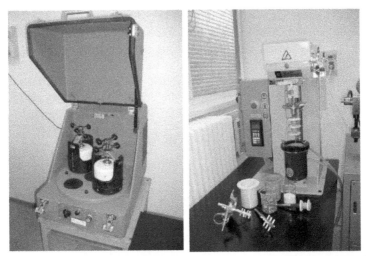

FIGURE 2.2   Mills used for hydroxyapatite preparation. (a) ball mill, (b) attritor mill.

2.2.2   STRUCTURAL INVESTIGATION OF HYDROXYAPATITE

The structural investigations were performed by scanning electron microscope (SEM, LEO 1540 XB) and transmission electron microscope (TEM, Philips CM-20). The eggshell structure (Fig. 2.3a) is relatively compact with average grain size about 3 μ m. The XRD measurement of the calcined eggshell confirmed mainly the CaO (JCPDS-PDF 0371497) phase (Fig. 2.3b).

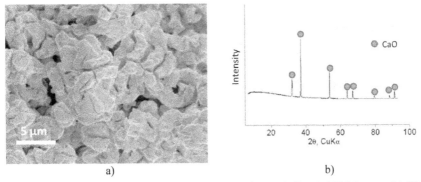

a)   b)

FIGURE 2.3   Structural investigation of calcinated egg shell. (a) SEM image, (b) XRD measurement.

The effect of various milling was observed after attritor and ball milling[17,27]. SEM image of structure prepared by attritor is shown in Fig. 2.4a. The structure produced by ball milling is shown in Fig. 2.4b. The SEM investigation confirmed that the attritor milling is more efficient in grain size reduction compared to ball milling. Thus, smaller particle size with homogeneous size distribution may be achieved with attritor milling. In both cases, the grains were agglomerated. The XRD measurements of powders prepared by two different milling showed different phases (Fig. 2.5). In the both cases, the powders are consisted of hydroxyapatite (HAp, JCPDSPDF 74-0565), calcite ($CaCO_3$, JCPDS-PDF 05-0586), calcium hydroxide ($Ca(OH)_2$, JCPDS-PDF 01-0653).

a)   b)

FIGURE 2.4   SEM images of milled egg shells. (a) attritor mill at 5 h, 4000 rpm, (b) ball milling at 10 h, 350 rpm.

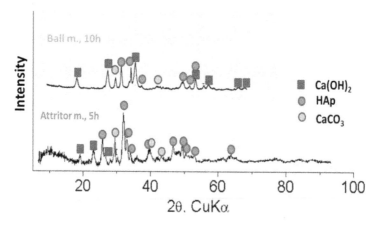

FIGURE 2.5   XRD measurements of milled powders.

The reduction of average particle size of milled powders from few hundred nm to 100 nm was observed after heat treatment at 900 °C during 2h. The agglomeration is still present, but the size of agglomerates is enhanced (Fig. 2.6). Smooth surfaces are evolving after heat treatment at 900 °C in the cases of ball milling.

a)   b)

FIGURE 2.6   SEM images of milled powder after 2h heat treatment at 900 °C. (a) attritor, (b) ball mill.

Figure. 2.7 shows X-ray diffractograms of the milled powders after heat treatment at 900 °C during 2h.Only two phases were observed; the main phase hydroxyapatite (HAp) and minor phase calcium oxide (CaO, JCPDS-PDF 037-1497).

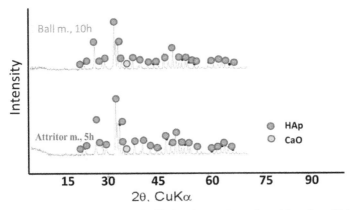

FIGURE 2.7   XRD measurements of milled and treated powders. (a) attritor, (b) ball mill.

In order to check the presence of trace elements in addition to the diffractograms energy dispersive X-ray microanalysis (EDS) measurements were performed on the milled powder samples. As the hydrogen content can not be analyzed by X-ray methods, furthermore the irregular and porous samples do not allow rigorous correction of the EDS spectra, the results can be considered as semiquantitative ones. The spectra were collected in a JSM-25 S-III SEM using Bruker Si(Li)EDS detector and Quantax system using area scan mode for averaging. The collection times were typically25-60 min (live lime) on a single 1 mm² area of each sample, then the data were turned into numerical composition values using the no-standard P/B

ZAF package of the Quantax system. The normalized mass percent output values of the P/B ZAF program are summarized in Table 2.2. Zirconia may be found in the samples prepared by attrition milling only, because of wearing of zirconia milling parts (balls, tanks). Other elements found by EDS are sodium and silicon (as shown in Tab.2.1).

TABLE 2.2   EDS Measurements of Milled and Heat Treated Powders

Elements at%	O	Na	Mg	Si	P	S	Cl	Ca	Zn	Zr	Ca/P
Attritor mill	42.05	0.08	0.45	0.06	16.01	0.1	0.03	41.01	0.06	0.16	1.98
Attritor, heat tr.	31.76	0.05	0.47	0.1	18.34	0.1	0.01	48.98	0.06	0.14	2.06
Ball mill	49.67	0.01	0.4	0.03	14.81	0.09	0.01	34.87	0.06	0.05	1.82
Ball m., heat tr.	41.25	0.06	0.47	0.07	16.85	0.08	0	40.99	0.04	0.17	1.88

These elements are characteristic trace elements of eggshell. Sulfur and chlorine are other contaminants. As observed by EDS the chlorine can be found only in attrition milled samples. Although the EDS results are semiquantitative ones, it is interesting to see that magnesium is found. Increasing concentration of Mg in HAp has the following effects on its properties: (i) decrease in crystallinity, (ii) increase in HPO_4^{2-} incorporation, and (iii) increase in extent of dissolution [28]. Mg is one of the main substitutes for calcium in biological apatites. Enamel, dentin and bone contain, respectively, 0.44, 1.23, and 0.72wt.% of Mg [28]; Mg-substituted HAp materials (denoted hereafter as Mg-HAp) are expected to have excellent biocompatibility and biological properties [29].

The phase composition of powders was studied by Fourier transform infrared spectroscopy (FTIR-Varian Scimitar FTIR spectrometer equipped with broad band MCT detector). The FTIR spectrum after attritor milling (Fig. 2.8a) resembles the characteristic spectral feature of bone mineral. The spectrum is dominated by the typical PO_4 bands of poorly crystalline apatite phase components of the triply degenerated $\upsilon_3 PO_4$ asymmetric mode at 1021 and 1087 cm^{-1} (shoulder), nondegenerated symmetric stretching mode of $\upsilon_1 PO_4$ at 962 cm^{-1} and components of the triplet of $u_4 PO_4$ bending mode at 599 and 562 cm^{-1}[17,27]. Carbonate bands are also observed at 1550–1350 cm^{-1} (υ_3), 873 cm^{-1} (υ_2) and 712 cm^{-1} (u_4). By analogy with bone mineral, the position of carbonate bands (1456, 1415 and 872 cm^{-1}) indicates the formation of a carbonated apatite with B-type substitution (in tetrahedral positions) [30]. The broad band of low intensity in the range 3000–3400 cm^{-1} can be attributed to traces of water incorporated to the structure, together wit the very weak, broad band around 1640 cm^{-1} of H–O–H bending mode. In the OH stretching vibration region, beside the υOH of HAp, surface OH band at 3644 cm^{-1} also appears, probably connected to CaO occurring on the surface. In the case of attritor milled and treated powder, the intensity of surface OH band also increases with increasing the temperature (Fig. 2.8a).

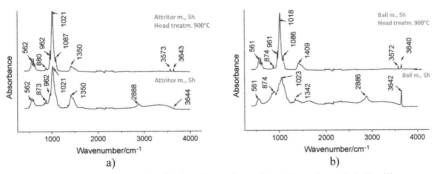

FIGURE 2.8   FITR spectra of milled and heated powders, (a) attritor, (b) ball mill.

FTIR spectrum of ball-milled powders shows a complex mixture of different calcium phosphate phases (Fig. 2.8b). This band decreases in matured bone apatite and in highly crystalline HAp[31]. The pure HAp phase with some carbonate substitution is formed only after 900 °C. However, surface OH bands are also present in the spectrum. After heat treatment, spectral features of the apatite phase become dominant and above 800–900 °C beside some peaks of carbonate (v_3 at 1466 and 1409 cm^{-1}, v_2 at 874 cm^{-1}, u_4 at 713 cm^{-1}) and surface –OH (nOH at 3640 cm^{-1}) typical bands of well-crystallized HAp can be observed (v_3PO_4 at 1086 and 1018 cm^{-1}, v_1PO_4 at 961 cm^{-1} and the triplet of u_4PO_4 at 626, 599 and 561 cm^{-1})[30]. No traces for acid phosphate (HPO_4^{2-} peak around 540–530 cm^{-1}), characteristic for immature bone mineral or incomplete apatite phase can be observed[32].

2.2.3   BIOLOGICAL PROPERTIES OF HYDROXYAPATITE

Sixteen 4-month-old New Zealand white rabbits with an average weight of 2.8 kg (range 2.5–3.0 kg) were used in this experiment, which was approved by the Institutional Animal Care and Use Committee of Gangneung-Wonju National University, Gangneung, Korea (no. 2010–16).

General anesthesia was induced by intramuscular injection of a combination of 0.4 mL ketamine (100 mg/mL; Ketara; Yuhan, Seoul, Korea) and 0.3 mL xylazine (10 mg/kg body weight; Rompun; Bayer Korea, Seoul, Korea). The cranium area was shaved and disinfected with povidone-iodine. From the nasal bone to the occipital protuberance, a longitudinal incision was made in the skull. Then a midline incision was created in the periosteum. Sharp subperiosteal dissection reflected the pericardium from the outer table of the cranial vault, exposing the parietal bones. A dental trephine bur was used under copious saline irrigation to create a bilateral full-thickness calvarial defect. Two 8-mm diameter defects were created, one on each side of the midline. The graft, HAp from eggshells, was placed on the cavarial defects. Some defects remained unfilled as the control. Assignment to each group for the corresponding defect was done randomly, and each group was composed of 10 animals (10 defects for each group). None of the animals received the same grafts in both calvarial defects. Eight animals were killed at 4 weeks and at 8 weeks. The dimensions of the calvarial specimens was $25 \times 12 \times 3$ mm at the largest, including both defects. They were fixed in 10% formalin[33,34].

The bones were dehydrated in ethanol and decalcified by using 5% formic acid for 2 weeks for histomorphometric evaluation. The right and left parietal bones were separated through the midline sagittal suture. Both segments were embedded to show the sagittal sections in the paraffin blocks. Then the sections were sliced and stained with Masson trichrome. The section showing the widest defect area was selected, and cuts from 50 µm before and after were also selected. Digital images of the selected sections were taken by using a digital camera (DP–20; Olympus, Tokyo, Japan). The images were analyzed by Sigma Scan Pro (SPSS, Chicago, IL).

TABLE 2.3   Histomorphometry Analysis After 4 and 8 Weeks

	Unfilled	HAp graft	Unfilled	HAp graft
(%)	4 weeks		8 weeks	
Total new bone	17.11 ±10.24	25.68 ±10.89	27.50 ±10.89	41.99 ±8.44
Residual graft	-	40.63 ±12.19	-	9.60 ±4.95

The results of the histomorphometry are presented in Table 2.3. Total new bone was $17.11 \pm 10.24\%$ in the control group at 4 weeks after the operation. It was $25.68 \pm 10.89\%$ in the HAp group at 4 weeks after the operation[33]. Residual graft was $40.63 \pm 12.19\%$ in the HAp group at 4 weeks after the operation (Fig. 2.9a). HAp had a low inflammatory response. Total new bone at 8 weeks after the operation was $27.50 \pm 10.89\%$ in the control group; it was $41.99 \pm 8.44\%$ in the HAp group (Fig. 2.9b).

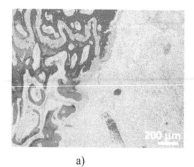
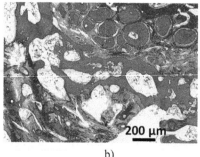

a)   b)

FIGURE 2.9   Histologic view of hydroxyapatite from egg shells after the operation. (a) After 4 weeks. (b) After 8 weeks.

HAp was significantly different compared with the unfilled control ($P < 0.038$). Well-organized lamella bony islands were formed in the HAp group (Fig. 2.9b). Most defect areas were filled with regenerated bone in the HAp group, and the remaining HAp particles were incorporated into the regenerated bone at 8 weeks after the operation.

After bilateral parietal bony defects formation (diameter: 8.0 mm), nanosized HAp was grafted. The control was unfilled defect (Fig. 2.10a, d). Results of bone regeneration evaluated by microcomputerized tomograms (μCT) at 4 and 8 weeks are shown in Fig.10b and Fig. 2.1°C. HAp showed much more bone formation compared to unfilled control group in both micro tomographic and histomorphometric analysis. The region of interest for each specimen was analyzed for bone mineral content (BMC) and bone mineral density (BMD).

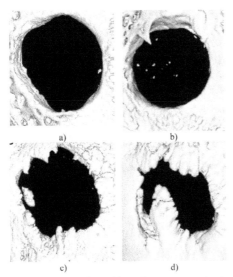

FIGURE 2.10   Bone regeneration evaluated by microcomputerized tomograms. a) unfilled control after 4 weeks, HAp graft after 4 weeks, c) unfilled control after 8 weeks, d) HAp graft after 8 weeks.

TABLE 2.4   μCT Analysis of Unfilled and HAp Grafts at 4 and 8 Weeks after Operation

	Unfilled	HAp graft	Unfilled	HAp graft
	4 weeks		8 weeks	
BMC (mg)	43.71 ± 3.28	190.28 ± 13.67	25.66 ± 4.36	187.77 ± 28.07
BMD (mg/cm³)	303.16 ± 18.82	635.21 ± 46.48	243.34 ± 10.02	1245.35 ± 182.95
TMC (mg)	22.43 ± 0.93	101.39 ± 15.62	12.32 ± 4.43	75.59 ± 33.92
TMD (mg/cm³)	418.06 ± 25.54	1158.61 ± 41.34	327.87 ± 8.31	2100.04 ± 45.24

The tissue mineral content (TMC) and tissue mineral density (TMD) were calculated by software (Table 2.4). All measured variables, such as BMC, BMD, TMC, and TMD, were significantly higher on μCT analysis in the HAp group than in the unfilled control group ($P < 0.001$ except for BMD at 4 weeks [$P< 0.002$] and TMC at 8 weeks [$P< 0.014$]). Bone regeneration rate of the nHA group at 8 weeks after surgery was 187.77% ± 28.07%, and it was significantly higher than that in the unfilled control group (25.66% ± 10.98%) ($P ± 0.046$). The nanosized HAp can increase initial cellular attachment compared with a plastic surface [35]. Calcium ions are important in osteoblast differentiation [36]. The degradation of nanosized HAp alters calcium/phosphate metabolism and activates the osteoblast via a specific calcium ion channel [37]. The HAp nanoparticles showed increased osteoblast attachment and higher calcium release than conventional sized HA particles[38].

2.3 POLYMER/HYDROXYAPATITE COMPOSITE IMPLANT MATERIAL

Ceramics are known for their good biocompatibility, corrosion resistance, and high compression resistance. Drawbacks of ceramics include, brittleness, low fracture strength, difficult to fabricate, low mechanical reliability, lack of resilience, and high density. Polymer composite materials provide alternative choice to overcome many shortcomings of homogenous materials mentioned above. A lot of polymers are used in various biomedical applications; such as polyethylene (PE), polyurethane (PU), polytetrafluoroethylene (PTFE), polyacetal (PA), polymethylmethacrylate (PMMA), polyethylene terephthalate (PET), silicone rubber (SR), polysulfone (PS), poly(lactic acid) (PLA), and poly(glycolic acid) (PGA). Each type of material has its own positive aspects that are particularly suitable for specific application. One of these applications is an orthopedic implant. One of the major problems in orthopedic surgery is the mismatch of stiffness between the bone and metallic or ceramic implants. In the load sharing between the bone and implant, the amount of stress carried by each of them is directly related to their stiffness. In this respect, the use of low-modulus materials such as polymers appears interesting; however, low strength associated with low modulus usually impairs their potential use. Since the fiber reinforced polymers, that is, polymer composite materials exhibit simultaneously low elastic modulus and high strength, they are proposed for several orthopedic applications [39]. The mechanical properties of various polymers are shown in Table 2.5.[39]

TABLE 2.5   Mechanical Properties of Typical Polymeric Biomaterials[39]

Material	Modulus (GPa)	Strength (MPa)
Polyethylene (PE)	0.88	35
Polyurethane (PU)	0.02	35
Polytetrafluoroethylene (PTFE)	0.5	27.5
Polyethylene terephthalate (PET)	2.85	61
Polyacetal (PA)	2.1	67
Polysulfone (PS)	2.65	75

When the diameters of polymer fiber materials are decreased from micrometers to submicrons or nanometers, there appear several amazing characteristics such as very large surface area to volume ratio, flexibility in surface functionalities and superior mechanical properties (stiffness and tensile strength) compared with any other known form of the material. A number of processing techniques such as drawing [40], template synthesis [41], phase separation [42], self-assembly [43], electrospinning [44], etc. have been used to prepare polymer nanofibers in recent years.

Cellulose, $(C_6H_{10}O_5)n$, is the major component in the rigid cell walls in plants, a linear polysaccharide polymer with many glucose monosaccharide units. As an acetate ester of cellulose, cellulose acetate (CA) is a biodegradable polymer. CA scaffolds have been used for growing "structurally mature" and "functionally competent" cardiac cell networks [45]. For ceramic–polymer hybrid systems, poor dispersion of ceramic powders in the polymer matrix and agglomeration of ceramic component has been a grave issue to the manufacturing of artificial scaffolds.

The use of nanotechnology has helped overcome these limitations. The synthesis of hybrid ceramic polymer nanofibers by means of electrospinning is a major breakthrough in biotechnology-related nanomanufacturing. Nanosized HAp prepared from eggshells and CA were combined to form novel hybrid 3D scaffolds mimicking the extracellular matrix (ECM) architecture.

2.3.1 PREPARATION OF POLYMER-HAP COMPOSITES

2.3.1.1 ELECTROSPINNING

During electrospinning process a high voltage (5–30 kV) is applied between a needle attached to a syringe and a target. The out flowing solution gets charged which takes it to the target. During the flight the solvent evaporates and polymer fibers develop. The fibers on the target are disordered, their diameter can reach 100 nanometer to few millimeters. The fibers can evolve thicker than the average diameter forming concentrated beads, which significantly decreases the specific surface area. The fiber diameter and the number of beads are the main characteristics of the final samples. In the electrospinning process there are given parameters which cannot be modified: humidity, temperature, pressure, atmosphere, molecular weight of the polymer. Other parameters such as viscosity, surface tension, conductivity, dielectric constant may be modified by selecting the appropriate polymer. The others parameters may be modified directly: flow rate, voltage, diameter of the needle and the distance between the needle and the target [46].

2.3.1.2 PREPARATION METHOD OF COMPOSITE

Nanosized HAp powder was synthesized from raw materials [16,17]. The as-received HAp powder was added to acetic acid forming a 9.38% wt./vol solution. This solution was then mixed with CA solution, which was CA powder dissolved in acetone to form 15% wt./vol composition (Fig. 2.11). The mixed solution (71–29 vol.% acetone/acetic acid) was introduced into a standard vertical electrospinning setup and was electro-spun by KD Scientific KDS202-CE electrospinning device with high voltage supply. The needles were common medical needles, which ending-points were polished perpendicular to the direction of the flow. The polymer was cellulose-acetate (CA) (22188; Mr~29000 {9004–35–7}; Fluka/Sigma-Aldrich), the applied solvents were acetone (Spektrum3D), isopropanol (Reanal), and acetic-acid (Spektrum3D), flow rate 9.6 mL/h, voltage 19 kV and distance from needle to collector 10 cm.

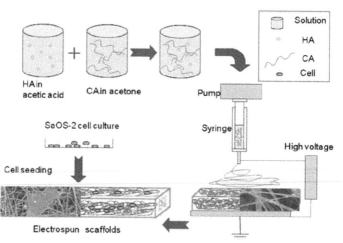

FIGURE 2.11   Schematic view of Polymer-HAp biocomposite preparation.[44]

2.3.2 STRUCTURAL PROPERTIES OF POLYMER-HAP COMPOSITE

Structural investigations showed that the diameter of the fibers is nearly uniform, about 500–1000 nm, but too many beads were formed and spheres developed with a few hundred microns in diameter (Fig. 2.12a). The EDS confirmed that the fibers contain the HAp, but the element mapping shows that the distribution of it is not homogenous. The EDS measurement showed C (found in CA), O (CA and HAp), P (HAp), Mg, S (HAp-eggshell trace element) Si (sample holder) and Al (sample cover) compounds. According to the Ca (Fig. 2.12b)) and P element maps (Fig. 2.12c)) the HAP presence in 1–2 micrometer sized clusters. The presence of Ca and P can be in 1–2 micron clusters. The distribution of O (Fig. 2.12d)) is homogenous, because the CA contains it too.

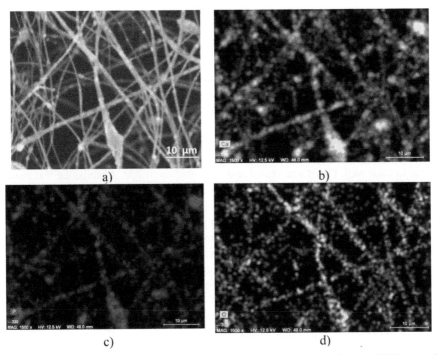

FIGURE 2.12 Structural investigation of CA- HAp composite. (a) SEM images, (b) Elemental map of Ca, c) Elemental map of P, d) Elemental map of O.

In our study, the mixing parameters were between 15 and 25 w/w% solid material (HAp+CA), increasing the concentration varied of Hap+CA resulted larger fiber diameter and fewer beads [47]. Using more than 30 w/w% and less than 5 w/w% it was impossible to produce fibers. Applying 20 w/w% acetic-acid electrospinning process generated uniform fiber diameter. Using any other w/w% acetic-acid caused larger fibers diameter and more beads. It could be concluded that the composition of the solution has the most striking influence on the morphology of the final mats.

2.3.3 BIOLOGICAL PROPERTIES OF POLYMER-HAP COMPOSITE

MTS array as applying for mazan (CellTiter 96® Aqueous One Solution Cell Proliferation Assay, Promega, Madison, WI) used the novel tetrazolium compound (MTS) and the electron coupling reagent, phenazine metho sulfate (PMS). MTS is chemically reduced by cells into formazan, which is soluble in tissue culture medium. The measurement of the absorbance of the formazan can be carried out using 96 well microplates at 492 nm. The assay measures dehydrogenase enzyme activity found in metabolically active cells.

The metabolic activity of cells was monitored using the MTS that is chemically reduced by metabolically active cells into formazan. The amount of formazan produced is an indicator of the cell viability. The measurement of formazan absorbance was performed in 96 well plates after several days of incubation. Standard curves were prepared by diluting a series of cell suspensions from 15.7 cells/mL to 157,000 cells/mL. An aliquot (1.0 mL) of each dilution was transferred to wells of a 24-well tissue culture plate in triplicate. Subsequently, 150 μL of the MTS solution was added to each suspension.

The plate was incubated at 37 °C in a humidified atmosphere containing 5% CO_2 in the dark for 4 h, after which 1.0 mL of the kit's solubilization/stop solution was added to each well. Plates was sealed and incubated overnight. Absorbance was read at 570 nm wavelength and also at 650 nm as a reference wavelength using the plate fluorimeter.

The scaffolds were sterilized by incubating with 70% ethanol for 30 min, seeded with human osteoblast-like cells (SaOS-2) and cultured for up to 14 days. The viability of the osteoblasts seeded to the electrospun scaffolds was determined first by using the MTS assay (Fig. 2.13). The amount of formazan produced is proportional to number of living cells in culture since the chemical reduction of formazin to a colored product is dependent on the number of viable cells. The MTS viability assay demonstrated that the cells exposed to these scaffolds maintain the ability to proliferate for up to 14 days that the experiment lasted for.

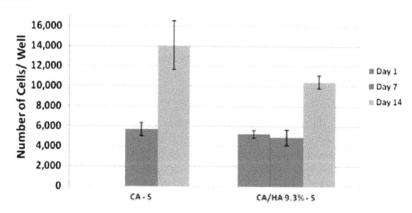

FIGURE 2.13   MTS cell viability assay testing results.[44]

To evaluate cell morphology on the scaffolds, samples were prepared for electron microscopy (EM) staining. Samples were washed 2 times with PBS and fixed with 2.5% gluteraldeyde [Sigma] for 2 h then washed with PBS. The cell-scaffold constructs were then attached to aluminum stubs, sputter-coated with gold, and then examined under a LEO-Gemini Schottky FEG scanning electron microscope.

The cells exposed to the nano scaffolds interacted with multiple fibers. Anchoring sites for cell attachment to the fibers were visualized by SEM. The nanoclusters of HAp mineral were consistently located at the edge of cells, which provides additional evidence that they act as anchoring sites for cell attachment to the fibrous hybrids. This improvement in cell adhesion and growth are deemed to the biological role of HAp. Figure 2.14 shows the morphology of the cells cultured on the scaffolds for 1, 7 and 14 days. The size and number of the cells are both increasing with time.

Our studies suggest that the morphology and structure of the CA-HAp composite scaffolds play important roles in facilitating cell spreading and differentiation and enhance apatite mineralization. Based on our observations, the electrospun CA scaffolds with nanosized HAp are considered as a promising candidate for bone tissue engineering application.

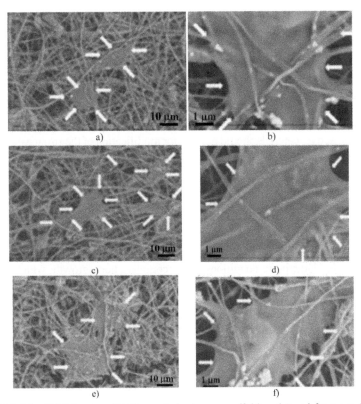

FIGURE 2.14 SEM images of cell morphologies on scaffold sculptured for up to 14 days. Yellow marks show the anchor age sites of cells. (a) CA-HAp day 1, (b) CA-HAp day 1 (high magnification), (c) CA-HAp day 7, (d) CA-HAp day 7 (high magnification), (e) CA-HAp day 14, (f) CA-HAp day 14 (high magnification).

2.4 BIOINERT TITANIUM/BIOACTIVE HYDROXYAPATITE IMPLANTS

Titanium (Ti) is most commonly used as orthopedic implant materials or bone substitute materials. Ti possesses the good biocompatibility and the sufficient mechanical properties for medical applications (Table 2.6) [48]. One negative property of Ti is a low abrasion resistance and minute Ti abrasion powders may cause inflammatory reactions [49]. Scaling treatment is only method for removal of plaque and dental calculus adheres, and it is a necessary treatment process to obtain good prognosis throughout the long-term maintenance of implant [50]. Therefore, for abutment division of implant, it is important to possess high abrasion resistance to keep the smoothness of the implant surface after scaling treatment [51]. Biomaterials must be nontoxic, noncarcinogenic, chemically inert, stable and mechanically strong enough to withstand the repeated forces of a lifetime. From this point of view, TiC is a very stable phase in comparison to pure Ti or Ti alloys. Titanium carbide (TiC) is an useful material for biomedical instruments because it possesses a range of desirable properties. The combination of very high hardness, high melting temperature, and excellent thermal and chemical stabilities makes TiC suited to a number of commercial applications. TiC is often used in abrasives, cutting tools, grinding wheels, and coated cutting tips [52].

Ti based implants are classified as bioinert. Bioinert refers to a material that retains its structure in the body after implantation and does not include any immunologic host reaction. Bioactive materials should be used for modification of the surface that occurs upon implantation. Bioactive refers to materials that direct chemical bonds with bone or even with soft tissue of a living organism. One of most used bioactive materials is a hydroxyapatite. The major inorganic constituent of bones and teeth is calcium phosphate, whose composition is similar to that of synthetic hydroxyapatite (HAp; $Ca_{10}(PO_4)_6OH)_2$. This similarity provides HAp based materials excellent bioactivities like bone bonding capability, osteoconductivity, and biocompatibility. The major disadvantage of HAp is its poor adhesion, poor mechanical integrity, high brittleness, degradation in acidic/basic conditions and incomplete bone growth, which restricts its application only in non load-bearing areas of the human body. The creation of nanocomposites of ceramic materials with particle size few ten nanometers can significantly improve the bioactivity of the implant and enhance the osteoblast adhesion.

TABLE 2.6 Mechanical Properties of Typical Metallic and Ceramic Biomaterials[39]

Material	Modulus (GPa)	Strength (MPa)
Stainless steel	190	586
Co–Cr alloy	210	1085
Ti-alloy	116	965

TABLE 2.6   *(Continued)*

Material	Modulus (GPa)	Strength (MPa)
Zirconia	220	820
Bioglass	35	42
Hydroxyapatite	95	50
Alumina	380	300

2.4.1 TIC BASED THIN FILMS

2.4.1.1 PREPARATION OF TIC/A:C THIN FILMS

Modern methods of vacuum deposition provide great flexibility for manipulating material chemistry and structure, leading to films and coatings with special properties. TiC/A:C nanocomposite thin films have been prepared by DC magnetron sputtering (Fig. 2.15) on silicon (001) substrate with 300 nm thick oxidized silicon sublayer. Films have been deposited at 200 °C in argon at 0.25 Pa. The input power of 99.999% purity carbon target (C) was kept constant at 150 W and the input power of the 99.995% purity titanium target (Ti) was changed between 15 and 50 W. The deposition rate was ~0.06 nm/s. The thickness of thin films was about 300 nm. The structural investigations have been performed on a Philips CM–20 Microscope using a 200 kV accelerating voltage. The elemental analysis of film composition was also performed in this microscope, which is equipped with a NORAN EDS (Energy Dispersive Spectrometer), with an HP-Ge detector.

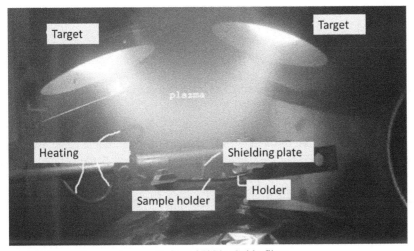

FIGURE 2.15   DC magnetron sputtering of TiC/A:C thin films.

2.4.1.2   STRUCTURAL INVESTIGATION OF TIC/A:C THIN FILMS

In our previous works, TiC based nanocomposites have been deposited between 25 °C and 800 °C. As it has been showed, the nanocomposite film deposited at 200 °C exhibited the best mechanical properties (nanohardness 18 GPa and elastic modulus 205 GPa) [53,54]. The film consisted of TiC crystallites separated by thin carbon matrix (Fig. 2.16a) as showed by TEM investigations. The crystallites have columnar structure with average width around 10–15 nm (Fig. 2.4.2b). The mechanical properties of films, namely hardness and elastic modulus may be compare with mechanical properties of bulk Ti implants (Table 2.6). In the case of TiC film, the 18GPahardness and 205 GPa elastic modulus value was measured[53,54]. TiC phase was the reinforcing phase to enhance the hardness of films. Highest H/E ratio ~ 0.094 and elastic recovery ~ 0.634 indicate that the deformation of the TiC nanocomposites arises mainly from elastic deformation of the C matrix which determines the elasticity of the asperities in a tribological contact.

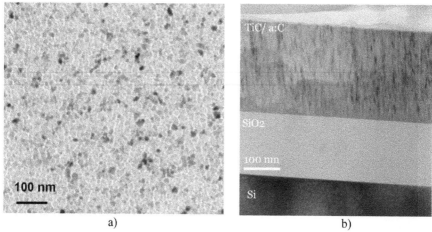

a)   b)

FIGURE 2.16   TEM investigation of TiC/A:C thin film prepared at 200 °C. (a) plan view TEM image, (b) cross-section TEM image.

Selected area electron diffraction (SAED) confirmed the stable cubic TiC phase (Fig. 2.17. JCPSFWIN 32–1383). According to Balden et al.,[55] the optimal content from the structure (crystallites separated by thin carbon matrix) of the doped C films is one with a low metal concentration within 1–20 at.%. EDS elemental analysis of the film composition resulted in ~62–72 at.% C and 23–26 at.% Ti composition of the films (Table 2.7).

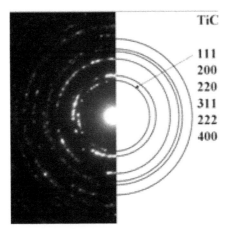

FIGURE 2.17   SAED of TiC/A:C thin film.

TABLE 2.7   Elemental Composition of TiC Based Film Measured by Two Different Methods

Element	Ti	C	O
EDS (at%)	26.8	72.6	0.6
AES (at%)	17	81	2

The film prepared at argon-contained ~30 at% Ti and ~70 at% C according to the EDS. The Auger Electron Spectroscopy (AES) depth profiling was applied to determine the in-depth compositions. This surface sensitive analysis obviously gave the concentration values in the surface close region as well, which might play a role in the interaction with the biological substances. To minimize the ion bombardment induced roughening Ar+ ions of 1 keV with angle of incidence (with respect to the surface normal) of 78° were applied, and the sample was rotated during sputtering. The morphology of films was studied by atomic force microscopy (AFM AIST-NT, Smart SPM 1010), in semicontact mode. AES analysis gave somewhat higher C and lower Ti concentrations, which can be accepted considering the difficulties of the C analysis[56]. The concentration of oxygen according to the EDS was below than the detectability limit, while AES detected oxygen in all of the 300 nm thickness of TiC film as shown in Fig. 2.18. The oxygen concentration was constant, around 2%, inside the layer, while it strongly increased at the outer surface as well as at the TiC/SiO$_2$ interface.

The carbon concentration followed similar trend; it was constant inside the layer, while it changed at the outer and inside interface. The carbon concentration, however, decreased at the surface (Fig. 2.18b). Thus the surface of the film consists Fig. 2.18b. AES spectra of carbon in TiC film of amorphous carbon, titanium carbide and titanium oxide.

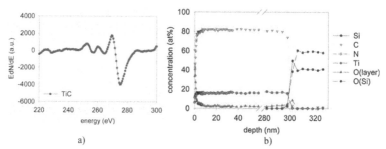

a)   b)

FIGURE 2.18   AES measurements of TiC/A:C thin film. (a) Spectrum of carbon, (b) Depth profile.

2.4.1.3 BIOLOGICAL PROPERTIES OF TIC BASED THIN FILMS

The in-vitro tests of biocompatibility and bioactivity of TiC coatings were studied on the adhesion, growth, maturation, viability, stress adaptation and potential immune activation of osteogenic cells in cultures on these materials. For the cell culture experiments, the samples have been sterilized with 70% ethanol, inserted into 24-well polystyrene cell culture plates (TPP, Switzerland; internal well diameter 15.6 mm). These studies were carried out on human osteoblast-like cell line MG–63 (European Collection of Cell Cultures, Salisbury, UK).

The cells have been cultured for 1, 3, and 7 days at 37 °C in a humidified air atmosphere containing 5% CO_2. For the cell culture experiments, glass slides and also the bottom of standard polystyrene cell culture dishes have been used as reference materials. The detailed description of cell seeding was described by Balazsi et al.[56,57]

The samples have been used for evaluation of cell number and their viability by the LIVE / DEAD viability/cytotoxicity kit for mammalian cells (Invitrogen, Molecular Probes, USA) according to the manufacturer's protocol. Briefly, nonfixed cells have been incubated for 5 to 10 min at room temperature in a mixture of two of the following probes: calcein AM, a marker of esterase activity in living cells, emitting green fluorescence (excitation/emission ~495/~515 nm), and ethidium homodimer-1, which penetrated into dead cells through their damaged membrane and produced red fluorescence (excitation/emission ~495/~635 nm).

TiC/A:C nanocomposite thin films did not lead to an increase in cell number after 1 day (Fig. 2.19a) in culture in comparison with the control microscopic glass coverslips [56]. However, on days 3 and 7 after seeding, the cell numbers on TiC/A:C surface have been found to be similar to PS and significantly higher than that on the microscopic glass coverslips (Fig. 2.19). Harcuba et al. [58] investigated the biological properties of Ti–6Al–4 V alloy after surface treatment by the electric discharge machining (EDM) process using MG63 osteoblast cells. These results could be compared with our results and showed that samples modified by EDM provide a better substrate for the adhesion and growth of human bone-derived cells than the alloy plasma-sprayed withTiO$_2$[58].

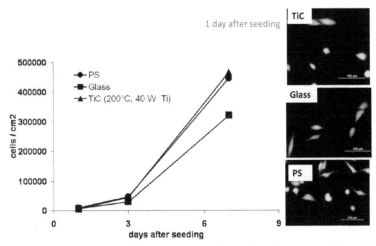

FIGURE 2.19 The growth dynamics of number of MG 63 cells on day 1, 3 and 7 after seeding. PS – polystyrene dishes, Glass – glass dishes.

The cell–matrix adhesion involved cell adhesion with many integrin or other adhesion receptors of an appropriate type, associated with cell differentiation, and also the formation of a large number of mature cell–matrix adhesion sites.[59] The microscopic investigations have been performed on cells stained by immunofluorescence staining. The β1-integrin adhesion receptors have been localized predominantly in the central region of the cells (Fig. 2.20). Vinculin, which formed dot-like structures, has been located on the whole surface of cells in the case of TiC/A:C (Fig. 2.4.7). However, on microscopic glass coverslips, vinculin has been situated mainly at the cell edges. The cells on TiC/A:C displayed larger and more numerous vinculin-containing focal adhesions than the cells on microscopic glass coverslips (Figs. 2.20 and 2.21).

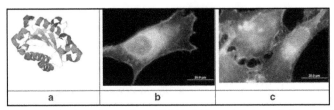

FIGURE 2.20 Investigation of integrin on TiC/A:C films. a) Schematic image of β1 integrin [60]. b) Microscopic image of β1 integrin on TiC/A:C films. c) Microscopic image of β1 integrin on glass coverslips.

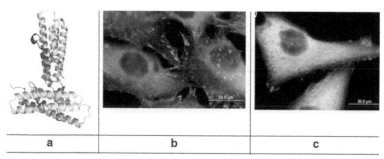

FIGURE 2.21   Investigation of vinculin on TiC/A:C films. (a) Schematic image of vinculin.[61] (b) Microscopic image of vinculin on TiC/A:C films. (c) Microscopic image of vinculin on glass coverslips.

Ti based implants are classified as bioinert. Bioinert refers to a material that retains its structure in the body after implantation and does not include any immunologic host reaction. Bioactive materials could be used for modification of the surface that occurs upon implantation. Bioactive refers to materials that direct chemical bonds with bone or even with soft tissue of a living organism. One of most used bioactive materials is a hydroxyapatite. The structure of bioinert TiC /A:C thin films covered by bioactive HAp is shown in Fig. 2.22.

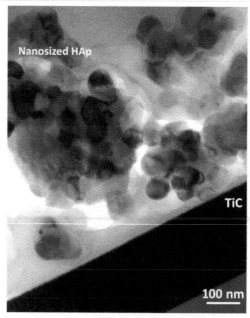

FIGURE 2.22   TEM images of bioinert TiC /A:C thin films covered by bioactive HAp.

2.5. CONCLUSION

Biomaterials used for implant should posses some important properties in order to long-term usage in the body without rejection. The creation of nanocomposites of ceramic materials with particle size few ten nanometers can significantly improve the bioactivity of the implant and enhance the osteoblast adhesion. The most used biomaterials are hydroxyapatite, polymer and titanium (Ti).

Hydroxyapatite (HAp) prepared from eggshells are a good candidate as biomaterial. Nanosized HAp prepared by attritor or ball milling had significantly higher bone formation than the unfilled control at 8 weeks after the operation. The histologic measurements showed that the remaining graft material was much lower in the HAp group. HAp from eggshells could be considered as an economic bone graft material. The results from this animal study cannot be extrapolated to clinical applications; subsequent toxicologic assessment and clinical trials would be necessary.

Polymer-HAp composites were successfully manufactured and applied in bone tissue engineering by employing the electrospinning technique. CA-HAp nano scaffolds were proved to promote favorable adhesion and growth of osteoblasts as well as to stimulate the cells to exhibit functional activity of bone cells. Overall, our studies suggest that both the morphology and structure of the CA-HAp composite scaffolds play important roles in facilitating cell spreading and differentiation and enhance apatite mineralization.

The sputtered TiC/A:C nanocomposite thin films were prepared as potential barrier coating for interfering of Ti ions from pure Ti or Ti alloy implants. Columnar TiC crystallites with 15–20 nm width have been embedded in 5 nm thin amorphous carbon matrix. MG63 osteoblast cells have been used for in-vitro study of nanocomposites. The 7 day lasting tests showed a higher value of cells on TiC/A:C nanocomposite surface. On the other hand, the cells on TiC/A:C film showed amore spreading tendency than the cells on control. The distribution of osteocalcin on microscopic glass coverslip did not reach the intensity of osteocalcin in cells on TiC/A:C films. The initial adhesion, subsequent growth and viability of human osteoblast-like MG63 cells in cultures on different substrates have been studied. From these measurements, no significant differences in the viability of MG63 cells have been found concerning all tested materials.

Based on our observations, the nanosized HAp prepared from eggshells by milling, electrospun CA-HAp nano scaffolds or TiC based thin films are considered as a promising candidate for bone tissue engineering application.

2.6 ACKNOWLEDGEMENTS

This work was supported by OTKA Postdoctoral grant Nr. PD 101453, the János Bolyai Research Scholarship of the Hungarian Academy of Sciences, OTKA 105355. The research leading to this result has received funding for European Community

Seven Framework Program FP7/2007–2013 under grant agreement Nr. 602398 (HypOrth).

The authors thank to Prof. Gouma useful cooperation related to polymer based composites, Dr. S.G. Kim, L. Bacáková, M. Vandrovcová for biological measurements, M. Menyhárd and S. Gurbán for AES measurements.

KEYWORDS

- **Biocompatibility**
- **Biomaterials**
- **Composites**
- **Electrospinning**
- **Hydroxyapatite**
- **Nanostructure**

REFERENCES

1. Boretos, J. W., & Eden, M. (1984). Contemporary biomaterials: Material and host response, clinical applications, new technology, and legal aspects Noyes Publications.
2. Ramakrishna, S., Mayer, J., Wintermantel, E., & Leong, K. (2001). Bio-medical applications of polymer-*composite* materials, A review. Compos Sci Technol, 61, 1189–1224.
3. Wise, D. L. (2000) Biomaterials engineering and devices. Humana Press, Berlin.
4. Park, J. B., & Bronzino, J. D. (2003). Biomaterials: principles and applications. Boca Rator, CRC Press.
5. Long, M., & Rack, H. J. (1998). Titanium alloys in total joint replacement. Biomaterials, 19, 1621–1639.
6. Okazaki, Y., & Gotoh, E. (2005). Comparison of Metal Release from Various Metallic Biomaterials in-vitro *Biomaterials,* 26, 11–21.
7. Wapner, K. L. (1991). Implications of metallic corrosion in total knee arthroplasty. *Clin Orthop Relat Res*, 1271, 12–20.
8. McGregor, D. B., Baan, R. A., Partensky, C., Rice, J. C., & Wilbourn, J. D. (2000). *Evaluation of the carcinogenic risks to humans associated with surgical and other foreign bodies a report of an IARC monographs program meeting. Eur J Cancer*, 36, 307–313.
9. Manivasagam, G., Dhinasekaran, D., & Rajamanickam, A. (2010). Biomedical Implants: Corrosion and its Prevention A Review. *India Recent Patents on Corr Sci*, 2, 1877–6108.
10. Black, J., & Hastings, G. W. (1998). *Handbook of Biomaterials Properties,* London UK Chapman and Hall.
11. Murugan, R., & Rao, K.P. (2002). Controlled Release of Antibiotic From Surface Modified Coralline Hydroxyapatite *Trends Biomat. Artif Organs*, 16, 43–45.
12. Rocha, J. H. G., Lemos, A. F., Agathopoulos, S., Valério, P., Kannan, S., Oktar, F.N., & Ferreira, J. M. F. Scaffolds for bone restoration from cuttlefish. *Bone*, 37, 850–857.
13. Herliansyah, M. K., Pujianto, E., Hamdi, M., Ide-Ektessabi, A., Wildan, M. W., & Tontowi, A. E. (2009). The Influence of Sintering Temperature on the Properties of Compacted Bovine Hydroxyapatite *Mat Sci Eng.,* C, (29), 1674–1680.

14. Herliansyah, K., Nasution, D. A., Hamdi, M., Ide-Ektessabi, A., Wildan, M. W., & Tontowi, A. E. (2007). Preparation and characterization of natural hydroxyapatite: A comparative study of bovine bone hydroxyapatite and hydroxyapatite from calcite. *Mater Sci. Forum,* 561–565, 1441–1444.
15. Herliansyah, M. K., Pujianto, E., Hamdi, M., Ide-Ektessabi, A., Wildan, M. W., & Tontowi, A. E. (2006). *Proceeding of ICPDM* X–31–IX–36.
16. Balázsi, C., Wéber, F., Kövér, Z., Horváth, E., & Németh, C. (2007). Preparation of Calcium Phosphate Bioceramics from Natural Resources *J Eur Ceram Soc* (27), 1601–1606.
17. Gergely, G., Wéber, F., Lukács, I., Tóth, A. L., Horváth, Z. E., Mihály, J., & Balázsi, C. (2010). Preparation and characterization of hydroxyapatite from eggshell, *Ceram Inter,* 36, 803–806.
18. Rodrıguez, R., Coreno, J., & Castano, V. M. (1996). Nanocomposites produced by growth of hydroxyapatite onto silica particles prepared by the sol-gel method. *Adv ComposLett,* 5, 25–28.
19. Arita, I. H., Wilkinson, D., Castano, V.M. (1995). Synthesis and Processing of Hydroxyapatite Ceramic Tapes with Controlled Porosity *J. Mater Sci. Med.,* 6, 19–23.
20. Wang, P. E., Chaki, T. K. (1993). Sintering Behavior and Mechanical Properties of Hydroxyapatite and Dicalcium Phosphate *J. Mater Sci. Med.,* 4, 150–158.
21. Liu, D. M. (1997). Fabrication of hydroxyapatite ceramic with controlled porosity. *J. Mater Sci. Med.,* 8, 227–232.
22. Riviera, E. M., Araiza, M., Brostow, W., Castan˜o, V. M., Diaz-Estrada, J. R., Hernandez, R., Rodriguez, R. (1999). Synthesis of Hydroxyapatite from Eggshells *Mat Lett,* 41, 128–134.
23. Hench, L. L. (1998). Sol Gel Materials for Bioceramic Applications J. Am Ceram Soc, 81, 1705–1728.
24. Cancedda, R., Giannoni, P., & Mastrogiacomo, M. (2007). A tissue engineering approach to bone repair in large animal models and in clinical practice. *Biomaterials,* 28, 4240–4250.
25. Ogose, A., Kondo, N., & Umezu, H. (2006). Histological Assessment in Grafts of Highly Phosphate (OSferion®) in Human Bone *Biomaterials,* 27, 1542–1549.
26. Valentini, P., Abensur, D., & Maxillary, D. (1997). Sinus Grafting with Anorganic Bovine Bone: A clinical report of long-term. *Int J Periodontics Restor Dent.,* 17, 233–241.
27. Gergely, G., Weber, F., Lukacs, I., Illes, L., Toth, A. L., Horvath, Z. E., Mihaly, J., & Balazsi, C. (2010). Nano Hydroxyapatite Preparation from Biogenic Raw Materials, *Cent Eur J Chem,* 8(2), 375–381.
28. LeGeros, R. Z. (1991). *Calcium phosphates in oral biology and medicine* Basel, Switzerland: Karger AG.
29. Suchanek, W. L., Byrappa, K., Shuk, P., Riman, R. E., Janas, V. F., & TenHuisen, K. S. (2004). Preparation of magnesium-substituted hydroxyapatite powders by the mechanochemical–hydrothermal method. *Biomaterials,* 25, 4647–4657
30. Rey, C., Renugopalakrishnan, V., Collins, B., & Glimcher, M. J. (1991). A resolution enhanced Fourier transform infrared spectroscopic study of the environment of the CO32- ion in the mineral phase of enamel during its formation and maturation. *Calcified Tissue Inter,* 49, 251–258.
31. Miller, L. M., Vairavamurthy, V., Chance, M. R., Mendelsohn, R., Paschalis, E. P., Betts, F., & Boskey, A. L. (2001). Insitu Analysis of Mineral Content and Crystallinity in Bone Using Infrared Micro Spectroscopy of the u4 PO_4^{3-} vibration *Biochimicaet Biophysica Acta,* 1527, 11–19.
32. Farmer, V. C. (1974). *The Infrared Spectra of Minerals* Bartholomew Press Dorkung Surrey.
33. Lee, S. W., Kim, S. G., Balázsi, C., Chae, W. S., Lee, H. O. (2012). Comparative Study of Hydroxyapatite from Eggshells and Synthetic Hydroxyapatite for Bone Regeneration *Oral Surg Oral Med Oral Phat Oral RadiolEndoeth,* 113(3), 348–355.
34. Lee, E. H., Kim, J. Y.,Kweon, H. Y., Jo, Y. Y., Min, S. K., Park, Y. W. (2010). A Combination Graft of Low-Molecular-Weight Silk Fibroin with Choukroun Platelet Rich Fibrin for Rabbit Calvarial Defect *Oral Surg Oral Med Oral Pathol Oral RadiolEndod,* 109, 33–38.

35. Shu, R., McMullen, R., & Baumann, M. J. (2003). Hydroxyapatite accelerates differentiation and suppresses growth of MC3T3-E1 osteoblasts. *J Biomed Mater Res A*, 67, 1196–1204.
36. Duncan, R. L., Akanbi, K. A., & Farach-Carson, M. C. (1998). Calcium Signals and Calcium Channels in Osteoblastic Cells *SeminNephrol*, 18, 178–190.
37. Turhani, D., Weissenböck, M., Watzinger, E., Yerit, K., Cvikl, B., Ewers, R., Thurnher, D. (2005). In-vitro Study of Adherent Mandibular Osteoblast Like Cells on Carrier Materials *Int J Oral MaxillofacSurg*, 34, 543–550.
38. Kweon, H. Y.,Lee, K. G., Chae, C. H., Balázsi, C., Min, S. K., Kim, J. Y., Choi, J. Y., Kim, S. G. (2011). Development of Nano Hydroxyapatite Graft with Silk Fibroin Scaffold as a New Bone Substitute *J Oral MaxilSurger*, 69(6), 1578–1586.
39. Stuart, M. (1991). *Orthopedic Composites International Encyclopedia of Composites* VCH Publishers New York.
40. Ondarcuhu, T., & Joachim, C. (1998). Drawing a Single Nanofiber over Hundreds of Microns, *Europhys Lett*, 42(2), 215–220.
41. Feng, L., Li, S., Li, H., Zhai, J., Song, Y., & Jiang, L. (2002). Super-Hydrophobic Surface of Aligned Polyacrylonitrile Nanofibers. *Angew Chem Int Ed*, 41(7), 1221–1223
42. Ma, P. X., & Zhang, R. (1999). Synthetic nano scale fibrous extracellular matrix. *J Biomed Mat Res*, 46, 60–72.
43. Liu, G. J., Ding, J. F., Qiao, L. J., Guo, A., Dymov, B. P., & Gleeson, J. T. (1999). Polystyrene block poly (2-cinnamoylethyl methacrylate), nanofibers Preparation, characterization, and liquid crystalline properties *Chem A European J*, 5, 2740–2749.
44. Gouma, P., Xue, R., Goldbeck, C. P., Perrotta, P., & Balázsi, C. (2012). Nano-hydroxyapatite Cellulose acetate composites for growing of bone cells. *Mat Sci Eng*, 32(3), 607–612.
45. Entcheva, E., Bien, H., Yin, L., Chung, C. Y., Farrell, M., & Kostov, Y. (2004). Functional cardiac cell constructs on cellulose-based scaffolding. *Biomaterials*, 25, 5753–5762.
46. Gouma, P. I., Ramachandran, K., Firat, M., Connolly, M., Zuckermann, R., Balazsi, C., Perrotta, P. L., & Xue, R. (2010). Novel bioceramics for bone implants. *Ceram Eng Sci Proceed*, 30(6), 35–44.
47. Tóth, M., Gergely, G., Lukács, I. E., Wéber, F., Tóth, A. L., Illés, L., & Balázsi, C. (2010). Production of polymer nanofibers containing hydroxyapatite by electrospinning, *Mat Sci Forum*, 659, 257–262.
48. Matsuno, H., Yokoyama, A., Watari, F., Uo, M., & Kawasaki, T. (2001). Biocompatibility and osteogenesis of refractory metal implants, titanium, hafnium, niobium, tantalum and rhenium *Biomaterials*, 22, 1253–1262.
49. Tamura, Y., Yokoyama, A., Watari, F., Uo, M., & Kawasaki, T. (2002). Mechanical Properties of Surface Nitrided Titanium for Abrasion Resistant Implant Materials. *Mater Trans.*, 43, 3043–3051.
50. Zhu, Y. H., Watari, F. (2007). Surface carbonization of titanium for abrasion resistant implant materials. *Dent. Mater J.*, 26, 244–252.
51. Mengel, R, Buns, C. E., Mengel, C., & Flores-de-Jacoby, L. (1998). An in-vitro study of treatment of implant surface with different instruments *Int. J. Oral Max. Implants*, 67, 91–96.
52. Koc, R., Meng, C., & Swift, G. A. (2000). Sintering Properties of Submicron TiC Powders From Carbon Coated Titania Precursor. J. Mater. Sci., 35, 3131–3141.
53. Sedlácková, K., Grasin, R., & Radnóczi, G. (2008). *New Research on Nanocomposites* Nova Science Publ New York.
54. Sedláčková, K., Ujvári, T., Grasin, R., Lobotka, P., Bertoti, I., & Radnoczi, G. (2008). C-Ti Nanocomposite Thin Films Structure Mechanical and Electrical Properties Vacuum 82, 214–216.

55. Balden, M., Cieciwa, B. T., Quintana, I., De Juan Pardo, E., Koch, F., Sikora, M., Dubiel, B. (2005). Metal doped carbon films obtained by magnetron sputtering *Surf Coat. Technol.*, 200, 413–417.

56. Balázsi, K., Lukács, I. E., Gurbán, S., Menyhárd, M., Bacáková, L., Vandrovcová, M., & Balázsi, C. (2013). Structural, Mechanical and Biological Comparison of TiC and TiCN Nanocomposites Films *J. Eur Ceram Soc*, 33(12), 2217–2221.

57. Balázsi, K., Vandrovcová, M., Bačáková, L., & Balázsi, C. (2013). Structural and biocompatible characterization of TiC/a: C nanocomposite thin films*Mat Sci Eng C*, 33(3), 1671–1675.

58. Harcuba, P., Bačáková, L., Stráský, J. Bačáková, M., Novotná, K., & Janeček, M. J (2012). Surface treatment by electric discharge machining of Ti-6Al-4 V alloy for potential application in orthopedics. *Mech. Behav Biomed Mater*, 7, 96–105.

59. Bacakova, L., Filova, E., Parizek, M., Ruml, T., & Svorcik, V. (2011). Modulation of cell adhesion, proliferation and differentiation on materials designed for body implants. *Biotechnol Adv.*, 29, 739–767.

60. http://sr.wikipedia.org/wiki/Integrin.

61. http://www.pdb.org/pdb/explore/explore.do?structureId-1qkr.

CHAPTER 3

LIQUID CRYSTALS AND CELLULOSE DERIVATIVES COMPOSITES

ADINA MARIA DOBOS, MIHAELA-DORINA ONOFREI, and SILVIA IOAN

ABSTRACT

This chapter summarizes recently developed researches concerning cellulose derivatives composites obtained by blending of cellulose derivatives or of cellulose derivatives with different polymers, as well as by processes involving nano-particles in the cellulose derivative matrix. Knowledge and understanding of the interactions manifested in these cellulose systems constitute essential elements for the conception and optimization of novel structures. Different characterization techniques allow a more complete evaluation of the mechanisms of multicomponent systems, of their fundamental interactions, such as hydrogen bonding, and the manner in which these interactions affect the final properties. This research reveals the relation between the molecular interactions and physical properties, which represents an important challenge from both scientific and industrial perspectives. On the other hand, the chapter shows that, like many other cellulose derivatives, hydroxypropylcellulose (HPC), is the most common ether of native cellulose, whose concentrated solutions possess optical properties characteristic to cholesteric liquid crystals. In crystalline state, on one hand, these types of cellulose present specific arrangements, while, on the other, these arrangements depend on the concentration of their solutions. In cellulose derivative composites, the absence or presence of liquid crystals properties is dependent on mixture composition, solution concentration and used solvent, according to the application in biomedical or electronic domain.

3.1 INTRODUCTION

Cellulose is a natural polymer used in a variety of applications. Like most polymers, cellulose is not completely crystalline, also containing disordered regions. Out of the variety of models that explain its dual nature, the fringed micelle pattern considers that chains pass through both crystalline and amorphous (disordered) regions, whereas other models attribute the disorder to imperfections, chain ends, and surface regions of the microfibril. Cellulose derivatives use is limited by their poor solubility in various solvents, which is primarily due to the hydrogen bonds formed between the hydroxyl groups from the anhydroglucose chain. In dissolved state, all hydroxyl groups are accessible to the reactant molecules, so that the cellulose structure can be chemically modified to improve their processability and performance for particular uses. To this end, various cellulose derivatives have been synthesized for the diversification of their characteristics. Thus, cellulose acetates are cellulose esters partially substituted at the C–2, 3 and 6-positions of the anhydro glucopyranose residue. They can be easily molded into different forms, such as membranes, fibers, and spheres. To conjugate the mechanical properties of the polymer with the intrinsic properties of cellulose acetates, hybrid organic/inorganic materials have been prepared.[1] The obtained composite materials present many intrinsic advantages,[2] such as low cost, availability, biodegradability and easy handling. At the same time, the modern technology makes possible to manufacture, from the existing suitable polymeric materials, different optical cellulosic components for a wide variety of applications, such as spectacle lenses, contact lenses, intraocular lenses, consumer products, instrumentations, etc. Generally, cellulose acetates are not used for corrective lenses, but are occasionally employed for plane lenses. High optical transparency, high moisture absorption and low dimensional stability characterize these derivatives of cellulose.[1,3] Recent researches describe different processes for obtaining membranes with optical and dielectric properties for biomedical applications. Cellulose acetates have been widely used for dialysis membranes, for example, in artificial kidneys, as membranes in plasmapheresis, and as drug delivery matrices for controlled release.[4,6] Mention should be here made of the researches on the integration of cellulose acetates onto silicon wafers by the standard microfabrication process, to add filtration capability on the chip. The membranes are biocompatible, showing good structural integrity and good adhesion to the substrate.[7,8] Moreover, cellulose acetate membranes are recommended for applications that require superior clarity, for example, for optical sensors.[9] Also, a pH-sensitive membrane, consisting of a polyester support covered with a thin layer of cellulose onto which a pH-indicator was covalently immobilized, has been developed.[10]

Cellulose acetate phthalate (CAP), a mixed ester of cellulose obtained through phthaloylation of cellulose acetate, is used in different domains as a pharmaceutical excipient, due to its pH dependent solubility in aqueous media. CAP enteric coatings are resistant to gastric acid and easily soluble in the slightly alkaline environment of the intestine. The pH-dependent solubility is mainly determined (among other properties of the mixed ester) by the substitution degree (DS), namely the average number of substituent groups linked by one unit of anhydro glucopyranose and by the molar fraction (acetyl and phtaloyl groups). These two structural characteristics are dependent on the synthesis conditions. The potential of this polymer to inhibit infections caused by several types of herpes virus, such as Herpes Simplex type 1, and by other sexually transmitted diseases, has been analyzed in-vitro.[11,13]

Also, CAP is a well-established safety record for human application, being used for enteric film coating of tablets and capsules.[14]

Cellulose ethers, such as methylcellulose, ethylcellulose, hydroxypropylcellulose, hydroxypropyl-methylcellulose, methylhydroxyethylcellulose possess a remarkable combination of important properties for biomedical and pharmaceutical applications, for example, as carriers for drug targeting,[15] vaccine bullets,[16] sustained release of drugs,[17,18] as materials for the disintegration of matrix tablets,[19] as electro-optical devices,[20,21] as transparent media for liquid crystalline display technology,[22] and for stabilization of dispersion polymerization.[23] On the other hand, the most common ether of native cellulose, whose concentrated solutions display optical properties typical for cholesteric liquid crystals, is hydroxypropylcellulose (HPC). In crystalline state, the HPC molecules are present in a helical arrangement. The structure is not dependent on the solvent alone, but also on concentration. Thus, the rheological behavior of HPC,[24,25] either pure or in mixtures with other polymers, involves the absence or presence of liquid crystals properties, as a function of composition, solution concentration and solvent,[26,28] on also considering to the applications areas, for example, food industry,[29] medicine,[30] etc. HPC is cheap and biocompatible, which makes it a valuable tool for applications in biological and medical fields, especially for the fabrication of macromolecular prodrugs.[31,33] Recently, cross-linked HPC derivatives have been shown to form elastomeric thin films.[34] At the same time, researchers have found chiral nematic order in natural materials, such as animal and plant tissues.[35,37] In this context, literature shows different stages in which these natural materials, similarly with cellulose crystallites and nanocrystallites, may be replicated in-vitro.[38,40]

In the present study, CAP and HPC have been preferred especially for the large range of their pharmaceutical and biomedical applications. Their miscibility is generally considered as a result of the specific interactions between polymer segments in casting solutions of organic solvents. Thus, literature shows that CAP/HPC composite films obtained in different organic solvents, such as ethanol, rapidly reduce the infectivity of several sexually transmitted disease pathogens, HIV-1 included.[41] The infections reduction mechanism involves conversion of these composite films into gel in the presence of water. These films were used in different biomedical applications, such as controlled release of drugs.[42,43]

Knowledge of the solution properties is important for handling/formulation, and also for a better understanding and control of polymer blending processes.

Extensive studies have been conducted on the effect of the used solvents, on the morphology and performance of derivative cellulose or derivative cellulose blends with various polymers in different applications, such as ultrafiltration membranes, reverse-osmosis membranes, gas-separation membranes, and pervaporation membranes. Addition of a second solvent to the casting solution increases the permeation flux of a membrane or improves the performance of nano-fibers.[44] Literature data reveal that the polymer structures, boiling point, content of solvent mixture, and intrinsic viscosity may affect performance in some applications.[45,46] Moreover, the electrospinning process of polymers in solvent mixtures employs an electrostatic potential to form fibers with different diameters. These fibrous mats with high specific surface area and nano-degree of porosity lend themselves to a wide range of applications.

3.2 INFLUENCE OF SELF-COMPLEMENTARY HYDROGEN BONDING ON SOLUTION PROPERTIES OF DERIVATIVE CELLULOSES IN SOLVENT/NON-SOLVENT MIXTURES, OVER A LARGE CONCENTRATION DOMAIN

Recently, the literature has pointed out intense interest in investigating new solvents, in particular for cellulose compounds.[47,48] The mechanism of cellulose dissolution is important in description of the strategies employed for the synthesis of cellulose derivatives, and in the future perspectives to produce some complex structures, including nano-composites, "smart" polymers that respond reversibly to external stimuli, and bio-compatible materials. In this respect, in relation with the molecular structure of cellulose (Schemes 1 and 2), both intra and intermolecular hydrogen bonding interactions, influencing various properties must be mentioned.[39,49] The intramolecular hydrogen bond interactions between O–2–H and O–5' of the adjacent glucopyranose unit and O–2–H and O–6' contribute to single-chain conformation and stiffness.

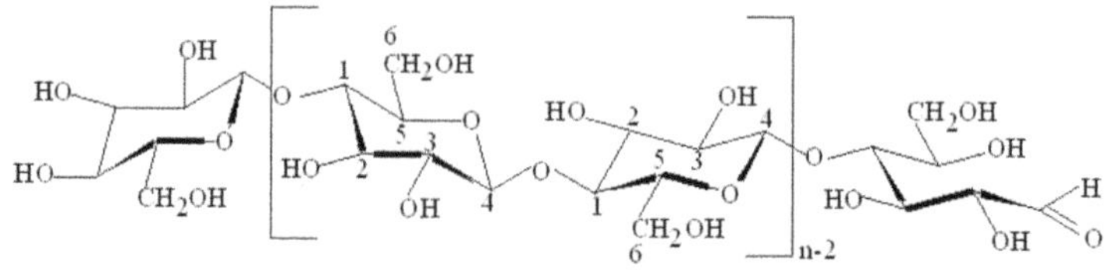

| Non-reducing end groups | Anhydroglucose unit | Reducing end groups |

SCHEME 1   Molecular structure of cellulose.

The intermolecular hydrogen bonding interactions in cellulose, responsible for the sheet-like nature of the native polymer, are localized between the hydroxyl groups, -OH group at the C–6' and C–2' positions of cellulose molecules adjacently present in the same lattice plane.

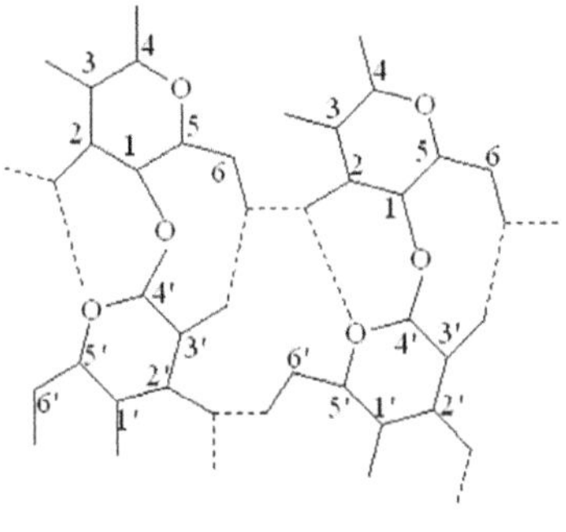

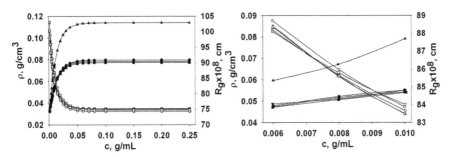

SCHEME 2   Hydrogen bonds in cellulose.

Literature has demonstrated the versatility of multiple hydrogen bonding interactions in the formation of derivative cellulosic structures.

Moreover, solvents contribute essentially to the modification of solution behavior (association, complex, micelle, and core-shell structure of the polymer chains), and also to the processes, which determine the morphology of cellulose derivatives. Most of the parameters that play an important role in the formation of cellulose films have been presented in detail in this chapter. Among these variables, one should mention the nature of cellulose and the substitution degree, the nature of the casting solvent mixtures, their composition and temperature.[50,53] Conformational properties of cellulose acetate phthalate (obtained from cellulose acetate with a 1.93 substitution degree according to Scheme 3)[54,55] in solutions of 2-methoxyethanol (2-Me)/water (W) (solvent/nonsolvent mixtures), are evaluated from the modification of coil density, (ρ – Eq. (1) or (2)), and gyration radius, (R_g – Eq. (3)), with concentration over both a dilute and a large domain of concentrations (Fig. 3.1).[56,58]

FIGURE 3.1   Variation of coil density and gyration radius, respectively, of CAP on the large domain (theoretical data, left hand side image) and in dilute domain (experimental data, right hand side image) of concentrations, at 25 °C, in different solvent mixtures,: (■, □) – 100/0 2-Me/W % v/v, (▲, △) – 50/50 2-Me/W % v/v, (▼, ▽) – 70/30 2-Me/W % v/v, (◆,◊) – 75/25 2-Me/W % v/v and (●,○) – 80/20 2-Me/W % v/v. – left part.[55]

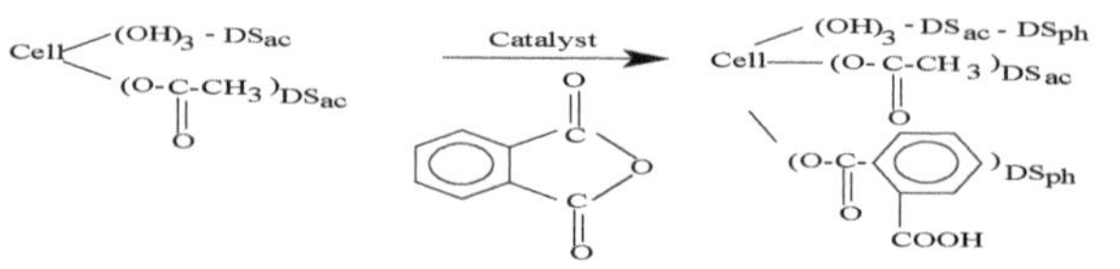

SCHEME 3   Molecular structure of cellulose acetate phthalate.[54,55]

$$\rho = \frac{c}{\eta_{sp}}\left(1.25 + 0.5\sqrt{56.4\eta_{sp} + 6.25}\right) \tag{1}$$

or

$$\rho = \frac{c*}{0.77^3}\left\{1 + \frac{[\eta]-[\eta]_\theta}{[\eta]_\theta}\left[1 - \exp\left(-\frac{c}{c*}\right)\right]\right\} \tag{2}$$

where: $\eta_{sp} = \dfrac{t}{t_0} - 1$ is specific viscosity, t and t_0 are the flow times of the polymer

solution and of the solvent, respectively, c* is the critical concentration at which the polymer coils begin to overlap each other, defined by equation 4, and $[\eta]_\theta$ is intrinsic viscosity in unperturbed state, defined by equation 5:

$$c* = \frac{0.77}{[\eta]} \tag{4}$$

$$[\eta]_\theta = \frac{[\eta]\left[1 - \exp\left(-\dfrac{c}{c*}\right)\right]}{\dfrac{0.77^3\rho}{c*} - \exp\left(-\dfrac{c}{c*}\right)} \tag{5}$$

Polymer coil density increases and radius of gyration decreases with increasing polymer concentration and, above a critical concentration in the semidilute domain, $c^+ \cong 8c*$, both acquire values corresponding to the unperturbed state of the polymer chain.

The thermodynamic parameters represented by intrinsic viscosities, ($[\eta]$), radii of gyration in perturbed, $R_{g,c=0}$, and unperturbed, $R_{g,c=\theta}$, states, the critical concentrations delimiting the dilute – semidilute domain ($c*$), semidilute unentangled – semidilute entangled domain ($c**$), and semidilute – concentrated domain ($c*** = c*\cdot(R_{g,c=0}/R_{g,c=\theta})^8$), presented in Table 3.1, show the influence of solvent mixtures.

TABLE 3.1   Critical Concentrations c^*, c^{**}, c^+, and c^{***}, Radius of Gyration in Perturbed and Unperturbed State and Intrinsic Viscosity for CAP in 2-Me/W (% v/v)[55]

2-Me/W	c^* (g/dL)	c^{**} (g/dL)	c^+ (g/dL)	c^{***} (g/dL)	$R_{g,c=0} \times 10^8$ (cm)	$R_{g,c=\theta} \times 10^8$ (cm)	$[\eta]$ (dL/g)
100/0	1.25	7.40	9.98	19.79	106.00	75.05	0.617
80/20	1.26	7.93	10.07	20.35	105.81	74.72	0.612
75/25	1.26	-	10.07	21.29	105.78	74.28	0.611
70/30	1.37	9.80	10.93	21.75	104.93	74.27	0.563

The perturbed and unperturbed dimensions of cellulose acetate phthalate decrease with increasing the water content, which also reflects modification of the solvation power of solvents. 2-methoxyethanol is a good solvent for CAP, and the addition of water reduces the quality of the solvent mixtures in 0–0.30 volume fractions of water. At the same time, 2-methoxyethanol is preferentially adsorbed (preferential adsorption coefficient, λ_1 shows positive values of 0.460, 0.685 and 0.526 for 80, 75, and 70% 2-Me contents, respectively) over the mentioned domains, and both solvents tend to minimize preferential adsorption at the extreme values of 2-methoxyethanol volume fractions (ϕ_1). Consequently, from preferential adsorption data, two distinct ranges of water composition can be distinguished, over which the solvent mixture evidences different interactive properties. Therefore, the polymer chains have the tendency to surround themselves with the thermodynamically most efficient solvent, at a given composition of the solvent mixture.

Generally, for nonassociating polymers appears that:[59,60]

$\eta_{sp} \propto c^{**1.25}$ in semidilute unentangled regime;

$\eta_{sp} \propto c^{**4.8}$ in semidilute entangled regime – if the polymer is in a theta solvent; and

$\eta_{sp} \propto c^{**3.7}$ in semidilute entangled regime if the polymer is in a good solvent.

Entanglement concentration, c^{**}, defined as the transition from the semidilute unentangled regime to the semidilute entangled one, can be measured using the change in slope at the onset of the entangled regime. Figure 3.2 illustrates the dependence of specific viscosity for CAP solutions in 2-Me/W solvent mixtures at 25 °C as a function of a dimensionless parameter, $c \cdot [\eta]$, defined as the coil overlap parameter, which provides an index of the total volume occupied by the polymer. A slope modification in these dependencies occurring at critical concentrations, c^{**}, which separates the semidilute unentangled/semidilute entangled domain. As a function of solvent mixture composition, these concentrations take the following values: 8.69, 9.55 and 10.00 g/dL for 100/0, 80/20 and 70/30% v/v 2-Me/W, respectively. In addition, the dependencies between specific viscosity and concentration are expressed by the following equations:

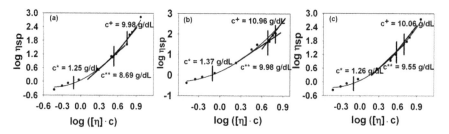

FIGURE 3.2 η_{sp} dependence on $c \cdot [\eta]$ for CAP in 2-methoxyethanol/water solvent mixtures at different compositions: (a) – 100/0 v/v 2-Me/W, (b) – 80/20 v/v 2-Me/W, (c) 70/30 v/v 2-Me/W.[55]

$$\log \eta_{sp} = 1.603 + 1.215 \cdot (c \cdot [\eta]) + 5.207 \cdot (c \cdot [\eta])^2 + 1.093 \cdot (c \cdot [\eta])^3 \text{ for} \tag{6}$$

100/0 v/v 2-Me/W

$$\log \eta_{sp} = 1.673 + 1.205 \cdot (c \cdot [\eta]) + 4.781 \cdot (c \cdot [\eta])^2 + 1.124 \cdot (c \cdot [\eta])^3 \text{ for} \tag{7}$$

80/20 v/v 2-Me/W

$$\log \eta_{sp} = 2.759 + 1.379 \cdot (c \cdot [\eta]) + 4.045 \cdot (c \cdot [\eta])^2 + 1.190 \cdot (c \cdot [\eta])^3 \text{ for} \tag{8}$$

70/30 v/v 2-Me/W

A lower quality of the solvent mixture leads to higher values for η_{sp} and also for c^{**} concentrations; increase of the water content leads to a decrease in the quality of the solvent mixtures, and increase of polarity by water addition leads to an increase in the number of intermolecular polymer associations. The η_{sp} power law exponent in semidilute entanglement regime increases, from 4.5 to 5.5 for 100/0 – 70/30 v/v solvent mixture compositions. In the dilute and semidilute unentanglement domain, η_{sp} increases with the increase of the 2-Me content, suggesting that 2-Me is a better solvent for CAP than 80/20 v/v 2-Me/W or 70/30 v/v 2-Me/W (Table 3.1 and Fig. 3.2).

Consequently, η_{sp} increases with solvent quality for dilute or semidilute unentangled chains, due to a stronger apparent repulsive force among chain segments, whereas η_{sp} decreases with increasing solvent quality of the chains in entangled regime. This phenomenon has been described by a two-parameter scaling relationship for polymers in entangled regime,[60] and the increase in η_{sp} with decreasing solvent quality in the entangled regime has been reported to be caused by a higher number of entanglement couplings in the poor solvent, which result from the intense attractive forces between the chain segments.[59,61] Increase of the water content over the

0.70 volume fraction displays significant hydrogen bonding with cellulose acetate phthalate, generating the entanglement process among other types of interactions generated by carboxyl groups. CAP, which contains carboxyl and hydroxyl groups, is able to develop hydrogen bonding, easily dissociating into a relatively low polarity solvent, such as 2-Me, and leading to strong intermolecular interactions with the addition of higher polarity water, which shows the influence of hydrogen bonding on the thermodynamic properties. Therefore, transition from the semidilute unentangled regime to the semidilute entangled one is determined by solvent polarity. The concentration corresponding to this transition increases from 8.69 g/dL in 100/0 v/v 2-Me/W to 9.98 g/dL in 70/30 v/v 2-Me/W. In addition, literature[62] establishes an optimal concentration of around 30 g/dL for obtaining nano-fibers. According to Table 3.1, this concentration is located at the supper limit of the semidilute – concentrated domain. In this context, it is stated that the solvent plays an important role in the electrospinning process and that the optimal concentration of CAP for obtaining nano-fibers is: 35%, 25% and 12.5% in 2-Me, 50/42.5/7.5 v/v/v 2-Me/acetone (Ac)/W, in 85/15 v/v Ac/water, respectively. These concentration values correspond to the entangled domain and decrease with decreasing solvent mixtures' quality.

Knowledge on conformational properties of cellulose derivatives in solution is important in various applications, such as those involving the production of films.[43,63] For the above illustrated system, the water content in the casting solutions favored the occurrence of roughness (according to the atomic force microscopy (AFM) images from Figs. 3.3–3.5), more visible for intermediary compositions of the mixed solvents, corresponding to a better preferential adsorption of 2-methoxyethanol from the solvent mixtures.

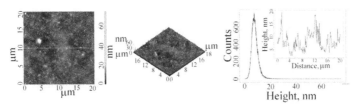

FIGURE 3.3   2D, 3D, AFM images, histogram and surface profile of CAP films obtained in 100/0% v/v 2-methoxyethanol/water solvent mixtures.[55]

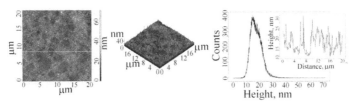

FIGURE 3.4   2D, 3D, AFM images, histogram and surface profile of CAP films obtained in 80/20% v/v 2-methoxyethanol/water solvent mixtures.[55]

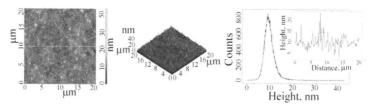

FIGURE 3.5   2D, 3D, AFM images, histogram and surface profile of CAP films obtained in 70/30% v/v 2-methoxyethanol/water solvent mixtures.[55]

For extreme water compositions, lower average roughness values and root-mean-square roughness values appear, the phenomenon being more intense at higher water contents (Fig. 3.6, Table 3.2).

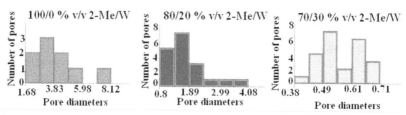

FIGURE 3.6   Pore number distribution attending to pore size as obtained from AFM image analysis.[55]

TABLE 3.2   Pore Characteristics, Including Pore Number (No.), Diameter (d, μm), Depth (dp, nm), Perimeter (p, μm) and Area (A, μm^2), and Surface Roughness Parameters, Including Average Roughness (Sa, nm), Root-Mean-Square Roughness (Sq, nm), Nodule Height from the Height Profile (nhp, nm), and Average Height from the Histogram (Ha, nm) of Membranes Prepared from Cellulose Acetate Phthalate Solutions in different 2-Me/W Solvent Mixtures (% v/v)[55]

2-Me/W	Pore characteristics					Surface roughness			
	No.	d	dp	P	A	Sa	Sq	nhp	Ha
100/0	9	4.1	12.5	12.7	13.0	2.0	2.9	17.8	8.1
80/20	18	1.8	13.0	5.8	2.6	3.4	4.3	26.3	17.8
70/30	23	0.6	7.2	1.7	0.2	1.6	2.5	23.0	9.9

This changing trend in morphology is due to the modification of polymer chain conformation in solution that can be speculated according to the applied area. In this context, modification of the rheological properties and some morphological aspects of cellulose acetate phthalate in 2-methoxyethanol/acetone/water, at different compositions of solvent mixtures, allowed the establishment of optimal composition of solvent mixtures for obtaining fibers with controlled diameters.[64] Figure 3.7 plots the modification of dynamic viscosity, η, *versus* shear rate, $\dot{\gamma}$, and water content.

Increasing water content leads to decrease in dynamic viscosity, concomitantly with increasing the Newtonian plateau and flexibility. At the same time, at constant values of shear rate, mentioned literature shows that dynamic viscosity varies insignificantly until an approx. 25% vol. water content, while, for a 27.5% vol. composition, dynamic viscosity increases, attaining approx. the same values for different shear rates. At higher water composition, decreasing of dynamic viscosity signifies rearrangement of macromolecules in solution.

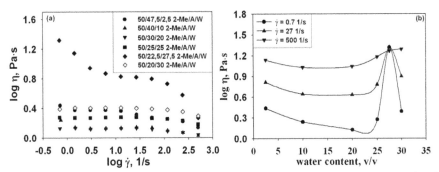

FIGURE 3.7   Logarithmic plot of dynamic viscosity for CAP in different mixing ratio of 2-Me/Ac/W as a function of: (a) shear rate – for a good visualization, the curves are detached with 0, 0.5, and 1 from bottom to top; (b) water content.[64]

The values of transition frequency from viscous to elastic domain and the values at which storage, G', and loss, G'', moduli are equal, slightly decrease with increasing the water content (Fig. 3.8). A sharp increase, up to a water content exceeding 25% vol., is also evidenced.

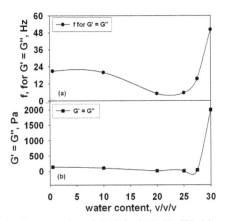

FIGURE 3.8   Viscoelastic properties of CAP in 2-Me/Ac/W: (a) oscillation frequency *vs.* water content; (b) G'= G'' *vs.* water content.[64]

The spinning process of CAP in acetone/water solvent mixtures leads to short fibers with small diameter. When 2-methoxyethanol was added to the systems, the diameter of the fibers becomes larger and the irregularities disappear. Fiber diameters were found to vary insignificantly with the water content, yet, at an approx. 25% water content, the fibers evidence a smaller diameter, when the viscosity of solutions and the boiling point of the solvents mixtures increase.

On the other hand, the AFM images (with 20×20 μm^2 scanning area) from Fig. 3.9 show the influence of casting solutions on films morphology. Table 3.3 lists the average values of pore characteristics and of the surface roughness parameters.

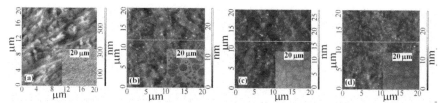

FIGURE 3.9   AFM images (2D and phase image (small inserted image)) at 2 0 x 20 μm^2 scan area for cellulose acetate phthalate films prepared from solutions in: (a) – 2-methoxyethanol; (b) – 50/47.5/2.5 v/v/v 2-methoxyethanol/acetone/water; (c) −50/22.5/27.5 v/v/v 2-methoxyethanol/acetone/water; (d) – 50/20/30 v/v/v 2-methoxyethanol/acetone/water.[64]

Increasing of the water content in the solvent mixtures determines modification of pores number and of their characteristics, so that, at approx. 25% vol., the pores number is maximum, while the area, perimeter and diameter are minimum.

TABLE 3.3   Pore Characteristics Including the Area, Average Perimeter, Diameter, Length, and Mean Width, and Surface Roughness Parameters Including Average Roughness (Ar), Root Mean Square Roughness (Rms), and Nodule Height from the Height Profile (Nhp) of Cellulose Acetate Phthalate Films Prepared from Solutions in 2-methoxyethanol/Acetone/ Water (Column 1), with 20 x 20 μm^2 Scanned Areas, Corresponding to the 2D AFM Images[64]

Solvent mixtures, v/v/v	Pore characteristics				Surface roughness		
	Number pores	Area (μm^2)	Perimeter (μm)	Diameter (μm)	Ar (nm)	Rms (nm)	Nhp (nm)
100/0/0	-	-	-	-	83.26	102.81	301
50/47.5/2.5	46	7.28	7.99	2.54	1.22	1.62	5.0
50/40/10	30	3.01	5.97	1.90	1.58	2.18	12.0
50/30/20	20	15.01	12.07	3.89	1.55	2.12	7.6
50/25/25	48	0.76	2.84	0.90	1.65	2.22	12.9
50/22.5/27.5	21	8.53	10.07	3.21	1.16	1.56	4.5
50/20/30	12	6.45	8.88	2.83	1.22	1.65	5.8

Modification of morphology is due to the modification in the chain conformation of the polymer, which is influenced by the quality of the mixed solvents.[45,63]
Also, it may be assumed that the association phenomena of 2-methoxyethanol, acetone or water over different composition domains of their mixtures may influence
the preferential adsorption of one of the solvents by the macromolecular chain.[65]

Therefore, conformational modifications generated by the interaction from
the system change the solubility of cellulose derivatives, and affect the
rheological and morphological properties.

3.3   SOLUTION PROPERTIES AND MICROSTRUCTURES OF CELLULOSE DERIVATIVE BLENDS

A successful alternative for the development of new polymeric materials is blending
of the already existing polymers, for attaining a balance among the desired properties exhibited by the individual components. Development of characterization
techniques has improved understanding of the mechanisms involved in polymers
mixing, of their fundamental interactions, and of the manner in which these interactions affect their final properties.[66] The relation between molecular interactions and
the physical and engineering properties continues to be an important challenge from
both scientific and industrial perspectives, due to the increasing economical impact
of polymer blends and alloys in many domains, directly affecting our everyday life.
The rheology, as well as the interface properties or morphology of polymer blends
involving cellulose materials have been the subject of numerous researches.[26,65,67,70]
As a consequence, a good knowledge on the rheological properties of these solutions is important for their handling and formulation, and also for better understanding and controlling the processes involved in cellulose acetate phthalate/hydroxy-
propyl cellulose blends for different applications.

In this context, literature shows that the microstructure and CAP/HPC-solvent
interactions are observed by rheological studies involving shear experiments, which
can evidence the lyotropic properties in different solvents at higher concentrations.[26]
Figures 3.10 and 3.11 plot the modification of dynamic viscosity, η, *versus* shear
rate, $\dot{\gamma}$, and concentration, c, respectively, for different compositions of the polymer
mixtures, including pure polymers, in 2-methoxyethanol. The concentration domain
is lower than that corresponding to the occurrence of crystal liquid properties –
where, at high shear rates, viscosity at high concentration is lower that viscosities
obtained for smaller concentrations.[26]

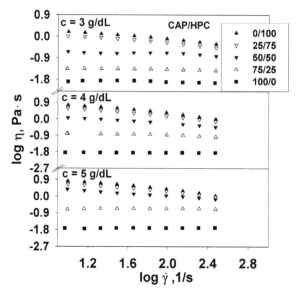

FIGURE 3.10 Log-log plots between dynamic viscosity and shear rate for CAP/HPC blends at different concentrations in 2-methoxyethanol.[69]

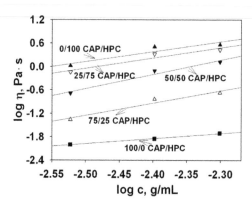

FIGURE 3.11 Log-log plots between dynamic viscosity and different concentrations in 2-methoxyethanol for CAP/HPC blends: (a) 0/100 wt./wt., (b) 25/75 wt./wt., (c) 50/50 wt./wt. (d) 75/25 wt./wt., (e) 100/0 wt./wt.[69]

In addition, a Newtonian behavior for CAP was observed for all concentrations, as well as a thinning behavior for HPC and CAP/HPC solutions at higher HPC compositions. Figure 3.11 shows that dynamic viscosity increases with increasing concentration, the maximum slope occurring at 50/50 wt./wt. CAP/HPC, in the absence of any liquid crystal phenomenon.[69] The dependence of viscosity on concentration at a shear rate of 1.6 s^{-1} is described by a power law $\eta \propto c^x$, with exponents of 2.27, 3.14, 3.69, 2.64 and 2.53 for 100/0, 75/25, 50/50, 25/75 and 0/100 wt./wt. CAP/HPC, respectively. These dependencies coincide with the theoretical prediction for semidiluted polymer solutions close to the entanglement domain, especially for 50/50 wt./wt. CAP/HPC. As already mentioned, dynamic viscosity is proportional with $c^{1.25}$ and $c^{3.7}$ in the semidilute unentangled regime for nonassociating polymers, as well as in the semidilute entangled regime, respectively.[60]

The interactions between chain segments, which reflect the existence of polymer unentanglements or entanglements and hydrogen bonding, are described by the activation energy, E_a (Eq. (9)):[71,72]

$$\ln \eta = \ln \eta_0 + \frac{E_a}{RT} \tag{9}$$

where $\eta_0 \propto e^{-\Delta S/R}$ represents a preexponential constant,[73] ΔS is the flow activation entropy, "R" is the universal gas constant and "T" is absolute temperature.

Generally, this property is directly related to the disengagement (E_{dis}, with positive contribution) of the associated chain formation, (E_{ass}, with negative contribution), according to Eq. (10):

$$E_a = E_{dis} + E_{ass} \tag{10}$$

Thus, for presented blends,[69] the activation energy $E_a > 0$, $E_{dis} > E_{ass}$, decreases with increasing the CAP content (see Fig. 3.12). The association phenomena generated by hydrogen bonding appear preponderantly in the HPC/2-Me system (as due to the presence of a higher number of hydroxyl groups and 2-Me, which is a poorer solvent for HPC than for CAP, as shown by viscometric data). The anhydroglucose units of HPC reduce chain flexibility in 2-methoxyethanol, so that the entropy of the systems is lower, resulting in the highest mean activation energy values $E_a = 26.81 \div 29.28$ kJ/mol, comparatively with the CAP pure components, for which $E_a = 6.48 \div 20.42$ kJ/mol. A lower value of E_a implies a lower energy barrier for the movement of an element in the fluid. In the here discussed case, this barrier is related to the interaction between chain segments, and is determined by polymer entanglements or by specific interactions, such as hydrogen bonding. A pseudoplastic behavior and consequently, a decrease in viscosity with increasing shear rate are observed both in HPC solutions and in their corresponding blends with CAP. The dependence of shear stress, σ, on shear rate partially obeys the power law relationship described by Eq. (11):

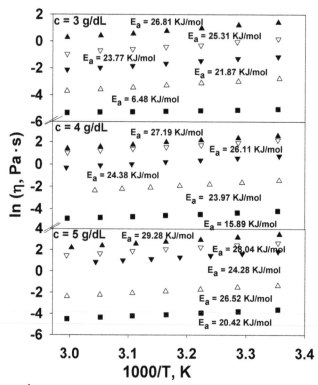

FIGURE 3.12 $\ln \eta$ vs. 1000/T for ▲- 0/100 CAP/HPC, ▽ – 25/75 CAP/HPC, ▼ – 50/50 CAP/HPC, △ – 75/25 CAP/HPC and ■ – 100/0 CAP/HPC blends (wt./wt.) in 2-methoxyethanol at different concentrations[69]

$$\sigma = K \cdot \dot{\gamma}^n \qquad (11)$$

where n and K are the flow behavior and consistency index, respectively.

It is estimated that $n = 1$ for a Newtonian behavior of fluids, $n < 1$ for a thinning behavior, and $n > 1$ for a thickening behavior.[74,75] Table 3.4 confirms the thinning behavior introduced by HPC in polymer blends; the flow behavior indices are sub-unitary for HPC and at a higher HPC content, and around the unit for CAP and at a higher CAP content in polymer blends, according to the shear thinning rheological behavior evidenced in Fig. 3.10. The solution consistency index, K, induced by pure HPC takes higher values than those induced by pure CAP, and decreases with decreasing the HPC content and solution concentrations in polymer blends. These results, influenced by both composition and solution concentrations of the polymer blends, are a consequence of the modification of polymer interactions in the system, such as hydrogen bonding interactions. Likewise, HPC helps to the specific molecular rearrangement in solution, being influenced by temperature and concentration.

TABLE 3.4 Flow Behavior Index, n, and the Consistency Index, K, ($Pa \cdot s^n$), for CAP/HPC Casting Solutions in 2-Me at Different Concentrations, c (g/dL), and Mixing Ratio (wt./wt.)[69]

CAP/HPC blends	c	n	K	c	N	K	c	n	K
0/100	3	0.66	3.55	4	0.49	20.89	5	0.47	25.12
25/75	3	0.67	2.29	4	0.51	10.23	5	0.55	13.80
50/50	3	1.04	0.15	4	0.66	0.43	5	0.66	5.01
75/25	3	0.95	0.05	4	0.92	0.20	5	1.01	0.20
100/0	3	0.98	0.01	4	0.99	0.02	5	1.03	0.02

Viscoelastic measurements significantly contribute to the knowledge and differentiation of polymer systems, completing the rheological studies developed in shear regime. Prior to the measurements of these CAP/HPC blends, suitable strain amplitude tests must be performed, to establish the domain of shear stress over which the storage and loss moduli are constant. Figure 3.13 exemplifies the variation of storage and loss modulus *versus* shear stress at a constant frequency of 1 **Hz**, for a 4 g/dL concentrated CAP/HPC blend.[69] Initially, G' and G" are constant, with G'< G", after which both moduli decrease at higher deformation, showing viscoelastic fluid properties.

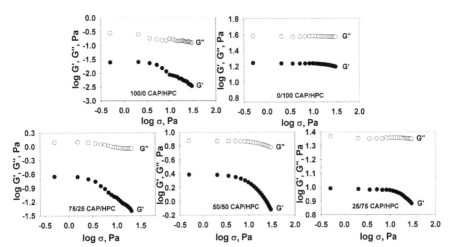

FIGURE 3.13 Log-log plots of storage and loss moduli *versus* shear stress for CAP/HPC blends in 2-Me at 4 g/dL concentration.[69]

At the same time, the constant zone is seen as dependent on the intermolecular association from the polymer blends. The aspect of the curves for both moduli is related to polymer topological entanglements and association phenomena.[76] Therefore, knowledge on the morphological and rheological properties as a function of shear stress provides insight into molecular interactions and surface organization. In addition, Fig. 3.13 permits the selection of shear stress on 2 Pa (from the linear viscoelastic domain *versus* shear stress) in viscoelastic measurements *versus* frequency. The tendency towards aggregation phenomena in casting solutions of CAP, HPC and their blends is reflected in Fig. 3.14, where moduli G' and G'' are presented as a function of frequency.[69] Over the low frequencies domain of 0.3–0.9 **Hz**, the storage and loss moduli are proportional to frequency – where the exponents for G' and G'' are between 2.1–0.9 and 1.1–0.8, respectively, maintaining the behavior characteristic to a viscoelastic fluid, where G'< G''.[77] These exponents decrease at lower values of polymer concentrations and at lower HPC compositions in the polymer blend. In addition, the frequencies corresponding to the crossover point, which delimits the viscous flow from the elastic one, and for which G'= G'', exhibit lower values for HPC in 2-Me, which increase with increasing the CAP content in polymer blends.

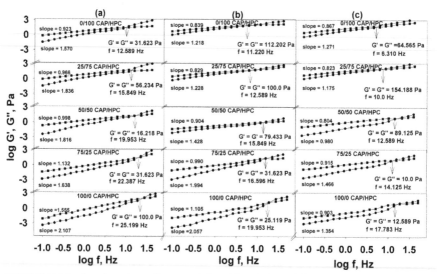

FIGURE 3.14   Log-log plots of storage (●) and loss (■) moduli as a function of frequency for CAP/HPC blends in 2-Me at (a) – 3, (b) – 4 and (c) – 5 g/dL concentration.[69]

All these aspects reflect the specific molecular rearrangements in the system, through modification of the mixing ratio of polymers, under the influence of hydrogen bonding interactions between the polymer components and solvent.

The rheological, as well as the morphological properties, are influenced by the hydrophobic characteristics of both CAP and HPC components (Fig. 3.15),[78,79] in which the disperse components take values of 38.8 mN/m and 34.3 mN/m, respectively, while the electron-acceptor interactions (of $\gamma_{sv}^+ = 0.59$ mN/m and $\gamma_{sv}^+ = 0.18$ mN/m, respectively) are lower than the electron-donor ones (of $\gamma_{sv}^- = 18.86$ mN/m and $\gamma_{sv}^- = 16.23$ mN/m, respectively) (Table 3.5). For 50/50 wt./wt. CAP/HPC films, the apolar component achieves maximum value, while the polar components, with electron donor and electron acceptor parameters, and total surface tensions have minimum values.

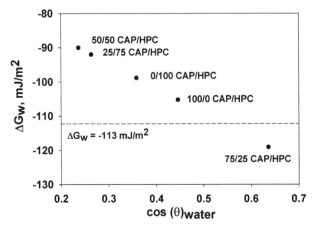

FIGURE 3.15 Surface free energy (ΔG_w) vs. water contact angle of CAP/HPC blends (wt./wt.) in 2-Me for 4 g/dL concentration. $\Delta G_w > -113$ mJ/m² correspond to more hydrophobic materials.[78]

TABLE 3.5 Surface Tension Parameters (mN/m) of CAP/HPC (wt./wt.) Blends in 2-Me for 4 g/dL Concentration, According to the Geometric and Acid/Base Method[78]

CAP/HPC	γ_{sv}^d	γ_{sv}^p	γ_{sv}^+	γ_{sv}^-	γ_{sv}
100/0	38.80	6.69	0.59	18.86	45.49
75/25	28.20	4.22	0.23	19.97	32.42
50/50	35.32	3.03	0.12	19.16	38.35
25/75	31.50	3.28	0.14	19.36	34.78
0/100	34.34	2.87	0.18	16.23	37.21

According to Table 3.6, which summarizes the data provided by AFM images (Figs. 3.16–3.20), variation in polymer blend compositions and concentration of the casting solutions in 2-Me favors a different surface morphology.[69]

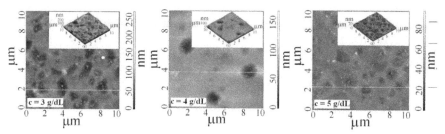

FIGURE 3.16 2D-AFM images at $10x10\ \mu m^2$ scan area, including the small images corresponding to 3D-AFM images, and cross-section profiles corresponding to the insert line in small 2D images, of CAP films obtained from solutions in 2-Me at different concentrations, 3 g/dL, 4 g/dL, and 5 g/dL.[69]

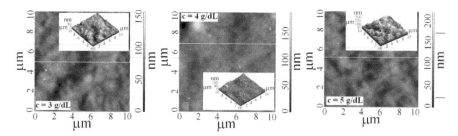

FIGURE 3.17 2D-AFM images at $10x10\ \mu m^2$ scan area, including the small images corresponding to 3D-AFM images, and cross-section profiles corresponding to the insert line in small 2D images, of HPC films obtained from solutions in 2-Me at different concentrations, 3 g/dL, 4 g/dL, and 5 g/dL.[69]

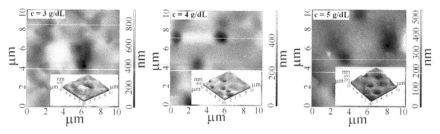

FIGURE 3.18 2D-AFM images at $10x10\ \mu m^2$ scan area, including the small images corresponding to 3D-AFM images, and cross-section profiles corresponding to the insert line in small 2D images, of 75/25 wt./wt. CAP/HPC films obtained from solutions at different concentrations, 3 g/dL, 4 g/dL, and 5 g/dL.[69]

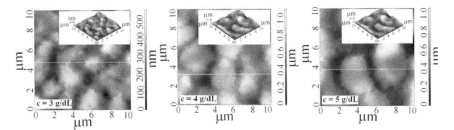

FIGURE 3.19 2D-AFM images at $10x10$ μm^2 scan area, including the small images corresponding to 3D-AFM images, and cross-section profiles corresponding to the insert line in small 2D images, of 50/50 wt./wt. CAP/HPC films obtained from solutions in 2-Me at different concentrations, 3 g/dL, 4 g/dL, and 5 g/dL.[69]

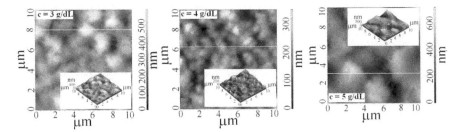

FIGURE 3.20 2D-AFM images at $10x10$ μm^2 scan area, including the small images corresponding to 3D-AFM images, and cross-section profiles corresponding to the insert line in small 2D images, of 25/75 wt./wt. CAP/HPC films obtained from solutions in 2-Me at different concentrations, 3 g/dL, 4 g/dL, and 5 g/dL.[69]

TABLE 3.6 Pore Characteristics (Area, Volume, Depth and Diameter) and Surface Roughness Parameters (Average Roughness, Sa, and Root Mean Square Roughness, Sq) for CAP/HPC Films Obtained from Casting Solutions in 2-Me at Different Concentrations, c, Corresponding to AFM Images[69]

CAP/HPC (wt./wt.)	c, (g/dL)	Pore characteristics				Surface roughness	
		Area (μm^2)	Volume (μm^2.m)	Depth (nm)	Diameter (μm)	Sa (nm)	Sq (nm)
100/0	3	0.68	27.49	77.41	0.85	11.58	15.76
75/25	3	0.44	164.72	474.31	0.74	99.25	134.15
50/50	3	1.30	514.37	–	1.26	70.57	88.64
25/75	3	1.45	663.35	–	1.35	66.55	84.43
0/100	3	–	–	–	–	12.72	16.22
100/0	4	1.13	41.24	89.23	1.14	7.43	11.67
75/25	4	1.09	162.39	411.07	1.17	53.74	77.14

TABLE 3.6   *(Continued)*

CAP/HPC (wt./wt.)	c, (g/dL)	Pore characteristics				Surface roughness	
		Area (μm^2)	Volume ($\mu m^2 .m$)	Depth (nm)	Diameter (μm)	Sa (nm)	Sq (nm)
50/50	4	4.55	4.23	–	2.39	153.25	188.58
25/75	4	0.38	95.79	–	0.69	41.62	51.65
0/100	4	–	–	–	–	11.74	16.13
100/0	5	1.77	62.16	57.87	1.33	4.27	5.59
75/25	5	1.74	129.13	206.93	1.47	50.50	66.21
50/50	5	7.66	5.26	–	3.09	151.71	184.49
25/75	5	5.52	2351.66	–	2.62	87.21	110.01
0/100	5	–	–	–	–	17.57	21.83

Different morphological aspects of the CAP film are the effect of chain conformation modification generated by the hydrogen bonding between the acetyl and hydroxyl groups, and also by the specific interactions, including hydrogen bonding, with 2-Me. Increase of casting solutions concentration determines modification of pores number and of their characteristics; thus, surface roughness decreases, both area and volume of pores increase, while pores depth decreases with increasing concentration.

The structure, formed after the slow evaporation of 2-Me at room temperature, evidences the importance of evaporation kinetics and concentration of the casting solution in making the different morphological aspects of the HPC films. The changes of molecular distribution in the corresponding films depend on solutions concentration. At a concentration of 3 g/dL, a large number of HPC molecules aggregate, showing their potential to form self-assembled structures generated by side-chain hydrogen bonding interactions. The dimensions of aggregates are in the 1.5 - 2.7 μm range, increasing with the concentrations of the casting solution over the 3–5 g/dL domain. Under some specific conditions, at higher concentrations, HPC films morphology gets modified, exhibiting a crystal liquid behavior and/or texture characteristics.[80,81]

Polymer–polymer miscibility is generally considered as a result of the specific interactions between polymer segments, which include donor–acceptor, dipole–dipole, hydrogen-bonding, ion–ion, acid–base, and ion–dipole interactions.[82] In CAP/HPC blends, with hydrophobic and electron-donor characteristics, miscibility is due especially to hydrogen bonding. 2-Me is known as a solvent with electron-donor properties, so that the morphological aspects of polymer blends films are a consequence of hydrogen bondings presence in the casting solution system, along with other kinds of apolar interactions; therefore, polymers dissolution is assured by the interfacial free energy of the polymers, on taking into account the polar as well as the apolar surface tension parameters.[83] A small content of HPC in the CAP/HPC blends generates pores with higher volumes and depths, and also an increased surface roughness, comparatively with the values corresponding to pure CAP; at a higher content of HPC, domains with large areas and lower average roughness appear. Both components are stabilized by hydrogen bonds interactions, which lead to areas whose diameter increases with increasing the concentration and composition of HPC. The occurrence of different areas, expressed as supernodular aggregates,[84] coincides with the thinning behavior evidenced by rheological data – caused by numerous hydrogen bonding interactions at higher HPC content. The next subchapter evidences that in some solvents, lyotropic mesophases usually have a characteristic critical concentration, where the molecules first begin to orient themselves into the anisotropic phase (which coexists with the isotropic one); the anisotropic or ordered phase increases with solution concentration in some domain of the biphasic region. At higher concentration, the solution becomes anisotropic.[85]

3.4   LYOTROPIC LIQUID CRYSTAL PHASES IN CELLULOSE DERIVATIVE COMPOSITES

On macroscopic scale, liquid crystalline phase of fluids polymers possesses long-range order. Generally, their molecules have asymmetric shapes to form liquid crystalline phases and mesogenic groups, linked either on the main chain with flexible segments, or on the side chain attached to the flexible main chain. Some polymers without mesogenic groups are also capable of forming liquid crystalline phases. These polymers have a rigid or semirigid backbone, arising from the steric effects or intramolecular hydrogen bonding, which restricts chain flexibility.

Based on a lattice polymer, Flory was the first to predict that linear rod-like polymer would form ordered phases in concentrated solutions.[86] According to this theory, at a given concentration, where the critical volume fraction of the polymer depends on the axial ratio, expressed as a length – diameter ratio, the rigid polymer solution would separate into two phases – isotropic and anisotropic.

Both entropy and enthalpy contribute to the stability of nematic mesophases, although the dominant factor is the entropy term, which is controlled by the size and shape of the molecule.[87,88] The original theory, which assumed that the polymer chains were completely rigid and rod-like, does not reflect the real situation. Besides, polydispersity, distribution of the substituted side groups as well as polymer-solvent interactions affect the flexibility of the polymer, so that even the most rigid polymers will have some degree of flexibility.

Formation of the liquid crystalline phase is highly dependent on specific polymer-solvent and polymer-polymer interactions. Transition from the isotropic to the ordered phase of lyotropic systems of semirigid polymers represents a balance of polymer-polymer and polymer-solvent interactions.[89,90] Not all cellulose derivatives form liquid crystalline phases at high concentrations; it is only those with appropriate solubility in a particular solvent that can form liquid crystalline solutions. In some cellulosic/solvent systems, increasing polymer concentration from the semidilute state, where already microgels appear, leads directly to the gel state, with no evidence of liquid crystallinity, even if the backbone of these derivatives has similar stiffness with that of the derivatives producing mesophases. Due to the strong polymer-polymer interactions cellulose derivatives are not sufficiently soluble in a particular solvent to achieve the concentrations necessary for mesophase formation. Literature shows that the lyotropic liquids crystalline phase formed by cellulose derivatives occurs in highly concentrated solutions, ranging from 20–70% by weight, higher than those of the rigid rods predicted by Flory's theory. The semiflexible nature of chains is due to substituents nature, substitution degree, substituents distribution for partially substituted cellulose derivatives, temperature and solvents. Thus, the stiffness of cellulose derivatives in solutions is a function of the steric interactions occurring between adjacent units, intrachain hydrogen bonding, polymer-polymer and polymer-solvent intermolecular interactions.[91]

On the other hand, liquid crystal polymer blends have been intensively investigated as to their unique electrical, optical, mechanical properties.[92] The self-aligning nature of the liquid crystal (LC) is used for obtaining organic polymer thin-film transistors in liquid crystal display devices.[93] Recently, a comprehensive study has been devoted to the cellulose triacetate-nitromethane system, to explore its phase separation for different concentration and temperature values.[94] The physical state of the polymer is identified within the coexistence phase limits on the phase diagram, which included three types of phase separation: amorphous, crystal, and liquid crystal. The limits of the regions determining the coexistence of the liquid crystal and of the partially crystal phase are found to be inside the region of amorphous liquid-liquid phase separation. For cellulose ester-solvent systems, this state diagram is the first experimental evidence for the possible coexistence of several phases with amorphous, liquid crystal, and crystal polymer ordering. The phase state of the system develops under the influence of temperature and concentration. At temperatures above the binodal and liquidus, all mixtures are homogeneous transparent solutions. If temperature is lowered and the configurative point moves to the metastable range, the system of various concentrations may have various morphologies, depending on polymer concentration and distance from the stability range. For low-concentrated solutions, under conditions of a rather high kinetic mobility of the molecules, amorphous phase separation is observed with lowering temperature. The solutions are visually characterized by opalescence and occurrence of precipitation. For medium-concentrated solutions with low viscosities, in time the system splits into two phases. A dense white precipitate is formed in the polymer-concentrated phase, the other phase appearing as an opalescent solution with suspended particles. Literature[95,98] shows that transition to LC state is possible under certain thermodynamic conditions, when the selective solvation of the different polymeric groups determines the extended helical conformations of macromolecules[94,99,102] and the appearance of liquid crystal ordering. In the same context, the structure, intra and intermolecular interactions, and conformations of macromolecules in cellulose derivative films – under conditions of the liquid crystal state formation during vapor sorption of some solvents – show that the number of intramolecular hydrogen bonds stabilizes the rigid helical conformation of the macromolecules.[103] The anisotropic structure is also preserved after desorption of the vapor sorbed solvent. The capability of cellulose derivative films to form a liquid crystal state in vapor solvent and to preserve the anisotropic structure after solvent desorption, is especially important for preparing new functional materials.

In crystalline state, the structure of polymers is influenced by solvent and concentration.[69,70] Fundamental research on the formation of banded textures in thin-film samples from lyotropic solutions subjected to shear is important, due to the large number of physical interactions here involved.[104,105] Surface anisotropy and the mechanical and optical properties of the polymer films,[106] together with their potential use as alignment layers for liquid crystal displays, make these systems particularly interesting and promising for new applications. Flow behavior is the most thoroughly studied rheological property. Some studies hypothesized the universal existence of three shear flow regimes to describe the viscosity of polymer liquid crystals namely: a shear thinning regime at low shear rates (Region I), a Newtonian plateau at intermediate shear rates (Region II), and another shear thinning regime at high shear rates (Region III). Region I, observed at low shear rates, shows shear thinning, exhibiting yield stress, as in some plastic materials. This region is characterized by distortional elasticity, associated with spatial variation in the director field (average local molecular orientation). Region II is a Newtonian plateau, reflecting a "dispersed polydomain" structure and Region III is a shear-thinning zone, showing viscoelastic behavior. In Regions I and II, the flow is not strong enough to affect molecular orientation while, in Region III, the flow field is very strong, so that the shear induces molecular orientation. Cellulose derivatives do not always cover the entire domain from Region I to Region III, because not every regime lies within the accessible shear rate range.[107]

In the preceding section, the specific interactions of CAP/HPC blends in 2-methoxyethanol are presented, such as hydrogen bonding, ion-ion pairing, and electron-donor and electron-acceptor complexation, which generates miscibility between components, have been discussed. In this context, the ATR-FTIR spectra of CAP, HPC and their blends (100/0, 75/25, 50/50 25/75 and 0/100 wt./wt. CAP/HPC), plotted in Fig. 3.21, show that equilibrium in polymer blends is assured by the hydrogen bonds.[69] A remarkably similar aspect of the spectra is observed for both polymers. Broad transmission bands are distinguished at 3421 cm^{-1} for HPC and 3431 cm^{-1}, respectively, for CAP, produced by stretching of the – OH groups, at 1723 cm⁻¹ for HPC and 1719 cm⁻¹ for CAP, produced by stretching of the C=O groups from the ester, carboxylic acid, and at 1252 cm⁻¹, respectively, for HPC and 1329 cm⁻¹ for CAP, produced by stretching of the C–O–C ester bond. The presence of hydrogen bond structures in blends are evidenced from peaks shape and intensity of the absorption band of the hydrogen stretching vibration.[108] The differences observed among the shape, broadening and shifting of the mentioned peaks for polymer blends suggest the existence of hydrogen bonding generated by -OH, C=O and C–O–C groups. However, the mentioned free and associated groups assure the equilibrium in these polymer blends *via* hydrogen bonds.

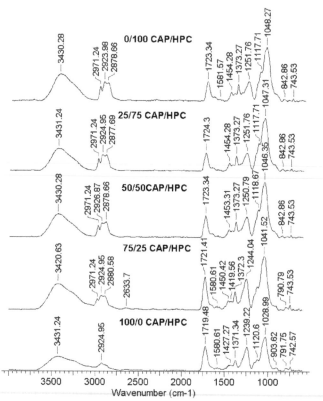

FIGURE 3.21   ATR-FTIR spectra of CAP, HPC and CAP/HPC wt./wt. blends.[69]

Quantitative measurements of weight loss are shown as TG/DTG plots (Fig. 3.22) for the same blends.[69] The highest mass loss was of 100/0 wt./wt. CAP/HPC, followed by 75/25, 50/50 25/75 and 0/100 wt./wt. CAP/HPC. It has been found out that the components of the blends decompose over different temperature ranges, the films loose water, and the curve for pure HPC shows an one-stage degradation step within the 300–400 °C range, with a maximum at 373 °C. The CAP and CAP/HPC blends showed a two-stage degradation step within the 210–298 °C and 280–560 °C ranges, respectively. The first weight loss is caused by the phthalic anhydride from CAP; with increasing the HPC content, this process is less evident, while the second weight loss is caused by the remaining cellulose. In addition, Table 3.7 shows that a percentage of 10 wt.% from the CAP and HPC mass is lost at 194 °C and 340 °C, respectively. For CAP/HPC blends, the temperatures at which 10 wt.% of their mass is lost represent intermediary values of the above-mentioned ones, increasing with increasing the HPC content.

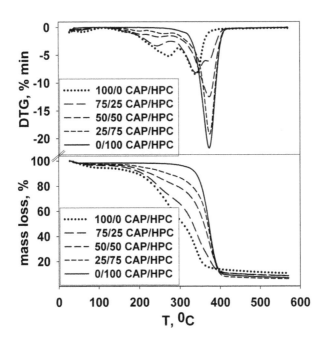

FIGURE 3.22 Experimental TG/DTG curves of CAP, HPC and CAP/HPC wt./wt. blends.[69]

TABLE 3.7   Degradation Stage, Temperature at which the Thermal Decomposition Begins (T_{onset}), Temperature at which the Degradation rate is Maximum ($T_{main\ peak}$), Temperature at End of the Process (T_{endset}), Weight Losses (W), and Temperatures Corresponding to 10 and 50 wt.% Weight Losses (T_{10}, T_{50}) from TG/DTG Curves for CAP, HPC and CAP/HPC wt./wt. Blends[69]

CAP/HPC blends	Degradation Stage, °C	T_{onset}, °C	$T_{main\ peak}$, °C	T_{endset}, °C	W, %	T_{10}, °C	T_{50}, °C
0/100	30–560	337	373	403	91.70	340	372
	residue				8.30		
25/75	30–140[(a)]	–	–	–	2.70		
	140–225[(b)]	–	–	–	3.28	270	369
	225–290[(c)]	–	–	–	6.10		
	290–560[(d)]	349	375	403	81.75		
	residue				6.17		
50/50	30–120[(a)]	–	–	–	3.20		
	120–224[(b)]	–	–	–	6.54	226	359
	224–280[(c)]	–	–	–	8.50		
	280–560[(d)]	325	374	405	72.86		
	residue				8.90		
75/25	30–120[(a)]	–	–	–	2.70		
	120–212[(b)]	–	–	–	7.10	213	333
	212–280[(c)]	215	246	271	18.23		
	280–560[(d)]	301	342	411	62.44		
	residue				9.53		
100/0	30–117[(a)]	–	–	–	5.63		
	117–185[(b)]	–	–	–	3.40	194	308
	185–298[(c)]	202	274	293	37.38		
	298–560[(d)]	312	340	366	43.05		
	residue				10.54		

(a) Degradation stage corresponding to the water loss.
(b) Degradation stage corresponding to the phthalic anhydride and acetic acid losses from CAP.
(c) Degradation stage corresponding to the phthalic anhydride losses from CAP.
(d) Degradation stage corresponding to the thermal decomposition of the remaining cellulose.

Thermogravimetric measurements show that CAP is less thermally stable than the HPC and CAP/HPC blends, as the anhydroglucose units increase the rigidity of the HPC chain.[69] These interactions impart specific properties to the polymer blends. In N,N-dimethylacetamide (DMAc), the anisotropic behavior appears under specific conditions of concentration and/or blend composition. Some studies have reported that the polymer structures, their mixing ratio and the used solvent influence the interactions from the systems and, consequently, the ordered domains in rheological behavior.[109,110] In this respect, Figure 3.23 presents the modification of dynamic viscosity, η, *versus* shear rate, $\dot{\gamma}$, for c = 20, 40, 60 wt. % at 25 °C, for pure CAP and pure HPC in DMAc. In addition, Fig. 3.24 shows the same modification of dynamic viscosity *versus* shear rate at different temperatures over the 25–45 °C domain for HPC in DMAc.

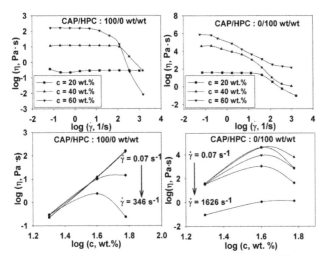

FIGURE 3.23 Log-log plots of dynamic viscosity *versus* shear rate at different concentrations and also *versus* concentrations at different shear rates for CAP and HPC samples in DMAc Logarithmic viscosity values of HPC in log η versus log $\dot{\gamma}$ are shifted upwards to 1 and 2 for 40 and 60 wt. %, respectively, for a better visualization[70]

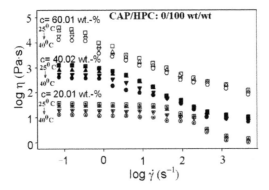

FIGURE 3.24 Logarithmic plot of viscosity as a function of shear rate for HPC in DMAc at different concentrations and temperatures. Logarithmic viscosity data points for 60 wt. % are shifted upwards to 1 for a better visualization[26]

As one can see, for CAP solutions with lower concentrations, a Newtonian behavior appears over the entire shear rate domain while, with increasing concentration to 60 wt. %, the thinning effect reduces dynamic viscosity. The shear experiments performed on lyotropic HPC solutions in DMAc reveal that the viscosity-shear rate dependence at different concentrations appears only in Regions II and III. As concentration increases, the Newtonian plateau becomes smaller, being shifted to lower shear rates. At low shear rate, dynamic viscosity increases with concentration for both cellulose derivatives, showing a maximum at intermediary concentration, in the 20–60 wt. % range, which means a transition from the isotropic to the anisotropic phase.

Ordering of macromolecules in both polymers at higher shear rate is associated with a viscosity lower than that of the isotropic solutions, revealing liquid crystal behavior. Below a certain critical shear rate and at a lower content of HPC, the isotropic solutions are Newtonian. Literature data show that the viscosity peak, observed for all lyotropic polymer solutions, is a decreasing function of shear rate, ascribed to several mechanisms.[70,111] A competition between the ordering induced by shear and that thermodynamically produced was suggested by Hermans,[111] while a correlation between maximum viscosity and the anisotropic phase appearance – valid only at lower shear rate – was suggested by Zugenmaier.[91]

For CAP/HPC blends in DMAc,[70] a distinct change in the rheological properties occurs at the critical concentration of 40 wt. %. Therefore, viscosity at higher shear rates takes lower values than at lower concentrations, at equivalent shear rates. Upon formation of the anisotropic phase, viscosity begins to decrease. Considering the dependence of viscosity on concentration for concentrations above 40 wt. %, the solution is in a fully liquid-crystalline mesophase. Below this concentration, the solution is biphasic. Similar results were obtained for HPC aqueous solutions[89,112] or CAP/HPC blends – in which Regions II and III have different extensions, the Newtonian behavior decreasing with increasing concentration and HPC content (Fig. 3.25).

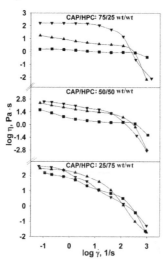

FIGURE 3.25   Log-log plots of dynamic viscosity *versus* shear rate for 75/25, 50/50 and 25/75 wt./wt. CAP/HPC blends in DMAc at different concentrations: (■) 20 wt. %; (▲) 40 wt. %; (▼) 60 wt. %[70]

Deviation of the CAP solution from the Newtonian behavior is also analyzed by the power law relationship between shear stress and shear rate (equation 11). Figures 3.26 and 3.27 show the values of the flow and consistency indices, obtained at different concentrations and blend compositions, from the slope and intercept of log shear stress *versus* log shear rate plots, over the domain of lower shear rates. It was observed that, as concentration increases, the resistance of the fluid to the applied rate of shear or force (called shear stress) decreases, causing a decrease of the flow index to values < 1, and an increase in pseudoplasticity. Decrease of the flow index occurs when the HPC content increases and temperature decreases. At the same time, the consistency index increases with increasing concentration and HPC content, and decreases with the increase of temperature.

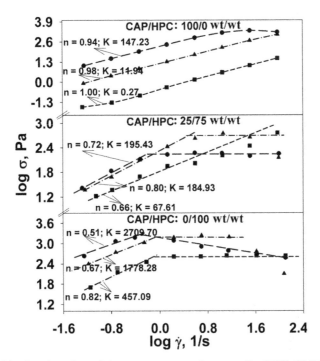

FIGURE 3.26 Log-log plots of shear stress *versus* shear rate for 100/0, 25/75, 0/100 wt./wt. CAP/HPC blends in DMAc at different concentrations: (■) 20 wt. %; (▲) 40 wt. %; (▼) 60 wt. %.[70]

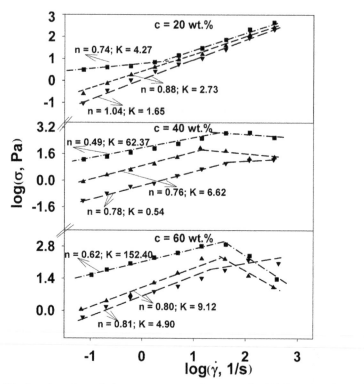

FIGURE 3.27   Log-log plots of shear stress *versus* shear rate for 50/50 wt./wt. CAP/HPC blends in DMAc at different concentrations and temperatures: (■) 25 °C; (▲) 35 °C; (▼) 45 °C.[70]

Figure 3.28 shows that the flow activation energies for CAP and HPC in DMAc, determined in the region of shear rate with Newtonian behavior, increase with concentration and are higher for HPC.

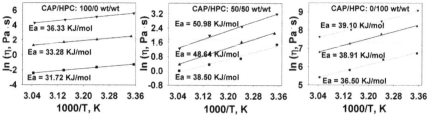

FIGURE 3.28   Arrhenius plots for CAP, HPC and 50/50 wt./wt. CAP/HPC blends in DMAc at (■) 20 wt. %, (▲) 40 wt. %, and (▼) 60 wt. % concentrations.[70]

Similarly with Fig. 3.13, Fig. 3.29 illustrates the variation of storage and loss moduli *versus* shear stress at a constant frequency of 1 Hz, for CAP, HPC and CAP/HPC blend solutions at 60 wt. % concentrations in DMAc. G' is constant and lower than G'', showing viscoelastic fluid properties for 100/0 and 75/25 wt./wt. CAP/HPC blends over a larger deformation region. Increase of the HPC content determines higher values of G' than of G'', generated by transition from the isotropic to the anisotropic phase.

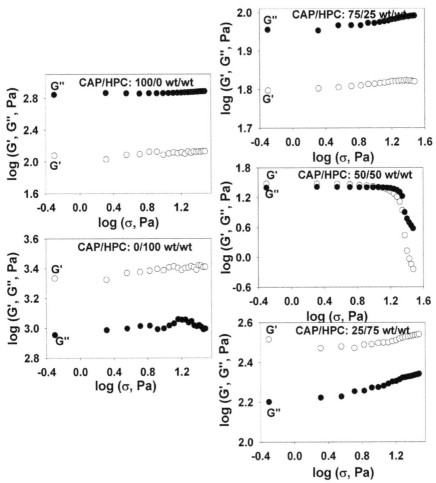

FIGURE 3.29   Log-log plots of storage and loss moduli *versus* shear stress for different compositions of CAP/HPC blends in DMAc at a concentration of 60 wt. %.[70]

The ordering tendency in casting solutions of CAP, HPC and their blends in DMAc is illustrated in Fig. 3.30, where moduli G' and G'' are presented as a function of frequency, in the same manner as in Fig. 3.14 (at lower concentrations, in 2-Me).

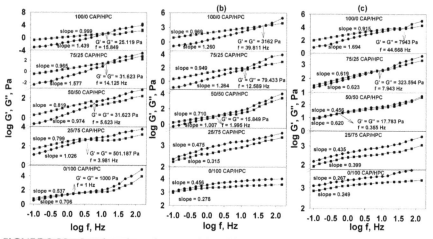

FIGURE 3.30   Log-log plots of storage (●) and loss (■) moduli as a function of frequency for different compositions of CAP/HPC blends in DMAc at a concentration of (a) – 20 wt. %; (b) – 40 wt. %; 60 wt. %.[70]

For all concentrations of CAP and CAP/HPC blends at higher content of CAP, over the low frequencies domain of 0.2–0.9 Hz, the storage and loss moduli are proportional to frequency – when the exponents for G' and G'' take values between 1.6–0.6 and 0.9–0.4, respectively, maintaining the characteristic of a viscoelastic fluid, where G' < G''.[76,77,113] These exponents decrease both with increasing polymer concentrations and at higher HPC compositions in the polymer blend. In addition, the frequencies corresponding to the crossover point, which delimits the viscous flow from the elastic one, and for which G' = G'', exhibit lower values for HPC in DMAc, and become higher with increasing the CAP content in polymer blends. A gel characteristic proves the presence of a liquid crystal phase, which appears for HPC, and a higher content of HPC at 40 wt. % and 60 wt. % concentrations. In these cases, the storage modulus, G', is always higher than the loss modulus, G'', over the entire frequency range, being characterized by $G' \propto f^{0.267-0.475}$ and $G'' \propto f^{0.249-0.399}$; these dependencies are observed when the lyotropic phase becomes predominant.

Literature shows that the liquid crystalline order in cellulose derivative solutions observed by rheological investigations is preserved in solid films by slow evaporation of the solvent.[114] Consequently, the microstructure of films depends on the drying conditions; also, films prepared under ambient conditions show polydomain structures with the helical axes of different chiral nematic domains pointing in different directions. Moreover, applying a magnetic field during drying increases the size of the chiral nematic domains and affect the orientation of the helical axes with respect to the film plane.[115] Studies on the microstructure of such chiral nematic solid films reveal parabolic focal conic defects, a symmetrical form of focal conic defects in which the line defects produce a pair of perpendicular, antiparallel, and confocal parabolas.[114] Depending on the substance and manner of sample preparation, the parabolas lie horizontally or vertically in the sample plane.[116] In vertical position, the parabolas intersect the top and bottom sample surfaces with their ends. The focal regions are located in midplane. The fact that these structures are captured in a solid film permits to study this highly symmetrical defect structure by other means than optical microscopy, for example, by atomic force microscopy (AFM). In addition, a chiral nematic system with a helical pitch large enough to be resolved by an optical microscope allows visualization of the individual structural layers.

Thus, AFM images from Figs. 3.31–3.34, for cellulose derivative films prepared under ambient conditions, show the polydomain structures with the helical axes of different chiral nematic domains pointing in different directions. Moreover, films obtained from CAP and HPC liquid crystalline solutions in both pure state and in mixture evidence some lights areas whose sizes depend on the HPC content and increase with increasing concentration. Intensity of the polarization colors varies cyclically, from zero up to a maximum brightness at 45 degrees, for HPC (Fig. 3.31 (b, b,' b")). A rotating stage and centration of samples in a polarized light microscope is a critical element for determining the quantitative aspects of the HPC liquid crystal. Centration of the objective make the center of the stage rotation coincides with the center of the field, for maintaining the specimen exactly in the center, when rotated.

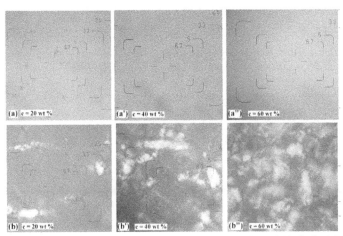

FIGURE 3.31   Optical microscopy images of CAP (a, a,' a") and HPC (b, b,' b") films obtained from solutions in DMAc at different concentrations.[70]

In the case of CAP/HPC blends, the number of methylene units (i.e., side chain length) inserted by increasing the HPC ratio in blends has a significant influence on the selective reflection characteristics of cholesteric liquid crystals.[117] Some formations of different sizes and intensities, namely droplets, appear. At a c = 20 wt. % concentration for all mixing ratios, the presence of the HPC liquid crystal is visible (white areas), without forming the droplets, such formations tending to appear starting from a 40 wt. % concentration (Figs. 3.32–3.34).

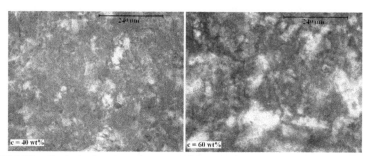

FIGURE 3.32   Optical microscopy images of films obtained from solutions of 25/75 wt./wt. CAP/HPC blends in DMAc at 40 wt. % and 60 wt. % concentrations.[70]

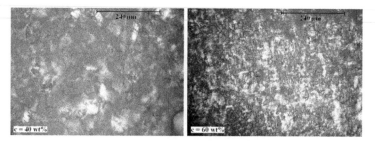

FIGURE 3.33   Optical microscopy images of films obtained from solutions of 50/50 wt./wt. CAP/HPC blends in DMAc at 40 wt. % and 60 wt. % concentrations.[70]

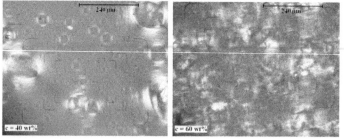

FIGURE 3.34   Optical microscopy images of films obtained from solutions of 75/25 wt./wt. CAP/HPC blends in DMAc at 40 wt. % and 60 wt. % concentrations.[70]

Atomic force microscopy studies, according to Figs. 3.33–37, [70] also evidence the occurrence of liquid crystal phases including these formations. Under particular conditions, HPC liquid crystalline films exhibit a characteristic structure, called "band texture,"[81,91] consisting of alternating bright and dark areas.

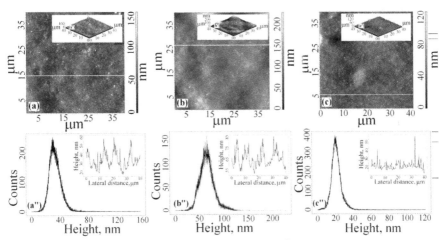

FIGURE 3.35 2D and 3D – AFM images at $40x40 \ \mu m^2$ scan area, and histograms – in which surface profiles were introduced as small plots, – of CAP films obtained from solutions in DMAc at different concentrations: 20 wt. % – (a, a,' a"); 40 wt. % – (b, b,' b"); 60 wt. % – (c, c,' c").[70]

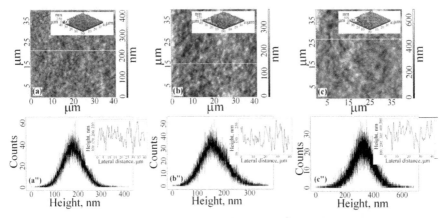

FIGURE 3.36 2D and 3D – AFM images at $40x40 \ \mu m^2$ scan area, and histograms – in which surface profiles were introduced as small plots, – of HPC films obtained from solutions in DMAc at different concentrations: 20 wt. % – (a, a,' a"); 40 wt. % – (b, b,' b"); 60 wt. % – (c, c,' c").[70]

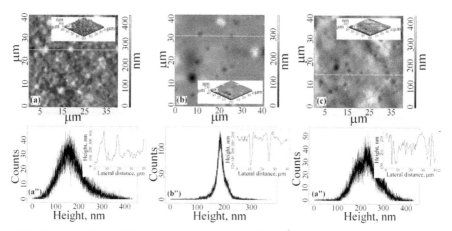

FIGURE 3.37   2D and 3D – AFM images at $40x40 \ \mu m^2$ scan area, and histograms – in which surface profiles were introduced as small plots, – of 75/25 wt./wt. CAP/HPC films obtained from solutions in DMAc at different concentrations: 20 wt. % – (a, a,' a"); 40 wt. % – (b, b,' b"); 60 wt. % – (c, c,' c").[70]

As the percent of the liquid crystalline counterpart decreases, the surface pattern is maintained, while its dimensions become higher and the "small" bands disappear. In the case of CAP/HPC blends in DMAc, hydrogen bonds appear only between HPC and DMAc, therefore band texture is larger and increases with decreasing the amount of HPC. A natural tendency is noticed for the semirigid segments of HPC, namely to self-align into ordered domains, thus lending high performance properties to these materials.[118] Also is observed that the LC phases are strongly dependent on the number of methylene units from the side chains of polymer blends.

On the other hand, although in optical microscopy images droplets cannot be seen at c = 20 wt. % concentration, in AFM images they are visible for all films, with the exception of the pure sample. They start to appear from c = 20 wt. %, 50/50 CAP/HPC wt./wt. blends, their dimensions being around 3.69μm. These values agree with those from literature,[119] obtained for hydroxypropylcellulose/polydimethylsiloxane blends, where the size of droplets varies between 7–15μm. With increasing concentration, domains without droplets appear, as well as areas with droplets in training. They become more visible for 25/75 CAP/HPC wt./wt. blends, increase of the HPC content and, implicitly, of the methylene units leading to a more compact network, caused by HPC/DMAc bonding. Thus, the size of droplets decreases with increasing concentration (Table 3.1). In addition, as shown by Table 3.1 and by the AFM images of films made from pure samples, the surfaces present low roughness, pores appearing only at 75/25 CAP/HPC wt./wt., their dimension increasing with increasing concentration. Both components are present on the surface after solvent evaporation, being stabilized by hydrogen bonds interactions, which leads to the formation of droplets/pores of different sizes and intensities. The periodicity of the average peak-to-valley height for these images is evident from the profile line plotted in Figs. 3.35–3.39.

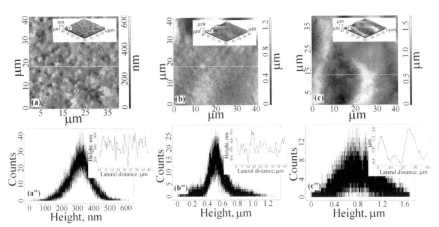

FIGURE 3.38 2D and 3D – AFM images at $40x40$ μm^2 scan area, and histograms – in which surface profiles were introduced as small plots, – of 50/50 wt./wt. CAP/HPC films obtained from solutions in DMAc at different concentrations: 20 wt. % – (a, a,' a"); 40 wt. % – (b, b,' b"); 60 wt. % – (c, c,' c")[70]

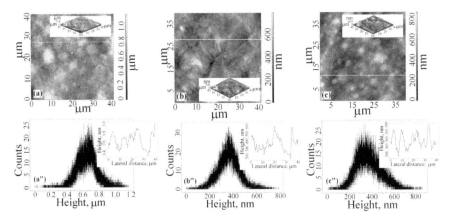

FIGURE 3.39 2D and 3D – AFM images at $40x40$ μm^2 scan area, and histograms – in which surface profiles were introduced as small plots, – of 25/75 wt./wt. CAP/HPC films obtained from solutions in DMAc at different concentrations: 20 wt. % – (a, a,' a"); 40 wt. % – (b, b,' b"); 60 wt. % – (c, c,' c").[70]

In the same context, some data on the rheological properties of epiclon-based polyimide/HPC blend solutions have been reported.[26] 2D topography AFM images ($40x40$ μm^2) of the corresponding films, prepared by mixing a

60 wt. % HPC solution with a 49 wt. % polyimide solution, shows that the hydrogen bonds between HPC and DMAc influence band texture, which increases with decreasing the amount of HPC (Fig. 3.40 (a) and (b)).

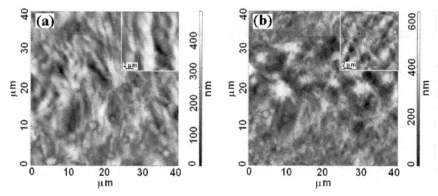

FIGURE 3.40   2D-AFM images for 30/70 wt./wt. polyimide/HPC (a) and 70/30 wt./wt. polyimide/HPC (b).[26]

As a result of the symmetry properties of the HPC liquid crystal solution, large domains of well-oriented polymer chains are formed during shear flow, while the defects are squeezed into small regions. Under specific shear flow conditions, the cholesteric liquid crystalline cellulose derivatives exhibit unwinding of the cholesteric helix and a cholesteric-to-nematic transition.[120] When shear is stopped, the system will first relax at a characteristic time to a transient state. In this state, the distortion energy is minimized and the orientational order is maintained, resulting a banded structure. The shear-induced anisotropy is affected by the inevitable relaxation of the chains, when the external field is removed. By introducing polyimide into the HPC solution, relaxation will take place collectively, due to the fact that the highly concentrated and aligned polymers cannot individually relax and, therefore, the inner stress induces a periodical contraction in the whole liquid crystalline polymer and different packing modes are observed. Structural relaxation after cessation of shear depends on the shear history of the mixture and on the dominant mechanism of stress relaxation. The band morphology of the blend is influenced by precursor and polyimide solution composition, solvent evaporation rate, film thickness, rate and duration of shear.[121] The effect of the chemical structure and composition on the viscoelastic properties is reflected on the orientation or mobility of segments in the shear field. The inherent long-range ordering tendencies of LC itself – and specifically its pattern-forming properties – open new perspectives to produce ordered polymer microstructures. The morphology of the ordered domains provides a means of "imaging," respectively potentially novel aspects of the pattern-forming LC states. Pure and applied researches related to the shear-induced morphology and structural relaxation after cessation of shear in polymer/LCP blends will become more and more important for developing high performance alignment layers used in display devices.

Consequently, the occurrence of droplets coincides with the thinning behavior evidenced by rheological data, both being caused by the more numerous hydrogen bonding interactions at higher HPC content. In this context, the results are in agreement with the literature ones, which show that cellulose derivatives form lyotropic mesophases.

All these results reflect the specific molecular rearrangements produced in the system through modification of the mixing ratio of polymers in certain solvents. As some biological polymers are insoluble in organic solvents or water, ionic liquids have attracted the attention of industry, especially because of the current need for creating a green chemistry environment. Thus, literature recommends 1–butyl-3–methylimidazolium acetate as a solvent for the study of sol/gel and liquid crystal transition of hydroxypropylcellulose (HPC).[25] According to the experimental data obtained from the parameters: relaxation time, hysteresis ratio, loss modulus, and also observing the LC textures *via* a polarization optical microscope, the liquid crystal transition concentration of HPC is slightly higher than sol/gel transition concentration, and increases with temperature which is also the case of common solvents.

Furthermore, the structural and molecular heterogeneity of cellulose derivatives, their high molecular masses, low diffusion coefficient values, high viscosities of even moderately concentrated solutions, and specific intermolecular and molecule-solvent interactions, as well as their behavior in different conditions raise numerous experimental problems. The investigations developed in this chapter describe the character of the structural transformations observed, starting from rheological and morphological data. Knowledge of phase separation kinetics mechanism which modifies the morphology of cellulosic systems allows designing of materials with desired properties for specific applications.

3.5 BIOCOMPATIBILITY OF DERIVATIVE CELLULOSE BLENDS

Typical research and application areas of polymeric biomaterials include tissue replacement, tissue augmentation, tissue support, and drug delivery. In many cases, the body needs only the temporary presence of a device/biomaterial, in which instantly biodegradable and certain partially biodegradable polymeric materials are used. Recent treatment concepts of scaffold based on tissue engineering principles differ from those based on standard tissue replacement and drug therapies, as the engineered tissue aims not only to repair but also to regenerate the target tissue. Cells have been cultured outside the body for many years; however, it has only recently become possible for scientists and engineers to grow complex three-dimensional tissue grafts to meet clinical needs. New generations of scaffolds based on synthetic and natural polymers are being developed and evaluated at rapid pace, aimed at mimicking the structural characteristics of the natural extracellular matrix.[122] The natural abundance and biodegradability of cellulose, together with its ability to provide unique properties through diversification of cellulosic structures determine a wide range of biomedical applications. In their native form, cellulosic materials have been widely used in the manufacture of optical products, such as hard contact lenses, due to their excellent clarity, good wettability and high gas permeability, textile fibers, molding powder sheets, optical membranes, etc. In addition, cellulose derivatives evidence excellent properties. Usually, these materials are molded and extruded into various consumer products, such as brush handles, tool handles, toys, steering wheels, or other items.

In view of a direct contact of the biomaterial with blood, a clear understanding of their interactions is a prerequisite. First of all, the material interacts instantaneously with blood constituents, which is critical in determining their potential side effects on the circulatory system and, eventually, on the whole organism.[123,124] Secondly, the interactions with blood can affect the in-vivo pharmacokinetic behaviors of the polymers and their ability to leave the blood compartment and enter other tissues. Blood has important physiological functions and a complex composition, being divided into two compartments, namely plasma – which contains proteins, lipids, salts – and specific cells, including red blood cells, white blood cells, and platelets, as well.

The artificial surfaces interact with blood platelets, initially causing platelet adherence and aggregation; when such foreign surfaces are placed in contact with the circulating blood, this interaction is believed to lead to thrombosis and thromboembolism, and to the removal of platelets from the circulation.[125] The surface characteristics of biomaterials, such as the hydrophilicity/hydrophobicity, roughness, and flexibility affect the cell-surface interactions, protein adsorption, behavior of cells adhesion and proliferation, and the host response, too. Therefore, cellular adhesion has a direct bearing on the thrombogenicity and immunogenicity of a specific material, predicting its blood compatibility and deciding the long-term use for a blood-contacting materials application.

On the other hand, adhesion of red blood cells, platelets or water to the cellulose derivative substrate plays an important role in biomedicine. For analyzing biocompatibility, the relations between the physicochemical properties of material surface and the adhesion of blood components should be known. Surface wettability, which is associated with surface free energy, has been often related to cell adhesion phenomena. In order to study the red blood cells, platelets, or water adhesion as a function of the substitution degree, it is preferable to compare a chemically – homologous series of polymers, for minimizing the contribution of specific interactions between the adherent cells and the chemical groups at the solid surface.

Cellulose derivatives compatibility with blood can be established by equation 12, where $W_{s,w}$, $W_{s,rbc}$, $W_{s,p}$, $W_{s,f}$, $W_{s,a}$, and $W_{s,IgG}$ describe the work of spreading of water, red blood cells, platelets, fibrinogen, albumin, and immunoglobulin G, respectively:[78,79,126]

$$W_s = W_a - W_c = 2 \cdot [(\gamma_{sv}^d \cdot \gamma_{lv}^d)^{1/2} + (\gamma_{sv}^+ \cdot \gamma_{lv}^-)^{1/2} + (\gamma_{sv}^- \cdot \gamma_{lv}^+)^{1/2}] - 2 \cdot \gamma_{lv} \quad (12)$$

where W_a and W_c are the work of liquid adhesion and cohesion, respectively. Also, the surface tension parameters listed in Table 3.5 (γ_{sv}^p, γ_{sv}^+, γ_{sv}^-, γ_{sv}^d, γ_{sv}), involved in the theoretical evaluation of liquids spreading, can be obtained by the acid/base method (Eqs. (13)–(16)),[127,130] using the known surface tension components of different liquids (γ_{lv}^p, γ_{lv}^-, γ_{lv}^+, γ_{lv}^d, γ_{sl}) given in Table 3.8.

$$1 + \cos\theta = \frac{2}{\gamma_{lv}} \cdot \left(\sqrt{\gamma_{sv}^d \cdot \gamma_{lv}^d} + \sqrt{\gamma_{sv}^+ \cdot \gamma_{lv}^-} + \sqrt{\gamma_{sv}^- \cdot \gamma_{lv}^+} \right) \quad (13)$$

$$\gamma_{sv}^p = 2 \cdot \sqrt{\gamma_{sv}^+ \cdot \gamma_{sv}^-} \quad (14)$$

$$\gamma_{sv} = \gamma_{sv}^d + \gamma_{sv}^p \quad (15)$$

$$\gamma_{sl} = \left(\sqrt{\gamma_{lv}^p} - \sqrt{\gamma_{sv}^p} \right)^2 + \left(\sqrt{\gamma_{lv}^d} - \sqrt{\gamma_{sv}^d} \right)^2 \quad (16)$$

where superscripts "d" and "p" indicate the disperse and polar component of the film surface tension obtained from the γ_{sv}^- electron-donor and γ_{sv}^+ electron-acceptor interactions, while γ_{sl} indicates the solid-liquid interfacial tension.

When the solid-liquid interfacial tension, γ_{sl}, takes negative values (Fig. 3.41), the interfacial free energy, ΔG_{sls}, has positive values (Eq. (17), Fig. 3.41) and rejection between the two surfaces of the same polymer, s, immersed in liquid, along with attraction of the liquid occurs:

$$\Delta G_{sls} = -2 \cdot \gamma_{sl} \quad (17)$$

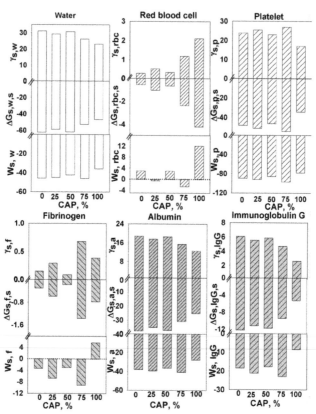

FIGURE 3.41 Solid-liquid interfacial tensions, γ_{sl}, interfacial free energy, ΔG_{sls}, and spreading work of CAP/HPC films, W_s, with water (w), read blood cell (rbc), platelet (p), fibrinogen (f), albumin (a), and immunoglobulin G (IgG).

TABLE 3.8 Surface Tension Parameters (mN m^{-1}) of Water and Some Biological Materials

Material	γ_{lv}^{d}	γ_{lv}^{p}	γ_{lv}^{+}	γ_{lv}^{-}	γ_{lv}
Water[127]	21.80	51.00	25.50	25.50	72.80
Red blood cell[126]	35.20	1.36	0.01	46.20	36.56
Platelet[126]	99.14	19.10	12.26	7.44	118.24
Fibrinogen[131]	37.60	3.89	0.10	38.00	41.50
Albumin[132,133]	26.80	35.70	6.30	50.60	62.50
IgG[134]	34.00	17.30	1.50	49.60	51.30

The hydrophilic/hydrophobic balance of the polymers can be described by the work of spreading of water, over the considered surface. In addition, when blood is exposed to a biomaterial surface, the life of the implanted biomaterials is decided by adhesion/cohesion of cells. Cellular adhesion to biomaterial surfaces could activate coagulation and the immunological cascades. Therefore, cellular adhesion has a direct bearing on the thrombogenicity and immunogenicity of a biomaterial, thus dictating its blood compatibility. The materials, which exhibit a lower work of adhesion, would show a lower extent of cell adhesion than those with a higher work of adhesion. Polymer interaction with red blood cells is mediated mostly by the hydrophobic interaction with the lipid bilayer (the red blood cell hydrophobic layer containing transmembrane proteins), the electrostatic interaction with the surface charges or/and the direct interaction with membrane proteins, depending on polymer characteristics. Figure 3.41 shows generally positive values for the work of spreading of red blood cells, $W_{s,rbc}$, and negative values for the work of spreading of platelets, $W_{s,p}$, suggesting a higher work of adhesion, comparatively with that of cohesion for the red blood cells, but a lower work of adhesion, comparatively with the work of cohesion for platelets. Blood platelets are essential in maintaining hemostasis, being very sensitive to changes in the blood microenvironment. Platelet aggregation is used as a marker for materials' thrombogenic properties, the polymer-platelet interaction being an important step for understanding their hematocompatibility.[135,136] Therefore, considering the exposure to blood platelets, the negative values of spreading work indicate that all compositions of cellulose derivative blends evidence cohesion; this result suggests that polymer blends do not interact with platelets, thus preventing activation of coagulation at the blood/biomaterial interface.

Also, an important problem in the evaluation of biocompatibility refers to the analysis of the competitive or selective adsorption of blood proteins at the biomaterial surface; predictions about these interactions can be formulated only by knowing exactly the structure of the biomaterial. Initially, the surface of an implanted material is mainly coated with albumin, immunoglobulins (especially immunoglobulin G (IgG)), and fibrinogen from plasma. These sanguine plasma proteins were selected for the study of the affinity of polymer blends towards physiological fluid media, due to their presence in the biological events from blood. Hence, Fig. 3.41 exhibits negative values of spreading work for all three plasma proteins, revealing that cohesion prevails, thus favoring a nonadsorbent behavior at the interface, as required by bio-applications. Also, all samples exhibit lower values of spreading work for albumin that, along with the rejection of platelets, emphasizes the important role they play in material-host interactions. On the other hand, CAP/HPC blends may be considered as being compatible with certain elements from the physiological environment (i.e., tissue, cells), since their interaction with the studied biological materials would cause no damage of the blood cells or change in the structure of plasma proteins. All these properties, along with the special microarchitecture of the CAP/HPC blends, recommend them as proper candidates for applications in cellular and tissue engineering.

Another significant observation relates to the interesting combination of CAP/HPC blends properties, such as suitable cohesion with sanguine plasma proteins and platelets, and small adhesion with red blood cells; these results show them as promising materials for blood-contacting devices (including vascular grafts, stents, pacemakers, extracorporeal circuits, etc.), even if long-term biocompatibility requires, however, administration of anticoagulant drugs (i.e., warfarin, heparin). At the same time, the studies performed have made possible the production of blood-compatible polymeric materials by preparing heparin-containing blends for biomedical fields.

The results reveal that, for tissue engineering, obtaining of some porous and interconnected 3D polymer networks is recommended. Thus, CAP/HPC blends can be used to accomplish the required properties for specific applications making them good tissue-engineered candidates. As a result, considering the traditional processes and the recently developed techniques, the improved ability to control the porosity and molecular microarchitecture of the CAP/HPC hydrophobic membranes will drive the research closer to its proposed goals.

This type of membranes – with surface topography and roughness as important factors in determining the response of cells to a foreign material – represents an excellent scaffold for applications in cellular and tissue engineering.[137,138] Mention should be made of the fact that fibroblasts and chondrocyte cells were shown to grow well in a 3D porous membrane, evidencing superior properties (specific molecular microarchitecture and controlled porosity) for tissue regeneration applications.

With these techniques, it is possible not only to specifically control individual and group pore architecture, but also to take the next step, namely to create microvascular features to improve integration within host tissues. Nevertheless, structural improvement and increased pore interconnectivity of porous scaffolds is claimed for the development of artificial blood vessels or peripheral nerve growth.

3.6 CONCLUSIONS

The chapter reviews especially our recent studies on the modification and applications of some cellulose derivatives from a nanotechnology consideration, involving biomedical applications. Generally, the information for each type of cellulose derivative includes aspects of synthesis, processing and properties in solid state and in solution, thus illustrating a variety of research directions. In this context, studies contribute to a better knowledge of the specific interactions that generate and modify the properties of cellulose derivatives, required by the applications in different domains. Information regarding the influence hydrogen bonding on properties of cellulose derivative in solvent/nonsolvent mixtures – over a large concentration domain, microstructures and appearance of lyotropic liquid crystal phases and biocompatibility, reveal important aspects necessary to diversify domain of their applications.

3.7 ACKNOWLEDGEMENT

This work was supported by a grant of the Romanian National Authority for Scientific Research, CNCS – UEFISCDI, project number PN-II-RU-TE-2012-3-143 (stage 2013).

KEYWORDS

- **Biocompatibility**
- **Blends**
- **Cellulose Derivatives**
- **Hydroxypropylcellulose (HPC)**
- **Liquid Crystals**
- **Lyotropic Liquid Crystal**
- **Nano-Particles**

REFERENCES

1. Gushikem, Y., & Toledo, E. A. (1999). In *Polymer Interfaces and Emulsions*, Esumi, K., (Ed.) Dekker: New York, Chapter 13, 509.
2. Splendore, G., Benvenutti, E. V. Y., Kholinc, Y. V., & Gushikem, Y. (2005). Cellulose Acetate-Al_2O_3 Hybrid Material Coated with N-Propyl-1, 4-diazabicyclo [2.2.2] Octane Chloride: Preparation, Characterization and Study of Some Metal Halides Adsorption from Ethanol Solution. J. Braz. Chem. Soc., 16, 147–152.
3. Jha, G. S., Seshadri, G., Mohan, A., & Khandal, R. K. (2007). Development of High Refractive Index Plastics. *E-Polymer*, 120, ISSN 1618–7229.
4. Paul, W., & Sharma, C. P. (2006). Polysaccharides: Biomedical Applications. In *Encyclopedia of Surface and Colloid Science*, Somasundaran, P., (Ed). Taylor and Francis: New York, 5056–5067.
5. Chen, J. C., & Li, Q., & Elimelech, M. (2004). In-situ Monitoring Techniques for Concentration Polarization and Fouling Phenomena in Membrane Filtration. *Adv. Collid Interface Sci.*, 107, 83–108.
6. Boussu, K. (1996). Influence of Membrane Characteristics on Flux Decline and Retention in Nanofiltration. PhD Dissertation, Department of Chemistry, Leuven, Belgium.
7. Mohamed, H., Szarowski, D. H., Spencer, M. G., Martin, D. L., Caggana, M. & Turner, J. N. (2005). Analysis of Cellulose Acetate Membranes for Lab-on-a-chip Applications. *Microscopy Microanalysis*, 11 (Suppl. 2), 1754–1755.
8. Ye, S. H., Watanabe, J., Iwasaki, Y., & Ishihara, K. (2002). Novel Cellulose Acetate Membrane Blended with Phospholipid Polymer for Hemocompatible Filtration System. *J. Membr. Sci.*, 210, 411–421.
9. Carofiglio, T., Fregonese, C., Mohr, G. J., Rastrellia, F., & Tonellatoa, U. (2006). Optical Sensor Arrays: One-pot, Multiparallel Synthesis and Cellulose Immobilization of pH and Metal Ion Sensitive Azo-dyes. *Tetrahedron*, 62, 1502–1507.

10. Werner, T., & Wolfbeis, O. S. (1993). Optical Sensor for the pH 10-13 Range Using a New Support Material. *Fresenius J. Anal. Chem.*, 346, 564–568.

11. Stone, A. (2002). Microbicides: A New Approach to Preventing HIV and Other Sexually Transmitted Infections. *Nat. Rev. Drug Disc*, 1, 977–985.

12. Manson, K. H., Wyand, M. S., Miller, C., & Neurath, A. R. (2000). Effect of a Cellulose Acetate Phthalate Topical Cream on Vaginal Transmission of Simian Immunodeficiency Virus in Rhesus Monkeys. *Antimicrob. Agents Chemother*, 44, 3199–3202.

13. Neurath, A. R., Strick, N., Li, Y. Y., & Debnath, A. K. (2001). Cellulose Acetate Phthalate, a Common Pharmaceutical Excipient, Inactivates HIV-1 and Blocks the Coreceptor Binding Site on the Virus Envelope Glycoprotein gp 120, *BMC Infectious Diseases*, 1, 17.

14. Goskonda, S. R., & Lee, J. C. (2000). Cellulose Acetate Phthalate. In *Handbook of Pharmaceutical Excipients*, 3rd ed., Kibbe, A. H. Ed., American Pharmaceutical Association: Washington, 99–101.

15. Abdel-Hamid, S. M., Abdel-Hady, S. E., El-Shamy, A. A., & El-Dessouky, H. F. (2006). Formulation of an Antispasmodic Drug as a Topical Local Anesthetic. *Int. J. Pharm.*, 326, 107–118.

16. Christie, R. J., Findley, D. J., Dunfee, M., Hansen, R. D., Olsen, S. C., & Grainger, D. W. (2006). Photopolymerized Hydrogel Carriers for Live Vaccine Ballistic Delivery. *Vaccine*, 24, 1462–1469.

17. Tokumura, T., & Machida, Y. (2006). Preparation of Amoxicillin Intragastric Buoyant Sustained-Release Tablets and the Dissolution Characteristics. *J. Controlled Release*, 110, 581–586.

18. Ludwig, A. (2005). The use of Mucoadhesive Polymers in Ocular Drug Delivery. Advanced Drug Delivery Reviews. *Adv. Drug Deliv. Rev.*, 57, 1595–1639.

19. Tanaka, N., Imai, K., Okimoto, K., Ueda, S., Tokunaga, Y., Ohike, A., Ibuki, R., Higaki, K., & Kimura, T. (2005). Development of Novel Sustained-Release System, Disintegration-Controlled Matrix Tablet (DCMT) with Solid Dispersion Granules of Nilvadipine. *J. Controlled Release*, 108, 386–395.

20. Rusig, I., Dedier, J., Filliatre, C., Godinho, M. H., Varihon, L., & Sixou, P. (1992). Effect of Degradation on Thermotropic Cholesteric Optical Properties of (2-Hydroxypropyl) Cellulose (HPC) Esters. *J Polym Sci Part A: Polym. Chem.*, 30, 895–899.

21. Costa, I., Filip, D., Figueirinhas, J. L., & Godinho, M. H. (2007). New Cellulose Derivatives Composites for Electo-Optical Sensors. *Carbohydr. Polym.*, 68, 159–165.

22. Amaral, A., de Carvalho, C. N., Brogueira, P., Lavareda, G., Melo, L. V., & Godinho, M. H. (2005). ITO Properties on Anisotropic Flexible Transparent Cellulosic Substrates under Different Stress Conditions. *Mater. Sci. Eng. B.*, 118, 183–186.

23. Chen, Y., & Yang, H. W. (1992). Hydroxypropyl Cellulose (HPC)-Stabilized Dispersion Polymerization of Styrene in Polar Solvents: Effect of Reaction Parameters. *J. Polym. Sci. Part A: Polym. Chem.*, 30, 2765–2772.

24. Chen, S., & Etzler, F. (1999). Rheological Characterization of Hydroxypropylcellulose Gels. *Drug Develop. Ind. Pharm.*, 25, 153–161.

25. Rwei, S. P., Lyu, M. S., Wu P. S., Tseng, C. H., & Huang, H. W. (2009). Sol/Gel Transition and Liquid Crystal Transition of HPC in Ionic Liquid. *Cellulose*, 16, 9–17.

26. Cosutchi, A. I., Hulubei, C., Stoica, I., & Ioan, S. (2010). Morphological and Structural-Rheological Relationship in Epiclon-Based Polyimide/Hydroxypropylcellulose Blend Systems. *J. Polym. Res.*, 17, 541–550.

27. Roschinski, C., & Kulicke, W. M. (2000). Rheological Characterization of Aqueous Hydroxypropyl Cellulose Solutions Close to Phase Separation. *Macromol. Chem. Phys.*, 201, 2031–2040.

28. Tolstykh, M. Y., Makarova, V. V., Semakov, A. V., & Kulichikhin, V. G. (2010). Rheological Properties and Phase Behavior of a Hydroxypropyl Cellulose Poly (ethylene glycol) System. *Polym. Sci., Ser. A*, 52, 144–149.

29. Belalia, R., Grelier, S., Benaissa, S., & Coma, V. (2008). New Bioactive Biomaterials Based on Quaternized Chitosan. *J. Agric. Food Chem.*, 56, 1582–1588.

30. Kumar, S., Chawla, G., & Bansal, A. K. (2008). *Pharm.* Spherical Crystallization of Mebendazole to Improve Processability. *Dev. Technol.*, 13, 559–568.

31. Heinze, Th., Liebert, T. F., Pfeiffer, K. S., & Hussain, M. A. (2003). Unconventional Cellulose Esters: Synthesis, Characterization and Structure-Property Relations. *Cellulose*, 10, 283–296.

32. Hussain, M. A., Liebert, T. F., & Heinze, Th. (2004). First Report on a New Esterification Method for Cellulose. *Polym. News*, 29, 14–17.

33. Sealey, J. E., Frazier, C. E., Samaranayake, G., & Glasser, W. G. (2000). Novel Cellulose Derivatives. V. Synthesis and Thermal Properties of Esters with Trifluoroethoxy Acetic Acid. *J. Polym. Sci. Part B: Polym. Phys.*, 38, 486–494.

34. Godinho, M. H., Filip, D., Costa, I., Carvalho, A. L., Figueirinhas, J. L., & Terentjev, E. M. (2009). Liquid Crystalline Cellulose Derivative Elastomer Films Under Uniaxial Strain. *Cellulose*, 16, 199–205.

35. Bouligand, Y. (1972). Twisted Fibrous Arrangements in Biological Materials and Cholesteric Mesophases. *Tissue Cell*, 4, 189–217.

36. Roland, J. C., Reis, D., Vian, B., & Roy, S. (1989). The Helicoidal Plant Cell Wall as a Performing Cellulose-Based Composite. *Biology of the Cell*, 67, 209–220.

37. Neville, A. C., & Levy, S. (1985). The Helicoidal Concept in Plant Cell Wall Ultrastructure and Morphogenesis. In *Biochemistry of Plant Cell Walls*, Brett, C. T., Hillman, J. R., (Eds.), Cambridge University Press: Cambridge, 99–124.

38. Klemm, D., Kramer, F., Moritz, S., Lindström, T., Ankerfors, M., Gray, D., & Dorris, A. (2011). Nanocelluloses: A New Family of Nature-Based Materials, *Angew. Chemie Int. Ed.*, 50, 5438–5466.

39. Habibi, Y., Lucia, L. A., & Rojas, O. J. (2010). Cellulose Nanocrystals: Chemistry, Self-Assembly, and Applications. *Chem. Rev.*, 110, 3479–3500.

40. Moon, R. J., Martini, A., Nairn, J., Simonsen, J., & Youngblood, J. (2011). Cellulose Nanomaterials Review: Structure, Properties and Nanocomposites. *Chem. Soc. Rev.*, 40, 3941–3994.

41. Stone, A. (2009). *Regulatory Issues in Micribicide Development*. WHO Document Production Services: Geneva.

42. Duncan, R. (2003). The Dawning Era of Polymer Therapeutics. *Nat. Rev. Drug Discov.*, 2, 347–360.

43. Kabanov A. V., & Okano, T. (2003) Challenges in Polymer Therapeutics: State of the Art and Prospects of Polymer Drugs. In *Polymer Drugs in the Clinical Stage*, Maeda, H., Kabanov, A. V., Kataoka, K., Okano, T. (Eds.). Kluwer Academic/Plenum Publishers: New York.

44. Kim, S. G., Kim, Y. L., Yun, H. G., Lim G. T., & Lee, K. H. (2003). Preparation of Asymmetric PVA Membranes Using Ternary System Composed of Polymer and Cosolvent. *J. Appl. Polym. Sci.*, 88, 2884–2890.

45. Qian, J. W., An, Q. F., Wang, L. N., Zhang, L., & Shen, L. (2005). Influence of the Dilute-Solution Properties of Cellulose Acetate in Solvent Mixtures on the Morphology and Pervaporation Performance of Their Membranes. *J. Appl. Polym. Sci.*, 97, 1891–1898.

46. Wang, D. M., Wu, T. T., Lin, F. C., Hou, J. Y., & Lai, J. Y. (2000). A Novel Method for Controlling the Surface Morphology of Polymeric Membranes. *J. Membr. Sci.*, 169, 39–51.

47. El Seoud, O. A., Nawaz, H., & Arêas, E. P. G. (2013). Chemistry and Applications of Polysaccharide Solutions in Strong Electrolytes/Dipolar Aprotic Solvents: An Overview. *Molecules*, 18, 1270–1313.

48. Gericke, M., Fardim, P., & Heinze, T. (2012). Ionic Liquids Promising but Challenging Solvents for Homogeneous Derivatization of Cellulose. *Molecules*, 17, 7458–7502.
49. Shanbhag, A., Barclay, B., Koziara J., & Shivanand, P. (2007). Application of Cellulose Acetate Butyrate-Based Membrane for Osmotic Drug Delivery. *Cellulose*, 14, 65–71.
50. Southward, R. E., & Thompson, D. W. (2001). Reflective and Conductive Silvered Polyimide Films for Space Applications Prepared via a Novel Single-Stage Self-Metallization Technique. *Material Design*, 22, 565–576.
51. Yuan, Y., Fendler, J. H., & Cabasso, I. (1992). Photoelectron Transfer Mediated by Size-Quantized Cadmium Sulfide particles in Polymer-Blend Membranes. *Chem. Mater.*, 4, 312–318.
52. Shim, I. W., Kim, J. Y., Kim D. Y., & Choi, S. (2000). Preparation of Rh-Containing Polycarbonate Films and the Study of Their Chemical Properties in the Polymer. *React. Funct Polym.*, 43, 71–78.
53. Shim, I., Wun, C. S., Noh, W. T., Kwon, J. Y., Cho Chae, D. Y., & Kim, K. S. (2001). Preparation of Iron Nanoparticles in Cellulose Acetate Polymer and Their Reaction Chemistry in the Polymer. *Bull. Korean Chem. Soc.*, 22, 772–774.
54. Olaru, N., & Olaru, L. (2005) Phthaloylation of Cellulose Acetate in Acetic Acid and Acetone Media. *Iranian Polym. J.* 14, 1058–1065.
55. Dobos, A. M., Onofrei, M. D., Stoica, I., Olaru, N., Olaru, L., & Ioan, S. (2013). Influence of Self-Complementary Hydrogen Bonding on Solution Properties of Cellulose Acetate Phthalate in Solvent/Non-Solvent Mixtures. *Cell. Chem. Technol.*, 1–2, 13–21.
56. Qian, J. W., & Rudin, A. (1992a). Prediction of Thermodynamic Properties of Polymer Solutions. *Eur. Polym. J.* 28, 725–732.
57. Qian, J. W., & Rudin, A. (1992b). Prediction of Hydrodynamic Properties of Polymer Solutions. *Eur. Polym. J.* 28, 733–738.
58. Qian, J. W., Wang, M., Han, D. L., & Cheng, R. S. (2001). A Novel Method for Estimating Unperturbed Dimension $[\eta]_\theta$ of Polymer from the Measurement of its $[\eta]$ in a Non-Theta Solvent. *Eur. Polym. J.* 37, 1403–1407.
59. McKcc, M. G., Elkins, C. L., & Long, T. E. (2004). Influence of Self-Complementary Hydrogen Bonding on Solution Rheology/Electrospinning Relationships. *Polymer*, 45, 8705–8715.
60. Colby, R. H., & Rubinstein, M. (1990). Two-Parameter Scaling for Polymers in Theta Solvents. *Macromolecules*, 23, 2753–2757.
61. Isono, Y., & Nagawasa, M. (1980). Solvent Effects on Rheological Properties of Polymer Solutions. *Macromolecules*, 13, 862–867.
62. Olaru, N., & Olaru, L. (2010). Electrospinning of Cellulose Acetate Phthalate from Different Solvent Systems. *Ind. Eng. Chem. Res.*, 49, 1953–1957.
63. Huang, D. H., Ying, Y. M., & Zhuang, G. Q. (2000). Influence of Intermolecular Entanglements on the Glass Transition and Structural Relaxation Behaviors of Macromolecules. 2. Polystyrene and Phenolphthalein Poly (ether sulfone). *Macromolecules*, 33, 461–464.
64. Necula, A. M., Dunca, S., Stoica, I., Olaru, N., Olaru, L., & Ioan, S. (2010). Morphological Properties and Antibacterial Activity of Nano Silver Containing Cellulose Acetate Phthalate Films. *Int. J. Polym. Anal. Charact*, 5, 341–350.
65. Necula, A. M., Stoica, I., Olaru, N., Doroftei, F., & Ioan, S. (2011). Silver Nanoparticles in Cellulose Acetate Polymers: Rheological and Morphological Properties. *J. Macromol. Sci., Part B: Phys.*, 50, 639–651.
66. Dutta, D., Fruitwala, H., Kolhi, A., & Weiss, R. A. (1990). Polymer Blends Containing Liquid Crystals: A Review. *Polym. Eng. Sci.*, 30, 1005–1018.
67. El-Zaher, N. A., & Osiris, W. G. (2005). Thermal and Structural Properties of Poly (vinyl alcohol) Doped with Hydroxypropyl Cellulose. *J. Appl. Polym. Sci.*, 96, 1914–1923.
68. Ulcnik Krump, M. (2006). Study of Morphology Influence on Rheological Properties of Compatibilized TPU/SAN Blends. *J. Appl. Polym. Sci.*, 100, 2303–2316.

69. Dobos, A. M., Onofrei, M. D., Stoica, I., Olaru, N., Olaru, L., & Ioan, S. (2012). Rheological Properties and Microstructures of Cellulose Acetate Phthalate/Hydroxypropyl Cellulose Blends. *Polym. Comp.*, 33, 2072–2083.

70. Onofrei, M. D., Dobos, A. M., Stoica, I., Olaru, N., Olaru, L., & Ioan, S. (2013). Lyotropic Liquid Crystal Phases in Cellulose Acetate Phthalate/Hydroxypropyl Cellulose Blends, *J. Polym. Env.* Proof: doi: 10.1007/s10924–013–0618–7.

71. De Vasconcelos, C. L., Martins, R. R., Ferreira, M. O., Pereira, M. R., & Fonseca, J. L. C. (2001). Rheology of Polyurethane Solutions with Different Solvents. *Polym. Int.*, 51, 69–74.

72. Ioan, S., Filimon, A., Hulubei C., Stoica, I., & Dunca, S. (2013). Origin of Rheological Behavior and Surface/Interfacial Properties of Some Semi-Alicyclic Polyimides for Biomedical Applications. Polym. Bull., 70, 2873–2893.

73. Gupta, K., & Yaseen, M. (1997). Viscosity-Temperature Relationship of Dilute Solution of Poly (vinyl chloride) in Cyclohexanone and in its Blends with Xylene. *J. Appl. Polym. Sci.*, 65, 2749–2760.

74. Choi, J. H., & Rha, C. K. (1998). Dependence of Power-Law Parameters on Polysaccharide Concentration Using Methylan. *Biotechnol. Tech.*, 12, 377–380.

75. Lue, A., & Zhang, L. (2009). Rheological Behaviors in the Regimes from Dilute to concentrated in Cellulose Solutions Dissolved at Low Temperature. *Macromol. Biosci.* 9, 488–496.

76. Cassagnau, Ph., & Melis, F. (2003). Non-linear Viscoelastic Behavior and Modulus Recovery in Silica Filled Polymers. *Polymer*, 44, 6607–6615.

77. Ferry, J. D. (1980). *Viscoelasticity Properties of Polymers*, Willey Interscience: New York.

78. Dobos, A. M., Onofrei, M. D., Olaru, N., Olaru, L., Ioanid, G. E., & Ioan, S. (2012). Surface and Morphological Properties of Interpolymer Complexes and Blends Based on Cellulose Acetate Phthalate/Hydroxypropyl Cellulose. Book of Abstracts, the 7th International Conference on Advanced Materials, ROCAM, Brasov, Romania, August 28–31, 2012, Bucharest University Press, 29.

79. Dobos, A. M., Stoica, I., Olaru, N., Olaru, L., Ioanid, G. E., & Ioan, S. (2012). Surface Properties and Biocompatibility of Cellulose Acetates. *J. Appl. Polym. Sci.* 1, 2521–2528.

80. Peuvrel, E., & Navard, P. (1991). Band Textures of Liquid Crystalline Polymers in Elongational Flows. *Macromolecules*, 24, 5683–5686.

81. Patnaik, S. S., Bunning, T. J., & Adams, W. W. (1995). Atomic Force Microscopy and High-Resolution Scanning Electron Microscopy Study of the Banded Surface Morphology of Hydroxypropyl Cellulose Thin Films. *Macromolecules*, 28, 393–395.

82. Rao, V., Ashokan, P. V., & Shridhar, M. H. (2000). Miscible Blends of Cellulose Acetate Hydrogen Phthalate and Poly (vinyl pyrollidone) Characterization by Viscometry, Ultrasound, and DSC. *J. Appl. Polym. Sci.*, 76, 859–867.

83. Van Oss, C. J., & Good, R. J. (1989). Surface Tension and the Solubility of Polymers and Biopolymers: The Role of Polar and Apolar Interfacial Free Energies. *J. Macromol. Sci. Part A: Chem.*, 26, 1183–1203.

84. Kesting, R. E. (1990). The Four Tiers of Structure in Integrally Skinned Phase Inversion Membranes and Their Relevance to the Various Separation Regimes. *J. Appl. Polym. Sci.*, 41, 2739–2752.

85. Gilbert, R. D., & Kadla, J. F. (1998). Polysaccharides-Cellulose. In *Biopolymers from Renewable Resources*, Kaplan, D. L. Ed., Springer Verlag: Berlin, Heidelberg, New York, Chapter III.

86. Flory, P. J. (1984). Molecular Theory of Liquid Crystals. *Adv. Polym. Sci.*, 59, 1–35.

87. Flory, P. J., & Ronca, G. (1979). Theory of Systems of Rodlike Particles. 1. Athermal Systems. *Mol. Cryst. Liq. Cryst.*, 54, 289–309.

88. Warner, M., & Flory, P. J. (1980). The Phase-Equilibrium in Thermotropic Liquid-Crystalline Systems. *J. Chem. Phys.*, 7, 6327–6332.

89. Appaw, C. (2004). Rheology and Microstructure of Cellulose Acetate in Mixed Solvent systems. PhD Dissertation, North Carolina State University.

90. Appaw, C., Gilbert, R., Khan, S., & Kadla, J. (2010). Phase Separation and Heat-Induced Gelation Characteristics of Cellulose Acetate in a Mixed Solvent System. *Cellulose*, 17, 533–538.

91. Zugenmaier, P. (1994). Polymer Solvent Interaction in Lyotropic Liquid Crystalline Cellulose Derivatives Systems. In *Cellulosic Polymers Blends and Composites*, Gilbert, R. D., Ed., Hanser, New York, Chap. 4, 71–94.

92. Kinder, L., Kanickia, J., & Petroffa, P. (2004). Structural Ordering and Enhanced Carrier Mobility in Organic Polymer Thin Film Transistors. *Synth. Met.*, 146, 181–185.

93. Chae, B., Lee, S. W., Lee, B., Choi, W., Kim, S. B., Jung, Y. M., Jung, J. C., Lee, K. H., & Ree, M. (2003). Sequence of the Rubbing-Induced Reorientations of Polymer Chain Segments in Nanofilms of a Well-Defined Brush Polyimide with a Fully Rodlike Backbone as Determined by Polarized FTIR Spectroscopy and Two-Dimensional Correlation Analysis. *Langmuir*, 19, 9459–9465.

94. Shipovskaya, A. B., Gegel, N. O., Shmakov, S. L., Shchyogolev, S., & Yu. (2012). Phase Analysis of the Cellulose Triacetate-Nitromethane System. Int. J. Polym. Sci., doi:10.1155/2012/126362.

95. Yunusov, B. Y., Khanchich, O. A., Shablygin, M. V., Nikitina, O. A., & Serkov, A. T. (1983). Structural Transformations at Solidifying Anisotropic Solutions of Cellulose Acetates in Trifluoroacetic Acid. *Vysokomol. Soedin. B*, 25, 292–294.

96. Meeten, G. H., & Navard, P. (1983). Gel Formation and Liquid Crystalline Order in Cellulose Triacetate Solutions. *Polymer*, 24, 815–819.

97. Aharoni, S. M. (1980). Rigid Backbone Polymers-8. Effects of the Nature of the Solvent on the Lyotropic Mesomorphicity of Cellulose Acetate. Molecular Crystals and Liquid Crystals, 56, 237–241.

98. Timofeeva, G. N, & Averianova, V. M. (1980). Studying Kinetics Phase Division in Student not Forming System Triacetate Cellulose-Nitrometan a Rheological Method, *Colloid J.* 42, 393–397.

99. Shipovskaya, A. B., & Timofeeva, G. N. (2003). Peculiarities of Phase Separation in the Variacetylated Cellulose Acetate-Mesophasogenic Solvent (Nitromethane) System. In *Struktura i Dinamika Molekuliarnykh Sistem*, vol. 10, part 2, 222–225, Kazan University Press, Kazan, Russia.

100. Shipovskaya, A. B., Shmakov, S. L., & Timofeeva, G. N. (2006). Phase Processes and Energetics of Spontaneous Change in the Dimensions of Acetate Fibers in Nitromethane Vapors. *Polym. Sci., Series A*, 48, 509–519.

101. Shipovskaya, A. B., & Timofeeva, G. N. (2008). Self-Organizing a Macropierkul of Cellulose Acetates. Under the Influence of Steams of Mezofazogen Solvents. In *Sinergetika v Estestvennykh Naukakh*, Gegel, N. O., Shchyogolev, S. Yu. (Eds.), Tvergu Press: Tver Cuty, Russia, 178–181.

102. Papkov, S. P. (1984). About a mutual combination liquid, liquid crystal and crystal phase balance in connection with a problem of morphology of natural polymers," *Vysokomolekulyarnye Soedineniya A*, 26, 1083–1089, 1984.

103. Lirova, B. I., & Lyutikova, E. A. (2012). Formation of the Liquid Crystal State of Cellulose Diacetate in Nitromethane Vapor: A Fourier IR Study. *Russian J Appl. Chem.*, 85, 1617–1621.

104. Israelachvili, J. N. (1985). *Intermolecular and Surface Forces with Applications to Colloidal and Biological Systems*, Academic Press: London, 480.

105. Godinho, M. H., Fonseca, J. G., Ribeiro, A. C., Melo, L. V., & Brogueira, P. (2002). Atomic Force Microscopy Study of Hydroxypropylcellulose Films Prepared from Liquid Crystalline Aqueous Solutions. *Macromolecules*, 35, 5932–5936.

106. Wang, J., & Labes, M. M. (1992). Control of the Anisotropic Mechanical Properties of Liquid Crystal Polymer Films by Variations in Their Banded Texture. *Macromolecules*, 25, 5790–5793.

107. Onogi, S., & Asada, T. (1980). Rheology and Rheo-Optics of Polymer Liquid Crystals. In *Rheology*, Astarita, G., Marucci, G., Nicolais, L. Eds., Plenum Press: New York, 1, 127–146.

108. Li, H., Shen, X., Gong, G., & Wang, D. (2008). Compatibility Studies with Blends Based on Hydroxypropylcellulose and Polyacrylonitrile. *Carbohydr. Polym.* 73, 191–200.

109. Necula, A. M., Olaru, N., Olaru, L., & Ioan, S. (2008). Influence of the Substitution Degree on the Dilute Solution Properties of Cellulose Acetate. *J. Macromol. Sci. Part B: Phys.* 47, 913–928.

110. Ioan, S., & Dobos-Necula, A. M. (2012). Silver Nanoparticles in Cellulose Derivative Matrix. In *Nanotechnology in Polymers*, Thakur V. K., Singha A. S. (Eds.), Studium Press LLC: USA, Chapt. 11, 196–253, ISBN: 1–933699–90–6.

111. Jr. Hermans, J. (1962). The viscosity of Concentrated Solutions of Rigid Rodlike Molecules (Poly γ-benzyl-l-glutamate in m-Cresol). *J. Colloid. Sci.*, 17, 638–648.

112. Ernst, B., & Navard, P. (1989). Band Textures in Mesomorphic Hydroxypropylcellulose Solutions. *Macromolecules*, 22, 1419–1422.

113. Tirtaatmadja, V., Tam, K. C., & Jenkins, R. D. (1997). Superposition of Oscillations on Steady Shear Flow as a Technique for Investigating the Structure of Associative Polymers. *Macromolecules*, 30, 1426–1433.

114. Roman, M., & Gray, D. G. (2005). Parabolic Focal Conics in Self-Assembled Solid Films of Cellulose Nanocrystals. *Langmuir*, 21, 5555–5561.

115. Edgar, C. D., & Gray, D. G. (2001). Induced Circular Dichroism of Chiral Nematic Cellulose Films. *Cellulose*, 8, 5–12.

116. Sein, A., & Engberts, J. B. (1996). F. N. Lyotropic Phases of Dodecylbenzenesulfonates with Different Counterions in Water. *Langmuir* 12, 2913–2923.

117. Huang, B., Ge, J. J., Li, Y., & Hou, H. (2007). Aliphatic Acid Esters of (2-Hydroxypropyl) Cellulose. Effect of Side Chain Length on Properties of Cholesteric Liquid Crystals. *Polymer*, 48, 264–26.

118. Kasajima, Y., Kato, T., Kubono, A., Tasaka, S., & Akiyama, R. (2008). Wide Viewing Angle of Rubbing-Free Hybrid Twisted Nematic Liquid Crystal Displays. *Jpn. J. Appl. Phys.*, 47, 7941–7942.

119. Lee, H. S., & Denn, M. M. (1999). Polymer Blends with a Liquid Crystalline Polymer Dispersed Phase. *Korea-Australia Rheol. J.*, 11, 269–273.

120. Andresen, E. M., & Mitchel, G. R. (2013). X-ray Rheology of Liquid Crystal Polymers. In *Rheology: Theory, Properties and Practical Applications*, Mitchell, G. R. Ed., Nova Science Pub. Incorporated, Chapter 15.

121. Yan, L., Zhu, Q., & Ikeda, T. (2003). Alignment behavior of liquid crystals on ethyl cellulose films with banded-texture structure. *Polym. Int.*, 52, 265–268.

122. Entcheva, E., Biena, H., Yina, L., Chunga, C. Y., Farrella, M., & Kostovc, Y. (2004). Functional cardiac cell constructs on cellulose-based scaffolding. *Biomaterials*, 25, 5753–5762.

123. Liu, Z. H., Janzen, J., & Brooks, D. E. (2001). Adsorption of Amphiphilic Hyperbranched Polyglycerol Derivatives onto Human Red Blood Cells. *Biomaterials*, 31, 3364–3373.

124. Jain, K., Kesharwani, P., Gupta, U., & Jain, N. K. (2010). Dendrimer Toxicity: Let's Meet the Challenge. *Int. J. Pharm.*, 394, 122–142.

125. Ware, J. A., Kang, J., DeCenzo, M. T., Smith, M., Watkins, S. C., Slayter, H. S., & Saitoh, M. (1991). Platelet Activation by a Synthetic Hydrophobic Polymer, Polymethylmethacrylate. *Blood*, 78, 1713–1721.

126. Vijayanand, K., Deepak, K., Pattanayak, D. K., Rama Mohan, T. R., & Banerjee, R. (2005). Interpenetring Blood-Bioma- terial Interactions from Surface Free Energy and Work of Adhesion, *Trends. Biomater. Artif. Organs*, 18, 73–83.

127. Rankl, M., Laib, S., & Seeger, S. (2003). Surface Tension Properties of Surface-Coatings for Application in Biodiagnostics Determined by Contact Angle Measurements. *Colloid Surf B: Biointer*, 30, 177–186.

128. Albu, R. M., Avram, E., Stoica, I., Ioanid, E. G., Popovici, D., & Ioan, S. (2011). Surface Properties and Compatibility with Blood of New Quaternized Polysulfones. *Journal of Biomaterials and Nanobiotechnology*, 2, 114–123.

129. Ioan, S., & Filimon, A. (2012). Biocompatibility and Antimicrobial Activity of Some Quaternized Polysulfones. In *A Search for Antibacterial Agents*, Bobbarala, V. (Ed). InTech: Rijeka: Chapt. 13, 249–274.

130. Ioan, S., Buruiana, L. I., Petreus, O., & Avram, E. (2011). Rheological and Morphological Properties of Phosphorus-Containing Polysulfones. *Polym. Plast. Technol. Eng.*, 50, 36–46.

131. Van Oss, C. J. (1990). Surface Properties of Fibrinogen and Fibrin. *J. Protein Chem.* 9, 487–491.

132. Kwok, S. C. H., Wang, J., & Chu, P. K. (2005). Surface Energy, Wettability, and Blood Compatibility Phosphorus Doped Diamond-Like Carbon Films. *Diamond Relat. Mate.*, 14, 78–85.

133. Agathopoulos, S., & Nikolopoulos, P. (1995). Wettability and Interfacial Interactions in Bioceramic-Body-Liquid Systems. *J. Biomed. Mater. Res. Part A*, 29, 421–429.

134. Van Oss, C. J. (2003). Long-range and Short-Range Mechanisms of Hydrophobic Attraction and Hydrophilic Repulsion in Specific and Aspecific Interactions. *J. Mol. Recognit.*, 16, 177–190.

135. Dobrovolskaia, M. A., Patri, A. K., Simak, J., Hall, J. B., Semberova, J., De Paoli Lacerda, S. H., & McNeil, S. E. (2012). Nanoparticle Size and Surface Charge Determine Effects of PAMAM Dendrimers on Human Platelets in-vitro. *Mol. Pharmaceutics*, 9, 382–393.

136. Michanetzis, G. P. A. K., Missirlis, Y. F., & Antimisiaris, S. G. (2008). *J.* Haemocompatibility of Nanosized Drug Delivery Systems: Has It Been Adequately Considered? *Biomed. Nanotechnol.* 4, 218–233.

137. Annabi, N., Nichol, J. W., Zhong, X., Ji, C., Koshy, S., Khademhosseini, A., & Dehghani, F. (2010). Advances in Bioartificial Materials and Tissue Engineering. *Tissue Eng. Part B*, 16, 371–383.

138. Martina, M., & Hutmacher, W. D. (2007). Biodegradable Polymers Applied in Tissue Engineering Research: a Review. *Polym. Int.*, 56, 145–157.

CHAPTER 4

BIOCOMPOSITES COMPOSED OF BIO-BASED EPOXY RESINS, BIO-BASED POLYPHENOLS AND LIGNOCELLULOSIC FIBERS

MITSUHIRO SHIBATA

ABSTRACT

In recent years, renewable resources-derived polymers (bio-based polymers) and composites (biocomposites) are attracting a great deal of attention because of the advantages of these polymers such as conservation of limited petroleum resources, possible biodegradability, the control of carbon dioxide emissions that lead to global warming. This chapter deals with the preparation, thermal and mechanical properties of the bio-based network polymers prepared by bio-based epoxy resins and bio-based polyphenols, and their biocomposites with lignocellulosic fibers. As bio-based epoxy resins, glycerol polyglycidyl ether (GPE), polyglycerol polyglycidyl ether (PGPE), sorbitol polyglycidyl ether (SPE) and epoxidized soybean oil (ESO) were used. As bio-based polyphenols, tannic acid (TA) which is a hydrolysable tannin and quercetin (QC) which is a flavonoid were used. Also, the polyphenols (TPG) prepared by the reaction of tung oil (TO) and pyrogallol (PG) and guaiacyl pyrogallol[4]arene (PGVNC) prepared by the reaction of PG and vanillin (VN) were also used. As lignocellulosic fibers, wood flour (WF) made from Sanbu cedar crushed into powders through 3 mm screen mesh and microfibrillated cellulose fiber (MFC) were used. The thermal and mechanical properties of the bio-based polymer networks and their biocomposites were investigated in detail by means of dynamic mechanical analysis (DMA), thermogravimetric analysis (TGA) and tensile test. The morphology of the fractured surface of the biocomposites was observed by field emission-scanning electron microscopy (FE-SEM). Consequently, the SPE cured with PGVNC showed the highest tan δ peak temperature (148 °C). The PGPE/TA/WF and GPE/TA/WF biocomposites with WF content 50–60 wt.% showed the highest Young's modulus (4–5 GPa). The SPE/TA/MFC biocomposites with MFC content 10 wt.% showed the highest tensile strength (80 MPa).

4.1   INTRODUCTION

Biocomposites which are composed of matrix resins and natural fibers such as wood and plant fibers have recently gained much attention due to their low cost, environmental friendliness, and their potential to compete with man-made fiber-reinforced composites.[1,3] Furthermore, the concept of using bio-based polymers as matrix resins for biocomposites is becoming increasingly important due to dwindling petroleum resources.[4,5] Such biocomposites composed of bio-based polymers and natural fibers are especially termed as green composites. For example, the green composites of bio-based polymers such as poly(lactic acid),[6,14] poly(hydroxyalkanoate),[15,18] and cellulose acetate[19,20] with lignocellulosic natural fibers such as flax, jute, hemp, kenaf, abaca, bamboo, and wood flour have been reported by several groups. However, the main problem encountered in using their bio-based polymers is its rather poor interfacial adhesion between the polar lignocellulose and the more hydrophobic characteristics of those polymeric matrices. The poor adhesiveness results in a poor strength, a relatively low stiffness and high moisture uptake. Another major shortcoming of this type of matrix is the relatively low fiber content, typically of less than 50–60 wt.%. One way to improve the poor adhesion is a modification of the interface of matrix and fiber.[6,8,9,12,15,16] However, the surface treatment normally increases both processing steps and its cost. Bio-based epoxy resin cured with polyphenol-based hardener should be a good candidate for a matrix resin of the green composite, because high loading of natural fiber is possible due to a low viscosity of the resin before curing, and superior interfacial adhesion is expected due to the hydrogen bonding interaction between lignocellulosic fiber and hydroxypropyl (or hydroxyethyl) moiety formed by the curing reaction of glycidyl (or epoxy) group with phenol. In this chapter, after the promising bio-based aliphatic epoxy resins and bio-based polyphenol hardeners are introduced, the thermal and mechanical properties of the biocomposites using their bio-based epoxy resins/hardeners and lignocellulosic fibers are reviewed based on our previous studies.[21,26]

4.2   BIO-BASED EPOXY RESINS AND HARDENERS

4.2.1   BIO-BASED EPOXY RESINS

Bio-based aliphatic epoxy resins such as glycerol polyglycidyl ether (GPE), polyglycerol polyglycidyl ether (PGPE), sorbitol polyglycidyl ether (SPE), epoxidized soybean oil (ESO), epoxidized linseed oil (ELO), diglycidyl ester of dimer acid (DGEDA) and limonene diepoxide (LMDE), etc. are industrially available in large volumes at a reasonable cost (Fig. 4.1). Glycerol is an abundant and inexpensive bio-based aliphatic polyols, which can be derived from triglyceride vegetable oil. The biodiesel boom of the recent decade led to a significant increase in biomass-derived glycerol.[27] At present, the worldwide glycerol production is around 1.2–1.4

million tons. This amount will further increase to 1.54 million tons in 2015.[28] However, the total global demand is less than 1 Mt. Thus, the market for petrochemical derived glycerol that is available via propene, allyl chloride and epichlorohydrin (ECH) no longer exists. The utilization of glycerol for the production of other intermediate chemicals and final materials will become very important in near future. Indeed, some of these products are close to commercialization or already introduced into the market. One such example is the Epicerol® Process introduces by Solvay in 2007 enabling the ECH synthesis from bio-derived glycerol.[29] Also, glycerol is an attractive renewable building block for synthesis of polyglycerols which have several uses in different field. Polyglycerols, especially diglycerol and triglycerol are the main products of glycerol etherification.[30] Therefore, GPE and PGPE which are synthesized by the reactions of glycerol and polyglycerol with ECH should be very promising bio-based epoxy resins. Although the industrially available GPE and PGPE have been used in textile and paper processing agents and reactive diluents, etc., their epoxy resins have not yet been applied to matrix resins for fiber-reinforced plastics.

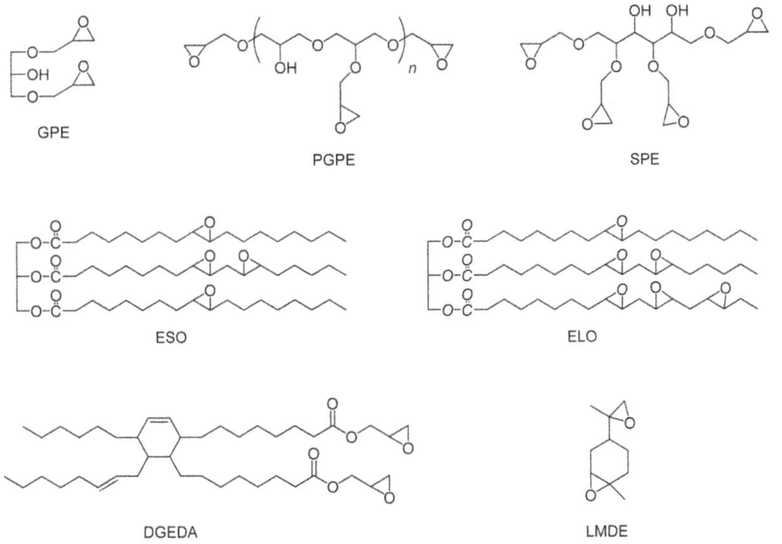

FIGURE 4.1 Bio-based aliphatic epoxy resins.

Sorbitol is also abundant and inexpensive bio-based aliphatic polyols, which can be produced by the catalytic hydrogenation of glucose derived from corn starch.[31,32] It is widely used in the food industry, not only as a sweetener but also as a humectant, texturizer, and softener. The SPE which is prepared by the reaction of sorbitol and ECH has been mainly used for tackifier, coatings, and paper and fiber-modifier, etc. Epoxidized vegetable oils such as ESO and ELO are manufactured by the epoxidation of the double bonds of vegetable oils with hydrogen peroxide, either in acetic acid or in formic acid,[33,35] and have mainly been used as plasticizer or stabilizer to modify the properties of plastic resins such as poly(vinyl chloride). Because ELO has a 30% more oxirane content than ESO does, the cured ELO has a higher cross-linking density, which results in a better performance.

Dimer acid-based DGEDA is a flexible epoxy resin produced from dimer acid and ECH. Although dimer acid can be also obtained from animal fats or vegetable oils, most of the dimer acids appearing in the market are synthesized from the crude tall oil provided as a byproduct of Kraft pulp.[36] The commercially available dimer acid usually contains monomer (1–5%) and trimer or more (14–16%) in addition to dimer. Limonene-based LMDE (1-methyl-4-(2-Methyl-2–oxiranyl)-7-oxabicyclo[4.1.0]heptane) is commercially produced by the reaction of limonene and peracetic acid, which is used as cationically curable resins and reactive diluents. (+)-d-Limonene is a popular monoterpene which is commercially obtained from citrus fruits.

However, their bio-based aliphatic epoxy resins had not been versatile materials because of inferior mechanical and thermal properties to the bisphenol-A or novolac-based epoxy resins. Therefore, the selection of hardener and the addition of natural fibers are important for the use of the bio-based aliphatic epoxy resins in wide applications. In the following section, we used GPE, PGPE, SPE, and ESO as bio-based aliphatic epoxy resins for the preparation of green composites. Their physical properties and suppliers are summarized in Table 4.1. The average number of epoxy groups per molecule of GPE, PGPE, SPE and ESO are 2.0, 4.1, 3.6, and 4, respectively.

TABLE 4.1   Physical Properties of the Bio-Based Aliphatic Epoxy Resins Used in this Study

Epoxy resin (Abbreviation)	Supplier (Trade name)	Epoxy functionality	Epoxy equivalent weight (g/eq.)	Viscosity (cps, 25 °C)
Glycerol polyglycidyl ether (GPE)	Nagase ChemteX, Corp. (DENACOL® EX-313)	2.0	140	150
Polyglycerol polyglycidyl ether (PGPE)	Nagase ChemteX, Corp. (DENACOL® EX-512)	4.1	169	1300
Sorbitol polyglycidyl ether (SPE)	Nagase ChemteX, Corp. (DENACOL® EX-614B)	3.6	172	5000
Epoxidized soybean oil (ESO)	Kao Chemical Co., Inc. (KAPOX® S-6)	4	239 (oxirane oxygen 6.7%)	-

4.2.2 BIO-BASED HARDENERS

Most of epoxy hardeners such as polyamines, polyphenols, and carboxylic acid anhydrides are derived from petroleum resources. The past studies on bio-based epoxy hardeners are less than that on the bio-based epoxy resins. Basically, bio-based polyamines, polyphenols, and carboxylic acid anhydrides can be used as epoxyhardener. We investigated a possibility of bio-based polyamines and polyphenols as hardeners of bio-based epoxy resins for biocomposites with lignocellulosic fibers considering an interfacial adhesion between matrix resin and fibers. Although the reaction of polyamine or polyphenol with epoxy resin generates hydroxy groups which can form hydrogen-bonding with lignocellulosic fibers, carboxylic acid anhydrides do not produce hydroxy groups but ester groups.

As a bio-based polyamine hardener, ε-poly(L-lysine) (PL) was investigated.[37] The PL is produced by aerobic bacterial fermentation using *Streptomyces albulus* in a culture medium containing glucose, citric acid, and ammonium sulfate.[38,39] The PL has been used as food preservatives,[40] while has not yet been applied to the industrial polymeric materials. PL differs from usual proteins in that the amide linkage is not between the α-amino and carboxylic groups in typical of peptide bonds, but is between the ε-amino and carboxy group. The pendant α-amino groups are expected to react with epoxy groups. When GPE or PGPE was cured with PL, a soft cured resin (GPE/PL or PGPE/PL) with a tensile modulus lower than 10 MPa and glass transition temperatures (T_g's) measured by DSC lower than 50 °C was obtained. Although PL is interesting as a hardener for bio-based epoxy/clay nanocomposites,[37] we did not use PL as a hardener for bio-based epoxy/natural fiber biocomposites because of the inferior mechanical and thermal properties.

As bio-based polyphenol hardeners, we investigated tannic acid (TA) and quercetin (QC) as are shown in Fig. 4.2.[21,24] Commercial TA is comprised of mixtures of gallotannins from sumac galls, Aleppo oak galls, or sumac leaves.[41] The chemical formula for commercial TA is often given as $C_{76}H_{52}O_{46}$ as shown in the figure. But, in fact it contains a mixture of related compounds. Its structure is based mainly on glucose ester of gallic acid. QC (3,3,'4,'5,7-pentahydroxyflavone) is one of the most abundant flavonoid found in glycosylated forms in plants such as onion, capers and tea. TA is industrially available from various Makers (for example, Fuji Chemical Industry, Co., Ltd. (Wakayama, Japan)), we used the reagent grade TA of Kanto Chemical Co., Inc. (Tokyo, Japan). QC can be obtained from plants via extraction of the quercetin glycosides followed by hydrolysis to release the aglycone and subsequent purification.[42] Although QC is used as an ingredient in supplements, beverages and foods, it has not been used as an ingredient of polymer materials to the best of our knowledge. When QC is used as an epoxy hardener, it is expected that the cured resin has a high T_g and superior adhesiveness to plant fibers because of the rigid and polar polyphenol moiety. We used the QC purchased from Sigma-Aldrich Japan Co. Ltd. (Tokyo, Japan).

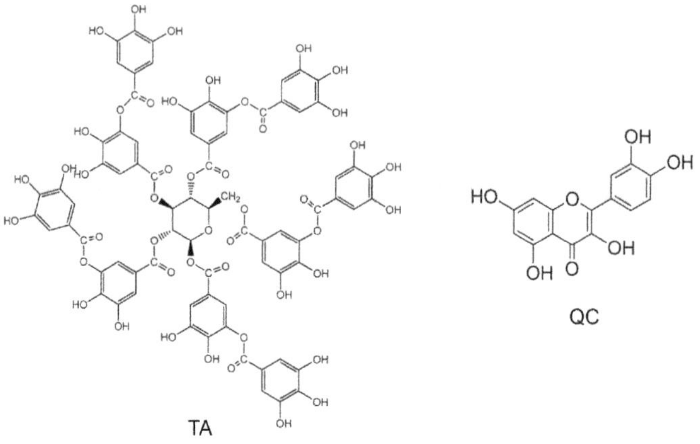

FIGURE 4.2   Structure of TA and QC.

Also, bio-based phenols such as pyrogallol (PG) and cardanol (CD) are promising raw materials for the preparation of bio-based phenolic epoxy hardeners (Fig. 4.3). PG is obtained by decarboxylation of gallic acid, which is a basic component of hydrolysable tannin. Utilization of PG as a raw material of the preparation of epoxy hardeners was described in the following section in detail. CD is a phenol meta-substituted with a flexible unsaturated hydrocarbon chain, which is derived from cashew nutshell liquid.[43] There have been many studies on the utilization of CD to phenol resins[44,46] and epoxy resins[46,49], and some of them have been already commercialized by Cardolite Corp. (Newark, NJ, USA) and Shanghai Meidong Biomaterials Co. Ltd (Shanghai, China), etc.

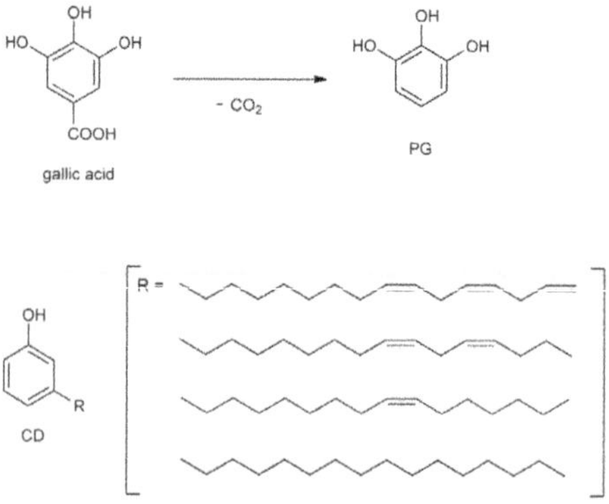

FIGURE 4.3   Synthetic scheme of PG by decarboxylation of gallic acid and the structure of CD.

4.3   REINFORCING LIGNOCELLULOSIC FIBERS

In the past studies, various natural fibers such as flax, jute, hemp, kenaf, sisal, abaca, pineapple leaf fiber, cotton, coir, bamboo, and wood flour have been investigated as reinforcing fibers for biocomposites. Their properties and the application to biocomposites are described in detail in some review articles.[50,52] We used wood flour (WF) made from Sanbu cedar crushed into powders through 3 mm screen mesh and microfibrillated cellulose (MFC). The reason for the use of WF is as follows: Among the natural fibers, the use of waste wood generated from forest-thinning and wrecking of wooden building, etc. is very important for solving the severe environmental problem.[53,54] Also, a massive outbreak of damaged cedar trees infected by *Cercospora sequoiae Ellis et Everhart* ("*Sugi-Mizogusare*" disease) which cannot be used as log and lumber is becoming a serious problem in Chiba, Japan. Fig. 4.4 shows the FE-SEM micrographs of the WF supplied from Kowa Technos, Co. Ltd. (Sammu, Chiba, Japan). The photograph at a low magnification shows that the WF particles are mainly composed of fibrous substance of *ca.* 0.2–1.0 mm in length. The average length and aspect ratio of the WF fibers were *ca.* 0.7 mm and 4.2, respectively. The photograph at a high magnification revealed that each particle is composed of fiber bunch with a rough surface.

1.00 mm   200 µm

FIGURE 4.4   FE-SEM images of WF.[22]

In recent years, MFC with the diameters in the range of 10–100 nm, which is obtained through a simple mechanical process which includes refining and high pressure homogenizing, has received significant research attention as a reinforcing fibers of polymers.[52,55,63] Yano et al. reported that the poly(lactic acid)/MFC biocomposites prepared by the method using acetone as a mixing solvent exhibit high strength and modulus.[59] Drzal et al. reported that the mechanical properties of the composites of MFC and the bisphenol F-type epoxy resin cured with a polyether amine are improved by the silane coupling treatment of the MFC.[60] It is important for the preparation of superior polymer/MFC composites to devise the dispersion method of MFC in the hydrophobic polymers and the surface treatment method of MFC. The merit of the use of bio-based epoxy resin system instead of petroleum-based epoxy resin system is that MFC can be easily dispersed in aqueous solution of GPE/TA or SPE/TA without a special surface modification of MFC. We used the MFC, trade name Celish KY-100G was supplied by Daicel Chemical Industries, Ltd. (Tokyo, Japan). This product is a 10 wt.% solid content in water suspension.

4.4   BIOCOMPOSITES COMPOSED OF BIO-BASED EPOXY RESIN AND TANNIC ACID (TA)

4.4.1   BIO-BASED EPOXY RESINS CURED WITH TA

The GPE, PGPE, SPE, and ESO were cured with TA in order to compare the properties of the cured bio-based epoxy resins.[21,23] Because TA is soluble in liquid GPE and SPE, the mixing of GPE/TA and SPE/TA is very easy. In case of PGPE/TA and ESO/TA, it is necessary to add ethanol to get homogenous solutions. The chemical formula for the commercial TA is often given as $C_{76}H_{52}O_{46}$ (Fig. 4.2). However, in fact it contains a mixture of related compounds. Also, it is supposed that all of the three hydroxy groups of PG moiety of TA are hard to react with epoxy groups. Therefore, the curing temperature, curing time and epoxy/hydroxy ratio were optimized for GPE/TA, SPE/TA and ESO/TA. First, curing temperature and time were changed at the fixed epoxy/hydroxy ratio of 1/1. As ESO with alicyclic epoxy groups has a lower reactivity than GPE and SPE with glycidyl groups, ESO/TA was cured at a higher curing temperature range (150–230 °C) than that of GPE/TA and SPE/TA (120–200 °C). The GPE/TA and SPE/TA had the maximal tan δ peak temperatures (73 and 95 °C), when cured at 160 °C for 3 h and 2 h, respectively. Also, ESO showed the highest tan δ peak temperature (57 °C), when cured at 210 °C for 2 h (Table 2). These results indicate that the control of curing temperature and time is very important for the bio-based epoxy curing system containing aliphatic and sugar-based moieties with relatively low heat resistance. When PGPE/TA with epoxy/hydroxy 1/1 was cured at the same curing temperature and time (160 °C and 3 h) as GPE/TA, the tan δ peak temperature was 77 °C. When the tan δ peak temperatures of the materials cured at epoxy/hydroxy 1/1 are compared, the higher order was SPE/TA (95 °C) > PGPE/TA (77 °C) > GPE/TA (73 °C) > ESO/TA (57 °C), as is supposed from the relationship between epoxy functionality and the shortest distance between the two cross-linked points.

TABLE 4.2 Optimization of Curing Temperature and Time for the GPE/TA, SPE/TA and ESO/TA with Epoxy/Hydroxy Ratio 1/1

Sample	Curing temperature (°C)	Curing time (h)	Tan δ peak temperature (°C)
GPE/TA	120	2	68
	140	2	70
	160	1	63
		2	72
		3	73
		4	66
		5	69
	180	2	77
	200	0	63
SPE/TA	120	2	74
	140	2	82
	160	1	89
		2	95
		3	89
		4	90
		5	90
	180	2	84
	200	2	65
ESO/TA	150	2	46
	170	2	52
	190	2	53
	210	1	55
		2	57
		3	55
		4	51
		5	51
	230	2	50

Next, epoxy/hydroxy ratio was changed at the curing temperature/time, which showed the highest tan δ peak temperature at epoxy/hydroxy 1/1. The tan δ peak temperatures of GPE/TA and SPE/TA increased with decreasing epoxy/hydroxy ratio (Table 4.3). The increase of T_g with TA content should be attributed to an increase of the content of highly hindered aromatic framework rather than an increase of crosslinking density. However, the cured resins at epoxy/hydroxy 1/1 showed the highest tensile strength and modulus. Judging from the trend of tensile properties, it is thought that the incorporation of TA component in the crosslinked structure is insufficient for the TA-rich compositions (GPE/TA 1/1.2–1/1.4, SPE/TA 1/1.2–1/1.4). Consequently, the epoxy/hydroxy ratio 1/1 was selected for both GPE/TA and SPE/TA, considering the balance of T_g and tensile properties. The epoxy/hydroxy ratio of PGPE/TA was fixed to 1/1, considering the result of GPE/TA, In case of ESO/TA, both the tan δ peak temperature and tensile properties increased with decreasing epoxy/hydroxy ratio. When the epoxy/hydroxy ratio was lower than 1/1.4, the ESO/TA mixture became so viscous that we could not prepare a void-free cured sample. Consequently, the condition of curing temperature 210 °C, curing time 2 h, and epoxy/hydroxy ratio 1/1.4 was selected for the curing of ESO/TA. When petroleum-based PN is used as a hardener of SPE, the tan δ peak temperatures for SPE/PN resins cured at epoxy/hydroxy ratios 1/0.8–1/1 were 78–81 °C, which were lower than those of SPE/TA resins cured at epoxy hydroxy ratios 1/1–1/1.4 (95–111 °C).

TABLE 4.3 Thermal and Mechanical Properties of the GPE, PGPE and SPE Cured with Various Hardeners

Sample	Epoxy/ hydroxy ratio	Curing condition (°C/h)	Tan δ peak temperature (°C)	Tensile strength (MPa)	Tensile modulus (MPa)	5 wt.% loss temperature (°C)
GPE/TA	1/1	160/3	73	36.5	2430	317
	1/1.2		79	31.2	2360	312
	1/1.4		91	32.1	2260	
PGPE/TA	1/1	160/3	77	63.5	2710	316
SPE/TA	1/1	160/2	95	60.6	1710	314
	1/1.2		109	25.3	1698	
	1/1.4		111	30.0	1275	
SPE/PN	1/0.8	170/3	81.0			
	1/0.9		80.6			
	1/1		78.1			346.3
ESO/TA	1/0.8	210/2	46	6.0	116	
	1/1		57	12.7	409	
	1/1.2		58	12.7	450	
	1/1.4		58	15.1	458	

4.4.2 PGPE/TA/WF AND GPE/TA/WF BIOCOMPOSITES

All the bio-based materials used in this study (PGPE, GPE, and TA) are water-soluble and hydrophilic substances. The average number of epoxy groups per molecule of PGPE and GPE is 4.1 and 2.0, respectively. Because the viscosity of GPE (150 cps at 25 °C) was much lower than that of PGPE (1300 cps at 25 °C), a mixture of GPE, TA, and WF can be compounded without solvent. However, it was necessary to add a solvent in case of a mixture of PGPE, TA, and WF. Since some precipitate was liberated when aqueous solutions of PGPE and TA were mixed, ethanol was used as a mixing solvent. The mixture of PGPE, TA, and WF or GPE, TA, and WF was cured at the condition of 160 °C for 3 h with epoxy/hydroxy ratio of 1/1, at which most balanced thermal and mechanical properties were attained for the cured products of GPE and TA in the previous section.[22] Although the biocomposites with WF content higher than 70 wt.% can be prepared, the obtained composites became brittle and the surface was rough. Figure 4.5 shows FE-SEM photographs of the fractured surfaces of PGPE/TA(1/1)/WF and GPE/TA(1/1)/WF composites with WF contents of 60 and 70 wt.%. It appeared that WF is tightly incorporated into the crosslinked epoxy resins and their interfacial adhesion is good. This result may be attributed to the fact that the polyphenol moiety of TA and lignocellulose moiety of WF resemble each other. There are some voids on the fractured surface of PGPE/TA(1/1)/WF, probably generated during the evaporation of ethanol when compared with GPE/TA/WF.

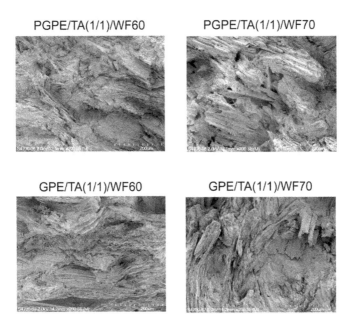

FIGURE 4.5   FE-SEM images of the fracture surfaces of PGPE/TA(1/1)/WF and GPE/TA(1/1)/WF biocomposites with WF contents of 60 and 70 wt.%.[22]

Figures 4.6 and 4.7 show the temperature dependency of storage modulus (E') and tan δ for PGPE/TA(1/1)/WF and GPE/TA(1/1)/WF measured by DMA, respectively. The E' at the rubbery plateau region over 80 °C for all the composites was much higher than that of control cured resins, suggesting a superior reinforcement effect due to the wood fibers. The tan δ peak temperature related to T_g for the composites was a little lower than that of the corresponding neat resins. The reason is not clear, but it is thought that some components of WF react with the epoxy resins, and/or that WF disturbs the crosslinking reaction.

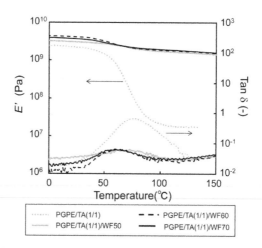

FIGURE 4.6   DMA curves of PGPE/TA(1/1) and PGPE/TA/WF(1/1) biocomposites.[22]

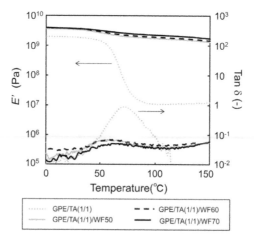

FIGURE 4.7   DMA curves of GPE/TA(1/1) and GPE/TA(1/1)/WF biocomposites.[22]

Figure 4.8 shows typical TGA curves of GPE/TA(1/1), GPE/TA(1/1)/WF60 and WF. Since the thermal decomposition temperature of WF was lower than that of GPE/TA(1/1), the GPE/TA(1/1)/WF composite exhibited two-step thermo-degradation. The 5% weight loss temperatures of all the composites are summarized in Table 4.4. Consequently, the 5% weight loss temperatures of all the composites were lower than those of the corresponding cured neat resins.

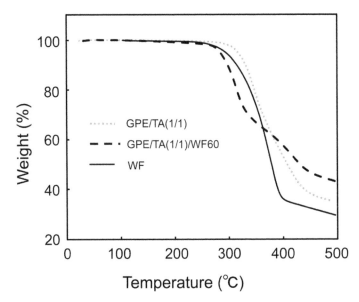

FIGURE 4.8   TGA curves of GPE/TA(1/1), GPE/TA(1/1)/WF and WF.[22]

TABLE 4.4   Tan δ Peak Temperatures and 5% Weight Loss Temperatures of all the Samples.

Resin	Epoxy/hydroxy ratio	WF content (wt.%)	Tan δ peak Temperature (°C)	5 wt.% loss temperature (°C)
PGPE/TA	1/1	0	77	316
	1/1	50	70	295
	1/1	60	63	294
	1/1	70	60	295
GPE/TA	1/1	0	73	317
	1/1	50	59	287
	1/1	60	61	284
	1/1	70	65	287

TABLE 4.4 *(Continued)*

Resin	Epoxy/hydroxy ratio	WF content (wt.%)	Tan δ peak Temperature (°C)	5 wt.% loss temperature (°C)
GPE/TA	1/0.6	0	60	322
	1/0.8	0	67	321
	1/1	0	73	317
	1/1.2	0	79	312
GPE/TA	1/0.6	60	42	290
	1/0.8	60	65	293
	1/1	60	61	284
	1/1.2	60	70	289
WF	-	100	-	296
TA	-	-	-	285

Figures 4.9 and 4.10 show the relationship between tensile properties and fiber content for PGPE/TA(1/1)/WF and GPE/TA(1/1)/WF, respectively. Although tensile modulus (4.3 GPa) of PGPE/TA(1/1)/WF was much higher than that of PGPE/TA(1/1) (2.7 GPa), the tensile strength of the composite was lower than that of PGPE/TA(1/1). On the other hand, both the tensile modulus and strength of GPE/TA(1/1)/WF were much higher than those of GPE/TA(1/1) (2.4 GPa and 37 MPa). Those values increased with WF content, became maximal values (5.1 GPa and 51 MPa) at WF content 60 wt.%, and were lowered at 70 wt.%. In general, although the tensile modulus of polymer/plant fiber biocomposites is higher than the control polymer, the strength is rather lower because of a poor interfacial adhesion. It is noteworthy that the tensile strength of GPE/TA is improved by the addition of WF without any interfacial modification. This result should be attributed to the superior interfacial adhesion between GPE/TA and WF. As the reason that the tensile strength of PGPE/TA(1/1)/WF did not increase, the following factors are considered. As tensile strength of PGPE/TA(1/1) is much higher than that of GPE/TA(1/1), the interfacial adhesion strength between PGPE/TA(1/1) and WF is not higher than the strength of PGPE/TA(1/1). The structural defects due to some voids are observed as was shown in Fig. 4.5. Also, the fact that the tensile modulus and strength of both PGPE/TA(1/1)/WF and GPE/TA(1/1)/WF composites with WF content 70 wt.% are lower than those of the composites with 60 wt.% suggests that the packing of matrix resin between the WF particles is relatively insufficient for composites with 70 wt.%.

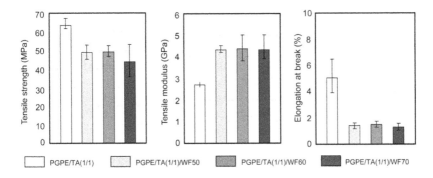

FIGURE 4.9 Tensile properties of PGPE/TA(1/1) and PGPE/TA(1/1)/WF biocomposites.[22]

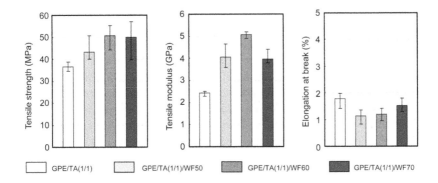

FIGURE 4.10 Tensile properties of GPE/TA(1/1) and GPE/TA(1/1)/WF biocomposites.[22]

As the tan δ peak temperature of GPE/TA(1/1)/WF was lower than that of GPE/TA(1/1) (Fig. 4.7), the epoxy/hydroxy ratio appropriate for GPE/TA/WF composites was investigated. Table 4.4 summarizes the tan δ peak temperature of GPE/TA and GPE/TA/WF60 prepared at epoxy/hydroxy ratios from 1/0.6 to 1/1.2. In case of the control GPE/TA, the tan δ peak temperature related to T_g increased with decreasing epoxy/hydroxy ratio. Considering that all of the three hydroxy groups of PG moiety of TA are hard to react with epoxy groups of GPE, it is supposed that an actual stoichiometric epoxy/hydroxy ratio should be lower than 1/1. Although GPE/TA(1/1.2)/WF60 exhibited the highest tan δ peak temperature among the GPE/TA/WF60 composites in a similar manner to the control resins, GPE/TA(1/0.8)/WF60 had a little higher tan δ peak temperature than GPE/TA(1/1)/WF60. In case of GPE/TA(1/0.8)/WF60, there is a possibility that the excess epoxy groups of GPE reacted with the hydroxy groups in WF at the curing temperature of 160 °C. As a result, GPE/TA(1/0.8)/WF60 had almost the same tan δ peak temperature as GPE/TA(1/0.8).

Table 4.4 also summarizes 5% weight loss temperature of GPE/TA and GPE/TA/WF60 prepared at epoxy/hydroxy ratios from 1.0/0.6 to 1.0/1.2. Regarding the control GPE/TA, the 5% weight loss temperature a little decreased with decreasing epoxy/hydroxy ratio. As TA itself has the lowest 5% weight loss temperature (285 °C), the presence of unreacted TA moiety in the GPE/TA with a higher TA content caused a decrease of the 5% weight loss temperature. In case of the composites with WF content 60 wt.%, GPE/TA(1/0.8)/WF60 exhibited the highest 5% weight loss temperature among the GPE/TA/WF composites.

Figure 4.11 shows the tensile properties of GPE/TA/WF60 composites prepared at various epoxy/hydroxy ratios. The GPE/TA(1/0.8)/WF60 showed the highest tensile modulus (5.22 GPa), strength (54.9 MPa) and elongation at break (1.35%), indicating that the best ratio of epoxy/hydroxy is *ca.* 1/0.8. In case of polypropylene (PP)/WF composites, it is known that the preparation of the PP/WF composite with WF content higher than 50 wt.% is not easy, and that the addition of maleic anhydride-grafted polypropylene (MAH-PP) improves the tensile properties. The tensile modulus and strength of PP/MAH-PP/WF (45/5/50) composite are reported to be 4.55 GPa and 40.4 MPa, respectively.[64] It is also known that that tensile modulus and strength of high-density polyethylene (HDPE)/WF (35/65) composite are 2.6 GPa and 15.6 MPa, and those of HDPE/poly(ethylene-*co*butyl acrylate-*co*-maleic anhydride)/WF (32.5/2.5/65) are 2.5 GPa and 18.6 MPa, respectively.[65] The GPE/TA/WF composites in which neither modifier nor compatibilizer is added have higher tensile modulus and strength than these petroleum-based plastics/WF composites.

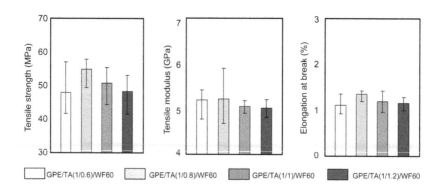

FIGURE 4.11 Influence of epoxy/hydroxy ratio on the tensile properties for GPE/TA/WF60 biocomposites.[22]

4.4.3 GPE/TA/MFC AND SPE/TA/MFC BIOCOMPOSITES

When GPE was used as a bio-based epoxy resin, direct mixing method of GPE with MFC (water content 90 wt.%) is possible, because GPE has much lower viscosity (150 cps, 25 °C) than SPE (5000 cps, 25 °C). For GPE/TA/MFC, the direct mixing method was compared with a water suspension method where 50% aqueous solution of GPE and TA in epoxy/hydroxy ratio 1/1 is mechanically mixed with MFC.[21] The obtained mixtures were subsequently freeze-dried, and finally pressure-molded at 160 °C for 3 h. Figure 4.12 shows the comparison of tensile properties of the GPE/TA(1/1)/MFC biocomposites with fiber content 10 wt.% (GPE/TA(1/1)/MFC10) prepared by both the methods. The GPE/TA(1/1)/MFC prepared by water suspension method had higher tensile strength and modulus than the GPE/TA(1/1)/MFC by direct mixing method. Figure 4.13 shows the FE-SEM images of fracture surfaces of both the composites. It is obvious that the composite by water suspension method has a better dispersion of MFC, and some aggregation of MFC is appeared for the composite by direct mixing method. In the following experiments, water suspension method was used for both the GPE/TA/MFC and SPE/TA/MFC because all the reagents except for MFC are water soluble.

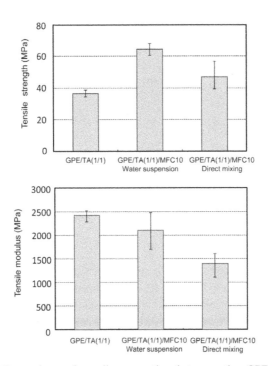

FIGURE 4.12 Comparison of tensile properties between the GPE/TA(1/1)/MFC10 biocomposites prepared by water suspension method and no solvent method.[21]

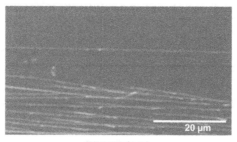

GPE/TA(1/1)

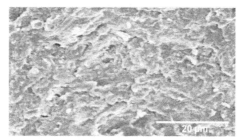

GPE/TA(1/1)/MFC10 Water suspension

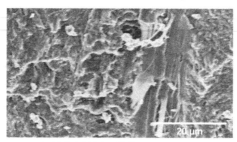

GPE/TA(1/1)/MFC10 Direct mixing

FIGURE 4.13 FE-SEM images of the fracture surfaces of GPE/TA(1/1) and the GPE/TA(1/1)/MFC10 biocomposites prepared by water suspension method and no solvent method.[21]

Figure 4.14 shows the relationship between tensile properties and fiber content for GPE/TA(1/1)/MFC and SPE/TA(1/1)/MFC. Tensile modulus increased with increasing fiber content for SPE/TA(1/1)/MFC, while the modulus did not improved for GPE/TA(1/1)/MFC. What GPE/TA(1/1) itself has much higher tensile modulus than SPE/TA(1/1) may be related to the difference of influence of fiber content on the modulus. The SPE/TA(1/1)/MFC10 (2660 MPa) had a 55% higher tensile strength than SPE/TA(1/1) (1710 MPa). Although SPE/TA(1/1)/MFC3 had a lower tensile strength than SPE/TA(1/1), the tensile strength of SPE/TA(1/1)/MFC increased with fiber content over the range of 3–10 wt.%, and leveled off at around 15 wt.%. It is supposed that critical fiber content where the fracture mode is changed from matrix control to fiber control is around 3 wt.%. When the fiber content is not more than 3 wt.%, the MFC in the composite had been pull out or broken at the maximal stress point. The SPE/TA(1/1)/MFC10 (78.6 MPa) had a 30% higher tensile strength than SPE/TA(1/1) (60.6 MPa). In case of GPE/TA(1/1)/MFC, tensile strength increased with the fiber content over the range of 0–10 wt.%, and then dropped at 15 wt.%.

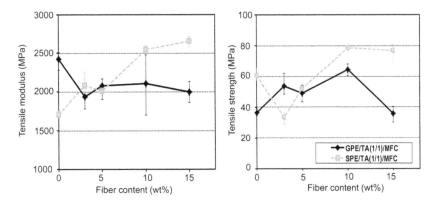

FIGURE 4.14   Tensile properties of GPE/TA(1/1)/MFC biocomposites with various fiber contents.[21]

Figure 4.15 shows FE-SEM images of the fractured surfaces of GPE/TA(1/1)/ MFC composites. The fractured surface of GPE/TA(1/1) is very smooth, indicating brittle crack propagation, as is shown in Fig. 4.13. On the other hand, uneven corrugation with the size of several tens micrometers was observed on the fractured surface of GPE/TA(1/1)/MFC3, indicating heterogeneous distribution of MFC. The uneven texture became more homogeneous and finer with increasing amount of MFC, and GPE/TA(1/1)/MFC10 had a rough surface, suggesting a complex fracture process involving both localized deformation of the matrix polymer as well as local interaction of finely dispersed MFC with cracks formed in the matrix. However, GPE/TA(1/1)/MFC15 showed bigger and more heterogeneous uneven texture than GPE/TA(1/1)/MFC10 did. This result suggests that some aggregation of MFC occurs for GPE/TA(1/1)/MFC15. Figure 4.16 shows FE-SEM images of the fractured surfaces of SPE/TA(1/1)/MFC composites. The heterogeneous texture observed for GPE/TA(1/1)/MFC3 and GPE/TA(1/1)/MFC5 was not appeared for SPE/TA(1/1)/ MFC composites. The relatively homogenous surface gradually became rougher with an increase of MFC content over the range from 3 to 15 wt.%. These results suggest that SPE/TA(1/1)/MFC composites have better dispersion of MFC than GPE/TA(1/1)/MFC composites do, and that the aggregation of MFC does not occur even at MFC content 15 wt.%.

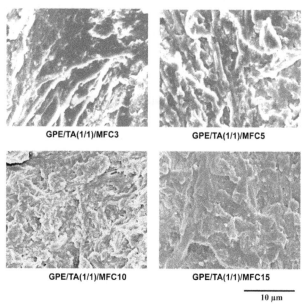

FIGURE 4.15   FE-SEM images of the fracture surfaces of GPE/TA(1/1)/MFC biocomposites with various fiber contents.[21]

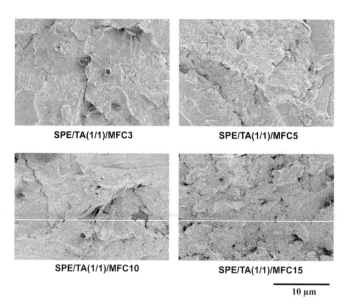

FIGURE 4.16   FE-SEM images of the fracture surfaces of SPE/TA(1/1)/MFC composites with various fiber contents.[21]

Figures 4.17 and 4.18 show the temperature dependency of E' and tan d for GPE/TA(1/1)/MFC and SPE/TA(1/1)/MFC measured by DMA, respectively. The E' at the rubbery plateau region over 100 °C increased with MFC content for both the composites, suggesting a good dispersion of MFC in the matrix is attained. The tan δ peak temperature related to T_g increased a little with MFC content for both the composites, indicating that there is some interaction between MFC and crosslinked epoxy resins. The composites composed of MFC and the bisphenol F-type epoxy resin (BPE) cured with polyether amine (PEA) is known to have a little lower tan δ peak temperature than the control cured epoxy resin (BPE/PEA), and that the peak temperature rises up to the value as high as the control by using the MFC surface-modified with 3–aminopropyltriethoxysilane (AMFC).[66] Table 4.5 summarizes the DMA data of various epoxy resins and their MFC composites. The E' at 30 and 130 °C and tan δ peak temperature of bio-based epoxy resin systems, GPE/TA(1/1) and SPE/TA(1/1), are comparable to those of petroleum-based epoxy resin system, BPE/PEA. Also, the biocomposites, GPE/TA(1/1)/MFC5 and SPE/TA(1/1)/MFC5 had higher E' at 130 °C than did BPE/PEA/MFC5 and BPE/PEA/AMFC5. In agreement with the results of DMA, the T_g measured by TMA rose with MFC content. The coefficient of thermal expansion's (CTE's) below T_g and above T_g somewhat increased with MFC content (Tables 4.5 and 4.6). The density of both the composites (1.3–1.2) a little decreased with MFC content. Considering that the density of cellulose fiber such as cotton is *ca.* 1.5, micro bubbles or voids are contaminated into the composites. Although we focused on the properties of the biocomposites prepared by a general procedure, a better result should be obtained by the addition of homogenizing operations such as sonication and subsequent vacuum degassing.

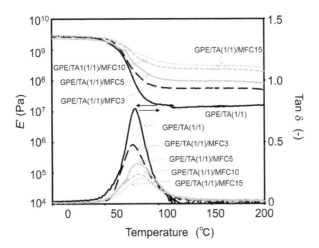

FIGURE 4.17   DMA curves for GPE/TA(1/1)/MFC composites with various fiber contents.[21]

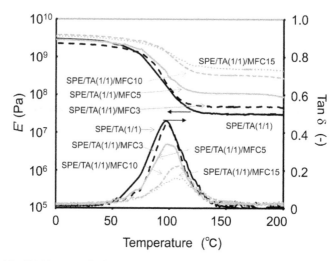

FIGURE 4.18   DMA curves for SPE/TA(1/1)/MFC composites with various fiber contents.[21]

TABLE 4.5   DMA Parameters of Various Epoxy Resin/MFC Composites

Sample	E' at 30 °C (GPa)	E' at 130 °C (MPa)	Tan δ peak temp. (°C)
GPE/TA(1/1)	2.6	14	73
SPE/TA(1/1)	2.7	33	95
BPE/PEA	2.6	9.7	85
GPE/TA(1/1)/MFC5	2.2	96	81
SPE/TA(1/1)/MFC5	2.6	113	99
BPE/PEA/MFC5	3.1	37	81
BPE/PEA/AMFC5	3.3	66	85

TABLE 4.6   Properties of GPE/TA(1/1) and GPE/TA(1/1)/MFC with Various MFC Contents

Sample	Density (g/cm³)	5 wt.% loss temp. (°C)	Tan δ peak temp. (°C)	T_g (°C) [TMA]	CTE (×10⁵ K⁻¹) a_1^{*1}	CTE (×10⁵ K⁻¹) a_2^{*2}
GPE/TA(1/1)	1.35	316.8	73	87.3	5.77	27.35
GPE/TA(1/1)/MFC3	1.26	312.6	76	89.8	6.86	39.38
GPE/TA(1/1)/MFC5	1.29	303.8	81	90.5	8.19	45.43
GPE/TA(1/1)/MFC10	1.29	300.8	82	95.0	9.10	38.94
GPE/TA(1/1)/MFC15	1.17	300.8	84	96.4	9.94	38.20

*¹ Coefficient of thermal expansion (CTE) between $(T_g - 35)$°C and $(T_g - 10)$°C.
*² Coefficient of thermal expansion (CTE) between $(T_g + 10)$°C and $(T_g + 35)$°C.

4.4.4   ESO/TA/MFC BIOCOMPOSITES

When the commercial MFC containing 90% water was directly added to a 50% ethanol solution of ESO and TA, some ESO-rich component was phase-separated and homogeneous cured material was not obtained. When the mixture was cured after freeze-drying, the relatively homogeneous ESO/TA/MFC composite with MFC content 5 wt.% was obtained. However, the composite had much lower tan δ peak temperature (42 °C) and tensile strength (4.0 MPa) than the control ESO/TA(1/1.4) (58 °C, 15.1 MPa). Therefore, the water in MFC aqueous suspension was substituted with ethanol and the obtained ethanol suspension of MFC was added to a mixture of ESO and TA, as is described in the experimental section. By this method, a homogeneous composite with a higher performance than the control-cured resin was obtained.[23]

Figure 4.19 shows the relationship between tensile properties and fiber content for ESO/TA(1/1.4)/MFC composites. Tensile modulus of the composites increased with increasing fiber content, and reached 1.33 GPa at MFC content 11 wt.%. On the other hand, elongation at break decreased with an increase of MFC content. Regarding tensile strength, although all the ESO/TA(1/1.4)/MFC composites had higher value than the control ESO/TA(1/1.4), a clear tendency between tensile strength and MFC content was not elucidated. As a result, highest tensile strength (26.3 MPa) was obtained when the MFC content is 9 wt.%.

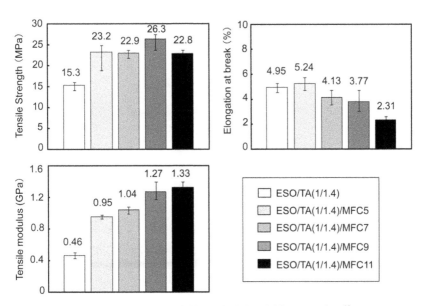

FIGURE 4.19   Tensile properties of ESO/TA(1/1.4)/MFC biocomposites.[23]

Figure 4.20 shows FE-SEM images of the fractured surfaces of ESO/TA(1/1.4) and ESO/TA(1/1.4) /MFC biocomposites with MFC content 9 and 11 wt.%. The surface of ESO/TA(1/1.4) was very smooth except for the fractured pattern. On the other hand, the surface of ESO/TA(1/1.4)/MFC9 is very rough, suggesting the MFC is homogeneously dispersed in the matrix polymer. However, the surface of ESO/TA(1/1.4)/MFC11 is heterogeneous and several microcracks were observed, suggesting that some aggregation of MFC occurs and the space between the aggregated fibrils is not fully filled out with the epoxy resin because of the high volume fraction of MFC. The difference of morphology is responsible for the fact that tensile strength of ESO/TA(1/1.4)/MFC11 is lower than that of ESO/TA(1/1.4)/MFC9.

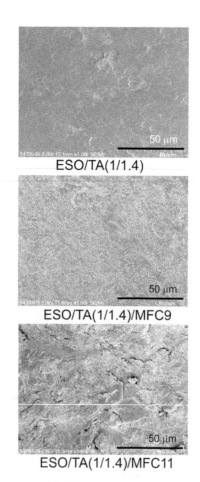

ESO/TA(1/1.4)

ESO/TA(1/1.4)/MFC9

ESO/TA(1/1.4)/MFC11

FIGURE 4.20   FE-SEM images of the fractured surfaces of ESO/TA(1/1.4), ESO/TA(1/1.4)/MFC9 and ESO/TA(1/1.4)/MFC11.[23]

Figure 4.21 shows the temperature dependency of E' and tan d for ESO/TA(1/1.4)/MFC composites. The E' at the rubbery plateau region over 100 °C increased with MFC content, suggesting a good dispersion of MFC in the matrix is attained. The tan d peak temperature corresponding to T_g increased a little with MFC content over the range from 5 to 9 wt.% (see also Table 7), indicating that there is some interaction between MFC and crosslinked ESO/TA. However, the tan d peak temperature of ESO/TA(1/1.4)/MFC11 was rather lower than that of ESO/TA(1/1.4) /MFC9. This result is attributed to the heterogeneous morphology of the former composite. Table 4.7 summarizes the results of TMA and TGA measurements of ESO/TA(1/1.4)/MFC composites. The T_g measured by TMA exhibited a similar tendency to the tan d peak temperature measured by DMA. The coefficient of thermal expansion's (CTE's) below T_g and above T_g somewhat increased with MFC content. Considering that crystalline cellulose has much lower CTE than ESO/TA(1/1.4),[67] it is thought that micro bubbles or voids are contaminated into the composites. Although 5 wt.% loss temperature measured by TGA for ESO/TA(1/1.4)/MFC decreased with increasing MFC content, their values were higher than that of dried MFC (315.1 °C).

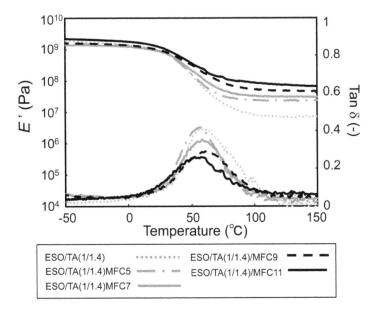

FIGURE 4.21   DMA curves for ESO/TA(1/1.4)/MFC biocomposites.[23]

TABLE 4.7 Properties of ESO/TA(1/1.4) and ESO/TA(1/1.4)/MFC with Various MFC Contents

Sample Abbreviation	5 wt.% loss temp. (°C)	Tan δ peak temp. (°C)	T_g [TMA] (°C)	CTE (10^{-5} K^{-1})	
				a_1^{*1}	a_2^{*2}
ESO/TA(1/1.4)	345.4	58	47.7	9.74	20.46
ESO/TA(1/1.4)/MFC5	329.5	58	48.6	10.98	24.54
ESO/TA(1/1.4)/MFC7	329.1	59	51.1	12.37	27.79
ESO/TA(1/1.4)/MFC9	324.9	61	56.7	14.28	27.06
ESO/TA(1/1.4)/MFC11	318.5	57	56.4	16.10	27.79

*1 Coefficient of thermal expansion (CTE) between $(T_g - 20)°C$ and $(T_g - 10)°C$.
*2 Coefficient of thermal expansion (CTE) between $(T_g + 10)°C$ and $(T_g + 20)°C$.

4.5 BIOCOMPOSITES COMPOSED OF SORBITOL POLYGLYCIDYL ETHER (SPE), BIO-BASED HARDENER AND WOOD FLOUR

4.5.1 SPE/QC/WF BIOCOMPOSITES

4.5.1.1 PROPERTIES OF CURED EPOXY RESINS

In order to optimize the curing condition of SPE and QC, the curing temperature and epoxy/hydroxy ratio were changed.[24] Table 4.8 summarizes the tan δ peak temperature measured by DMA and 5% weight loss temperature of SPE/QC cured at various conditions. When the curing temperature was changed from 150 °C to 190 °C for SPE/QC at a typical epoxy/hydroxy ratio of 1/1, the SPE/QC cured at 170 °C had the highest tan δ peak temperature (78.4 °C) and 5% weigh loss temperature (335.4 °C). Because SPE is an aliphatic epoxy resin, it is presumed that some thermal degradation starts to occur at around 190 °C. Therefore, the curing temperature was fixed to 170 °C. When the epoxy/hydroxy ratio was changed from 1/0.8 to 1/1.2 at the curing temperature of 170 °C, the SPE/QC 1/1.2 had the highest tan δ peak temperature (85.5 °C) and 5% weigh loss temperature (342.5 °C). This result suggests that four hydroxy groups of QC with five hydroxy groups per molecule actually reacted with SPE. Although it is not clear why SPE/QC 1/0.8 had a higher tan δ peak temperature than SPE/QC 1/0.9–1/1.1, it is supposed that cationic homopolymerization of SPE occurs by the action of acidic hydroxy proton at δ-position of carbonyl group of QC. Consequently, the epoxy/hydroxy ratio of 1/1.2 and curing temperature of 170 °C were selected as the optimized curing condition for SPE/QC. Table 4.8 also sum-marizes the thermal properties of the cured resins of SPE/PN, diglycidyl-ether of bisphenol A (DGEBA)/QC and DGEBA/PN. As a result of optimi-zation of the epoxy/hydroxy ratio for SPE/PN, SPE/PN(1/0.8) showed the highest tan δ peak temperature (81.0 °C), which was still lower than that of SPE/QC(1/1.2) (85.5 °C). Although we did not fully optimize the epoxy/hy-droxy ratio for DGEBA/QC and DGEBA/PN, DGEBA/QC(1/1) and DGE-BA/QC(1/1.2) showed higher tan δ peak temperature than DGEBA/PN(1/1). These results indicate that QC is a superior epoxy hardener to produce the cured resin with a high T_g.

TABLE 4.8 Thermal Properties of Epoxy Resins Cured with QC and PN at Various Conditions

Sample	Epoxy/hydroxy ratio	Curing Temperature (°C)	Tan δ peak temperature (°C)	5 wt.% loss temperature (°C)
SPE/QC	1/1	150	58.7	330.1
		170	78.4	335.4
		190	75.3	329.5
SPE/QC	1/0.8	170	84.9	342.2
	1/0.9		79.8	344.9
	1/1		78.4	335.4
	1/1.1		79.5	335.8
	1/1.2		85.5	342.5
SPE/PN	1/0.8	170	81.0	
	1/0.9		80.6	
	1/1		78.1	346.3
DGEBA/QC	1/1	170	130.2	385.2
	1/1.2		145.1	407.4
DGEBA/PN	1/1	170	90.8	395.5

4.5.1.2   PROPERTIES OF SPE/QC/WF BIOCOMPOSITES

Figure 4.22 shows the temperature dependency of E' and tan δ for SPE/QC(1/1.2), SPE/PN(1/0.8), SPE/QC(1/1.2)/WF and SPE/PN(1/0.8)/WF biocomposites mea-sured by DMA. The tan δ peak amplitude for the biocomposites became weaker with increasing WF content, indicating that amorphous content of the biocompos-ites certainly decreased with WF content. The E' curve of at the rubbery plateau region over 120 °C for all the biocomposites was much higher than those of con-trol SPE/QC(1/1.2) and SPE/PN(1/0.8), suggesting a superior reinforcement effect due to the wood fibers. The tan δ peak temperatures (SPE/QC(1/1.2)/WF20, 30,

40:106.2, 112.7, 107.2 °C) related to T_g for the SPE/QC(1/1.2)/WF biocomposites were significantly higher than that of SPE/QC(1/1.2) (85.5 °C). This trend is marked contrast to the fact that SPE/PN(1/0.8)/WF30 had a lower tan δ peak temperature (69.9 °C) than that of SPE/PN(1/0.8) did (81.0 °C). Similar lowering of tan δ peak temperature of WF biocomposite relative to the corresponding cured neat resin had been also observed for GPE/TA/WF biocomposites as was reported by our group.[22] Also, the E' of the SPE/QC(1/1.2)/WF biocomposites declined at around 90–120 °C due to the glass transition, and then again decreased at around 180–200 °C, probably due to the disappearance of specific interaction between WF and the cured SPE/QC resin. Figure 4.23 shows FT-IR spectra of WF, QC and a mixture of QC/WF 1/1 (w/w) prepared by mixing in THF and drying at 40 °C for 24 h. The band at 1621 cm⁻¹ for QC due to C=C stretching vibration at C–2 and 3 did not shift for QC/WF. In contrast, the band at 1675 cm⁻¹ for QC due to unsaturated carbonyl (C=O) stretching vibration significantly shifted to a lower wavenumber region for QC/WF (1659 cm⁻¹), indicating that there is a hydrogen bonding interaction between unsaturated carbonyl group of quercetin moiety and hydroxy group of lignocellulose component of WF. This interaction is based on the resonance structure of QC generating highly polarized carbonyl group as is shown in Fig. 4.24.

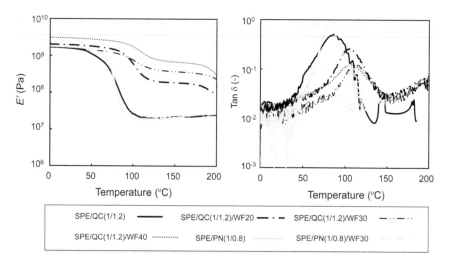

FIGURE 4.22 DMA curves of SPE/PN(1/0.8), SPE/QC(1/1.2), and SPE/PN(1/0.8)/WF and SPE/QC(1/1.2)/WF biocomposites.[24]

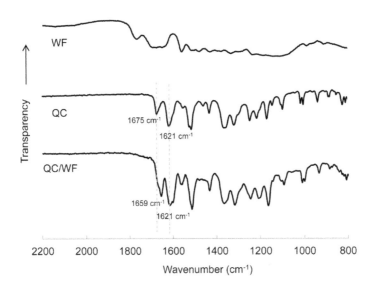

FIGURE 4.23 FT-IR spectra of WF, QC and QC/WF. [24]

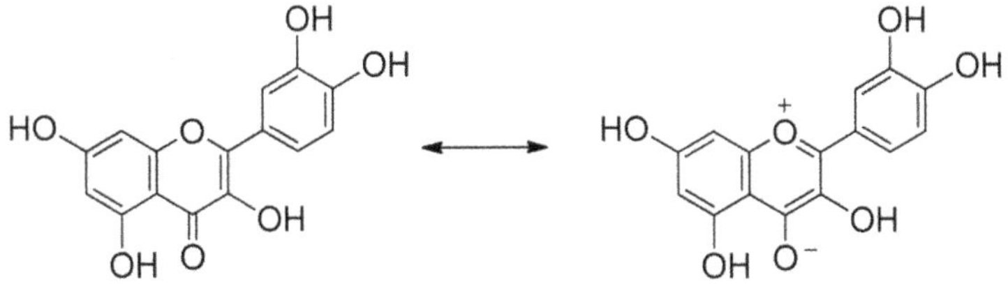

FIGURE 4.24 Resonance structure of QC. [24]

Figure 4.25 shows typical TGA curves of SPE/PN(1/0.8), SPE/QC(1/1.2), SPE/QC/(1/1.2)WF biocomposites and WF. Since the thermal decomposition temperature of WF was lower than that of SPE/QC(1/1.2), the SPE/QC(1/1.2)/WF composite exhibited two-step thermo-degradation. The 5% weight loss temperatures of SPE/QC(1/1.2), SPE/QC(1/1.2)/WF20,30,40 and WF were 342.5, 311.4, 300.8, 299.2 and 295.5 °C, respectively. The SPE/QC showed a comparable 5% weight loss temperature to SPE/PN(1/0.8) (346.3 °C) in agreement with the fact that both QC and PN are aromatic polyphenols. In addition, the cured resin of DGEBA and

QC, both of which are aromatic compounds had a superior 5% weight loss temperature (407.4 °C) as is shown in Table 4.8.

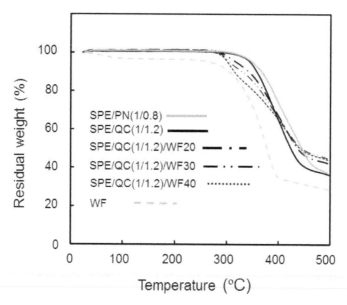

FIGURE 4.25   TGA curves of SPE/PN(1/0.8), SPE/QC(1/1.2) and SPE/QC(1/1.2)/WF biocmposites.[24]

Figure 4.26 shows the relationship between tensile properties and fiber content for SP/QC(1/1.2)/WF composites. The tensile modulus of SPE/QC(1/.12)/WF biocomposites increased with increasing WF content, and SPE/QC(1/1.2)/WF40 had a higher tensile modulus than SPE/PN(1/0.8). However, tensile strength and elongation at break of the biocomposite were lower than those of SPE/QC(1/1.2). Figure 4.27 shows the FE-SEM images of the fractured surface of SPE/QC(1/1.2)/WF biocomposites. The micrograph of WF shows that the fiber length and width of WF are ca. 0.2–0.4 mm and 40–200 mm, respectively. It appeared that the composites are fractured at the interface between WF and the cured resin. As SPE/QC(1/1,2) itself has a high tensile strength (43 MPa), a considerably high interfacial adhesiveness between WF and SPE/QC is necessary to obtain the biocomposite with a higher tensile strength than the cured resin.

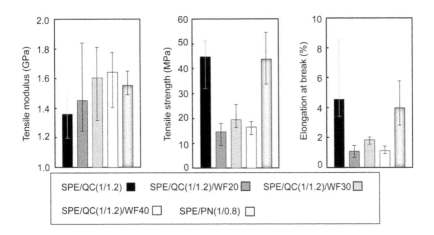

FIGURE 4.26   Tensile properties of SPE/PN(1/0.8), SPE/QC(1/1.2) and SPE/QC(1/1.2)/WF biocomposites.[24]

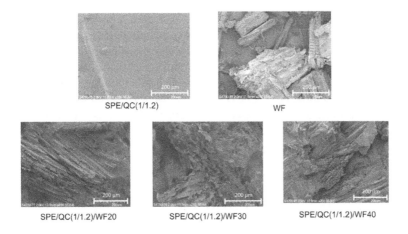

FIGURE 4.27   FE-SEM images of WF and the fracture surfaces of SPE/QC(1/1.2) and SPE/QC(1/1.2)/WF biocomposites.[24]

4.5.2   SPE/TPG/WF BIOCOMPOSITES

4.5.2.1   PREPARATION AND CHARACTERIZATION OF TPG

Vegetable oils such as soybean oil and tung oil (TO) and bio-based phenols such as cardanol (CD) and pyrogallol (PG) are promising raw materials for the preparation of flexible bio-based phenolic epoxy hardeners. TO is a triglyceride extracted from the seeds of the tung tree (*Aleurites fordii*), in which approximately 80% of the fatty acid chains is α-eleostearic acid, that is, 9-*cis*,11,13-*trans*octadecatrienoic acid.[68,69] Therefore, TO with the conjugated triene moiety shows a characteristic reactivity which is not seen in the convention soybean oil and linseed oil, etc. From the past studies, it was found that the reaction of soybean oil and phenol in the presence of a super acid such as trifluoromethanesulfonic acid or tetrafluoroboric acid produce a complex mixture of phenolated soybean oils oligomerized by Diels-Alder reaction.[70,71] In contrast, the reaction of TO and phenol smoothly proceed in a mild acidic condition without the formation of oligomerized materials to produce a desired TO-phenol resin.[72,75] PG is obtained by decarboxylation of gallic acid which is a basic component of hydrolysable tannin. In the past studies, PG-formaldehyde resin[76] and TO-PG resin[77] (TPG) have been successfully synthesized and applied for a thermosetting wood adhesive and a positive photoresist developed by alkaline solutions, respectively. We carried out the preparation and structural analysis by ^{1}H NMR spectroscopy of TPG and used TPG as an epoxy-hardener.[25] The reaction of TO and PG in the presence of *p*-toluenesulfonic acid in dioxane at 80 °C for 3 h gave TPG as a brown viscose liquid in 37% yield (Fig. 4.28). The fact that a considerable amount of TPG is lost during the repeated washing with hot water for the removal of unreacted PG is a reason for the low yield. So, there is a possibility that the yield is improved by the optimization of purification method. As PG has a high reactivity at both the 4- and 6-positions to an electrophile, it is supposed that crosslinking reaction should occur in the reaction with a multifunctional reagent such as TO. Actually, the reaction at a higher temperature than 80 °C or the use of other acid catalysts such as hydrochloric acid and borontrifluoride diethyl etherate resulted in a formation of gelatinous materials. The obtained TPG was soluble to ethanol, acetone, ethyl acetate, tetrahydrofuran, diethyl ether, *N,N*-dimethylformamide and dimethylsulfoxide, and insoluble to water, chloroform and hexane.

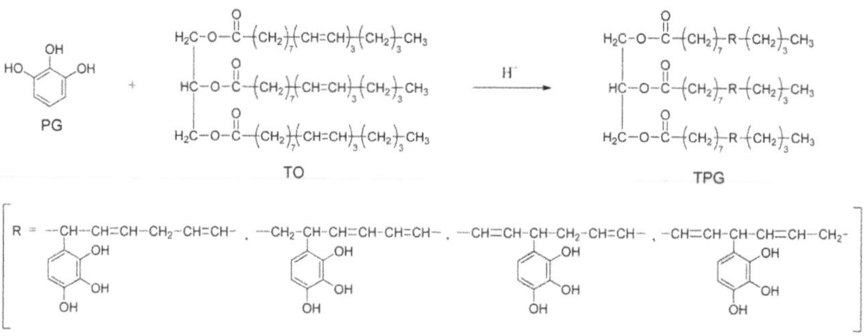

FIGURE 4.28   Synthetic scheme of TPG.[25]

Figure 4.29 shows FT-IR spectra of TO, TPG and PG. The band at 3375 cm⁻¹ due to O-H stretching vibration and that at 1623 cm⁻¹ due to benzene ring framework stretching vibration in addition to the band at 1714 cm⁻¹ due to C=O stretching vibration and those at 2950–2840 cm⁻¹ due to sp³C-H stretching vibration were observed for the spectrum of TPG, indicating that pyrogallol moiety and tung oil moiety certainly bonded. Also, the fact that the bands at 993 cm⁻¹ and 732 cm⁻¹ due to =C-H out-of plane bending vibrations of *trans* and *cis*-olefinic moieties, respectively observed for TO considerably diminished for TPG, suggesting that the addition reaction of PG to the olefinic moieties of TO certainly proceeded.

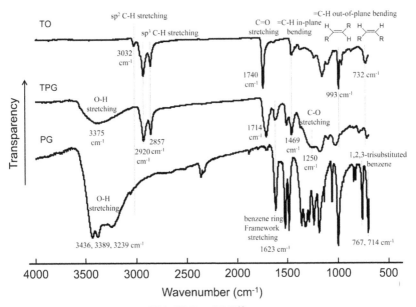

FIGURE 4.29 FT-IR spectra of TO, TPG and PG.[25]

Figure 4.30 shows the ¹H-NMR spectra of TO and TPG measured in d_6-acetone. The ¹H signal of methine proton of glyceride unit (H$_a$ and H$_{a'}$) in TPG and TO was observed at 5.29 ppm (s) and 5.27 ppm (s), respectively. The integral values of other proton signals were evaluated relative to those of H$_a$ and H$_{a'}$ signal (1H). Because we could not specify the phenolic hydroxy groups of PG unit in TPG, the H-D exchange reaction was performed by the addition of D$_2$O in a NMR tube. As a result, the ¹H signals from 7.92 to 6.71 ppm (6.8H) in d_6-acetone disappeared in the spectrum of TPG in d_6-acetone/D$_2$O, indicating 2.3 pyrogallol units are added to a TO triglyceride moiety. This number is in agreement with the integral values of ¹H signals at 6.45 ppm (d, 2.3H, H$_f$, $J = 8.3$ Hz) and 6.34 ppm (d, 2.2 H, H$_g$, $J = 8.3$ Hz) in the pyrogallol ring of TPG. The fact that coupling constant of two protons (H$_f$ and H$_g$)

of the pyrogallol ring is 8.3 Hz indicates that electrophilic substitution reaction of the TO-derived carbocation occurred at 4-position of pyrogallol (1,2,3-trihydroxy-benzene), because the product obtained by the reaction at 5-position should have the coupling constant at around 3 Hz. The PG-substituted methine proton (H$_i$) is also observed at 3.65 ppm (m, 2.3H). The number of olefins of TO per triglyceride is estimated to be 7.6 from the integral value of the olefinic ^{1}H signals at 6.45–5.41 ppm relative to that of H$_a$.. Similarly, the number of olefins of TPG is estimated to be 4.3 from the integral value of the olefinic ^{1}H signals at 6.07–5.35 ppm relative to that of H$_a$. From their values, the number of diminished olefins of TPG relative to TO is estimated to be 3.3, which is a little higher than the degree of addition of PG (2.3). This discrepancy may be attributed to the possibility that some components of TO with lower olefinic number are eliminated by the purification. Four structural formulae (R) of TPG in Fig. 4.28 are capable as the structures of the pyrogallol-substituted hexadiene moiety of TPG, considering the stability of the carbocation of reaction intermediate, if the horizontally flipped structures of R are omitted. In addition, there is a possibility that the original *cis-trans* configuration of triene part of TO is transformed to other configurations by the migration of π-bond in the car-bocation intermediates. Among the olefinic proton signals of TPG, the signal at a lower magnetic field (6.07 ppm) is assigned to inner protons (-CH=C*H*-C*H*=CH-) of conjugated diene moiety, and that at a higher magnetic field (5.35 ppm) is related to the protons of isolated olefin moiety. However, we could not assign the olefinic proton signals of TPG more precisely because many structural and configurational isomers are contained.

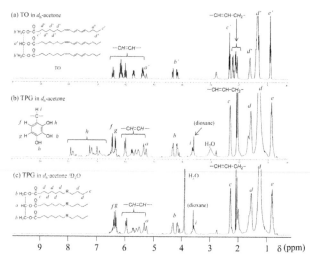

FIGURE 4.30 ^{1}H-NMR spectra of TO and TPG in d_6-acetone, and TPG in d_6-acetone / D$_2$O.[25]

4.5.2.2 PROPERTIES OF SPE/TPG/WF BIOCOMPOSITES

The onset and peak temperatures of the exothermic curve on the first heating DSC thermogram for the SPE/TPG compound with a standard epoxy/hydroxy ratio of 1/1 were 142.3 and 192.9 °C, respectively. Based on the DSC data, the curing temperature of the SPE/TPG(1/1) was changed between 150 and 190 °C. The tan δ peak temperature (43.0, 43.5 and 53.5 °C) measured by DMA increased with an increase of curing temperature (150, 170 and 190 °C). Also, the 5% weight loss temperature (344.3, 344.8 and 361.1 °C) increased with an increase of curing temperature. When the mixture was cured at a temperature higher than 190 °C, the cured material considerably colorized. The curing temperature was fixed to 190 °C, considering the stability of SPE/TPG and wood flour which is subsequently added.

Figure 4.31 shows DMA curves of the SPE/TPG(1/1)/WF biocomposites cured at 190 °C. The E' at the rubbery plateau region over 50 °C for the composites was much higher than that of SPE/TPG, suggesting a superior reinforcement effect due to the wood fibers. The tan δ peak temperature related to T_g for the composites (WF40:45.6 °C; WF50:45.7 °C; WF60:44.5 °C) was a little lower than that of the corresponding neat resins (53.5 °C). The reason is not clear, but it is thought that hydroxy groups of WF reacted with epoxy groups of SPE and the stoichiometry of epoxy and hydroxy is deviated. A similar decline of T_g by the addition of WF was also observed for the GPE/TA/WF biocomposites.[22] Figure 4.32 shows TGA curves of WF, SPE/TPG(1/1) and SPE/TPG(1/1)/WF composites. Since the thermal decomposition temperature of WF was lower than that of SPE/PGT, the SPE/PGT/WF composite exhibited two-step thermo-degradation, and the 5% weight loss temperature decreased with increasing WF content (0 wt.%: 361.1 °C, 40 wt.%: 294.7 °C, 50 wt.%: 286.3 °C, 60%: 279.6 °C).

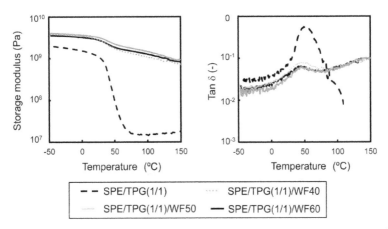

FIGURE 4.31   DMA curves of SPE/TPG(1/1) and SPE/TPG(1/1)/WF biocomposites.[25]

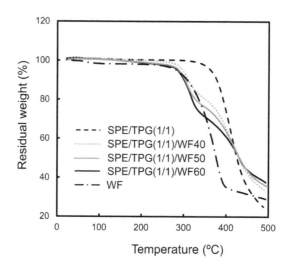

FIGURE 4.32   TGA curves of SPE/TPG(1/1), SPE/TPG(1/1)/WF biocomposites and WF.[25]

Figure 4.33 shows the tensile properties for SPE/TPG(1/1)/WF composites. The tensile modulus of SPE/TPG(1/1)/WF increased with increasing WF content in the range of 0–50 wt.%. However, the tensile modulus of SPE/TPG(1/1)/WF60 was lower than that of SPE/TPG(1/1)/WF50 in agreement with the influence of WF content on the E' measured by DMA. Also, the tensile strength of the composites with WF content 40–50 wt.% was a little higher than the corresponding neat resin (SPE/TPG(1/1)). The fact that the improvement of tensile strength is not so high as that of tensile modulus is related to the decease of elongation at break for the WF biocomposites. In the previous our study on GPE/TA/WF and SPE/QC(1/1.2)/WF biocomposites, the tensile strength considerably decreased by the addition of WF.[22,25] When TPG was used as an epoxy-hardener, the tensile strength of the WF composite did not decrease.

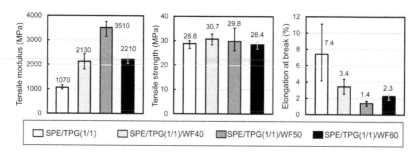

FIGURE 4.33   Tensile properties of SPE/TPG(1/1) and SPE/TPG(1/1)/WF biocomposites.[25]

Figure 4.34 shows SEM images of WF and the fractured surfaces of SPE/TPG(1/1) and SPE/TPG(1/1)/WF composites. The micrograph of SPE/TPG(1/1) showed no phase separation, indicating that SPE is homogeneously cured with TPG. The micrograph of WF shows that the fiber length and width of WF are ca. 0.2–0.4 mm and 40–200 mm, respectively. All the micrographs of SPE/TPG(1/1)/WF biocomposites show that WF is tightly incorporated into the crosslinked epoxy resin and their interfacial adhesion is good. The fact that tensile strength did not decrease by the addition of WF is related to the good affinity of SPE/TPG(1/1) and WF. The good affinity is inferred from what TO is widely used as a coating material for woody surface and the structure of pyrogallol moieties of TPG resembles that of lignin of WF.

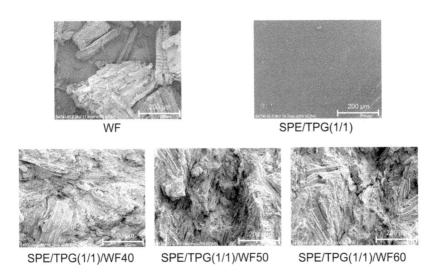

FIGURE 4.34   FE-SEM images of WF and the fracture surfaces of SPE/TPG(1/1) and SPE/TPG(1/1)/WF biocomposites.[25]

4.5.3   SPE/PGVNC/WF BIOCOMPOSITES

4.5.3.1   PREPARATION AND CHARACTERIZATION OF PGVNC

The T_g's measured by DMA for SPE/TA(1/1) and SPE/QC(1/1.2) were 95 °C and 86 °C, respectively, whose values were much lower than that of the cured conventional DGEBA (>100 °C). Therefore, when industrially available and inexpensive SPE is combined with a bio-based phenolic hardener, the hardener having a lower hydroxy value and a higher aromatic content than TA should be used. Vanillin (VN) is contained in essential oil of clove or vanilla, and is also prepared from bio-based eugenol or guaiacol. In this section, the synthesis of a bio-based novolac is examined by the reaction of PG and VN.[26] As a result of the spectral analyzes of the obtained reaction product, it was found that pyrogallol-vanillin calixarene (PGVNC) mainly composed of guaiacyl pyrogallol[4]arene is formed. Although aryl pyrogallol[4]arenes synthesized by the reactions of pyrogallol with benzaldehyde, *p*-methylbenzaldehyde and *p*-methoxybenzaldehyde, etc., are known compounds, their calixarenes have attracted little attention because of the very poor solubility in common solvents.[78,79] In contrast, the PGVNC can be easily used as building blocks of polymeric materials because of the good solubility to some organic solvents.

The reaction of PG and VN in the presence of *p*-toluenesulfonic acid gave PGVNC as a pale purple powder in 51% yield (Fig. 4.35). The obtained PGVNC was soluble to tetrahydrofuran, *N,N*-dimethylformamide and dimethylsulfoxide, and insoluble to water, methanol, ethanol, acetone, chloroform, ethyl acetate and hexane. Fig. 36 shows a FD-MS spectrum of PGVNC. A strong peak of *m/z* 1040 ($C_{56}H_{48}O_{20}$:4PG+4 VN−4H$_2$O) corresponds to a calix[4]arene, guaiacyl pyrogallol[4] arene. Other peaks at *m/z* 520, 398, 262 and 126 correspond to ($C_{28}H_{24}O_{10}$: 2PG+2 VN−2H$_2$O), ($C_{22}H_{22}O_7$: PG+2 VN−2O), ($C_{14}H_{14}O_5$: PG+VN−O) and ($C_6H_6O_3$: PG), respectively. Although the peak of *m/z* 520 corresponds a calix[2]arene, the structure would not be possible because of highly strained structure. From the MS spectrum of PGVNC, we could not decide whether the peaks other than 1040 are the fragment peaks of guaiacyl pyrogallol[4]arene or due to the compounds contained in PGVNC.

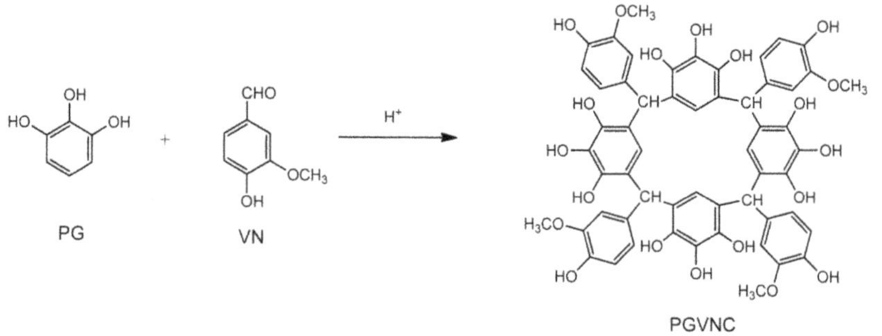

FIGURE 4.35   Synthetic scheme of PGVNC and molecular structure of SPE.[26]

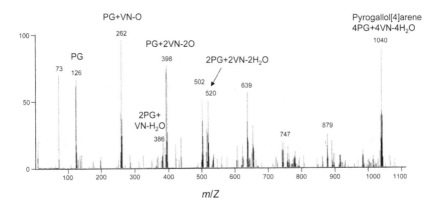

FIGURE 4.36   Mass spectrum of PGVNC.[26]

Figure 4.37 shows the ¹H-NMR spectrum of PGVNC measured in d_6-DMSO. Although novolac derivatives prepared from phenols and aldehydes generally show complex spectral patterns because of different substitution positions of benzene ring and degree of polymerization, PGVNC showed a simple spectral pattern, suggesting the formation of a single product with a symmetrical structure. Almost the main ¹H-signals can be assigned to the protons of guaiacyl pyrogallol[4]arene. The ¹H signals at 7.89 (s, 4H), 7.70 (s, 2H), 7.63 (s, 2H), 7.43 (s, 4H) and 7.19 (s, 4H) are assigned to the protons of phenolic hydroxy groups of PGVNC, because the ¹H signals disappeared by the H-D exchange by the addition of D_2O. The fact that two separated hydroxy protons are observed at 7.70 (s, 2H) and 7.63 (s, 2H) indicates that there are two pairs of hydroxy groups with a different conformation. The ¹H signals of guaiacyl group were observed at 6.33 (d, 4H, H_b, $J = 8.0$ Hz), 6.17 (s, 4H, H_d), 6.10 (d, 4H, H_c, $J = 8.0$ Hz). The ¹H-signals of pyrogallol ring were observed at 6.06 (s, 2H, H_a) and 5.58 (s, 2H, H_a), indicating that there are two kinds of pyrogallol rings with a different conformation, and that thermodynamically stable *rctt* (*cis-trans*) isomer is preferentially formed. Although the ¹H NMR spectral data of aryl pyrogallol[4]arenes have not yet been reported because of the very poor solubility, the reported NMR data of acylated aryl pyrogallol[4]arenes in $CDCl_3$ resemble those of PGVNC.[78] The stable conformation of the PGVNC with *rctt* configuration was calculated by MM2. Fig. 38 shows the calculated structure of the *rctt* PGVNC. The four-pyrogallol units in the calixarene ring were divided into two groups with two pyrogallol rings at almost perpendicular direction and other two pyrogallol rings nearly in horizontal position. The stretching direction of two perpendicular pyrogallol rings is opposite. One pyrogallol ring is upper standing and the other is upside down. The four side gauiacyl groups are also divided two groups with two neighboring guaiacyl groups at left side, while other two-guaiacyl groups locating at right side. For acylated *p*-methylphenyl pyrogallol[4]arene and acrylated *p*-methoxyphenyl pyrogallol[4]arene, similar structure is confirmed by the X-ray crystal structure analysis.[78]

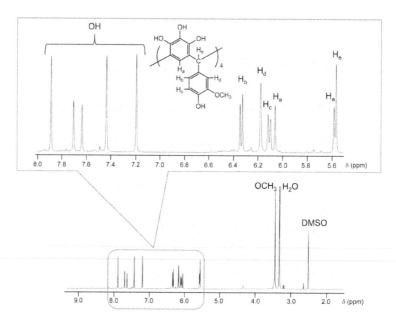

FIGURE 4.37   ¹H NMR spectrum of PGVNC in CDCl₃.[26]

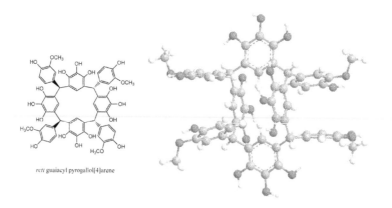

FIGURE 4.38   The structure of *rcct* guaiacyl pyrogallol[4]arene optimized by MM2.[26]

4.5.3.2   PROPERTIES OF SPE/PGVNC

When PGVNC is used as an epoxy hardener, it is supposed that all the phenolic hydroxy groups are hard to react with epoxy groups because of the steric hindrance. So, the epoxy/hydroxy ratio (1/1.14, 1/1.76, 1/2.65, 1/3.97) and curing temperature (150, 170, 190 °C) were optimized for the curing system of SPE and PGVNC. Table 4.9 summarizes the tan δ peak temperature measured by DMA and 5% weight loss temperature of SPE/PGVNC cured at various conditions. When the epoxy/hydroxy ratio was changed at the fixed curing temperature of 170 °C which is a standard curing temperature of epoxy resin, SPE/PGVNC(1/2.65) had the highest tan d peak temperature, although 5% weight loss temperature decreased a little with decreasing epoxy/hydroxy ratio. When the curing temperature was changed between 150 and 190 °C at the fixed epoxy/hydroxy ratio of 1/2.65, the cured resin at 190 °C showed the highest tan δ peak temperature (148.1 °C) and 5% weight loss temperature (319.2 °C). This result suggests that 6.0 of 16 hydroxy groups of PGVNC are reacted with epoxy groups of SPE. For example, this number corresponds to the sum of two hydroxyl groups of four guaiacyl groups and four sets of one hydroxyl group per one pyrogallol unit in the guaiacyl pyrogallol[4]arene. Also, when the curing temperature is 190 °C, the tan d peak temperature of SPE/PGVNC 1/2.65 was higher than that of SPE/PGVNC(1/1.76). We did not investigate the curing temperature higher than 190 °C, considering the stability of wood flour which is subsequently added. When SPE was cured with PN at 190 °C, SPE/PN(1/1) had a higher tan δ peak temperature and 5% weight loss temperature than that of SPE/PN 1/2.65. This result is reasonable, considering that all the hydroxy groups of PN can react with the epoxy groups of SPE in contrast to the case of SPE/PGVNC.

TABLE 4.9   Tan δ Peak Temperature Measured by DMA and 5% Weight Loss Temperature Measured by TGA for SPE/PGVNC and SPE/PN Cured at Various Conditions.

Sample	Epoxy/hydroxy ratio	Curing temperature (°C)	Tan δ peak temperature (°C)	5% weight loss temperature (°C)
SPE/PGVNC	1/1.14	170	104.1	329.0
	1/1.76	170	119.2	323.7
	1/2.65	170	133.1	316.8
	1/3.97	170	123.6	301.0
	1/2.65	150	103.2	317.5
	1/2.65	170	133.1	316.8
	1/2.65	190	148.1	319.2
	1/1.76	190	130.1	319.0
SPE/PN	1/1	190	78.1	346.3
	1/2.65	190	66.2	285.5

Figure 4.39 shows the temperature dependency of E' and tan δ for SPE/PGVNC(1/2.65) and SPE/PN(1/1) cured at 190 °C. The tan δ peak temperature of SPE/PGVNC(1/2.65) (148. 1 °C) was much higher than that of SPE/PN(1/1) (78.1 °C). Also, the E' of SPE/PGVNC(1/2.65) was higher than that of SPE/PN(1/1) over the temperature range from 0 to 200 °C. The fact that SPE/PGVNC has high glass transition temperature and rigidity should be attributed to the pyrogallol[4] arene structure. Figure 4.40 shows the comparison of tensile properties of SPE/PGVNC(1/2.65) and SPE/PN(1/1). The SPE/PGVNC showed a higher tensile modulus than SPE/PN(1/1) in agreement with the result of DMA. However, tensile strength and elongation at break for SPE/PGVNC(1/2.65) were lower than those of SPE/PN(1/1), indicating a more brittle character due to the rigid calixarene structure.

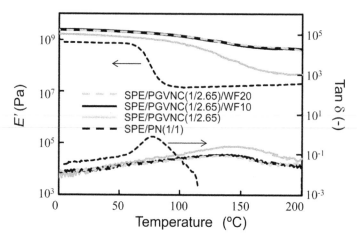

FIGURE 4.39 DMA curves of SPE/PN(1/1), SPE/PGVNC(1/2.65) and SPE/PGVNC(1/2.65)/WF biocomposites.[26]

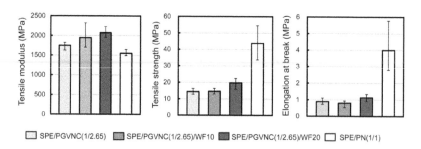

FIGURE 4.40 Tensile properties of SPE/PN(1/1), SPE-PGVNC(1/2.65) and SPE/PGVNC(1/2.65)/WF biocomposites.[26]

4.5.3.3   PROPERTIES OF SPE/PGVNC/WF BIOCOMPOSITES

Figure 4.39 also shows the temperature dependency of E' and tan δ for SPE/ PGVNC(1/2.65)/WF composites measured by DMA. The E' at the rubbery plateau region over 150 °C for the composites was much higher than that of SPE/ PGVNC(1/2.65), suggesting a superior reinforcement effect due to the wood fibers. The tan δ peak temperature related to T_g for the composites was a little lower than that of the corresponding neat resins. The reason is not clear, but it is thought that hydroxy groups of WF reacted with epoxy groups of SPE and the stoichiometry of epoxy and hydroxy is deviated.

Figure 4.40 also shows the tensile properties for SPE/PGVNC(1/2.65)/WF composites. The SPE/PGVNC(1/2.65) composites showed higher tensile modulus than SPE/PGVNC(1/2.65) in agreement with the result of storage modulus by DMA. Tensile strength also improved by the addition of WF. Figure 4.41 shows SEM images of WF and fracture surfaces of SPE/PGVNC(1/2.65) and SPE/PGVNC(1/2.65)/ WF20. The photograph of WF shows that the fiber length and width of WF are ca. 0.2–0.4 mm and 40–200 mm, respectively. Although some voids due to vaporization of THF were observed in the microphotograph of SPE/PGVNC(1/2.65), the cured resin itself is homogeneous, suggesting that SPE was homogeneously cured with PGVNC. In case of SPE/PGVNC(1/2.65), it appeared that WF was tightly incorporated into the crosslinked epoxy resins and their interfacial adhesion is good. This result may be attributed to the fact that the structures of guaiacyl and pyrogallol moieties of PGVNC resemble that of lignin of WF. The fact that tensile strength and elongation at break did not decrease by the addition of WF should be related to the good affinity of SPE/PGVNC and WF.

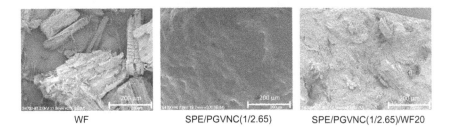

WF	SPE/PGVNC(1/2.65)	SPE/PGVNC(1/2.65)/WF20

FIGURE 4.41   FE-SEM images of WF and fracture surfaces of SPE/PGVNC(1/2.65) and SPE/PGVNC(1/2.65)/WF20.[26]

Figure 4.42 shows TGA curves of WF, SPE/PGVNC(1/2.65), SPE/PN(1/1), SPE/PGVNC(1/2.65)/WF composites. Since the thermal decomposition temperature of WF was lower than that of SPE/PGVNC(1/2.65), the SPE/PGVNC(1/2.65)/

WF composite exhibited two-step thermo-degradation, and the 5% weight loss temperatures decreased with increasing WF content (0 wt.%: 319.2 °C, 10 wt.%: 312.1 °C, 20 wt.%: 301.8 °C, 100%: 293.2 °C). The 5% weight loss temperature of SPE/PGVNC(1/2.65) (319.2 °C) was lower than that of SPE/PN(1/1) (346.3 °C). The feed ratio of 1/2.65 for SPE/PGVNC was selected based on the highest tan δ peak temperature. In order to get the cured material with higher thermal stability, the epoxy/hydroxy ratio should be approached to 1/1, as is obvious from Table 4.9. Regarding the biodegradability, it is supposed that SPE/PGVNC and SPE/PGVNC/ WF are fairly resistant to both aerobic and anaerobic biodegradation because their materials contain highly crosslinked aromatic structure, which is similar to that of lignin.[80,81]

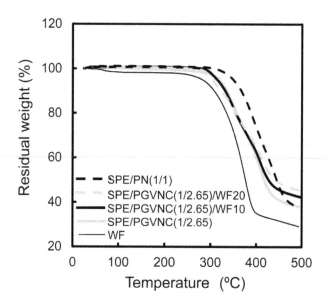

FIGURE 4.42   TGA curves of SPE/PN(1/1), SPE/PGVNC(1/2.65), SPE/PGVNC(1/2.65)/ WF biocomposites and WF.[26]

4.6   CONCLUSION

The biocomposites composed of bio-based epoxy resins (GPE, PGPE, SPE, and ESO), bio-based polyphenol hardeners (TA, QC, TPG, and PGVNC), lignocellulosic fibers (WF and MFC) were prepared and their thermal and mechanical properties were investigated. Tan δ peak temperatures and tensile properties of all the biocomposites are summarized in Table 4.10. Among the bio-based epoxy resins cured with TA, SPE/TA(1/1) showed the highest tan δ peak temperature (95 °C), and

PGPE/TA(1/1) showed the highest tensile strength and modulus (63.5 MPa and 2.71 GPa). Among the SPE cured with various bio-based polyphenol hardeners, SPE/PGVNC(1/2.65) showed the highest tan δ peak temperature (148 °C), and PGPE/TA(1/1) showed the highest tensile strength and modulus. Regarding the biocomposites with WF, although the tensile modulus increased with increasing WF, tensile strength rather decreased by the addition of WF for all the WF biocomposites except for GPE/TA(1/1)/WF. In case of GPE/TA(1/1)/WF biocomposites, the tensile modulus and strength were improved by the addition of WF. The GPE/TA(1/0.8)/WF60 had the highest tensile modulus (5.22 GPa) among the all biocomposites, and had a superior tensile strength (54.9 MPa). The tan δ peak temperature of all the WF biocomposites except SPE/QC(1/1.2)/WF biocomposites were lower than those of the corresponding cured neat resins. When QC was used as a hardener, the tan δ peak temperature considerably increased, probably due to a hydrogen bonding interaction between WF and QC. Although the maximal fiber content of MFC biocomposite (ca. 10 wt.%) was much lower than that of WF biocomposite (ca. 60 wt.%), the tensile strength and modulus increased with increasing MFC content. Tan δ peak temperature also increased a little with increasing MFC content. Among all the biocomposites, SPE/TA(1/1)/MFC10 showed the highest tensile strength (78.6 MPa) and a superior tan δ peak temperature (108 °C). As a whole, it can be said that WF biocomposites are suitable to get the materials with a higher tensile modulus, and that MFC biocomposites are suitable to get the materials with a higher tensile strength and T_g than the corresponding cured neat resins.

TABLE 4.10 Comparison of Tan δ Peak Temperature and Tensile Properties for all the Green Composites.

Sample	Fiber content (wt.%)	Epoxy/ hydroxy ratio	Curing temperature (°C) / time (h)	Tan δ peak temperature (°C)	Tensile strength (MPa)	Tensile modulus (GPa)
GPE/TA/WF	0	1/1	160/3	73	36.5	2.43
	60	1/1	160/3	61	50.7	5.06
	60	1/0.8	160/3	65	54.9	5.22
GPE/TA/MFC	10	1/1	160/3	82	64.4	2.11
	15	1/1	160/3	84	35.8	2.00
PGPE/TA/WF	0	1/1	160/3	77	63.5	2.71
	60	1/1	160/3	63	49.1	4.38
SPE/TA/MFC	0	1/1	160/2	95	60.6	1.71
	10	1/1	160/2	108	78.6	2.55
ESO/TA/MFC	0	1/1.4	210/2	58	15.1	0.46
	9	1/1.4	210/2	61	26.3	1.27

TABLE 4.10   *(Continued)*

Sample	Fiber content (wt.%)	Epoxy/ hydroxy ratio	Curing temperature (°C) / time (h)	Tan δ peak temperature (°C)	Tensile strength (MPa)	Tensile modulus (GPa)
SPE/QC/WF	0	1/1.2	170/3	86	44.7	1.36
	30	1/1.2	170/3	113	19.5	1.60
SPE/TPG/WF	0	1/1	190/3	54	28.8	1.07
	50	1/1	190/3	46	29.8	3.51
SPE/PGVNC/ WF	0	1/2.65	190/3	148	14.4	1.75
	20	1/2.65	190/3	135	19.7	2.08

KEYWORDS

- **Bio-based Epoxy Resins**
- **Bio-based Polyphenols**
- **Biocomposites**
- **Microfibrillated Cellulose**
- **Wood Flour**

REFERENCES

1. Bledzki, A. K., & Gassan, J. (1999). Composites Reinforced with Cellulose Based Fibers. *Prog. Polym. Sci.*, 24, 221–274.
2. John M. J., & Thomas, S. (2008). Biofibers and Biocomposites. *Carbohydr. Polym.* 71, 343–364.
3. Mohanty, A. K., Misra, M., & Hinrichsen, G. (2000). Biofibers, Biodegradable Polymers and Biocomposites: An Overview. *Macromol. Mater. Eng.*, 276/277, 1–24.
4. Faruk, O., Bleszki, A. K., Fink, H. P., & Sain, M. (2012). Biocomposites Reinforced with Natural Fibers: (2000–2010). *Prog. Polym. Sci.*, 37, 1552–1596.
5. Koronis, G., Silva, A., & Fontul, M. (2013). Green Composites: A Review of Adequate Materials for Automotive Applications. *Compos. Part B Eng.*, 44, 120–127.
6. Shibata, M., Ozawa, K., Teramoto, N., Yosomiya, R., & Takeishi, H. (2003). Biocomposites Made from Short Abaca Fiber and Biodegradable Polyesters. *Macromol. Mater. Eng.*, 288, 35–43.
7. Oksman, K., Skrifvars, M., & Selin, J. F. (2003). Natural Fibers as Reinforcement in Polylactic Acid (PLA) Composites. *Compos. Sci. Technol.*, 63, 1317–1324.

8. Plackett, D. (2004). Maleated Polylactide as an Interfacial Compatibilizer in Biocomposites. *J. Polym. Environ.* 12, 131–138.

9. Teramoto, N., Urata, K., Ozawa, K., & Shibata, M. (2004). Biodegradation of Aliphatic Polyester Composites Reinforced by Abaca Fiber, *Polym. Degrad. Stabl.* 86, 401–409.

10. Mathew, A. P., Oksman, K., & Sain, M. (2005). Mechanical Properties of Biodegradable Composites from Poly Lactic Acid (PLA) and Microcrystalline. *J. Appl. Polym. Sci.*, 97, 2014–2025.

11. Seriwzawa, S., Inoue, K., & Iji, M. (2006). Kenaf-Fiber-Reinforced Poly (lactic acid) Used for Electronic Products. *J. Appl. Polym. Sci.*, 100, 618–624.

12. Lee, S. H., & Wang, S. (2006). Biodegradable Polymers/Bamboo Fiber Biocomposite with Bio-Based Coupling Agent. *Compos. Part A: Appl. Sci. Manuf.*, 37, 80–91.

13. Gamstedt, E. K., Bogren, K. M., Neagu, R. C., Akerho, L. M. M., & Lindstrom, M. (2006). Dynamic-Mechanical Properties of Wood-Fiber Reinfoced Polylactide: Experimental Characterization and Micromechanical Modeling. *J. Thermoplast. Compos. Mater.* 19, 613–637.

14. Shibata, M. (2009). Poly (lactic acid)/Cellulosic Fiber Composites. In Biodegradable Polymer Blends and Composites from Renewable Resources, Yu, L. (Ed.), Wiley & Sons: Hoboken, 287–301.

15. Mohanty, A. K., Khan, M. A., & Hinrichsen, G. (2000). Surface Modification of Jute and its Influence on Performance of Biodegradable Jute-Fabric/Biopol Composites. *Compos. Sci. Technol.* 60, 1115–1124.

16. Shibata, M., Takachiyo, K., Ozawa, K., Yosomiya, R., & Takeishi, H. (2002). Biodegradable Polyester Composites Reinforced with Short Abaca Fiber. *J. Appl. Polym. Sci.*, 85, 129–138.

17. Zini, E., Focarete, M. L., Noda, I., & Scandola, M. (2007). Bio-Composite of Bacterial Poly (3-hydroxybutyrate-*co*-3-hydroxyhexanoate) Reinforced with Vegetable Fibers. *Compos. Sci. Technol.*, 67, 2085–2094.

18. Coats, E. R., Loge, F. J., Wolcott, M. P., Englund, K., & McDonald, A. G. (2008). Production of Natural Fiber Reinforced Thermoplastic Composites through the Use of Polyhydroxybutyrate-Rich Biomass. *Bioresour. Technol.*, 99, 2680–2686.

19. Mohanty, A. K., Wibowo, A., Misra, M., & Drzal, L. T. (2004). Effect of Process Engineering on the Performance of Natural Fiber Reinforced Cellulose Acetate Biocomposites. *Compos. Part A: Appl. Sci. Manuf.*, 35, 363–370.

20. Choi, J. S., Lim, S. T., Choi, H. J., Hong, S. M., Mohanty, A. K., Drzal, L. T., Misra, M., & Wibowo, A. C. (2005). Rheological, Thermal, and Morphological Characteristics of Plasticized Cellulose Acetate Composite with Natural Fibers. *Macromol. Symp.* 224, 297–308.

21. Shibata, M., & Nakai, K. (2010). Preparation and Properties of Biocomposites Composed of Bio-Based Epoxy Resin, Tannic Acid, and Microfibrillated Cellulose. *J. Polym. Sci. Part B: Polym. Phys.*, 48, 425–433.

22. Shibata, M., Teramotro, N., Takada, Y., & Yoshihara, S. (2010). Preparation and Properties of Biocomposites Composed of Glycerol-Based Epoxy Resins, Tannic Acid, and Wood Flour. *J. Appl. Polym. Sci.*, 118, 2998–3004.

23. Shibata, M., Teramotro, N., & Makino, K. (2011). Preparation and Properties of Biocomposites Composed of Epoxidized Soybean Oil, Tannic Acid, and Microfibrillated Cellulose. *J. Appl. Polym. Sci.*, 120, 273–278.

24. Shibata, M., Yoshihara, S., Yashiro, M., & Ohno, Y. (2013). Thermal and Mechanical Properties of Sorbitol-Based Epoxy Resin Cured with Quercetin and the Biocomposites with Wood Flour. *J. Appl. Polym. Sci.*, 128, 2753–2758.

25. Shibata, M., Teramotro, N., Yoshihara, S., & Itakura, Y. (2013). Preparation and Properties of Biocomposites Composed of Sorbitol-Based Epoxy Resin, Tung Oil-Pyrogallol Resin, and Wood Flour. *J. Appl. Polym. Sci.*, 129, 282–288.

26. Shimasaki, T., Yoshihara, S., & Shibata, M. (2013). Preparation and Properties of Biocomposites Composed of Sorbitol-Based Epoxy Resin, Pyrogallol-Vanillin Calixarene, and Wood Flour. *Polym. Compos.* 33, 1840–1847.

27. Haas, M. J., McAloon, A. J., Yee, W. C., & Foglia, T. A. (2006). A Process Model to Estimate Biodiesel Production Costs. *Bioresour. Technol.* 97, 671–678.

28. Katryniok, B., Paul, S., Belliére-Baca, V., Rey, P., & Dumeignil, F. (2010). Glycerol Dehydration to Acrolein in the Context of New Uses of Glycerol. *Green Chem.* 12, 2079–2098.

29. Martin, A., CHecinski, M. P., & Richter, M. (2012). Tuning of Diglycerol Yield and Isomer Distribution in Oligomerization of Glycerol Supported by DFT-Calculations. *Catal. Commun.* 25, 130–135.

30. Ayoub, M., & Abdullah, A. Z. (2013). Diglycerol Synthesis via Solvent-Free Selective Glycerol Etherification Process over Lithium-Modified Clay Catalyst. *Chem. Eng. J.* 225, 784–789.

31. Cazetta, M. L., Celligoi, M. A. P. C., Buzato, J. B., Scarmino, I. S., & Da Silva, R. S. F. (2005). Optimization Study for Sorbitol Production by *Zymomonas Mobilis* in Sugar Cane Molasses. *Process Biochem.* 40, 747–751.

32. Mishra, D. K., Lee, J. M., Chang, J. S., & Hwang, J. S. (2012). Liquid Phase Hydrogenation of D-Glucose to D-Sorbitol over the Catalyst (Ru/NiO-TiO$_2$) of Ruthenium on a NiO-Modified TiO$_2$ Support. *Catal. Today*, 185, 104–108.

33. Meffert, A., & Kluth, H. (1989). Process for the Preparation of Modified Triglycerides. *United States Patent*, 4 886–893.

34. Park, S. J., Jin, F. L., & Lee, J. R. (2004). Synthesis and Thermal Properties of Epoxidized Vegetable Oil. *Macromol. Rapid Commun.* 25, 724–727.

35. Swern, D., Billen, G. N., Findley, T. W., & Scanlan, J. T. (1945). Hydroxylation of Monounsaturated Fatty Materials with Hydrogen Peroxide. *J. Am. Chem. Soc.*, 67, 1786–1789.

36. Wheeler, D. H., White, J., & Mills, G. (1967). Dimer Acid Structures. The Thermal Dimer of Normal Linoleate, Methyl 9 cis, 12 cis Octadecadienoate. *J. Am. Oil Chem. Soc.* 44, 298–302.

37. Takada, Y., Shinbo, K., Someya, Y., & Shibata, M. (2009). Preparation and Properties of Bio-Based Epoxy Montmorillonite Nanocomposites Derived from Polyglycerol Polyglycidyl Ether and ε-Polylysine. *J. Appl. Polym. Sci.* 113, 479–484.

38. Kahar, P., Iwata, T., Hiraki, J., Park, E. Y., & Okab, E. M. (2001). Enhancement of ε-Polylysine Production by Streptomyces Albulus Strain 410 Using pH Control. *J. Biosci. Bioeng.* 91, 190–194.

39. Kahar, P., Kobayashi, K., Kojima, M., & Okabe, M. (2002). Production of ε-Polylysine in an Airlift Bioreactor. *J. Biosci. Bioeng.* 93, 274–280.

40. Kamioka, H. (1993). Use of polylysine pharmaceuticals in processed foods. *New Food Ind.*, 35, 23–31.

41. Salunkhe, D. K., Chavan, J. K., & Kadam, S. S. (1989). Dietary Tannins: Consequences and Remedies, CRC Press, Boca Raton, 10–17.

42. Harwood, M., Danielewska-Nikiel, B., Borzelleca, J. F., Flamm, G. W., Williams, G. M., & Lines, T. C. (2007). A Critical Review of the Data Related to the Safety of Quercetin and Lack of Evidence of in-vivo Toxicity, Including Lack of Genotoxic/Carcinogenic Properties. *Food Chem. Toxicol.*, 45, 2179–2205.

43. Sharma, V., & Kundu, P. P. (2006). Addition Polymers from Natural Oils A Review. *Prog. Polym. Sci.*, 31, 983–1008.

44. Yadav, R., Devi, A., Tripathi, G., & Srivastava, D. (2007). Optimization of the Process Variables for the Synthesis of Cardanol-Based Novolac-Type Phenolic Resin Using Response Surface Methodology. *Eur. Polym. J.*, 43, 3531–3537.

45. Santos, R. S. S., Souza, A. A., Paoli, M. A., & Souza, C. M. L. (2010). Cardanol–Formaldehyde Thermoset Composites Reinforced with Buriti Fibers: Preparation and Characterization. *Compos. Part A: Appl. Sci. Manuf.*, 41, 1123–1129.

46. Devi, A., & Srivastava, D. S. (2007). Studies on the Blends of Cardanol-Based Epoxidized Novolac Type Phenolic Resin and Carboxyl-Terminated Polybutadiene (CTPB), I. *Mater. Sci. Eng. A*, 458, 336–347.

47. Kim, Y. H., An, E. S., Park, S. Y., & Song, B. K. (2007). Enzymatic Epoxidation and Polymerization of Cardanol Obtained from a Renewable Resource and Curing of Epoxide-Containing Polycardanol. *J. Mol. Catal. B: Enzym.* 45, 39–44.

48. Campaner, P., D'Amico, D., Longo, L., Stifani, C., & Tarzia, A. (2009). Cardanol-Based Novolac Resins as Curing Agents of Epoxy Resins *J. Appl. Polym. Sci.* 114, 3585–3591.

49. Yadav, R., & Srivastava, D. (2009). Synthesis and Properties of Cardanol-Based Epoxidized Novolac Resins Modified with Carboxyl-Terminated Butadiene Acrylonitrile Copolymer. *J. Appl. Polym. Sci.*, 114, 1670–1681.

50. Bledzki, A. K., & Gassan, J. (1999). Composites Reinforced with Cellulose Based Fibers. *Prog. Polym. Sci.*, 24, 221–274.

51. Mohanty, A. K., Misra, M., & Hinrichsen, G. (2000). Biofibers, Biodegradable Polymers and Biocomposites: An Overview. *Macromol. Mater. Eng.*, 276/277, 1–24.

52. John, M. J., & Thomas, S. (2008). Biofibers and Biocomposites. *Carbohydr. Polym.* 71, 343–364.

53. Yang, H. S., Kim, D. J., & Kim, H. J. (2003). Rice Straw Wood Particle Composite for Sound Absorbing Wooden Construction Materials. *Bioresour Technol.*, 86, 117–121.

54. Ye, X. P., Julson, J., Kuo, M., Womac, A., & Myers, D. (2007). Properties of Medium Density Fiberboards Made from Renewable Biomass. *Bioresour. Technol.*, 98, 1077–1084.

55. Berglund, L. (2005). Cellulose-Based Nanocomposites. In *Natural Fibers, Biopolymers and Biocomposites*, Mohanty, A., Misra, M., Drzal, L. T. (Eds.), CRC Press: Boca Raton, Florida, 828.

56. Yano, H., & Narahara, S. (2004). Bio-Composites Produced from Plant Microfiber Bundles with a Nanometer Unit Web-Like Network. *J. Mater. Sci.*, 39, 1635–1638.

57. Nakagaito, A. N., & Yano, H. (2005). Novel High-Strength Biocomposites Based on Microfibrillated Cellulose Having Nano-Order-Unit Web-Like Network Structure. *Appl. Phys. A: Mater. Sci. Process*, 80, 155–159.

58. Lönnberg, H., Fogelström Berglund, M. A. S. A. S. L., Malmstöm, E., & Hult, A. (2008). Surface Grafting of Microfibrillated Cellulose with Poly (ε-caprolactone) Synthesis and Characterization *Eur. Polym. J.*, 44, 2991–2997.

59. Iwatake, A., Nogi, M., & Yano, H. (2008). Cellulose Nanofiber-Reinforced Polylactic Acid. *Compos. Sci. Technol.*, 68, 2103–2106.

60. Lu, J., Askeland, P., & Drzal, L. T. (2008). Surface Modification of Microfibrillated Cellulose for Epoxy Composite Applications. *Polymer*, 49, 1285–1296.

61. Suryanegara, L., Nakagaito, A. N., Yano, H. (2009). The Effect of Crystallization of PLA on the Thermal and Mechanical Properties of Microfibrillated Cellulose-Reinforced PLA Composites. *Compos. Sci. Technol.*, 69, 1187–1192.

62. Nakagaito, A. N., Fujimura, A., Sakai, T., Hama, Y., & Yano, H. (2009). Production of Microfibrillated Cellulose (MFC) Reinforced Polylactic Acid (PLA) Nanocomposites from Sheets Obtained by a Papermaking-Like Process. *Compos. Sci. Technol.* 69, 1293–1297.

63. Okubo, K., Fujii, T., & Thostenson, E. T. (2009). Multi-scale Hybrid Biocomposite: Processing and Mechanical Characterization of Bamboo Fiber Reinforced PLA with Microfibrillated Cellulose. *Compos. Part A: Appl. Sci. Manuf.*, 40, 469–475.

64. Nygard, P., Tanem, B. S., Karlsen, T., Brachet, P., & Leinsvang, B. (2008). Extrusion-Based Wood Fiber PP Composites: Wood Powder and Pelletized Wood Fibers. A Comparative Study. *Compos. Sci. Technol.* 68, 3418–3424.

65. Panthapulakkal, S., & Sain, M. (2007). Agro-Residue Reinforced High-Density Polyethylene Composites: Fiber Characterization and Analysis of Composite Properties. *Compos. Part A: Appl. Sci. Manuf.* 38, 1445–1454.

66. Lu, J., Askeland, P., & Drzal, L. T. (2008). Surface Modification of Microfibrillated Cellulose for Epoxy Composite Applications. *Polymer*, 49, 1285–1296.

67. Nishino, T., & Matsuda, I., & Hirano, K. (2004). All-Cellulose Composites. *Macromolecules*, 37, 7683.

68. Oyman, Z. O., Ming, W., & Linde, R. (2005). Oxidation of Drying Oils Containing Non-Conjugated and Conjugated Double Bonds Catalyzed by a Cobalt Catalyst. *Prog. Org. Coatings* 54, 198–204.

69. Blayo, A., Gandini, A., Nest, J. F. L. (2001). Chemical and Rheological Characterizations of Some Vegetable Oils Derivatives Commonly Used in Printing Inks. *Ind. Crop. Prod.*, 14, 155–167.

70. Nanaumi, K., Horiuchi, T., Nomoto, M., Inoue, M. (January 10, 1995). Method of Preparing Vegetable Oil-Modified Phenolic Resin and Laminate Produced by Using the Same. U. S. Patent 5, 380–789.

71. Ionescu, M., & Petrović, Z. S. (2011). Phenolation of Vegetable Oils. *J. Serbian Chem. Soc.*, 76, 591–606.

72. Yoshimura, Y. (1984). Reactions of Phenols with Tung Oil. *J. Appl. Polym. Sci.*, 29, 1063–1069.

73. Yoshimura, Y. (1984). Polymerization of Tung Oil by Reaction of Phenols with Tung Oil. *J. Appl. Polym. Sci.*, 29, 2735–2747.

74. Ziebarth, G., Singer, K., Gnauck, R., & Raubach, H. (1989). Untersuchungen des Reaktionsverlaufes und der Produkte der säurekatalysierten Reaktion von Phenolen mit Holzöl, 1. Phenol und Holzöl. *Angew. Makromol. Chem.*, 170, 87–102.

75. Singer, K., Ziebarth, G., Schulz, G., Gnauck, R., Raubach, H. (1989). Untersuchungen des Reaktionsverlaufes und der Produkte der säurekatalysierten Reaktion von Phenolen mit Holzöl, 2. Vergleichende Untersuchungen an Verschiedenen Phenolen. *Angew. Makromol. Chem.*, 170, 103–114.

76. Garro-Galvez, J. M., & Riedl, B. (1997). Pyrogallol-Formaldehyde Thermosetting Adhesives. *J. Appl. Polym. Sci.*, 65, 399–408.

77. Yu, S., Gu, J., Fang, W., & Fu, X. (1989). Tung Oil-Pyrogallic Acid Resin (TPA) and its Application to a Positive Photoresist. *J. Photopolym. Sci. Technol.*, 2, 51–56.

78. Yan, C., Chen, W., Chen, J., Jiang, T. Y., & Yao, Y. (2007). Microwave Irradiation Assisted Synthesis, Alkylation Reaction, and Configuration Analysis of Aryl Pyrogallol [4] arenes. *Tetrahedron*, 63, 9614–9620.

79. Beer P. D., & Tite, E. L. (1988). New Hydrophobic Host Molecules Containing Multiple Redox-Active Centres. *Tetrahedron Lett.* 29, 2349–2352.

80. Pessala, P., Keränen, J., Schutz, E., Nakari, T., Kurhu, M., Ahkola, H., Knuutinen, J. S., Herve, S., Paasivirta, J., & Ahtiainen, J. (2009). Evaluation of Biodegradation of Nonylphenol Ethoxylate and Lignin by Combining Toxicity Assessment and Chemical Characterization. *Chemosphere*, 75, 1506–1511.

81. Fernandes, T. V., Klaasse Bos, G. J., Zeeman, G., Sanders, J. P. M., & van Lier, J. B. (2009). Effects of Thermo-Chemical Pre-Treatment on Anaerobic Biodegradability and Hydrolysis of Lignocellulosic Biomass. *Biores Technol.* 100, 2575–2579.

CHAPTER 5

BIOCOMPOSITE STRUCTURES AS SOUND ABSORBER MATERIALS

NAZIRE DENIZ YILMAZ and NANCY B. POWELL

ABSTRACT

Biocomposites, provided that they are produced in porous form, that is, unconsolidated structure, act as noise control elements in a wide range of applications as they present a cost-effective, light-weight, and environmentally friendly alternative to conventional sound absorbers. This chapter presents an overview of biocomposites as rigid, porous sound absorbers. Sound absorption mechanisms that take place in porous biocomposites are explained. Methods of measuring sound absorption performance are presented. Some models to predict sound absorption capacity are described. Based on these models, factors that affect sound absorption behavior of biocomposites are given. An overview of biocomposite sound absorbers developed by researchers is reviewed. Suggestions for future research are listed.

5.1 INTRODUCTION

Advances in new technologies are often accompanied by noise pollution, besides air, soil, and water pollution.[1] To give an example, transportation is a major source of noise pollution with the ever-increasing number of more powerful and larger vehicles on the road. Vehicle passengers are affected by the noise generated by vehicles as much as the people outside the car. In addition to affecting the comfort of the passengers, it has negative effects also on the driver such as fatigue and distraction; hence, reduces the safety of the occupants. Not only the comfort and safety of passengers, but also the quality perception of the vehicle is deteriorated by unwanted sound.[2]

While progress in technologies has resulted in higher standards of living, development of advanced materials should be carried out with responsible environmental practices. In this respect, ever-tightening regulations, together with growing public awareness, force the manufacturing industry to select environmentally friendly materials and processes.[3] Biocomposites, which contain bio-based or biodegradable components, offer a viable alternative to their conventional counterparts: glassfiber based synthetic polymer composites. Automotives is a promising market segment for fiber-reinforced biocomposites with increasing product quantity, quality and variety. More than 40 automobile components including trunk and hood liners, floor mats, carpets, padding and door panels are conventionally made of fibrous structures and composites. This fact presents the significant potential for the use of bio-fibers as substitution for conventional petro-based fibers.[4] Bio-fiber based composites, that is, biocomposites have already found commercial uses by major vehicle producers since the 90s, in automobile components including door linings and panels, package shelves, and seatback linings,[5] for all of which, noise control is a requirement.

In today's conditions, environmentally friendly industrial practices cannot be carried out at the expense of quality performance.[6] Within this context, acoustic biocomposites should be able to compete with conventional sound absorbers such as glassfiber composites.[3] Glassfiber presents some critical disadvantages in terms of human and environment ecology, including being unsafe to handle and posing health risks when inhaled, in addition to being nonrecyclable.[2,7,8] Due to these afore mentioned drawbacks, bio-fibers are gaining increased attention in a variety of engineering fields to replace glassfibers.[9] Natural plant fibers offer some advantages compared to glassfibers. The specific gravity of plant fibers ($\sim$1.5 g/cc) is lower than that of glassfibers ($\sim$2.5 g/cc). If used in transportation, this, in turn, leads to lower gas consumption, that is, higher mileage per gallon and lower greenhouse gas emissions. Other advantages can be listed as lower cost, lower weight, and better heat insulation and noise reduction characteristics.[1]

There are three major methods to reduce unwanted noise. Primary methods consider modifications at noise and vibration sources. Secondary methods include alterations along the sound propagation path, and tertiary methods engage in sound receivers. Primary methods are restrained by economical and technical parameters to a great extent; while tertiary methods have to deal with each receiving person separately. This situation renders the secondary methods relatively advantageous in a number of applications.[3,10] The secondary methods concerning the control of airborne noise include the use of sound barrier and absorbers.[11] This chapter is focused primarily on sound absorbers.

Sound absorbers are porous materials. In this context, sound absorber biocomposites should be allowed to have pores as shown in Fig. 5.1. Noise is attenuated in tortuous channels of pores present in the porous materials due to viscosity and heat conductivity of the medium.[12,13] Porous sound absorbers can be classified into three groups: cellular, granular and fibrous materials.[10] Among sound absorbers, fibrous materials are promising materials for noise reduction applications. Fibrous materials are advantageous in that they absorb more sound over a broader frequency range compared to other materials.[14] Fibrous materials may also be more environmentally friendly in terms of production and after-service life practices.[1,15]

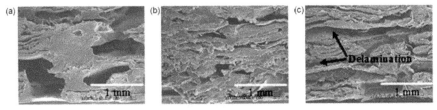

FIGURE 5.1 SEM images of unconsolidated sound absorber biocomposites from mechanically split corn husk (MSH) and PP at different MSH concentrations at (a) 35 wt.%, (b) 55 wt.%, and (c) 75 wt.%, respectively (From Huda, S.; Yang, Y. Industrial Crops and Products, 2009.[16] With permission from Elsevier).

Among parameters of porous materials, air flow resistivity, porosity, and tortuosity are the main factors that affect sound absorption.[13,17] In fibrous materials, flow resistivity increases with decreasing pore dimensions and fiber diameter. In addition to fiber size, fiber orientation,[18] web density, porosity, tortuosity,[19] mean pore size, pore size distribution, and absorber surface characteristics also affect flow resistivity.[20] Fiber reinforced composites offer some advantages compared to conventional sound absorber materials, including the economical price of the raw materials, effient thermo-processing, and lower specific weight.[21]

A thorough knowledge of sound propagation through fibrous materials is of prime importance for evaluating the noise absorption capacities of biocomposites, which are designed to serve as noise control elements in a wide range of applications. Sound absorber biocomposites are mostly produced by natural fiber nonwovens bonded by some means to produce three-dimensional rigid materials.

In this work, the term "biocomposite" refers to a material made up of distinct parts such as fibers or resin either of which is of biological origin. As the topic is related to noise control in terms of sound absorption, most of the biocomposite examples given in the open literature are in their unconsolidated form as they include pores to allow for sound wave dissipation.

This chapter presents an overview of biocomposites as rigid, porous sound absorbers. Sound absorption mechanisms that take place in porous biocomposites are explained. Methods of measuring sound absorption performance are presented. Some models to predict sound absorption capacity are described. Based on these models, factors that affect sound absorption behavior of biocomposites are given. An overview of biocomposite sound absorbers developed by researchers is reviewed. Suggestions for future research are listed.

5.2   NOISE CONTROL

Noise can be defined as any kind of undesired sound. It is the specific circumstances and attitudes of those who are exposed to the sound, which makes the distinction between noise and other sounds. However, loudness is never of secondary relevance for annoyance by noise. Sound pressure levels exceeding 85 dB may cause temporary or permanent damages in the hearing organ. Levels exceeding 60 dB can negatively affect blood circulation and metabolism. Annoyance,[10] fatigue,[22] sleep disturbance,[23] interference with speech,[22] and decrease in school and work performance[23] are some of the other unwanted effects of noise. There are regulations on noise levels in working environments to limit exposure of workers.[14] Consequently, in order to prevent the unwanted effects of noise, and to meet regulations, noise control measures have to be taken.

For sound, or noise, to be produced, three components are needed: a sound source, a medium, and a detector. The sound source is a vibrating body that produces a mechanical movement or sound wave. The medium, such as air, transfers the mechanical wave. The detector, such as an ear, detects the sound wave.[24]

ASTM describes sound absorption as *"the process of dissipating sound energy"* in ASTM C 634.[24] Every sound wave is subject to continuous reduction by certain dissipative processes whether it is propagating through air or porous materials. However, the sound dissipation during propagation through porous materials is much stronger than it is through air.[10]

5.2.1 POROUS SOUND ABSORBING MATERIALS

In porous absorbers, sound propagates through an interconnected pore network resulting in sound energy dissipation. Absorbers are only effective at mid to high frequency range, and this is the frequency range, which the ear is most sensitive to.[14]

Porous sound absorbing materials are used for the control of automotive acoustics, room acoustics, industrial noise control, and recording studio acoustics. Sound absorbers are generally used to decrease the undesirable sound reflection from hard, rigid and interior surfaces in order to reduce the reverberant noise levels.[10] Porous absorbers may be classified as cellular, granular, or fibrous materials.

Cellular materials include foams from polymers such as polyurethane and increasingly from metals, like aluminum.[10] In order to be an effective absorber, the foam should have an open-pore structure, that is, pores should be interconnected to allow for airflow from one to the other face of the foam. Although there are concerns about fire hazard and release of toxic combustion byproducts and the difficulty in recycling, polyurethane foams are widely used as sound absorbers.[2,17]

Wood-chip panels, porous concrete and pervious road surfaces are some examples of granular absorbers. Granular materials can be consolidated with binders or used in the unconsolidated form.[17] There is also an interest in producing granular absorbers from recycled materials such as used tires, waste foam,[14] and rubber particles.[25,26] An example of this is seen in Fig. 5.2.

(a)   (b)

(c)   (d)

FIGURE 5.2   Sound absorber biocomposites of pine sawdust and recycled rubber particles and polyurethane binder at different concentration and thickness levels. Reinforcement material is 50% pine sawdust and 50% recycled rubber particles for a, b and c, 70% pine sawdust and 30% recycled rubber particles for d, Polyurethane binder is 15% for a, 20% for b, c, and d; thickness 40 mm for a, b and d, 30 mm for c. (From Borlea A.; Rusu, T. U.; Ionescu, S.; Nemeş, Romanian Journal of Materials, 2012.[26] With Permission from Foundation for Materials Science and Engineering – "Serban Solacolu").

Fibrous materials may be composed of glass, mineral or organic fibers in the form of mats, boards or preformed elements. Whereas granular materials can achieve a broadband absorption of 80% at the most, fibrous absorbers can give higher absorption over a wider range of frequencies approaching 100% dissipation.[14]

Fibrous and granular absorbers may be produced by bonding the fibers or granules chemically with the use of a binder, mechanically or thermally.[10,27] It is a common practice to cover porous absorbers with a thin perforated sheet such as highly perforated panels of metal, wood or gypsum. The reason to do so is to give a more pleasant appearance, to protect them from damage and to prevent the particles harmful to humans from polluting the air and the surrounding.[10] The next section briefly explains sound absorption mechanisms that take place in porous materials.

5.2.2 SOUND ABSORPTION MECHANISMS

The absorption of sound mainly results from the dissipation of acoustic energy due to viscosity and heat conductivity of the medium. A number of dissipation mechanisms have been proposed by some authors.[10,13,28,29] Attenborough and Ver[17] cite the friction between the solid body (fiber) of the absorber and the fluid moving in it (air) as the main cause of sound attenuation. Similarly, Cox and D'antonio[14] refer to the friction due to viscosity of air as the primary cause and thermal conduction as the secondary cause of the sound energy loss. Fahy[13] explains the sound absorption phenomena on a molecular level as a combination of viscosity, thermal diffusion and relaxation processes which take place in the boundary layers next to "pore" surfaces. The next section gives information about sound absorption measuring techniques.

5.2.3 MEASURING SOUND ABSORPTION

Sound absorption coefficient of absorbing materials may be measured according to standard test methods ASTM E 1050-08, ASTM C384-04 (2011) and ASTM C 423-09a.[30,32] Normal-incidence sound absorption coefficient (NAC) may be measured according to ASTM E 1050-08, the Standard Test Method for Impedance and Absorption of Acoustical Materials Using A Tube, Two Microphones and A Digital Frequency Analysis, or ASTM C384-04(2011) Standard Test Method for Impedance and Absorption of Acoustical Materials by Impedance Tube Method. Random-incidence sound absorption coefficient may be measured according to ASTM C 423-09a Standard Test Method for Sound Absorption and Sound Absorption Coefficients by the Reverberation Room Method.

ASTM C 423-09a includes the size and construction of the sound absorber and a reverberation room. However, the test may be costly and time consuming for an early performance estimation of noise reduction capability of composites. Furthermore, it requires the test sample surface to be in massive dimensions as the minimum required area of the porous absorber specimen is 5.57 m^2.[33] While samples in small sizes are sufficient for ASTM C384-04, the testing standard requires sound absorption for each frequency to be measured separately which may take a very long time. This makes the test method ASTM E 1050-08 very feasible, taking into consideration that all data for numerous frequency points are measured simultaneously, and small dimensions of specimens are used: a specimen diameter of 100 mm is needed for a frequency range 50–1600 Hz, whereas a diameter of 29 mm is used for 500–6400 Hz range, similar to samples shown in Fig. 5.1. Due to its practicality, ASTM E 1050-08 test method will be briefly described here. In order to obtain knowledge on ASTM C384-04 and ASTM C423-09a, one may refer to the mentioned standards' specifications.

Normal-incidence sound absorption coefficient (NAC), α_n, can range from 0 (no absorption) to 1 (total absorption). The formulation defines α_n as follows;

$$\alpha_n = 1 - |\Re|^2,$$ (1)

where $\Re$ is reflection coefficient.[3] It is clear from Eq. (1) that the amount of sound absorption decreases when the reflection ratio increases. The reflection coefficient can be given as follows,

$$\Re = \frac{z_1 - Z_0}{z_1 + Z_0}.$$ (2)

In the expression above, z_1 represents the acoustical surface impedance of the porous material and Z_0 is the acoustical impedance of free air.[3] By merging Eqs. (1) and (2), the following equation can be obtained:

$$\alpha_n = 1 - \left(\frac{z_1 - Z_0}{z_1 + Z_0}\right)^2$$ (3)

Here, in the case of measurement in the impedance tube, where α_n is measured with the material backed by a hard wall as shown in Fig 5.2, in accordance with standard ASTM E 1050-08, the acoustical surface impedance of the porous material, z_1, takes the following value:

$$z_1 = Z_0 \coth(kl),$$ (4)

where k is the wave number and l is the thickness of the material. The Eq. (4) is true, provided that the hard wall backing material has a surface impedance of $z_2 = \infty$. As understood from Eq. (2), the more the difference between the impedance of free air and the surface impedance of the porous material, the greater becomes the reflection coefficient.[3] The schematic diagram of the impedance tube testing system is shown in Fig. 5.3 and the measurement system is shown in Fig. 5.4.

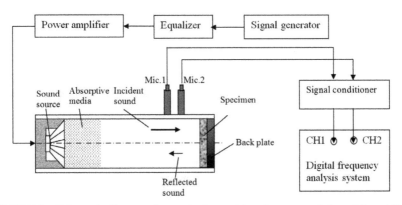

FIGURE 5.3   Schematic diagram of acoustical material testing system (adapted from ASTM E 1050-12[30] and Bruel & Kjaer[34]).

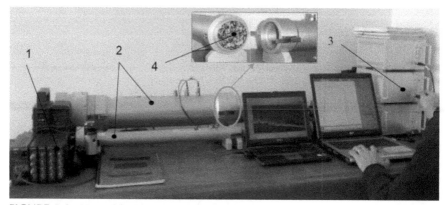

FIGURE 5.4 Normal-incidence sound absorption measurement. (1: white noise generator and acquisition system, 2: large and small diameter impedance tube, 3: signal amplifier, 4: sound absorber biocomposite sample (From Borlea A.; Rusu, T.U; Ionescu, S.; Neme, Romanian Journal of Materials, 2012.[26] With Permission from Foundation for Materials Science and Engineering – "Serban Solacolu").

A practical tool to express noise reduction capacities of sound absorbers is the noise reduction coefficient (NRC). NRC of a material is the average of the sound absorption coefficient values at 250, 500, 1000 and 2000 Hz frequencies.[32]

As shown in the aforementioned expressions, acoustic impedance, z, is a very important material parameter in terms of the noise reduction performance of porous materials. Acoustical impedance is the ratio between the sound pressure, p, and the particle vibration velocity, $\mathbf{u}_x$, as presented by Eq. (5),[10]

$$\frac{p}{u_x} = -z \tag{5}$$

Acoustical impedance determines a porous material's sound absorption capability as shown in Eqs. (2) and (3). The acoustical impedance includes two components, resistance and reactance as shown in Eq. (6),

$$z = r + i\chi, \tag{6}$$

where z stands for the impedance, r represents the resistance, which is a real quantity, and χ denominates the reactance, which is an imaginary quantity.[35]

Among the physical parameters, flow resistance is the most critical factor determining the sound absorptive properties of porous materials. A good number of researchers have used air flow resistivity to model sound absorption.[18,20,28,36] Even though different authors may use different denomination, terms as used in ASTM C522-03 Standard Test Method for Airflow Resistance of Acoustical Materials are adopted here and defined below accordingly.[37]

Flow resistance, R, in mks acoustic ohms (Pa·s·m^{-3}), is the pressure drop across a specimen divided by the volume velocity of airflow through the specimen. Specific flow resistance, r, in mks rayls (Pa·s·m^{-1}), is the product of the flow resistance of a specimen and its area. It is equivalent to the pressure difference across the specimen divided by the linear velocity of flow measured outside the specimen. Flow resistivity, r_0, in mks rayl/m (Pa·s·m^{-2}), of a homogeneous material, is the quotient of its specific flow resistance divided by its thickness. The flow resistance, R, the specific flow resistance, r, and the flow resistivity, r_0, of porous materials can be given as Eqs. (7)–(9),[3,37]

$$R = \frac{p}{u}, \tag{7}$$

$$r = \frac{p \times S}{u}, \tag{8}$$

$$r_0 = \frac{p \times S}{l \times u}, \tag{9}$$

where R is the flow resistance, r is the specific flow resistance, r_0 is the flow resistivity (denoted by σ in Delany and Bazley[28], Ξ in Mechel[18]) S is the area, in m^2, l is the thickness, in m, of the porous material, and u is the volumetric velocity of the fluid in m^3/s. Even though all these terms are concerned with steady flow and are not valid for sound with frequencies above a few hundred Hz for some sound absorbers,[13] they have been adopted by the majority of the researchers for the sake of simplicity.

The specific flow resistance is linearly related to the material thickness provided that the material is uniform. Thus, if the specific flow resistance is divided by the thickness, it will give the flow resistivity, in mks rayl/m, which is characteristic of the material independent of the thickness.[29]

5.2.4 MODELING SOUND ABSORPTION

There is an intensive amount of research and modeling efforts dedicated to understanding the sound absorption properties of fibrous structures. Early modeling efforts date back to the nineteenth century. Lord Rayleigh[38] is the first to try to explain sound propagation with a "capillary pore" model. In his model, the fibrous structure is represented by a solid structure with parallel cylindrical pores running vertical to the surface of the material. This model has been of great interest, and numerous researchers have contributed in further developing this model. The most important contributors include Morse and Bolt,[39] who introduced effective mass and flow resistance, Zwikker and Kosten,[40] who introduced structure factor and dependence on sound frequency, and Biot,[41] who introduced solid frame flexibility.

Another set of models view fibrous materials as solid cylinders in a fluid medium. These models may be a representation closer to fibrous structures; however, they require more intense mathematical computation. These models include either parallel arrays or stacks of cylinders. The models in this group also may be categorized in discrete or continuous models. Beranek,[42] Kawasima,[43] Attenborough,[44] and Mechel[18] are among the researchers who develop "solid-cylinders-in-fluid-medium" models.

Quasi-homogeneous models are the most basic models, which describe sound propagation in fibrous materials. With the integration of other models, a number of structure parameters are used to modify the plane wave equation to account for the effects of porosity, tortuosity and flow resistivity during propagation in porous materials.[18]

Due to the difficulty of representing the complex fibrous geometry with the aforementioned theoretical models, a number of empirical models have been introduced. The most important one of these models can be given as Delany and Bazley's power laws,[28] which predict the sound absorption behavior of fibrous materials with flow resistance, wave propagation number and sound frequency information.

5.2.5 MODELING SOUND ABSORPTION IN FIBROUS MATERIALS

Most of the useful models for predicting acoustical properties of porous materials can be categorized in two groups: microstructural theoretical models and phenomenological empirical models.[14]

Theoretical models generally deal with the behavior of sound waves in microstructural elements of porous material. However, it is confoundedly complicated to explain the acoustical behavior of many sound absorbers based on theoretical models due to structural and geometrical complexities of porous materials.[13] As a consequence, in order to predict sound absorption behavior of fibrous materials, empirical models with a macroscopic point of view have been developed. Nevertheless, these empirical models, which do not deal with the microstructure, may only contribute with a limited guidance during the design phase of an absorber.[14]

Among these empirical models, one has encountered general acceptance for the last several decades: the empirical model, which was presented by Delany and Bazley.[28] They investigated the sound absorption behavior of many fibrous materials with porosity close to unity for a large range of frequency. Based on their measurements, they found sound absorption indicators, the quantities of the wave number, k, and the characteristic impedance, Z_c, to mainly depend on the frequency, f (or angular frequency, ω), and the flow resistivity, r_0, of the material, as explained below.

The flow of a fluid in a circular cylinder of diameter a flowing at a steady Poiseuille flow can be given by

$$\frac{\partial p}{\partial x} = -\frac{128\mu u}{\pi a^4},$$ (10)

where p represents pressure in Pa, μ stands for dynamic viscosity of air in kg·m^{-1}·s^{-1}, u denominates the volume flow rate in m^3·s^{-1}, and a symbolizes the diameter of the channel in m.

Assuming that the number of parallel tubes is n per unit cross-sectional area in the porous material, the relation becomes,

$$\frac{\partial p}{\partial x} = -\frac{32\mu u'}{ha^2}$$ (11)

where u' stands for the average volume flow rate over cross-sectional area, and $h=(n/4)\pi a^2$, where a represents the pore diameter, gives porosity. The ratio of the pore radius to the boundary layer, η, $(\omega \rho_0 a^2/4 \mu)^{1/2}$ defines the acoustic characteristic of the sound absorber as explained by Fahy[13]. Here, can be replaced for $8 \mu/r_0 h$ in this ratio, η, using Eq. (11) to give

$$\eta = \sqrt{\frac{8\omega\rho_0}{hr_0}},$$ (12)

where ω, is the angular frequency, ρ_0 is the density of air. The porosity, h, can be ignored, as $h^{1/2}\approx1$ for most sound absorbers. This gives the nondimensional parameter $(\omega\rho_0/r_0)$ or:[13]

$$X = \frac{\rho_0 f}{r_0},$$ (13)

where X is the dimensionless sound absorber variable, which is denoted by E by some other researchers.[13,18]

Delany and Bazley[28] obtained sound absorption indicators, the values of the wave number, k, and the characteristic impedance, Z_c, through the following expressions:

$$Z_c = \rho_0 c_0 \left[1 + 0.0571X^{-0.754} - i0.087X^{-0.732}\right],$$ (14)

$$k = \frac{\omega}{c_0}\left[1 + 0.0978X^{-0.700} - i0.189X^{-0.595}\right],$$ (15)

where c_0 is the sound propagation velocity in air. X became a universal descriptor of fibrous sound absorbers as it collapsed all the sound absorption data for the 70 different fibrous materials that Delany and Bazley[28] measured.[18] The boundary suggested for the validity of the model is

$$0.01 < X < 1.0. \tag{16}$$

At low frequencies, Delany and Bazley's[28] formulas produce unfeasible results such as negative values for the real part of surface impedance of a hard-backed porous layer.[17] Several researchers modify these formulas to give more accurate predictions, such as Mechel[18]. Mechel's[18] modified formulas are as follows:

$$Z_c = \rho \, {}^\circ C_0 (1 + 0.081 X^{-0.699}) - i0.191 X^{-0.556}, \tag{17}$$

$$k = (1 + 0.136 X^{-0.641}) - i0.322 X^{-0.502} \tag{18}$$

for $X < 0.025$.

$$Z_c = \rho \, {}^\circ C_0 (1 + 0.0563 X^{-0.725}) - i0.127 X^{-0.655}, \tag{19}$$

$$k = \frac{\omega}{c_0} (1 + 0.103 X^{-0.716}) - i0.322 X^{-0.663} \tag{20}$$

for $X > 0.025$.

Yilmaz[3] developed a model with the frequency, f, thickness, l, and air flow resistivity, r_0, variables, presented in Eq. (21). The investigated materials included numerous fibrous composite materials from polypropylene, poly (lactic acid), hemp fibers and glassfibers in single, or multifiber, untreated, compressed, alkalized, heated and different layer-sequenced forms. The model estimates gave a goodness-of-fit, R^2, value of 0.97 and the boundary suggested for the validity of the model is a basis weight of 1.13 to 2.36 kg· m^{-2}, fiber diameter of 16.3×10^{-6} to 33.3×10^{-6} m, a material thickness of 3.9×10^{-3} to 1.31×10^{-2} m, porosity of 0.62 to 0.92, and a frequency range of 500 Hz – 5 kHz.

$$\alpha_n = sin(-5.27 \times 10^{-1} + 2.54 \times 10^{-4} f + 1.72 \times 10^{-6} r_0 + 3.54 \times 10^4 l) \tag{21}$$

The theoretical and empirical models provide guidance during the design of optimum sound absorbers in deciding which characteristics should be manipulated and how. Based on theoretical and empirical models, the factors, which directly or indirectly affect sound propagation through fibrous materials can be given as fiber properties (fiber size, fiber shape, etc.), material properties (flow resistance, density, thickness, and so on), process parameters and sound frequency. The following section investigates the factors that affect sound absorption.

5.3   THE FACTORS THAT AFFECT SOUND ABSORPTION

The first major step in noise control is to determine the source of the noise before designing the sound absorber. To give an automotive-related example where biocomposites find applications; the engine, the tire-road interaction and the wind are the three major factors that affect the acoustics of the passenger department of a vehicle. Hence, the floor coverings, the headliner, and the hood insulation are the critical contributions to the acoustic performance of vehicles.[2] The other most important characteristic to be sought in the design of absorption materials are the noise absorption capacities in the audible frequency range of interest. For example, medium range (1200–4000 Hz) is of interest for the interior of passenger vehicles.[45] Also, cost effectiveness and other conditions should be considered including durability in hostile environments such as high temperature, contamination, and high speed turbulent flow, etc.[17] Recyclability, lightweight, thermal comfort and contribution to passive safety systems are also desirable characteristics of noise absorbers in vehicles.[2]

Different researchers give importance to different material parameters as influencers of sound absorption. Among the material parameters, which are applicable to porous absorbers, Bies and Hansen[29] cite only the flow resistance/resistivity, Cox and D'antonio[14] consider flow resistivity and porosity, whereas Fahy,[13] and Attenborough and Ver[17] refer to porosity, and structure factor (tortuosity) in addition to flow resistance/resistivity as the primary parameters that affect the sound absorption properties.

Several authors give various parameters as factors affecting the sound absorption of fibrous structures. Cox and D'antonio[14] report that sound absorption efficiency of fibrous structures can be achieved by manipulating:

- Material density
- Fiber composition
- Fiber orientation
- Fiber dimensions.

Banks-Lee et al.[36] reported that the mass per unit area, thickness and porosity of the material and fiber fineness to be of significant importance to the airflow resistance and sound absorption of needled fibrous materials.

The factors affecting sound absorption behavior of fibrous materials, biocomposites in particular, are classified as fiber parameters, macroscopic physical parameters, process parameters of the porous material production and treatments as shown in Table 5.1 and explained below. It is important to note, beforehand, that the parameters to be described are not independent from each other.[3]

TABLE 5.1   Parameters that Control Sound Absorption of Fibrous Structures

Category	Sub-category	Parameter
Material parameters	Fiber parameters	Fiber type
		Fiber size
		Fiber shape

TABLE 5.1   *(Continued)*

Category	Sub-category	Parameter
	Macroscopic Physical Parameters	Flow resistivity and resistance
		Impedance
		Thickness
		Density
		Porosity
		Tortuosity
		Fiber orientation
		Composition
		Air gap
Process parameters	Production parameters	Web forming
		Web bonding technique
	Treatment parameters	Chemical treatment
		Physical treatment

Source: Yilmaz (2009).[3]

5.3.1   FIBER PARAMETERS

5.3.1.1   FIBER TYPE

A wide variety of fibers are used in biocomposite sound absorbers, including conventional synthetic fibers, conventional plant fibers, exotic plant fibers, animal fibers, reclaimed fibers of chemical and natural origin and engineered compostable fibers like poly (lactic acid), as given in Table 5.2. The characteristics of the constituent fibers also have an important effect on sound absorption.

The effect of fiber type on sound absorption is hard to detect as it is often accompanied by differences in fiber size and shape as seen in Figs. 5.5 and 5.6. Fiber type determines the relationship between the fiber size and flow resistivity.[14] The following relationships have been reported:

$$r_{0,g} = \frac{3.2\mu(1-h)^{1.42}}{a^2}, r_{0,P} = \frac{7.58\rho^{1.404}}{10^{12}a^2}, r_{0,W} = \frac{490\rho^{1.61}}{10^6 a} \tag{22}$$

where $r_{0,g}$, $r_{0,p}$, and $r_{0,w}$ represent flow resistivity values of glassfiber, polyester[14] and wool[46] webs in mks rayl/m, respectively; μ stands for viscosity of air which is 1.84×10^{-5} kg· m^{-1} s^{-1}, h denominates porosity, a is fiber radius in m, and ρ represents density of the fiber mat in kg· m^{-3}.

TABLE 5.2 Summary of Research on Biocomposite Sound Absorbers in Reverse Chronological Order

Fiber/material	Production method	Investigated parameters	Measured parameters	Frequency range (Hz)	Thickness (mm)	Fiber diameter (μm)	Max. NAC	Publication
PP, PLA, glass-fiber, hemp	Air laying, needle-punching, thermal treatment	Heat treatment	NAC, air flow permeability	500–5000	3.90–13.1	9–42	0.99	Yilmaz et al.[8]
PES, PP, cotton, wool, jute, rice straw, sawdust, jute,	Needle-punching	Thickness, cover plate, air gap, composition	NAC	100–6300	2.53–22.6	N.S.	0.99	Seddeq et al.[56]
PP, PLA, glass-fiber, hemp	Air laying, needle-punching	Compression	NAC, air flow permeability	500–5000	7.91–13.1	9–42	0.99	Yilmaz et al.[9]
PP, hemp	Air laying, needle-punching, alkalization	Alkalization	NAC, air flow permeability	500–6400	10.61–12.53	32–42	0.99	Yilmaz et al.[20]
PP, Bamboo strips	Laying stacking compression molding	Blend ratio, thickness, fiber type, fiber orientation	NAC, Noise Reduction Coefficient (NR)	0–3000	1.16–10.12	75	0.80	Huda et al.[48]
Coir fiber, wood particle debris, phenolic resin	Needle-punching, resin bonding	Blend ratio, needle-punching, fiber placement	NAC, Noise Reduction Coefficient (NR)	125–4000	N.S	N.S	0.99	Yao et al.[57]
PU binder, pine sawdust, recycled rubber	Resin bonding	Blending ratio, thickness, material type,	NAC	50–10,000	20–40	1–4* mm	0.92	Borlea et al.[26]
PP, PLA, glass-fiber, hemp	Air laying, needle-punching	Porosity, fiber type and size, layer sequence	NAC, air flow permeability	500–6400	11.45–12.68	9–42	0.99	Yilmaz et al.[58]

TABLE 5.2 (Continued)

Fiber/material	Production method	Investigated parameters	Measured parameters	Frequency range (Hz)	Thickness (mm)	Fiber diameter (μm)	Max. NAC	Publication
PES, formaldehyde, recycled PS, woodchip, furnace slag, municipal waste, power plant ash	Granulation	Resin bonding	Material type	10–3,150	2.5–10* mm	1–2* mm	0.91	Bratu et al.[51]
Flax tow	Grinding, washing, microwave, molding	Grinding, microwave treating, molding, thickness	NAC	100–4000	2–10	N.S	0.82	El Hajj et al.[54]
Recycled pulp, luffa fibers, yarn waste	Wet laying, cold pressing	Blend ratio, material type	NAC	500–4800	N.S	N.S	0.13	Karademir et al.[52]
PP, Jute, PES	Carding, needle-punching	Material density, number of layers	Sound insulation	N/S	2.6–51	8.7	N/A	Sengupta[59]
Jute, bamboo, banana, jute	Carding, needle-punching	Fiber type	NAC	100–1600	4.9–6.4	N.S	0.20	Thilagtvath[21]
PP, hemp, rapeseed straw, beech and flax	Extrusion granulating, compression molding	Fiber type	NAC	1000–6500	N/S	N.S	0.32	Markiewicz et al.[49]
PP, mechanically split corn husks, jute	Spunbonding, molding	Fiber type, blend ratio,	NAC	300–3000	3.2	1.3	0.42	Huda and Yang[60]

TABLE 5.2 *(Continued)*

Fiber/material	Production method	Investigated parameters	Measured parameters	Frequency range (Hz)	Thickness (mm)	Fiber diameter (μm)	Max. NAC	Publication
PP, PLA, glass-fiber, hemp	Air laying, needle-punching, thermal treatment	Heat treatment, needle-punching	Sound transmission loss, NAC	200–6400	9–42	N.S	0.99	Yilmaz et al.[1]
PP, PLA, glass-fiber, hemp	Air laying, needle-punching, thermal treatment, alkalization	Alkalization, heat treatment, needle-punching	Sound transmission loss, NAC	200–6400	13–94	N.S	0.99	Yilmaz et al.[4]
PP, corn husk fiber, jute	Carding, wet laying, thermal bonding	Enzyme treatment, fiber type	NAC	800–3000	3.2	18.8	0.60	Huda and Yang[55]
PP, chicken quill, jute	Grounding, carding, molding	Thickness, blend ratio	NAC	800–3000	4.4	N.S	0.51	Huda and Yang[53]
Kenaf, jute, waste cotton, recycled PES, flax, off-quality PP	Carding, air laying, needle-punching	Production method, fiber type	NAC	500–3200	7.36–19.03	N.S	0.99	Parikh et al.[22]
PP, cotton, hemp, flax	N.S.	Fiber type, fiber size,	NAC	400–5000	4.1–30	N.S	0.95	Nick et al.[45]

N/A: Not applicable, N.S: Not stated, *: granule diameter, PES: polyester, PP: polypropylene, PU: polyurethane, PS: polystyrene, PLA: polylactic acid.

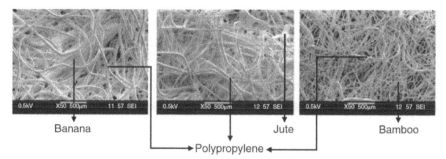

FIGURE 5.5   SEM images of needle-punched composites from PP/Banana, PP/jute and PP/Bamboo fibers (From Thilagavathi, G.; Pradeep, E.; Kannaian, T.; Sasikala, L. Journal of Industrial Textiles, 2010. [21] With permission from Sage Publications).

FIGURE 5.6   Surface morphology of plant fibers (a) hemp fiber –magnification 500x (Yilmaz et al., 2012: DOI: 10.1007/s12221–012–0915–0)[9], (b) flax fiber –magnification 500x (From El Hajj, N.; Mboumba-Mamboundou, B.; Deilly, R.-M.; Aboura, Z.; Benzeggagh, M.; Queneudec. Industrial Crops and Products, 2010.[54] With Permission from Elsevier), (c) corn husk fiber –magnification not stated (From Huda, S.; Yang, Y. Macromoleculer Materials, 2008.[55] With permission from Wiley VCH.).

Studies related to sound absorption properties of biocomposites include structures made up of different fibers. Surface properties of fibers and their cross-sections also play an important role. Accordingly, Nick et al.[45] found greater absorption for cotton-polypropylene (PP) blend fibrous material for automotive applications compared to the flax-polypropylene and hemp-polypropylene blends. This was probably due to the inherent superior fineness of cotton fibers as compared to flax and hemp.

Jayaraman et al.[47] examined the effect of kenaf fiber inclusion, which is a natural bast fiber, on the absorption of sound in fibrous absorbers. The addition of kenaf had a negative effect on the noise reduction performance compared to polyester and reclaimed polyester fibers, however, this effect is less pronounced in high frequencies. This negative effect may also be due to natural coarseness of kenaf fiber compared to synthetic fibers.

Parikh et al.[22] developed composites in various weight ratios of natural and synthetic fibers including kenaf, jute, waste cotton, and flax with recycled polyester and off-quality polypropylene and compared to absorbers of conventional fibers, that is, 70% polyester and 30% polypropylene. They reported that each of the natural fibers contributed to noise reduction because of their absorptive properties in comparison with the conventional material. Furthermore, adding a soft cotton underpad was found to greatly enhance the sound absorption properties of the nonwoven floor coverings.

Huda et al.[48] produced unconsolidated light-weight (0.312 g/cm^3) composites by laying fine bamboo strips on a PP web and by a subsequent compression molding process. They reported better mechanical and noise reduction capabilities for the mentioned composites compared to jute-based composites.

Markiewicz et al.[49] produced composites including PP and lignocellulosic fillers and measured their sound absorption performance in the 1000–6500 Hz frequency range. They reported the hemp filler addition allowed for significant increase in noise reduction over 3000 Hz, whereas rapeseed straw, beech and flax filler added to PP suppressed sound in the 3000–4000 Hz range.

Brencis et al.[50] presented a research study with an aim to develop a sound absorber from gypsum foam reinforced by fibrous hemp. They claimed that the gypsum, Gypsies rock, a local resource in Latvia, can have performance characteristics comparable to other state-of-the-art thermal and sound insulation materials. Additionally, gypsum poses an important fire-resistance characteristic. Fragility, which is the disadvantage of the gypsum material, claimed to be avoided with the use of plant fibers, such as hemp, as a reinforcement element.

Bratu et al.[51] studied composite materials including pellets from plastic bottles, sawdust, and ash from plant and sterile municipal wastes in a polymer type organic matrix in different blend ratios. The effects of the blend ratio and the type of the waste material on the sound absorption performance were investigated. They reported that use of sawdust and woodchips were advantageous in terms of noise reduction compared to the other recycled materials.

Karademir et al.[52] prepared biocomposites through a wet laying process from recycled corrugated boards with addition of 30% yarn waste and 15% luffa fibers. They found that the addition of luffa fibers and yarn waste led to an increase in sound absorption together with an increase in air permeability at the expense of tensile strength.

Among the very few examples of biocomposites containing materials of animal origin, Huda and Yang examined the sound absorption performance of ground chicken quill based PP composite and compared it with jute-based PP composites.[53] They reported that the chicken quill based composites resulted in better sound absorption performance in 500–2200 Hz frequency range as shown in Fig. 5.7.

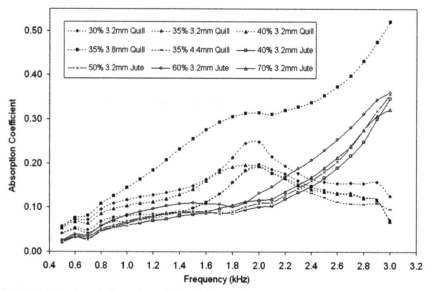

FIGURE 5.7   Sound absorption of PP-chicken quill composites compared to PP- jute mats at different thicknesses and blend ratios (From Huda, S.; Yang, Y. Composites Science and Technology, 2008.[53] With Permission from Elsevier.).

5.3.1.2   FIBER SIZE

Finer fibers lead to more effective sound absorption due to the greater number of fibers per volume, more contact area, and more tortuous channels.[61] Furthermore, the presence of finer fibers decreases the chances of pore connectivity[62] and increases the flow resistivity. Additionally, finer fibers can be vibrated easily compared to coarser ones; and cause acoustic energy dissipation.[61] Jayaraman et al.,[61] Lee and Joo,[62] and Koizumi et al.[63] reported higher sound absorption with finer fibers. Koizumi et al.[63] found substantial increase in absorption properties with incorporation of micro denier fibers. Technical fibers generally have Poisson distribution of fiber diameter which may be taken into account during modeling.[18] One disadvantage with plant fibers is that they generally are coarser than petroleum-based fibers due to the fact that a technical plant fiber is comprised of a great number of elementary fibers as seen in Fig 5.6.

5.3.1.3   FIBER SHAPE

Different fiber shapes result in different surface areas; and different surface areas, in turn, lead to different viscous and thermal effects.[14] Irregular cross section of fibers also increases the sound absorption due to increased surface area.[61] Greater fiber surface area allows greater sound absorption friction between fibers and air.[64] Watanabe et al.[64] and Narang et al.[65] report a direct correlation between sound absorption and fiber surface area. In the frequency range 1125–5000 Hz, fibers with serrated cross sections absorb more sound compared to ones with a round cross sectional area. Hur et al.[66] explain that sound absorption increases with specific surface area of fiber with an increase of relative density and friction of the pore wall. Accordingly, Taşcan and Vaughn[67] report greater sound transmission loss in fibrous structures of polyester fibers with deep grooves compared to that made from round polyester fibers.

In this regard, biocomposites are advantageous. Plant fiber component of biocomposites generally has increased surface areas caused by their inherent irregular shapes as shown in Fig 5.6 in contrast to smooth surfaces of man-made fibers. This might be the reason for the general acceptance that bio-fibers act as good noise reduction elements.

5.3.2   MACROSCOPIC PHYSICAL PARAMETERS OF BIOCOMPOSITES

The topological complexity of most acoustical materials necessitates the characterization of these materials based on gross properties.[13] The physical parameters which affect the acoustic properties of fibrous structures include flow resistance, thickness, porosity, weight of the structure, tortuosity, surface impedance, composition, the geometrical shape of the material, and the presence of air gap or cover screen.[61]

5.3.2.1   FLOW RESISTANCE AND FLOW RESISTIVITY

As explained above in Measuring Sound Absorption section, flow resistance is the most critical factor that determines the sound absorptive properties of porous materials. Ingard[35] introduces the normalized specific flow resistance as follows:

$$r_n = \frac{r}{\rho c}, \tag{23}$$

where r_n is the normalized specific flow resistance, ρ is density of the fluid, and c is the speed of sound, and ρc is the impedance of the fluid. The S.I. unit of specific flow resistance, r, is $kg \cdot m^{-2} \cdot s^{-1}$, of the density, ρ, is $kg \cdot m^{-3}$, and of the speed of the sound, c, is $m \cdot s^{-1}$; thus, the normalized specific steady flow resistance, r_n, of the porous material, is nondimensional.[35]

There is a close relationship between the flow resistivity and the density of the porous web. Flow resistivity increases in high densities, as air permeability, which is the reciprocal of air flow resistance, reduces when fiber packing density increases with an increase in the pressure drop.[68] Ballagh[46] found the following relationship for woolen fibrous webs:

$$r_0 = 16\rho_w^{1.61} \tag{24}$$

where r_0 is the flow resistivity in mks rayls/m and ρ_w is the density of web in kg· m^{-3}.

Yilmaz[3] generated an empirical model to explain how the fiber diameter and web density effect r_0, based on samples including layered webs of untreated, compressed and alkalized biocomposites of hemp, poly (lactic acid) (PLA), polypropylene (PP) and glass fibers in single or multifiber versions with a goodness-of-fit, R^2 value of 0.90:

$$r_0 = 15,651 + \frac{5.39 \times 10^{-4}}{\hat{a}^2 \rho_f^{1.6}} \rho_w^{1.6}, \tag{25}$$

where is the weighted root mean square fiber diameter of the composite in meters, ρ_f is the weighted average fiber density and, ρ_w is the density of the composite structure. As seen from the expression, air flow resistivity is inversely proportional to the square of the fiber size and is proportional to the quotient between the density of the biocomposite to its constituent fibers. The boundary suggested for the validity of the model is a basis weight of 1.13 to 1.57 kg·m^{-2}, fiber diameter of 16.3×10^{-6} to 33.3×10^{-6} m, a material thickness of 7.91×10^{-3} to 1.31×10^{-2} m, and porosity of 0.84 to 0.92. The agreement of the theoretical predictions with the experimental data is given in Fig. 5.8.

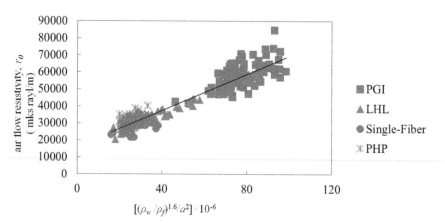

FIGURE 5.8 Comparison of statistical model estimates for Eq. (25) vs. actual values for air flow resistivity values of PGI: Three-layered PP-glassfiber intimate blend webs, LHL: PLA/Hemp/PLA sandwich structure, PHP: PP/Hemp/PP sandwich structure, Single-fiber: Three layered single-fiber webs of PP and PLA.[3]

The flow resistivity is proportional to the coefficient of shear viscosity, μ, of the fluid (i.e., air) involved and is reported to be inversely proportional to the square of the pore size of the material when the microstructure is considered. As the material is fibrous, the flow resistivity increases with the decreasing fiber size in accordance with Eq. (25).[35] However, the relationship between the fiber size and the flow resistivity depends on the fiber type.[14] Other than density and fiber diameter, flow resistivity is determined also by porosity, tortuosity, pore size distribution[19] and fiber orientation.[18]

Sound absorption increases with increasing flow resistivity up to a point then starts to decrease for higher resistivity values. If the resistivity is too low, there is a small amount of fibers to interfere with the sound wave to cause energy loss,[69] whereas when the flow resistance is too high, the material acts as a reflector rather than an absorber. High-frequency sound absorption requires relatively lower flow resistance.[68] Consequently, the absorption coefficient curve shifts to a lower frequency range as the flow resistance increases.

There is an optimum flow resistance for each frequency range. However, in practice, the specific flow resistance, r, of an absorber is typically in the range of 2 pc, as reported by Ingard.[35] Fahy[13] recommends a specific flow resistance of 3 pc, whereas Attenborough and Ver[17] report the range of 1–2 pc to be an optimal choice. Ballagh[46] finds a specific flow resistance of 1000 mks rayls, which is approximately 2.5 pc, gives optimum absorption of wool absorber. Fahy[13] reports that most of the sound absorbers have flow resistivity values between 2×10^3 to 2×10^5 Pa·s·m^{-2} or mks rayl/m.

5.3.2.2 THICKNESS

The resistance to air flow is achieved through the depth of the material. The thicker the material, the higher is the sound absorption. Generally, when the thickness of the material matches one tenth of the wavelengths of the incidence sound, effective sound absorption is achieved. At a resonance frequency of one-quarter wavelength of the incidence sound, peak sound absorption occurs. The necessity for the significant thickness to wavelength ratio renders porous materials inefficient sound absorbers at low frequency due to their greater wavelengths. This ratio is extremely small at low frequencies as the wavelength may reach values that are in the order of 10 meters.[14] To give some examples, the wavelength of 10-Hz sound is 34.3 m and 100-Hz sound is 3.43 m, whereas the wavelength of 1000-Hz sound is 34.3 cm.

Parikh et al.[22] tested velour-surface fibrous materials made of recycled fibers and PP/PES bicomponent fibers to see the effects of thickness, mass per area, and the production method on the sound absorption. The results showed that increase in thickness and mass per area leads to an increase in NAC as expected. A decrease in sound absorption accompanying the decrease in thickness was also reported by Jayaraman et al.,[47] el Hajj et al.[54] and Yilmaz et al.[20]

NAC (normal incidence sound absorption coefficient) increases as the thickness increases. However, for every frequency there is an upper limit of the thickness, a quarter of a wavelength, beyond which NAC decreases slightly.[17] This fact is contrary to the belief that NAC should improve continuously as the thickness increases.[3] As shown in Fig. 5.9, with increasing thickness, the frequency where the peak absorption takes place decreases, which is often desirable.[70]

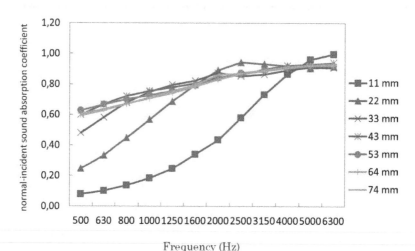

FIGURE 5.9   Effect of thickness on NAC of a needle-punched fibrous web of polypropylene.[3]

5.3.2.3   POROSITY

In order for a material to absorb sound, it must be porous, so that the sound waves can move through the material.[17] Porosity is the ratio of open space volume to the total volume of the porous material. Only connected pores, which are accessible to air flow, should be included in porosity calculation.[14]

The gravimetric measurement of the porosity requires the knowledge of the volume of the porous material and the density of the fibers. The porous material is weighed and fabric density is calculated. The porosity is calculated by using the following equation:

$$h = 1 - \frac{\rho_w}{\rho_f} \tag{26}$$

where h is porosity, r_w is density of fibrous material, r_f is the density of the fiber. Reciprocal of porosity, massivity, Ψ, is the ratio of fiber volume to the total volume of the porous material:[18]

$$\Psi = \frac{\rho_w}{\rho_f} = 1 - h \qquad (27)$$

Another method of measuring porosity involves saturating the porous materials with a kind of fluid such as water or mercury and determining the porosity from the relative weights of saturated and unsaturated samples. The disadvantage with this method is the deformation of the pore structure due to the introduction of the fluid.[17]

A dry method of porosity determination is based on the measurement of the change in pressure within a container, which contains the porous material, with addition of a small volume of air. It is advantageous as it only measures the connected pores, through which sound propagates.[17]

The range of porosity is higher than 0.95 for mineral and glass wools and porous plastic foams. The porosity values of fibrous mats have been reported to range between 0.83 and 0.95 by Cox and D'antonio.[14] Porosity determines the ratio between the average particle velocity in the channels and on the cross-sectional average particle velocity of the absorber material.[13] A number of models use the assumption that the porosity is close to unity including that of Delany and Bazley.[28] The effect of porosity should be included in air flow resistivity terms that takes a significant role in predictive models as the quotient between the density of the biocomposite to its constituent fibers, , given in Eq. (25) corresponds to massivity, Ψ, which is porosity subtracted from unity.

5.3.2.4 TORTUOSITY

The ratio between the passage ways through the pores and the thickness of the porous material is called tortuosity.[61] Tortuosity gives the extent of the deviation of the pores from the normal of material thickness.[71] The sound absorption performance of porous materials is generally affected by tortuosity which does not allow the sound waves to follow straight paths.[72] The air that is forced to follow a tortuous path suffers accelerations which cause momentum transfer from air to the material.[35] The value of tortuosity determines the high frequency behavior of sound absorbing porous materials.[61]

Tortuosity can be measured by an electrical conductivity method.[17] In this method, the porous material is saturated with a conducting fluid such as a brine solution, and the electrical resistivity of the saturated porous material and the solution alone is measured and compared.

$$F = \frac{R_{es}}{R_{ef}}, \qquad (28)$$

where F is the formation factor, and R_{ef} and R_{es} are the electrical resistance values of the fluid alone and the porous material saturated with fluid, respectively, which have the same dimensions, in ohm units. As the electrical resistance is proportional to the length, L, and inversely proportional to the area, S, of the conducting material the ratio between the resistances; formation factor, F, should be tortuosity, T_s, divided by porosity, h, as shown in the following equations:[17]

$$R \propto \frac{L}{S},\tag{29}$$

$$F = \frac{T_s}{h}\tag{30}$$

Tortuosity contributes to the 'structure factor,' Γ, along with the effect of inner structure of porous materials.[40] The structure factor, Γ, (χ in [18]) accounts for an increase in the inertial mass density of the air. In other words, the irregularity of the structure causes an addition of induced mass to the density of air. For fibrous materials, structure factor, Γ, is generally between 1.2 and 2.3,[13] but it is assumed to be unity in modeling studies by numerous researchers including Allard and Champoux,[73] Ballagh[46] and Cox and D'antonio.[14] High structure factor leads to low propagation speed, which increases the effective thickness of the absorber; and thus, used in sound absorber design.[13] Biocomposites are advantageous in terms of tortuosity due to the fact that the irregular shapes of constituent plant fibers prevent a straight flow of sound waves through the thickness of the absorbent material.

5.3.2.5   FIBER ORIENTATION

Fibers in fibrous structures have some level of orientation in machine- or cross-direction and parallel to material surface. Hence, fibrous materials are inherently anisotropic and the propagation constant, k, and surface impedance, z_1, changes according to the direction the sound waves propagate. Sound waves must be permitted to enter the material in order for the material to be able to absorb the sound rather than reflect it.[17] Fibrous mats with fibers arranged vertically to the surface, such as needle-punched products, allow sound to enter the material.[69] Accordingly, fibers arranged perpendicular to the material surface may be placed to face the sound source.

5.3.2.6   COMPOSITION

The use of multilayer absorbers has become increasingly important in noise reduction.[74] It is possible to tailor the materials for maximum absorption for the broadest frequency range by layering them.[2]

NAC values for various frequency ranges can be enhanced by changing the density and composition of the fibrous structure. This area needs more investigation on the inner structure of multilayer absorbers.[74] Multi-layer absorbers achieve higher sound absorption than the mono-layer absorbers with the same thickness.[17] Ingard[35] reported a significant drop in the critical frequency above which maximum absorption was achieved when the fibrous absorber consisted of layers with different flow resistivities. The critical frequency was significantly higher for a single-layer material with the same total thickness and flow resistance.

Ingard[35] reported that the sound absorption is greater when the flow resistivity, r_0, increases from the surface toward the rigid backing except for low frequencies below 150 Hz. This is in agreement with the statement that the sound waves should be able to penetrate the porous absorber in order for sound attenuation to occur. Ackermann et al.[75] reported that the smoothness and evenness of the surface of the absorbers facing the flow keeps frictional losses low and this allows higher sound absorption.

Yilmaz et al.[4] investigated the effect of sequence of two layers of poly (lactic acid) (PLA) and one layer of hemp fibrous webs on NAC values. They found that the biocomposite with the hemp layer facing the sound source higher sound absorption than the fabric with the hemp layer facing back plate or the hemp layer between the other two PLA layers. In fact, the composites including hemp layers facing the sound source or the back plate were actually the same materials, but just reversed. Hemp layer had greater pore sizes due to the coarser hemp fibers compared to man-made PLA fiber layer, so they allowed more sound waves to penetrate into the material. Consequently, it is possible to achieve better sound absorption only by changing the direction of the same composite absorber.

Using a different approach, Atalla et al.[76] produced nonhomogeneous absorbers from rock wool and glass fiber which include patches with different flow resistivity values in the same single layers. They found that surrounding patches interact together and better performances are obtained compared to homogeneous materials. This approach may also be applied to biocomposites.

5.3.2.7   AIR GAP

It is a common practice to leave an air gap between the absorber material and the hard backing wall in applications such as suspended acoustical ceilings, although the practical use is relatively limited.[68] The highest NAC will be achieved when the distance between the absorber and the wall is odd multiples of a quarter wavelengths for the sound frequency of concern. This phenomenon is due to the fact that the incident waves and the reflected waves will have a phase difference of 180°. Conversely, when the air gap is a multiple of half wavelengths, the airspace becomes totally ineffective as the incident and the reflected waves will be in phase.[17] Air gap behind the material increases NAC substantially in the low frequency range at the cost of high frequencies.[68] This effect is seen in Fig. 5.10. Jayaraman[61] found that the air gap caused an increase in the absorption in the frequency range between 500 and 4500 Hz. He did not find a significant difference in NAC values for fibrous mats with 5-mm air gap compared to those with 10-mm air gap, whereas the maximum peak is at a lower frequency, that is, a greater quarter wavelength for the greater depth of the gap. There are useful design charts in Attenborough and Ver[17] to predict the NAC of absorbers separated from the wall with an air gap with the knowledge of air gap distance, flow resistivity of the absorber and sound frequency.

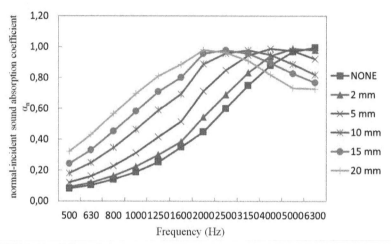

FIGURE 5.10   Effect of air gap on normal-incidence sound absorption of needle-punched polypropylene mat with 11 mm thickness.[3]

5.3.2.8   COVER SCREEN

Various forms of cover screens are used both to protect the porous absorbers from damage[77] such as loss of fibers,[14] to fine-tune their absorptive performance to meet practical demands,[13] and to give a more esthetical appearance.[10] Cover screens may be in the form of mineral wool felt sprayed on plastic, steel wool, mineral wool or glass fiber cloth; wire mesh cloth; or thin perforated metal. They are characterized by their specific flow resistance, r, and their mass per unit area.[17] The addition of a cover screen increases the sound absorption at low frequencies substantially, if it is not in contact with the porous material and able to vibrate freely. However, this may be at the cost of attenuation at high frequencies if flow resistance, r, of the screen is higher than 1 ρc.[35] If the cover screen has very low porosity[77] or is in contact with the porous absorber, the result will be decreased high frequency absorption with unaltered low frequency absorption[14]. The thinness and the lightness of the film increase the absorption.[75] Rebillard et al.[78] reported that a heavy film behaves like it is in contact with the porous absorber whether or not it is so. Jayaraman et al.[47] found the presence of PVC film on the side facing the sound source had a positive impact on NAC values in the frequency range below 4500 Hz. In this case, the curve has a completely different shape, similar to a bell shape rather than a typical "S" curve in

500–6400 Hz frequency range due to the decrease in higher frequencies. The maximum absorption was recorded to take place at 2200 Hz. When the PVC film was placed at the backside, it caused a slight increase in NAC values.

Similarly, Seddeq et al.[56] reported a shift to the lower frequency value of maximum sound absorption from 6300 Hz to 2250 Hz of jute mats with the incorporation of a perforated sawdust board as a cover plate. In terms of the absorption coefficient for frequency values higher than 3500 Hz a drastic decrease was observed, as "S" curve became a "bell" curve similar to that reported by Jayaraman et al.[47] When a cover plate and a back air space were incorporated to the porous absorber at the same time, multipeaked graphs of sound absorption was observed as seen in Fig. 5.11.

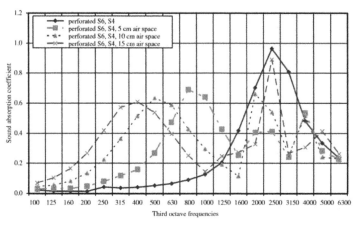

FIGURE 5.11   Effect of perforated cover plate and air gap on sound absorption coefficient of biocomposites (S6: compression molded sawdust board, S4:needle-punched jute mat) (From Seddeq, H. S.; Aly, N. M.; Ali, M. A.; Elshakankery, M. H. Journal of Industrial Textiles, 2013.[6] With permission from Sage Publications).

5.3.3   PROCESS PARAMETERS

Process parameters during absorbent material formation have an important impact on sound absorption due to their effects on the characteristics of the absorbent material. Process parameters have been classified into production and treatment parameter categories here. Different production and treatment processes used in biocomposite sound absorber formation are given in Table 5.2.

5.3.3.1   PRODUCTION

The fiber component of biocomposites is mainly manufactured by web formation and bonding methods before combining with other components. Web formation can be classified into three categories:
- Drylaid system (carding or airlaying)
- Wetlaid system (similar to paper production from pulp)
- Polymer-laid system (spunbonding, meltblowing, etc.).[79]
- There are also three classes of web bonding processes:
- Chemical bonding (use of binders)
- Thermal bonding (calendaring, through-air blowing, or ultra-sonic impact)
- Mechanical bonding (needling, stitching, water-jet entangling).[27]

Among web forming types, Jayaraman et al.[47] report higher sound absorption for fibrous structures formed by airlaying compared to carded ones irrespective of the fiber content. This finding was agreed by the findings of Parikh et al.[22]. This might be due to relatively random placement of fibers, and thus, higher tortuosity, higher number of pores with smaller sizes, higher number of fiber-to-fiber contact points, and gradient in porosity due to gravity.

Among web bonding methods, for needle-punching, the factors which affect noise reduction properties are given as the number of needling passes,[1,36] and punching density[80].

Genis et al.[80] found that the absorption coefficient reaches its maximum at material density $r_w = 100$ kg·m^{-3} for thermo-bonded polypropylene, and punching density $P = 28$ cm^{-2} for needle-punched polypropylene and polyamide materials in the sound frequency range of 63 to 8000 Hz, for fiber diameters between 10 to 40 μm, and material thicknesses of 3 to 20 mm. They found the absorption coefficient of their needle-punched samples to have more dependency on the frequency range compared to thermo-bonded ones; thus they have a narrower absorption efficiency frequency range. The absorption in needle-punched materials was also more dependent on the diameter of fibers compared to thermally bonded webs. Jayaraman et al.[47] did not report a significant difference between needled and needled plus thermally bonded fibrous structures.

5.3.3.2   TREATMENT

Different treatments may be applied to biocomposite according to the effect required. Examples of some chemical treatments include alkalization, enzymatic treatments, flame-retardant and antimicrobial agent applications. Physical treatments include compression and thermal treatments.

Among physical treatments, compression and thermal treatment play a very important role as they form the basis of molding process, which is a must for most of the noise control composites. Compression of a fibrous mat deteriorates its sound absorption properties according to Castagnede et al.[81] For a given homogeneous porous layer, compression is followed by a decrease in terms of porosity and thickness, and at the same time by an increase of tortuosity and resistivity. Jayaraman et al.,[47] Yilmaz et al.,[4] and Yilmaz et al.[58] also found a decrease in sound absorption with compression. The finding of Yilmaz et al.[20] is presented in Fig. 5.12.

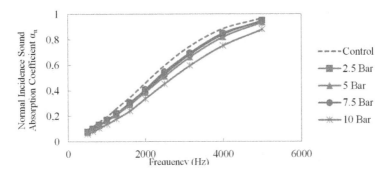

FIGURE 5.12   Effect of compression on the average normal-incidence sound absorption of PLA/Hemp/PLA sandwich biocomposites (Yilmaz et al., 2012: DOI: 10.1002/app.34712).[20]

Yilmaz et al.[4,8] investigated the effects of thermal treatment on noise control capability. Yilmaz et al.[8] treated three-layered sandwich structures of PLA/Hemp/PLA at temperature points of 125, 145, 165 and 185°C. An increase in air flow resistivity and a decrease in NAC were found. The decrease in NAC reached a substantial extent when the melting point of the constituent thermoplastic fiber, PLA, is reached as seen in Fig. 5.13. Yilmaz et al.[4] investigated the effects of thermal treatment on noise reduction performance of three-layered PP/Hemp/PP sandwich structures at 150 and 185°C (see Fig. 5.14). They reported a slight increase in NAC for the lower frequency range for 150°C treated composite and a drastic decrease in NAC for the biocomposite treated at 185°C as shown in Fig. 5.12. Similar to the finding of Yilmaz et al.[8], temperature exceeding the melting point of the thermoplastic component was reported to be very deteriorating for sound absorption performance. The slight increase in the lower region which is experienced for the 150°C treated sample might be due to increase airflow resistance. Accordingly, by fine-tuning the parameters of thermal treatment, the noise reduction capability of the biocomposite may be preserved or it can even be enhanced.

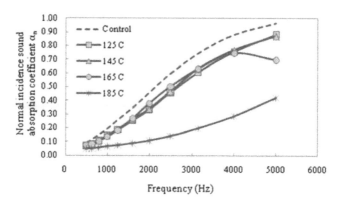

FIGURE 5.13 Effect of thermal treatment on the average normal-incidence sound absorption of PLA/Hemp/PLA sandwich biocomposites (Yilmaz et al., 2013:[8] DOI: 10.1177/1528083712452899).

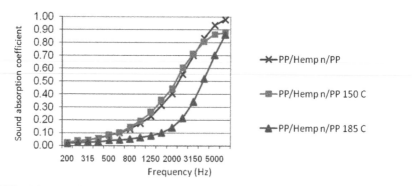

FIGURE 5.14 Sound absorption coefficients of untreated, and heat treated PP/Hemp /PP at 150°C and 185°C (Yilmaz et al., 2008).[4]

Among chemical treatments, Jayaraman et al.[47] studied the effect of fire retardancy treatment on sound absorption. They found that the treatment had a positive impact on the sound absorption of kenaf fibers. Yilmaz et al.[9] applied alkalization treatment on PP/hemp/PP to detect the effect of the treatment on sound absorption. As known, the alkalization process partially removes lignin, pectin and hemicelluloses present in the fibers and leads to a separation between the fibrils of the natural technical fibers. This results in finer fibers and a rougher surface, which might enhance sound attenuation. However, alkalization did not increase the sound absorption coefficient as expected. In contrary, the NAC values decreased with the decrease in the mass of the material due to loss of hemp fiber constituents and possibly the distortion of the pores during the wet treatment as given in Fig. 5.15.

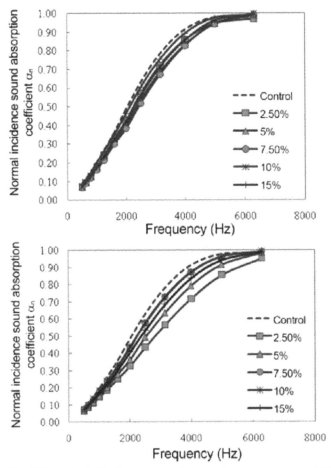

FIGURE 5.15 Effect of alkalization on sound absorption performance of PP/Hemp/PP webs. (a) Treated at room temperature. (b) Treated at boiling temperature. (Yilmaz et al., 2013.[9] DOI: 10.1007/s12221-012-0915-0).

Among wet treatments, Huda and Yang[55] extracted fibers from cornhusks by al-kalization and further treated a part of them with cellulose and xylanase enzymes to produce finer fibers with the elimination of extracellulosic materials. They blended cornhusk fibers with PP fibers then formed a web by carding and thermally bonded the web to produce a composite structure. They obtained better sound absorption from composites consisting enzyme treated cornhusk fibers compared to those that were untreated as shown in Fig. 5.16.

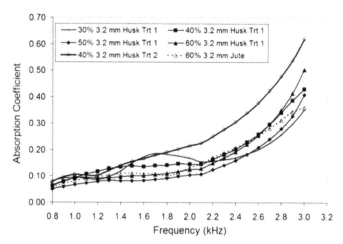

FIGURE 5.16   Effect of alkalization (trt 1), and subsequent enzymatic (trt 2) treatments on sound absorption of corn husk – PP composites (From Huda, S.; Yang, Y. Macromolecular Materials, 2008.[55] With permission from Wiley VCH).

5.4   CONCLUDING REMARKS

Fibrous materials act as noise control elements in a wide range of applications as they present a cost-effective, light-weight, and environmentally friendly alternative to conventional materials. Fibrous materials should be designed as composite structures to address the demands in terms of esthetics and performance characteristics such as moldability, durability and enhanced noise reduction capacity. Designing of an effective composite sound absorbing noise control element requires a good understanding of sound propagation in fibrous materials. This chapter has presented an overview of natural fiber based biocomposites as rigid porous sound absorbers. Sound absorption mechanisms that take place during sound wave propagation through fibrous media have been given. An overview of empirical models which may be applicable to explain sound propagation in biocomposites has been explained. Based on the models, factors that affect sound absorption behavior of biocomposites have been investigated referring to previous studies.

Most of the research and modeling effort for understanding the sound absorption characteristics of fibrous materials have been devoted to conventional petro-based materials. More study is needed with sound absorbers of biocomposites, which offer a safer production and service life with effective utilization of scarce resources.

Natural fibers, engineered environmentally friendly polymers, and recycled materials can be given as examples of "green" materials to be used during production of noise absorber biocomposites based on previous studies. There is especially very limited knowledge pertaining to the acoustical properties of biocomposites consisting of engineered biopolymers or agricultural byproduct materials. Future research efforts devoted to the understanding of these materials may enhance our understanding of sustainable materials as noise reduction elements.

KEYWORDS

- **Biocomposites**
- **Fiber reinforced Composites**
- **Noise Control**
- **Noise Pollution**
- **Sound Absorption Materials**
- **Sound Absorption Mechanism**

REFERENCES

1. Yilmaz, N. D., Banks-Lee, P., & Powell, N. B. (2008). In *Nonwovens Technical*, International Nonwovens Technical Conference Proceedings. Houston, TX.
2. Wilson, A. (2006). Engineered Nonwovens used for Automotive Acoustic Insulation. *Tech. Text. Intern.* 15, 11–14.
3. Yilmaz, N. D. (2009). Acoustical Properties of Biodegradable Nonwovens. PhD Dissertation, North Carolina State University, Raleigh, NC.
4. Yilmaz, N. D., Banks-Lee, P., & Powell, N. (2008). In *Nonwoven Enhancements,* Proceedings of Nonwoven Enhancements Conference, Houston, TX.
5. Polzer, C., & Heiling, R. (2005). Cars Made of Plants. [Cited 07/11 2011]. Available from http://www.dw.de/germany-gears-up-for-cars-made-of-plants/a-1631426.
6. Olesen, P. O., & Plackett, D. V. (1999). In *Natural Fibers Performance Forum*, Kopenhagen, Denmark.
7. Mohanty, A. K., Misra, M., Drzal, L. T., Selke, S. E., Harte, B. R., & Hinrichsen, H. (2005). Natural Fibers, Biopolymers, and Biocomposites: An Introduction. In *Natural Fibers, Biopolymers, and Biocomposites*, Mohanty, A. K., Misra, M., & Drzal, L.T. (Eds.), CRC Press: Boca Raton.
8. Yilmaz, N. D., Powell, N. B., Banks-Lee, P., Michielsen, S. (2013). Multi-Fiber Needle-Punched Nonwoven Composites: Effects of Heat Treatment on Sound Absorption Performance. *J. Ind. Text.* 43, 231–246.
9. Yilmaz, N. D., Powell, N. B., Banks-Lee, P., & Michielsen, S. (2012). Hemp-fiber Based Nonwoven Composites: Effects of Alkalization on Sound Absorption Performance. *Fib. Polym.* 13, 915–922.
10. Kuttruff, H. (2007). *Acoustics: an Introduction,* Taylor & Francis: New York.
11. Wyerman, B. (2003). Basics of acoustical materials, Acoustical Materials Workshop, SAE NVC.
12. Ng, Y. H., & Hong, L. (2006). Acoustic Attenuation Characteristics of Surface-Modified Polymeric Porous Microspheres. *J. Appl. Polym.* Sci., 102 1202.
13. Fahy, F. (2001). *Foundations of Engineering Acoustics,* Academic Press: San Diego, Calif.

14. Cox, T. J., & D'antonio, P. (2004). *Acoustic Absorbers and Diffusers*, Taylor & Francis: New York, NY.
15. Mukhopadhyay, S. K., & Partridge, J. H. (1999). Automotive Textiles, *Textile Progress*. 29, 1/2.
16. Huda, S., & Yang, Y. A. (2009). Novel Approach of Manufacturing Light-Weight Composites with Polypropylene Web and Mechanically Split Cornhusk. *Ind. Crops Products*. 30, 17–23.
17. Attenborough, K., & Ver, I. L. (2006). Sound-absorbing materials and sound absorbers. In *Noise and Vibration Control Engineering*, Istvan, L. V., Leo L. (Eds.), Wiley: Hoboken, NJ.
18. Mechel, F. P. (2002). *Formulas of Acoustics,* Springer: Berlin, Germany.
19. Mohammadi, M. (1998). Heat Barrier Properties of Heterogeneous Nonwoven Materials. PhD Dissertation, North Carolina State University, Raleigh, NC.
20. Yilmaz, N. D., Michielsen, S., Banks-Lee, P., & Powell, N. B. (2012). Effects of Material and Treatment Parameters on Noise Control Performance of Compressed Three-Layered Multi-Fiber Needle-Punched Nonwovens. *J. Appl. Polym. Sci.* 123, 2095–2106.
21. Thilagavathi, G., Pradeep, E., Kannaian, T., & Sasikala, L. (2010). Development of Natural Fiber Nonwovens for Application as Car Interiors for Noise Vontrol. *J. Indust. Text*. 39, 267–278.
22. Parikh, D. V., Chen, Y., & Sachinvala, N. D. (2006). Sound Dampening by Velour Nonwoven Systems in Automobiles. *AATCC Rev.*, 6, 40–44.
23. Finegold, L. S., Muzet, A. G., & Berry, B. F. (2006). "Sleep disturbance due to transportation noise exposure," in *Handbook of Noise and Vibration Control,* Crocker, M. J. Ed., John Wiley: Hoboken, N. J.
24. American Society for Testing and Materials (2008). *ASTM C 634–08 standard terminology relating to building and environmental acoustics.*
25. Hong, Z., Bo, L., Guangsu, H., & Jia, H. A. (2007). Novel Composite Sound Absorber with Recycled Rubber Particles. *J. Sound Vib.* 304, 400–406.
26. Borlea, A., Rusu, T. U., Ionescu, S., & Nemeş, O. (2012). Determination of the Sound Absorption Properties of Some New Composite Materials Obtained from Wastes. *Revista Romana de Materiale- Romanian J. Mater.*, 42, 405–414.
27. Massenaux, G. (2003). Introduction to nonwovens. In *Nonwoven Fabrics*, Albrecht, W., Fuchs, H., Kittelmann, W. (Eds), Wiley VCH Weinheim.
28. Delany, M. E., & Bazley, E. N. (1970). Acoustical Properties of Fibrous Absorbent Materials. *Appl. Acoust.* 3, 105–116.
29. Bies, D. A., & Hansen, C. H. (2003). *Engineering Noise Control: Theory and Practice. 3rd ed.*, Spon Press: New York.
30. American Society for Testing and Materials (2008). *ASTM E 1050-08 standard test method for impedance and absorption of acoustical materials using a tube, two microphones and a digital frequency analysis system.*
31. American Society for Testing and Materials (2011). *ASTM C384-04 Standard Test Method for Impedance and Absorption of Acoustical Materials by Impedance Tube Method.*
32. American Society for Testing and Materials (2008). *ASTM C 423 standard test methods for sound absorption and sound absorption coefficients by the reverberation room method.*
33. Bischel, M., Roy, K., & Greenslade, J. V. (July, 2008). In *Euro Noise*, Proceeding of Euro-Noise, [Paris] France.
34. Bruel & Kjaer. (2009). Product data impedance tube kit (50 Hz-6.4 kHz) type 4206. [Cited 12/3 2009]. Available from http://www.bksv.com/doc/bp1039.pdf.
35. Ingard, K. U. (1994). *Notes on Sound Absorption Technology,* Noise Control Foundation: Poughkeepsie, N.Y.
36. Banks-Lee, P., Peng, H., & Diggs, L. (1992). In *Nonwovens Technical*, International Nonwovens Technical Conference Proceedings, US.

37. American Society for Testing and Materials (2008). *ASTM C 522-03 standard test method for airflow resistance of acoustical materials.*

38. Lord Rayleigh. (1945). *Theory of Sound* (two volumes), Dover Publications: NY, New York, 1877, reissued.

39. Morse, P. M., & Richard H. B. (1944). Sound waves in rooms. *Rev. of Modern Phys.*, 16, 69–150.

40. Zwikker, C., & Kosten, C. W. (1949). *Sound Absorbing Materials.* New York: Elsevier Publishing.

41. Biot, M. A. (1956). Theory of Propagation of Elastic Waves in Fluid-Saturated Porous Solid. *J. Acoust. Soc. Am.*, 28, 168–178.

42. Beranek, L. L. (1947). Acoustical Properties of Homogeneous, Isotropic Rigid Tiles and Flexible Blankets. *J Acous. Soc. Am.*, 19, 556–568.

43. Kawasima, V. (1960). Sound Propagation in a Fibrous Block as a Composite Medium. *Acustica.* 10, 208.

44. Attenborough, K. (1971). The Influence of Microstructure on Propagation in Porous Fibrous Absorbents. *J. Sound Vib.* 16, 419–442.

45. Nick, A., Becker, U., & Thoma, W. (2002). Improved Acoustic Behavior of Interior Parts of Renewable Resources in the Automotive Industry. *J. Polym. Environ.* 10, 115–118.

46. Ballagh, K. O. (1996). Acoustical Properties of Wool. *Appl. Acoust.*, 48, 101–120.

47. Jayaraman, K. A., Banks Lee, P., & Jones, M. In. (2005). In *Nonwovens Technical*, International Nonwovens Technical Conference Proceedings, St. Louis, MO, September, 19–22.

48. Huda, S., Reddy, N., & Yang, Y. (2012). Ultra-light-weight Composites from Bamboo Strips and Polypropylene Web with Exceptional Flexural Properties. *Compos. Part B-Eng*, 43, 1658–1664.

49. Markiewicz, E., Boryslak, S., & Paukszta, D. (2009). Polypropylene-Lignocellulosic Material Composites as Promising Sound Absorbing Materials. *Polymery.* 54, 430–435.

50. Brencis, R., Skujans, J., Iljins, U., Ziemelis, I., & Osits, N. (2011). In *14th International Conference on Process Integration, Modelling and Optimization for Energy Saving and Pollution Reduction*, Press 2011:14th International Conference on Process Integration, Modelling and Optimization for Energy Saving and Pollution Reduction. Florance, Italy.

51. Bratu, M., Ropota, I., Vasile, O., & Muntean, M. (2011). Research on the Absorbing Properties of Some New Types of Composite Materials. *Revista Romana de Materiale- Romanian J. Mater.* 41, 147–154.

52. Karademir, A., Aydemir, C., & Yenidogan, S. (2011). Sound Absorption and Print Density Properties of Recycled Sheets Made from Waste Paper and Agricultural Plant Fibers. *African J. Agricultural Res.*, 6, 6073–6081.

53. Huda, S., & Yang, Y. (2008). Composites from Ground Chicken Quill and Polypropylene Source. *Compos. Sci. Technol.*, 68, 790–798.

54. El Hajj, N., Mboumba-Mamboundou, B., Deilly, R. M., Aboura, Z., Benzeggagh, M., & Queneudec, M. (2011). Development of Thermal Insulating and Sound Absorbing Agro-Sourced Materials from Auto Linked Flax-Tows. *Ind. Crops Products.* 34, 921–928.

55. Huda, S., & Yang, Y. (2008). Chemically Extracted Cornhusk Fibers as Reinforcement in Light-weight Poly (propylene) Composites. *Macromol. Mater. Eng.*, 293, 235–243.

56. Seddeq, H. S., Aly, N. M., Ali, M. A., & Elshakankery, M. H. (2013). Investigation on Sound Absorption Properties for Recycled Fibrous Materials. *J. Ind. Tex.*, 43, 56–73.

57. Yao, J., Hu, Y., & Lu, W. (2012). Performance Research on Coir Fiber and Wood Debris Hybrid Boards. *Bioresources.* 7, 4262–4272.

58. Yilmaz, N. D., Banks-Lee, P., Powell, N.B., & Michielsen, S. (2011). Effects of Porosity, Fiber Size and Layering Sequence on Sound Absorption Performance of Needle Punched Nonwovens. *J. Appl. Polym. Sci.* 121, 3056–3069.

59. Sengupta, S. (2010). Sound reduction by needle-punched nonwoven fabrics. *Ind J. Fiber Text. Res.*, 35, 236–242.
60. Huda, S., & Yang, Y. (2009). Feather Fiber Reinforced Light-Weight Composites with Good Acoustic Properties. *J. Polym. Environ.* 17, 131–142.
61. Jayaraman, K. A. (2005). Acoustic Absorptive Properties of Nonwovens. MS Thesis. North Carolina State University, Raleigh, NC.
62. Lee, Y. E., & Joo, C. W. (2003). Sound Absorption Properties of Recycled Polyester Fibrous Assembly Absorbers. *AUTEX Res. J.*, 3, 78–84.
63. Koizumi, T., Tsujiuchi, N., & Adachi, A. (2002). The Development of Sound Absorbing Materials using Natural Bamboo Fibers. In *High Performance Structures and Composites. High Performance Structures and Materials*, Brebbia, C. A., De Wilde, W. P., & Witpress: Belgium, (2002).
64. Watanabe, K., Minemura, Y., Nemoto, K., & Sugawara, H. (1999). Development of High Performance All-Polyester Sound-Absorbing Materials. *JSAE Rev.*, 20, 357–362.
65. Narang, P. P. (1995). *Material Parameter Selection in Polyester Fiber Insulation for Sound Transmission and Absorption.* Applied Science Publishers: Barking, Essex, 45.
66. Hur, B. Y., Park, B. K., Ha, D. I., & Um, Y. S. (2005). Sound Absorption Properties of Fiber and Porous Materials. *Mater. Sci. Forum.*, 475–479, 2687–2690.
67. Taşcan, M., & Vaughn, E. A. (2008). Effects of Total Surface Area and Fabric Density on the Acoustical Behavior of Needlepunched Nonwoven Fabrics. *Text. Res. J.* 78, 289–296.
68. Coates, M., & Kierzkowski, M. (2002). Acoustic Textiles Lighter, Thinner and More Sound-Absorbent. *Tech. Text. Intern.* 11, 15–18.
69. Taşcan, M. (2005). Acoustical Properties of Nonwoven Fiber Network Structures. PhD Dissertation, Clemson University, Celmson, SC.
70. Dent, R. W. (1983). *The Sound Absorption Properties of Felt*, Nonwovens Conference Papers: UMIST 1983, 257.
71. Wassilieff, C. (1996). Sound Absorption of Wood-Based Materials. *Appl. Acoust.*, 48, 339–356.
72. Wright, M. C. M., (ed.) (2005). *Lecture Notes on the Mathematics of Acoustics,* Imperial College Press: London.
73. Allard, J. F., & Champoux, Y. (1992). New Empirical Equations for Sound Propagation in Rigid Frame Fibrous Materials. *J. Acoust. Soc. Am.*, 91, 3346–3353.
74. Lee, F. C., & Chen, W. H. (2003). On the Acoustic Absorption of Multi-Layer Absorbers with Different Inner Structures. *J. Sound Vib.* 259, 761–777.
75. Ackermann, U., Fuchs, H. V., & Rambausek, N. (1998). Sound Absorbers of a Novel Membrane Construction. *Appl. Acoust.*, 25, 197–215.
76. Atalla, N., Panneton, R., Sgard, F. C., & Olny, X. (2001). Acoustic Absorption of Macro-Perforated Porous Materials. *J. Sound Vib.* 243, 659–678.
77. Chen, W. H., Lee, F. C., & Chiang, D. M. (2000). On the Acoustic Absorption of Porous Materials with Different Surface Shapes and Perforated Plates. *J. Sound Vib.* 237, 337–355.
78. Rebillard, P., Allard, J. F., Depollier, C., Guignouard, P., Lauriks. W., Verhaegen, C., & Cops, A. (1992). The Effect of a Porous Facing on the Impedance and the Absorption Coefficient of a Layer of Porous Material. *J. Sound Vib.* 156, 541–555.
79. Wilson, A. (2007). Development of the Nonwovens Industry. In *Handbook of Nonwovens*, Russell, S. J. Ed., CRC press LLC: Boca Raton, FL.
80. Genis, A. V., Kostyleva, E. Y., Andrianova, L. N., & Martem'yanov, V. A. (1990). Comparative Evaluation of Acoustical Properties of Heat-Bonded and Needlepunched Fibrous Materials Prepared from Polymer Melts. *Fib. Chem.*, 21, 479–482.
81. Castagnède, B., Aknine, A., Brouard, B., & Tarnow, V. (2000). Effects of Compression on the Sound Absorption of Fibrous Materials. *Appl. Acoust.*, 61, 173–182.

CHAPTER 6

BIOCOMPOSITES FOR INDUSTRIAL NOISE CONTROL

A. R. MOHANTY and S. FATIMA

6.1 INTRODUCTION

With growing concerns towards the environment, designers are looking forward to the use of environmental friendly materials for product development. Though the aim is not to replace completely the metals or synthetic materials by these environment friendly materials. For the former have high strength and long life. However in few applications the use of composites made up of a combination of natural fiber based materials and a binder have many advantages. Like they are environmental friendly, less weight, economical and in few places are also available abundantly in nature. Natural based fiber materials like flax, ramie, hemp, banana fiber, coir, jute, cotton can be used to produce biocomposites which have many industrial applications. These biocomposites are being used to manufacture machinery enclosures, furniture, highway crash barriers, geo-textiles, as reinforcement in concrete, apparels, bags, yarns, carpets and building materials. However, they have a strong potential to be used for noise control applications as well.

Human beings are uncomfortable at exposure to high sound levels for a long duration. Such undesired sound are known as noise. Thus designers of machines, building architects, city planners, automobile designers and environmentalist have to take due consideration in design of machines, building, city planning and so forth, so that the human being is comfortable in the respective situations. Moreover due to stiff global competition among manufacturers, everyone is striving to make their products quieter than the other. Keeping the above in consideration, this chapter introduces to the basics of sound along with the associated terminology. Then it focuses on the physical, mechanical, thermal and acoustical properties of such materials. Followed by a description of their use in various applications like architectural acoustics, home appliances, machinery enclosure and automobiles for noise control.

6.2   BASICS OF SOUND

Sound heard by human being is essentially a longitudinal pressure wave traveling in air, which is incident on the eardrum. A human being gets a sensation of hearing depending upon the intensity and frequency content of the pressure wave incident on the eardrum. The sound heard by a human being is characterized both by its frequency content and amplitude level. The human being can hear in the audible frequency range from 20 Hz to 20,000 Hz. Human ear has a very high dynamic range and one can hear from 20 mPa to 10 MPa of sound pressure. Since the dynamic range of hearing is high, for convenience in representing such large variation in pressures a logarithmic scale is preferred over a linear scale to represent the sound pressure. Sound pressure level (SPL) is thus represented in decibel (dB) as given in Eq. (1),

$$SPL(dB) = 20\log_{10}\left(\frac{p_{rms}}{p_{ref}}\right) \tag{1}$$

where in air p_{ref} has a value of 20 µPa.

A sound level increase by 6 dB corresponds to doubling of the sound pressure. And an increase of 10 dB is significantly higher, thus while performing SPL measurements of any sound source it is recommended that the source must be higher than the background noise level by at least 10 dB, so that the background noise does not contaminate the sound source noise. In other words the background must be 10 dB less than the source. It must also be noted that in a free-field measurement (no reflecting surfaces present) the SPL decreases by 6 dB for very distance doubled from the source. Typical values of sound pressure levels both in dB and pascals are shown in Fig. 6.1[1].

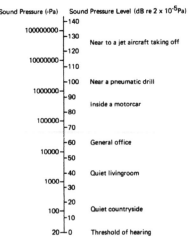

FIGURE 6.1   Typical sound pressure level (with permission from Bruel & Kjaer).

However the perception of sound heard by a human being is quite different than that measured by an instrument. Table 6.1 shows the subjective effect of changes in noise levels, as heard by a human being[2]. Thus, a noise reduction of less than 3 dB will not be appreciated much. At least a level difference of 5dB must be brought about for any perceivable change.

TABLE 6.1   Subjective effect of Changes in Noise Levels (with permission from Bruel & Kjaer)

Change in Level dB	Subjective Effect
3	just perceptible
5	clearly perceptible
10	twice as loud

Sound waves contain many frequencies. To determine these frequencies a frequency analysis is usually done by analog and digital means. Filters of varying and constant bandwidth are normally used. The frequency bands of the filters are known as octave band or 1/n octave band, depending on the upper and lower frequency limits of the band. For a 1/n octave band the relationship between the upper and lower frequency band are given by Eq. (2).

$$f_u = 2^{\frac{1}{n}} f_l \tag{2}$$

The audible frequency range of 20 Hz to 20,000 Hz is usually represented in such octave bands since human beings are better able to distinguish sound in octaves. For a 1/3 octave band which is of a finer frequency resolution than the octave band the value of n in Eq. 1.2 is taken as 3. The center frequency of the octave band is the geometric mean of the lower and upper frequency band. For example, for a 1/3 octave band of 1000 Hz, the lower frequency limit is 710 Hz and the upper frequency limit is 1420 Hz.

Human beings also hear differently at different frequencies, thus for hearing at normal sound pressure levels a weighting curve is used to represent the actual perception of SPL by the human being. And this weighting curve is known as the A-weighting. The weighting values for different frequency octave bands are available in the literature[2]. Such weighted SPL is represented in dBA.

The sound power of a sound source is the energy emitted by it and is independent of the acoustical properties of the surroundings. Sound power measured in watts is usually also represented in a dB scale. As per international standards the sound power reference is taken as 10^{-12} W, thus a source emitting 1 W of sound energy has a sound power level of 120 dB. One watt is a very small amount of energy, however it is large quantity from a hearing point of view. Figure 6.2 gives the values of typical sound power levels[1].

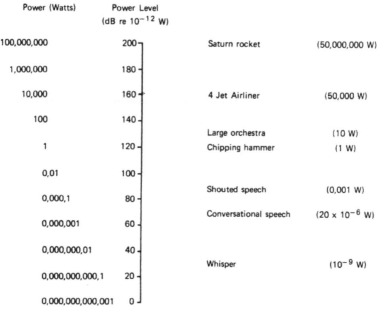

FIGURE 6.2   Typical Sound Power Levels (with permission from Bruel & Kjaer).

The sound power passing through a unit area is defined as sound intensity. Sound intensity is a vector quantity and has a direction associated with it. The sound intensity is also represented in a dB scale, where the reference sound intensity is taken as 10^{-12} W/m². The above acoustical quantities of sound pressure, sound power and sound intensity are all functions of the frequency of the sound wave.

Despite the care in designing and planning high levels of sound are unavoidable due to an inferior design, manufacturing and planning. Thus a noise control expert always comes in as a retrofit measure to bring the sound pressure levels to a comfortable level. There are three major aspects, which need to be understood while controlling the noise level. They are mainly the source of sound, then the path of sound propagation and finally the receiver, which is usually the human ear. Sound being elastic waves requires a medium for their propagation, the speed and nature of sound propagation depends upon the nature of medium and its elastic properties. Noise control materials can be used for treatment at various places in the path of propagation of sound for reducing the noise levels. Traditionally noise control materials like fiber glass, open cell foam, heavy concrete, thick steel plates and so forth have been used for noise control. In this chapter we focus on the use of biocomposites made up of natural materials for noise control applications. In the subsequent sections their physical, thermal, mechanical and acoustical properties are described, and then the procedure of use of such materials in the noise control of home appliances, machineries, automobile and buildings are described with few specific case studies.

6.3   BIOCOMPOSITES AS NOISE CONTROL MATERIAL

Traditionally noise control materials have three distinct properties: sound absorbing, sound blocking and sound damping. The very first step in noise control is to identify and characterize the noise source in term of its radiated sound pressure level and frequency. Designers make an effort to ensure that the products and associated components do not radiate high levels of noise. The noise which is received by the receiver propagates in the medium between the source and the receiver as elastic waves. Thus the sound energy is transferred between the source and the receiver by such waves. In order to reduce this transfer of energy, acoustical treatment can be done in the path between the source and the receiver by using the above noise control materials. For industrial noise control, materials like glass wool, elastomers and heavy sheet metal are used. Biocomposites made of natural materials are an excellent replacement for the above materials since they are abundantly available in nature and are economic to process. In order to use biocomposites for noise control their two important acoustical properties need to be known, normal specific sound absorption coefficient and the sound transmission loss. Another important property of the material, which is used to control the vibration of the structure generating noise is its damping factor. There exist standards by which all the above properties can be measured in the laboratory.

   In recent years, there is a growing interest in the development of new materials, which enhance optimal utilization of natural resources, and particularly, of renewable resources and the research on the natural fiber reinforced composite is on the forefront due to its significant properties like economical, biodegradable, recyclable and ecofriendly. For instance, in the year 2013 for 1 US dollar around 25 kg of raw jute fiber can be procured in India. Naturally, the composites reinforced with natural fibers like jute, sisal, banana and coir thus have been subjected to intense study for their low density and low cost application in contrast to synthetic reinforced composites. These materials have potential to replace traditional noise synthetic noise control materials because of their comparable acoustical and mechanical properties.

6.4   JUTE

Natural fibers have recently been used for making composite materials and they offer several advantages over synthetic materials. While these natural fibers can be extracted from many sources such as sisal, jute, coir, flax, hemp, pineapple and banana[3]; jute has been promoted as the most readily available, environment friendly, abundant, economic and bio-renewable source. It is specifically cultivated in large quantities in the eastern part of India and in Bangladesh[4]. It is a lignin-cellulose fiber which is composed primarily of the plant materials; cellulose (major component of plant fiber) and lignin (major components wood fiber). It falls into one of the bast fiber category (fiber collected from bast or skin of the plant) along with kenaf, industrial hemp, flax (linen), ramie and so forth.

Jute is used in various forms for noise control applications. The raw jute fiber after cleaning are used to produce jute yarn by a spinning process. The jute yarn is then weaved to make jute textile or cloth. Stacks of jute yarn laid in a random or a definite sequence are pressed under temperature to produce jute felt. The jute felt/fiber in turn can be chemically treated with a bonding agent usually natural rubber latex as a resin and pressed under certain temperature to form jute-based biocomposite panels. In few instances the raw fibers after appropriate processing can be chopped and used as fills in noise control blankets and pads. In jute mills where jute-based textiles are manufactured, during the trimming operations of these textiles many waste trims are produced. These trims can be used as acoustical fills as well, for noise control. All the above forms of jute derivatives are shown in Fig. 6.3 in some form or the other can be used for noise control purposes.

FIGURE 6.3   Jute and its biocomposite derivatives for noise control applications.

6.5   PROPERTIES OF JUTE

In order to apply these jute-based biocomposites in various industrial applications for noise control some of its important properties like mechanical strength, fire retardant properties, acoustical properties, chemical stability at extreme temperatures need to be known and understood, among many other important physical properties. Here a brief description of the properties is presented along with some of the values of the properties measured at the various experimental facilities available at the Indian Institute of Technology Kharagpur.

In general, jute fibers have an aspect ratio (length/diameter) above 1000 and thus can be easily woven and can be spun into coarse and strong threads. These fibers are mostly used for fishnets, sacks, bags, ropes and as a filling for mattresses and cushions. In general, bast fibers (skin fiber collected from plants) have good thermal and acoustical insulation properties. There are, however, few drawbacks associated with the application of jute fibers. The primary one is lack of consistency in fiber quality due to presence of hydroxy and other polar groups in various constituents.

Another one is high moisture absorption, which brings the dimensional instability to the composites. The compatibility between matrix and fiber is poor, which require surface or alkali treatment.

Limitation in performance of jute-based fiber composites can be greatly improved through chemical modification techniques[5].

6.5.1 PHYSICAL PROPERTIES

Biocomposites made up of flax, hemp, coir and jute have potential for application in automobile, building and machineries for noise control purposes. In order to use such materials their physical, mechanical and acoustical properties are needed. Many literature exist where some of these physical properties of biocomposites can be found[6,7].

6.5.1.1 DENSITY

The density of biocomposite plays a significant role in weight reduction. For energy savings, designer are striving to manufacture light weight and high strength components. However it is to be noted that the density of the biocomposite have a significant influence on their noise reduction capabilities. Jute fibers are heavier than water and have a density in the range of 1200 to 1400 kg/m^3.

6.5.1.2 WATER ABSORPTION

The water absorption of jute-based biocomposite samples were measured according to ASTM D 570[8]. This test is carried out to determine the amount of water absorbed and performance of the materials in water environments. This is very similar to the moisture absorption test. Three square shaped samples of 3.0 cm × 3.0 cm of 5% natural rubber based jute composite were kept in an oven at a temperature of 100°C. Then weight of the three samples were measured. After that these samples were kept in a Petri dish full of distilled water. Then after 24 h and 48 h, the weight of the samples were taken and the water absorption was calculated using the Eq. (3).

$$\text{Water absorption,} \qquad Wa = \frac{W_1 - W_2}{W_1} \times 100 \qquad (3)$$

where W_1 is original weight of the sample and W_2 is final weight of the sample after 24, 48 and 120 h.

The results of water absorption test are shown in Table 6.2. The percentage of the water absorption gradually increases with time and then it reaches a saturation point and an average of 114% of water is absorbed at the end of 120 h.

TABLE 6.2   Percentage of Water Absorption of Jute Composite

Sample No.	Initial wt. of sample (gm)	24 h	48 h	120 h	144 h	192 h
1	3.44	39.59	85.63	107.02	106.05	106.78
2	3.44	39.94	78.00	118.88	119.30	118.22
3	3.68	36.11	77.91	118.19	118.05	119.78

6.5.1.3 MOISTURE ABSORPTION

The moisture absorption of jute composite samples were also measured according to ASTM D 570[8]. This test was carried out to determine the amount of moisture absorbed and the performance of the material in humid environment. Four square shaped samples of 3.5 cm × 3.5 cm of 5% natural rubber latex jute composite were tested. The samples were kept in an oven at temperature 100°C for one hour. Then the weights of the samples were measured. After that the samples were kept in a desiccator containing saturated solution of NaCl. The relative humidity of saturated Sodium chloride is 74.87 ± 0.12 at room temperature of 35°C. The percentage of moisture absorption of samples after 24, 48, and 120 h were calculated according to Eq. (4) as used in the water absorption test.

$$\text{Moisture absorption,} \quad Ma = \frac{W_1 - W_2}{W_1} \times 100 \tag{4}$$

where W_1 is original weight of the sample and W_2 is final weight of the sample after 24, 48 and 120 h.

The results of moisture absorption test are shown in Table 6.3. The percentage of the moisture absorption gradually increases and then it reaches a saturation point and an average of 10.05% of moisture is absorbed approximately after 120 h.

TABLE 6.3   Percentage of Moisture Absorption of Latex Jute Composite

Sample No.	Wt. of Sample (gm)	24 h	48 h	120 h	144 h	192 h
1	3.23	9.84	10.15	10.33	10.56	10.89
2	3.42	9.49	10.05	10.11	10.34	10.71
3	3.00	9.51	9.55	10.01	10.52	10.90
4	3.81	9.29	9.55	9.78	10.06	10.25

6.5.1.4 WATER SWELLING

ASTM D570 standard covers method of determination of swelling in water of jute composite[8]. This test was done to determine the hydrophilic nature of 3 cm × 2.7 cm × 0.4 cm jute-based natural rubber composite. The average of the three values obtained for the change in thickness expressed as a percentage of the original average thickness is reported as the swelling value.

Thickness of swelling in water was measured as per Eq. (5),

Swelling Thickness, $Ts= \dfrac{T_2-T_1}{T_1} \times 100$ (5)

where T_1 is initial thickness and T_2 is the final thickness after water absorption.

Table 6.4 gives the percentage of thickness swelling of samples in water at 24 h. The saturation point occurs at 24 h and thickness swelling is approximately 100%.

TABLE 6.4   Thickness of Swelled Jute Composite in Water after 24 Hour

Sample No.	Initial thickness (mm)	Final thickness (mm)	Swelled Thickness (%)
1	4.01	8.10	101.99
2	4.02	8.01	99.25
3	4.03	8.02	99.01

6.5.1.5   BIODEGRADATION

This test was done to evaluate microbial biodegradation activities of jute composite by burying in soil[9]. Sample dimensions: 10 cm × 5 cm × 0.4 cm of the jute composite was used for the test. Weight of 10 pieces of rectangular shaped samples were measured and buried in soil mixture containing garden soil, cow dung and sand (2:1:1, w/w). After 15 and 30 days the samples were removed, cleaned properly and finally kept in oven at 100°C till a constant weight was obtained. The weight loss of the samples after 15 days was 3.74% and 30 days was 6.28%.

6.5.2   THERMAL PROPERTIES

The biocomposite materials have wide engineering applications ranging from moist to dry atmosphere, from clean to dusty environment and from low to high temperature, thus it becomes important for the noise control specialist to know the effect of temperature on the various physical and chemical properties of the biocomposite. Here some of the thermal properties of the jute-based biocomposite are presented.

6.5.2.1   THERMAL CONDUCTIVITY

The thermal conductivity of jute felt measured using the heat flow meter technique at the facilities of the Indian Institute of Technology Kharagpur as per ASTM C 518 is 0.064 W/m-K in the temperature range from 50°C to 80°C[10]. The thermal diffusivity of the jute felt measured using the laser flash technique as per ASTM E1461 standard was found to be 0.259 mm^2/s[11].

6.5.2.2   THERMOGRAVIMETRIC ANALYSIS

Thermogravimetric test is used to determine the temperature at which the thermal degradation starts and determines the usable temperature range of treated jute felts. Thermogravimetric analysis (TGA) of raw and natural rubber treated jute felts was carried out in nitrogen gas atmosphere at a heating rate of 10°C/min from 45°C to 60°C temperature using a Differential Thermal Analyzer (DTA). Using the first derivative of the TGA line, a DTG curve was obtained to identify the start, peak, and end temperature. From TGA it was found that thermal stability of raw jute felt is till 260.92°C whereas treated jute felt is till 269.3°C. This indicates that jute and its derivative can be used in applications up to a temperature of 260°C, which can be further enhanced with suitable chemical treatment.

6.5.2.3   FLAMMABILITY ASPECTS OF JUTE

Flammability of jute-based biocomposite is quantified by measuring the three parameters of limiting oxygen index, flame propagation and smoke density.

6.5.2.4   LIMITING OXYGEN INDEX TEST

Limiting Oxygen Index (LOI) test was carried out to measure the minimum volume concentration of oxygen that will just support flaming combustion of the jute biocomposite in a flowing mixture of oxygen and nitrogen. Oxygen concentration reported is the volume percent in a mixture of oxygen and nitrogen. Test was performed as per ASTM D 2863-97 standard[12]. The jute composite specimens for the LOI measurements were 152.4 mm ´ 5 mm ´ 4 mm in size. The volume percent of oxygen required for combustion is given in Table 6.5.

TABLE 6.5   Limiting Oxygen Index of Different Materials Indicating Oxygen Volume Percent

Material	Natural Rubber	Cellulose	Wool	Natural Rubber treated Jute Composite
LOI (%)	18.5	19.0	25.0	30.0

6.5.2.5   FLAME PROPAGATION

The rate of flame spread was measured as per federal motor vehicle safety system (FMVSS) standard[13]. Specimen of 152.4 mm ´ 5 mm ´ 4 mm in size was exposed horizontally at its one end to a small flame for 15 seconds. The distance and time duration of burning or the time to burn between two specific marks were measured. The burn rate B, was expressed as the rate of flame spread in mm/min given in Eq. (6).

$$B = 60 \times \frac{(L)}{(T)} \tag{6}$$

where B, L and T are burn rate in mm per minute, length of the flame traveled in mm and time in second for the flame to travel L mm, respectively.

From Table 6.6, 2.5% natural rubber latex based jute composite shows a lesser flame propagation rate than 5% natural rubber latex jute composite. Further, with a 1% sodium phosphate (Na_3PO_4; as a fire retardant) treatment on 5% natural rubber latex composite, the flame propagation rate of the jute composite reduces by almost half.

TABLE 6.6   Flame Propagation Test Results of Natural Rubber Latex Jute Composite

Material	Length (mm)	Flame propagation rate (mm/min)
2.5% natural rubber based jute composite	100	15.69
5% natural rubber based jute composite	100	20.56
5% natural rubber based jute composite + 1% sodium phosphate (fire retardant)	100	9.77

6.5.2.6   SMOKE DENSITY

The smoke density for a sample having dimension of 120 mm × 100 mm × 4 mm in size was measured by using a smoke density chamber as per ASTM D 2843–04[14]. The smoke generated (flaming mode) in the process of burning of sample was measured by the change of the light intensity. This test was useful for measuring and observing the relative amounts of smoke obscuration produced by burning or decomposition of the material. Smoke density rating which represents the total amount of smoke present in the chamber for 4 min was measured by Eq. (7).

$$\text{Smoke density rating} = \frac{A}{T} \times 100 \tag{7}$$

where, A and T are the area under the light absorption versus time curve and total area of the curve, respectively.

Table 6.7 shows that the smoke density rating of the natural rubber latex based jute composite is less than that of the traditionally used noise control material, fiber glass.

TABLE 6.7   Smoke Density of Natural Rubber Based Jute Composite

Material	Smoke density rating (%)	Max. light absorption (%)
2.5% natural rubber based jute composite	11.36	7.2
5% natural rubber based jute composite	9.89	6.9
Fiber glass	20.55	24.7

6.5.3   ACOUSTICAL PROPERTIES

Some part of the sound energy incident on a noise control material gets reflected, some get absorbed and the rest get transmitted, as shown in Fig. 6.4. This phenomenon of acoustical interaction of the sound energy with the material is very strongly dependent on the physical structure and physical properties of the material. The sound absorption coefficient is defined as the amount of energy absorbed by the noise control material to the energy incident on the material. The materials which have high sound absorbing coefficients are usually used as sound absorbers. A good barrier material "blocks" the sound and reflects it back to the incident medium. A good sound absorbing material has a poor reflecting capability. The sound transmission coefficient of an acoustical panel is the fraction of sound power in the incident airborne that appear in the transmitted airborne wave on the opposite or rear side of the acoustical panel. These coefficients are a function of the frequency of the sound wave, and thus need to be known for the noise control materials in the frequency or a frequency band at which noise has to be controlled[15,16]. Traditional sound absorbers are open cell porous polyurethane foam, fiber glass and naturally occurring materials like coir, cotton, hemp and jute. Traditionally used sound barrier material for industrial noise control are heavy concrete, steel, lead and so forth.

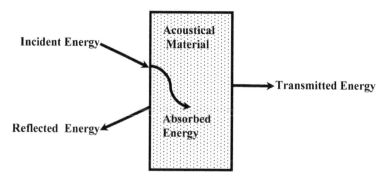

FIGURE 6.4   Phenomenon of sound interaction with an acoustical material.

The mechanism of sound absorption in a material is due to the viscous loss between the sound pressure waves while interacting with the walls of the pores in the material. The sound energy is thus dissipated as heat at the pores. At low frequencies this energy exchange is isothermal and at high frequencies it is adiabatic[17]. The sound absorption coefficient of a material depends upon its acoustical impedance at the surface. Several experimental techniques exist for the measurement of the same[18,19]. The fiber size, porosity, tortuosity, flow resistivity affect the impedance of the material and some of these values for the jute derivatives are measured by experiments and some of them can be estimated from analytical formulations. Many a times during numerical simulations of computer aided engineering models using techniques like the finite element method and the boundary element method the knowledge of impedance of these biocomposite materials are required to estimate the noise reduction obtained[20]. The sound absorption coefficients of recycled fibrous materials and effects of thickness, surface facing and compression on the sound absorption coefficients of some fibrous material are available in the literature.[21,22]

The research on natural fibers shows that Egyptian cotton can be used as an acoustical material in different forms[23]. It is an effective, cheap product and possesses a high sound reducing capability. Also, it is easy to use and creates no health risks compared to common commercial acoustical materials. Work has been done on natural fibers such as jute, coir and sisal where the structure-property relationship of these fibers including fracture modes has been determined. Attempts to incorporate them in polymers and characterization of these new composites, with and without subjecting them to environmental conditions, have also been reported[24]. Relationship between the sound absorption coefficients of a cover made of woven cotton fabric with its intrinsic parameters has been determined[25]. The effect of air space, behind and/or between the layers of new sample made of local textile material (100% cotton) which is produced from a specially woven structure on the absorption coefficient has been studied for use as a sound absorbing curtain[26]. Studies show that natural fiber composites are likely to be environmentally superior to glass fiber composites in most cases for the following reasons: (1) natural fiber production has a lower environmental impact compared to glass fiber production; (2) natural fiber composites have higher fiber content for equivalent performance, reducing the more polluting base polymer content; (3) the light-weight natural fiber composites improve fuel efficiency and reduce emissions, especially in automobile applications; and (4) end of life incineration of natural fibers results in recovered energy and carbon credits[27].

Sound proofing properties such as absorption coefficient and transmission loss index of natural organic multilayer coir fiber has been studied[28]. The effect of perforated size and air gap thickness on acoustical properties of coir fiber has also been studied. Comparison of acoustical properties between coir fiber and oil palm fiber has also been reported. The results obtained show that the coconut coir fiber gives an average noise absorption coefficient of 0.50. It shows a good noise absorption coefficient for higher frequencies but less for the lower frequencies. The oil palm fiber gives an average noise absorption coefficient of 0.64. The oil palm fiber shows a good noise absorption coefficient for higher frequency region compared to lower frequency region. Both fibers have a high potential to be used as sound absorber materials. The potential of using coconut coir fiber as sound absorber is also there. The effects of porous layer backing and perforated plate on sound absorption coefficient of sound absorber using coconut coir fiber were studied and implemented using woven cotton cloth as a layer type porous material in car boot liners in automobile industry[29]. Another type of natural sound absorbing material such as industrial tea-leaf-fiber waste material for its sound absorption properties has been investigated[30]. The experimental data indicate that a 1 cm thick tea-leaf-fiber waste material with backing provides sound absorption which is almost equivalent to that provided by six layers of woven textile cloth.

The technology based on the synthetic fiber composites made up of glass, Kevlar or carbon has played a vital role in noise reduction applications in the aerospace industry since 1950. The advancement in the composites design after reaching the aerospace requirements is targeted for the general industrial and domestic sectors. In contrast, the increased usage of electrical and mechanical appliances at home and industries has created a concern for noise pollution. Even though synthetic composites possess specific properties like light-weight, high strength-to-weight ratio and stiffness, they are not much applicable for the industrial and domestic sectors due to the high cost of the raw materials. Further these materials are harmful when kept exposed in the open environment. Though recently rice hull has been added to open cell polyurethane foam and its acoustical properties evaluated[31]. On the basis of these aspects, the new direction in industrial application on sound proofing materials based on natural biocomposites reinforced with fibers of sisal, coir, jute, etc. is steadily developing in the past few years.

There are many theoretical models available for predicting the sound absorption coefficients of sound absorbing materials[32]. Though the sound absorption coefficient can be measured, the knowledge of certain acoustical parameters help the material developer to estimate the sound absorption coefficients well before the material is made. Some such parameters are density, fiber size, porosity, tortuosity, flow resistivity and characteristic lengths.

6.5.3.1   FIBER SIZE

By using a scanning electron microscope, the jute fiber distribution is obtained as shown in Fig. 6.5. Based on the statistical averaging at the various places of different fibers, the effective diameter of the jute fiber was estimated. This is used to estimate the density and porosity of the sample by gravimetric analysis.

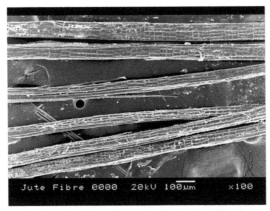

FIGURE 6.5   Photomicrograph of jute fibers distribution by SEM.[33]

6.5.3.2 POROSITY

Porosity is a measure of the air volume in the material. The porosity of jute (H) is estimated using Eq. (8)

$$H = 1 - \frac{\rho}{\rho'} \tag{8}$$

where, ρ is the density of the jute sample and ρ' is the density of the jute fiber[34]. The more the pores, the more is the interaction of the acoustical waves with the material at the pores and thus better is the sound absorption. Though an optimum amount of porosity is required, for if the material is highly porous it will let the sound wave pass without any interaction at the pores. Polyurethane foams are good absorbers of sound, but then they have to be of the open cell type and not the closed cell type.

6.5.3.3 FLOW RESISTIVITY

The flow resistivity is defined by the ratio of the static pressure difference (ΔP) to the product of the velocity V and the thickness l of the porous sample, as given in Eq. (9). Its unit is Ns/m⁴

$$\sigma = \frac{\Delta P}{Vl} \tag{9}$$

For measuring flow resistivity a sample of 25.4 mm thickness was used. With increasing the flow resistivity the sound absorption increases till an optimum value of the resistivity, beyond which the material reflects and the sound absorption decreases.

6.5.3.4   TORTUOSITY

Tortuosity ensures that the sound waves in a porous material do not travel in a straight path but interact with the pores in a tortuous path thus having maximum energy transfer at the pores. The tortuosity (α_∞) is determined by using the empirical formula in terms of porosity (H) as given by Eq. (10)[35].

$$\alpha_\infty = 1 + \frac{1-H}{2H}$$

(10)

6.5.3.5   CHARACTERISTIC LENGTHS

There are two measures of characteristic length which are of use. The viscous characteristic length is related to the size of the interconnection between two pores of the porous sound absorbing material. The thermal characteristic length is related to the diameter of the pore of the interconnecting channels. The thermal characteristic length is given as the ratio of the total inner surface area to the volume of the pore.

Here, the viscous and thermal characteristic lengths were estimated based on Allard's model and its extensions to Biot theory[35]. Assumption was made that jute fibers are cylindrical. Viscous characteristic length ($\wedge$) depending on the frame geometry and is given by Eq. (11)

$$\wedge = \frac{1}{2\pi r l}$$

(11)

where $2\pi r l$ is total perimeter of fiber per unit volume of material, is diameter of fiber and total length of fiber per unit volume of material defined by Eq. 12

$$l = \frac{\rho'}{\pi r^2 \rho}$$

(12)

ρ and ρ' are density of the sample and density of fiber, respectively.

Allard showed that for the material with porosity close to unity, $\wedge' = 2\wedge$. For material having pores of triangular cross section, $\wedge' = 1.14\wedge$ where $\wedge'$ is thermal characteristic length[35].

The above properties of these jute felt and jute fibers are shown in Table 6.8. The range of diameter of jute fibers which was measured by scanning electron microscope is 50–90 μm. By the statistical averaging of diameter at different locations, the effective diameter of single jute fiber is 68 μm. Density is about 1084.4 kg/m³.

TABLE 6.8   Acoustical Parameters of jute samples[33]

Material	Porosity (H)	Flow resistivity (σ) Ns/m⁴	Tortuosity (α_∞)	Characteristic lengths μm	
				$\wedge$	$\wedge'$
Jute felt	0.91	33190.84	1.05	1.51	3.02
Jute fiber	0.69	20087.72	1.22	5.28	6.02

6.5.4   *SOUND ABSORPTION COEFFICIENT*

Normal specific sound absorption coefficient of the materials has been determined by using an impedance tube, two microphones, dual channel frequency analyzer and the Indian Institute of Technology Kharagpur developed MATPRO software[36]. Test has been done as per ASTM E-1050 standard[19]. The impedance tube fabricated at the Indian Institute of Technology Khaargpur is used to measure the normal specific sound absorption coefficient of the biocomposite materials is shown in Fig. 6.6. Noise reduction coefficient (NRC), a simple quantification of absorption of sound by an acoustical material was calculated by averaging the four values of acoustical normal specific absorption coefficient at specified octave band levels. The NRC value lies anywhere between 0 and 1. Higher the NRC value better is its sound absorbing property. Figure 1.7 gives a comparison of the normal specific sound absorption coefficient of some of the sound absorbing materials used for noise control. Each of the material in Fig. 6.7 is of 25 mm thickness with a rigid backing. The normal specific sound absorption coefficient of all the materials increases with frequency. Fiberglass and wool have high sound absorption coefficients. Gypsum board has a low sound absorption coefficients since it is denser with less porosity. The natural materials like jute, cotton and coir also have relatively high sound absorption coefficient. The polyurethane open cell has a sound absorption coefficient higher than that of gypsum though less than the porous natural materials and fiberglass.

FIGURE 6.6   Impedance tube setup.

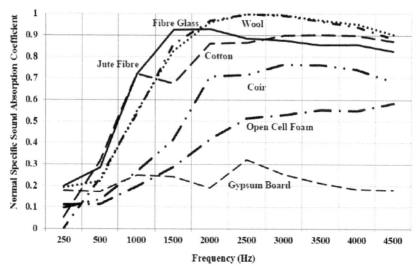

FIGURE 6.7   Sound Absorption of Noise Control Materials.

Table 6.9 gives the values of the normal specific impedance with both the real and imaginary part of a 50 mm jute felt with a density of 117.2 kg/m³ as a function of frequency from 100 Hz to 1000 Hz. The jute felt was provided with a rigid backing. The data in the Table 6.9 can be used in numerical simulations[20].

TABLE 6.9   Normal Specific Impedance of 50 mm Thickness Jute Felt

Frequency (Hz)	100	200	300	400	500	600	700	800	900	1000
Real Specific impedance	-0.0507	1.7120	0.8777	0.8560	0.9469	0.8905	1.0387	1.0909	1.1382	1.1969
Imaginary Specific impedance	-0.2700	-4.4004	-2.4079	-1.8327	-1.3750	-1.1178	-0.8590	-0.6663	-0.5084	-0.3880

6.5.4.1   EFFECT OF AIR GAP ON SOUND ABSORPTION

The sound absorption coefficient improves with frequency since at high frequencies the wavelengths reduce and for a given sample the velocity is a maximum at the interface. Better sound absorption occurs when the incident velocity is at a maximum at the interface. Thus to improve the low frequency sound absorption either the thickness of the sound absorbing material can be increased or an air gap between the material and the rigid backing can be provided. Usually in many applications where low frequency absorption is to be improved an air gap is usually provided as shown in Fig. 6.8. Figure 6.9 shows the effect of air gap on the sound absorption coefficient of a 50.8 mm jute fiber with a density of 348.2 kg/m^3.

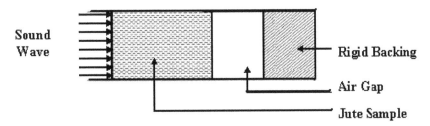

FIGURE 6.8   Configuration of using jute fiber with air gap in between a rigid backing.

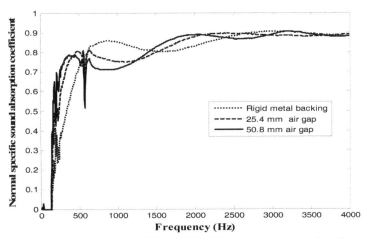

FIGURE 6.9   Sound absorption coefficient of jute fiber sample with varying air gap.

6.5.4.2   EFFECT OF CHEMICAL SURFACE TREATMENT ON SOUND ABSORPTION

Usually in certain applications when air flows over the jute fibers at grazing incidence, at times the fibers may come out of the jute felt with the airflow. In order to prevent the fibers from separating natural rubber is sprayed on to the jute felt surface. However, this coating of natural rubber may close some of the pores of the jute felt and thus decreases its sound absorbing capability. Thus rubber spraying has to be done with caution. Table 6.10 shows that untreated jute felt (density 117 kg/m^3)

has higher NRC value (0.85) as compared to 1% natural rubber latex jute composite (density 219 kg/m³).

TABLE 6.10 NRC Values of Untreated and NR Sprayed Jute Felts (400 gsm)

Treatment (400 gsm jute felt)	Density (kg/m³)	NRC
untreated	117	0.85
1% natural rubber	219	0.80
5% natural rubber	311	0.78

6.5.4.3 ACOUSTICAL TRANSMISSION LOSS

The acoustical transmission loss of an isotropic material can be estimated by Eq. (13), which characterizes the sound blocking capabilities of sound barrier materials. This equation states that for every doubling of the frequency the transmission loss increases by 6 dB. For a uniform density of the material, the transmission loss also increases by 6 dB. This is true below the first coincident frequency of the panel made out of such materials. However for composite materials Eq. (13) is not applicable since the mass density is not uniform. It is recommended to measure the transmission loss in such case. The transmission loss (TL) of materials can be determined experimentally and standards exist for the same[37,38]. A view of the setup used to measure the TL of the biocomposite panels is shown in Fig. 6.10. The two microphone sound intensity probe is used to measure the sound intensity with and without the sample whose transmission loss is to be determined. A random noise source is placed at the bottom of the box[39].

FIGURE 6.10 Transmission loss being measured.

$$TL = 20 \log_{10}(mf) - 42 \, [\text{dB}] \tag{13}$$

where, m is surface density in kg/m² and f is frequency in Hz.

The coincident frequency, f_c for a panel can be estimated by Eq. (14). At the coincident frequency there is drop in the value of the transmission loss of the material.

$$f_c = \frac{Eh^3c^3}{2\pi} \sqrt{\frac{\rho_s h}{12(1-v^2)}} \tag{14}$$

where E is the modulus of elasticity of the material, h is the thickness of the panel, c is the speed of the longitudinal sound wave in the material, r_s is the density of the material and v is the poison's ratio.

The natural rubber based jute composites were prepared as per the following procedure[39]. Jute felt of 400 gsm specimens were dried in an oven for 1hour to remove the water content in the specimen. The jute felt were treated with 1% NaOH (alkali) solution for 1 h. This alkali treatment was used to remove the impurities in the specimens. These alkali treated jute felts were again washed by water till they became alkali free. The washed jute felts were dried in an oven at 80°C for an hour. The dried felt was then dipped in 1% (by volume) natural rubber solution for 1 h. Excess rubber latex was drained off and the rubber treated jute felts were dried in a dry room for 1 h. Jute-based natural rubber latex composite was prepared by pressing ten pieces of natural rubber treated jute felts in a hydraulic press at 140°C with a load of 8 ton for 15 min. Similarly 2.5% natural rubber, 5% natural rubber and 10% natural rubber jute composites were prepared keeping all other parameters same. In all the sample preparations, natural rubber was used as bonding agent between the interfaces of the fibers.

The measured TL of all the samples are given in Table 6.11[36]. Usually a single number is preferred to represent the transmission characteristic of a barrier material known as the sound transmission class (STC)[37]. The STC ratings of the measured natural rubber latex treated jute composite of 5 mm thickness is shown in the last line of Table 6.9. These jute base composite panels of 5 mm thickness have comparable STC ratings to that of 3 mm Aluminum plate. The mass density of the panels have a strong influence on the TL. Higher the surface density higher will be the TL.

TABLE 6.11 Transmission Loss in dB of Natural Rubber (NR) Treated Jute Composite (152.4 mm × 152.4 mm)

Frequency (Hz)	1% NR	2.5% NR	5% NR	10% NR	15% NR
63	14.2	13.6	15.6	14.3	14.6
125	21.4	18.7	20.8	21.0	21.4
250	29.0	28.7	29.1	29.0	29.1
500	35.3	35.1	36.1	35.9	35.8
1000	40.5	39.9	42.5	43.7	43.2
2000	47.5	46.6	48.3	50.3	49.4
4000	59.1	56.2	60.2	58.7	56.6
STC Rating	37	38	39	40	40

6.5.5   DAMPING

Mechanical structures may vibrate during their operations. The air molecules near the structure thus receive the vibration energy and oscillate, and an air-borne sound wave is thus generated. To reduce the intensity of these sound waves, the vibration of the structure needs to be reduced. There exits three important modes of vibration control, namely the stiffness control, the mass control and the damping control. At the resonant frequency of the structure the vibration level and thus the airborne sound can be brought down by increasing the damping, and at frequencies below the resonant frequency the vibration is brought down by controlling the stiffness and at frequencies above the resonant frequency the vibration of the structure is brought down by controlling the mass[16].

Damping in a material is usually quantified by few of these related terms like, damping factor, loss factor, decay rate and damping capacity[40,41]. There exit standards for measuring damping[42]. The damping of the materials also vary with temperature and using a Dynamic Mechanical Analyzer (DMA) the same can be measured.

Experimental modal analysis has been done on composite plates with polymer base and coconut fiber to determine the frequency response function between the vibration response and the excitation force. From the measured frequency response function using the half power method the damping factor of the material can be determined[43,44]. It has been reported that polyester composites with 15% coconut coir fiber by volume have high damping ratio[44]. Polypropylene composites with various natural fibers like kenaf fibers, wood flour, rice hulls and newsprint fibers were developed. The variation of damping with temperature from −60°C to 120°C which were measured using a DMA has been reported. In this measurement a heating rate of 2°C/minute was applied[45]. Experimental investigations have been made to determine the sound absorption coefficient and the damping loss factors of natural fiber (flax) reinforced polyethylene honeycomb core panels. It is reported that the material constitution, fiber lengths and orientations yield to different behavior of the honeycomb cores[46].

6.5.6 MECHANICAL PROPERTIES

In engineering applications of jute-based composite panels the knowledge of their tensile and flexural strength is required. Several researchers have measured and reported the mechanical properties of biocomposites including jute and its derivatives[47,49]. The Tensile test was measured as per the ASTM D638 standard and the flexural strength was measured as per the ASTM D790 standard[50,51]. It has been observed that by a pretreatment of raw jute fibers with ultra violet light a better strength is obtained[5]. Figure 6.11 shows the SEM view of a untreated and UV treated failed jute fiber. A facture mode can be seen in Fig. 6.11(a) due to poor bonding between the fiber matrix and the natural rubber latex resin. The bonding is better between the fiber and the matrix in the case of pretreatment. However the improvement in mechanical strength has no significant change in the acoustical transmission loss of the jute composite panels as long as there is no change in the density of the panel.

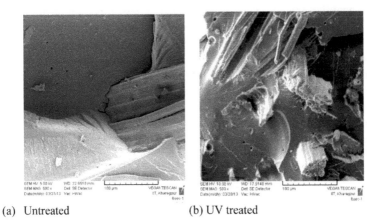

(a) Untreated   (b) UV treated

FIGURE 6.11   SEM micrograph of untreated and UV pretreated jute fiber in the matrix of the jute composite.

6.6   INDUSTRIAL APPLICATIONS FOR NOISE CONTROL

The jute derivatives in the form of fiber, felt, woven textile, composite panels have wide engineering applications for noise control. In architecture and building acoustics, noise control materials are used for treatment in the walls and ceiling for improving the reverberation time and speech intelligibility[52]. Due to consumer awareness and global competition among manufacturers of home appliances like vacuum cleaners, refrigerator, washing machines, room air-conditioners and the like there is a constant endeavor to improve the product quality by implementing many noise and vibration reduction technologies in their products. In order to control noise of a product the noise sources has to be ranked. Nowadays many experimental techniques exist for such noise source identification like the sound pressure level mapping, sound intensity method and the acoustical holography method[17]. There is a strict vehicle pass-by noise regulation in each country, which has to be adhered to the automobile manufacturers. Every automobile manufacturer is putting in efforts to bring about noise, vibration and harshness (NVH) reduction in the vehicles they manufacture. Reports of many automobile companies using biocomposite materials for NVH reduction are available[6,53,54]. The authors have helped few manufacturers in reducing the noise of their products by using such jute-based green technology, few such cases have been reported[55,56].

6.6.1 ARCHITECTURAL NOISE CONTROL

In the construction of auditorium, classroom, office buildings, residential and commercial buildings architects pay attention to the acoustic quality of such spaces. The acoustic qualities of such spaces are evaluated by few important measured / design parameters like noise Criterion (NC) rating, reverberation time and speech interference level (SIL). While designing such spaces the architects and engineers aim to obtain desirable value of the above parameters. The acoustical property of the building materials plays a significant role to obtain the target parameters. In this section the use of jute and its derivatives for improvement of the acoustical quality of the architectural space is presented.

The reverberation time of an architectural space is defined as the time taken for a sound level in a room to decrease by 60 dB[1]. The reverberation time is given by Eq. (15),

$$RT = \frac{0.161V}{A} \qquad (15)$$

where RT is the reverberation time defined as the time taken for a sound to decay by 60 dB after the sound source is suddenly switched off.

V is the volume of the auditorium in m³.

A is the total absorption of the auditorium in m²-sabins.

The absorption unit of 1 m²-sabins represents a surface capable of absorbing sound at the same rate as 1 m² of a perfectly absorbing surface, for example an open window.

However for practical purposes people use generally the Eyrings's formula for reverberation time is used which is given in Eq. (16) below

$$RT = \frac{0.161V}{-S\ln(1-\alpha)} \qquad (16)$$

where

$$\alpha = \frac{\alpha_1 s_1 + \alpha_2 s_2 + ... + \alpha_n s_n}{s_1 + s_2 + ... + s_n}$$

is the mean absorption coefficient of the room.

$S = s_1 + s_2 + ... + s_n$ is the areas of the various materials.

$\alpha_1, \alpha_2, ..., \alpha_n$ is the respective absorption coefficient.

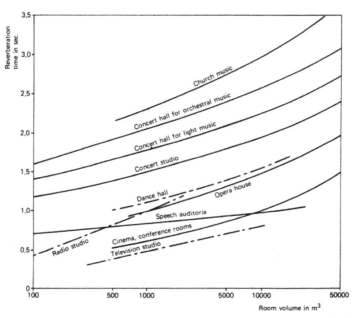

FIGURE 6.12 Typical variation of reverberation time with volume for auditoria to have good acoustical properties (with permission from Bruel & Kjaer).

Figure 6.12 gives the typical variation of overall reverberation time as a function of room volume[1]. By an international standard the reverberation time of an architectural space can be measured and improvement of the reverberation can be made by having the appropriate amount of absorption on the wall surface[57]. The speech interference level (SIL) is the arithmetic average of the sound pressure level in octave band centered at 500, 1000, 2000 and 4000 Hz and the preferred Speech Interference Level (PSIL) is the arithmetic average of the SPL in octave bands of 500 Hz, 1 kHz and 2 kHz. The PSIL as a function of distance from the receiver in presence of background noise is shown in Fig. 6.13[2]. Apart from using reverberation time and speech interference level to characterize the acoustics of a room the noise criterion (NC) curve is widely used in engineering. The NC curves are a set of sound pressure level in the various octave bands. To have a particular NC rating of the room the measures SPL in octave band should be less than or equal to the SPL given for the particular NC rating. A typical NC Curve is shown in Fig. 6.14 and the recommended NC values for various environments are given in Table 6.12.

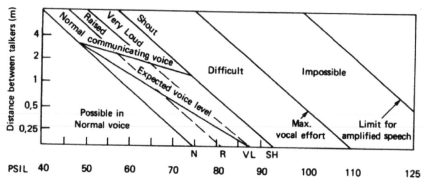

FIGURE 6.13   Communication limits in the presence of background noise (with permission from Bruel & Kjaer).

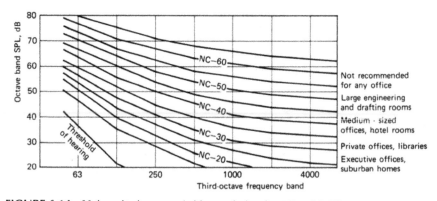

FIGURE 6.14   Noise criteria curves (with permission from Bruel & Kjaer)

TABLE 6.12 Recommended NC Values for Various Environments (with permission from Bruel & Kjaer)

Environment	Range of NC Levels likely to be acceptable
Factories (heavy engineering)	55–75
Factories (light engineering)	45–65
Kitchens	40–50
Swimming baths and sports areas	35–50
Department stores and shops	35–45
Restaurants, bars, cafeterias and canteens	35–45
Mechanised offices	40–50
General offices	35–45
Private offices, libraries, courtrooms and schoolrooms	30–35
Homes, bedrooms	25–35
Hospital wards and operating theatres	25–35
Cinemas	30–35
Theatres, assembly halls and churches	25–30
Concert and opera halls	20–25
Broadcasting and recording studios	15–20

6.6.2   ARCHITECTURAL APPLICATION OF BIOCOMPOSITES

Jute, cotton, coir, flax and their derivatives can be used as acoustical absorber in walls and ceilings of buildings. Such application of the biocomposites helps in controlling the reverberation time and the SIL. A typical case study using jute is presented here. Raw jute fiber, chopped jute strands and jute felt as shown in Fig. 6.3 can be used for wall and ceiling treatment. These jute derivatives can be held against a wooden board and faced with a perforated sheet or kept inside a jute cloth packet. The sound absorbing properties of such sample varies with thickness as discussed earlier. These jute derivatives can also be used in office partition and cubical ceilings. Since these materials cannot be put by themselves onto a wall they have to be kept inside a jute cloth packet which can be either glued or stapled to the wall. Some of these jute cloth packets can be stapled or screwed against a hard backing surface as shown in Fig. 6.15.

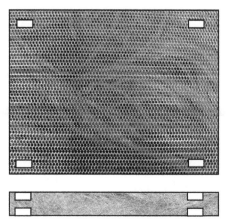

FIGURE 6.15   Jute fiber filled perforated panel, which can be glued or stapled.

Natural rubber based jute fiber composite panels have adequate transmission loss and can also be used as partitions in office cubicles. The transmission loss of such panels can also be measured as per ASTM E90 standard[38]. They also have adequate amount of flexural strength and can be used in such application. A typical used of jute fiber based panel with a perforated facing of 47% open are ratio used in our office cubical for noise reduction is shown in Fig. 6.16

FIGURE 6.16   Jute fiber based office cubicle partition.

Typical board size are 600 mm × 600 mm with a thickness of 50 mm. Figure 6.17 shows such boards being used in a room ceiling for controlling the reverberation time. Such jute boards can also be fixed on auditorium walls for improving the acoustics. These boards can be either screwed or nailed into the backing wall at the corners. These materials have good thermal stability properties and are fire retardant themselves, thus can be safely used[56].

FIGURE 6.17   Jute-based ceiling boards used in a room.

These boards have been used in ceilings of rooms for more than three years and have no signs of degradation in their physical condition. Depending upon the requirements, these acoustical boards can be made of larger dimensions for installation in the walls of auditoriums, cinema complexes, shopping malls, hospital, library and airports.

6.6.3   HOME APPLIANCES NOISE CONTROL

Jute-based composites can be used in the noise control of several home appliances like window air-conditioner, refrigerator, clothes washer and dryer, vacuum cleaner, dishwasher, etc. In many instances noise reduction of around 10 dB has been obtained by using the jute derivatives. In few other instances the noise reduction in terms of SPL has been hardly achieved, however there has been a significant improvement in the sound quality of the sound after the noise treatment. For instance a metallic tonal noise has been suppressed to a pleasant broadband noise. In fact currently noise control engineers are designing products with superior sound quality[58]. Case studies of noise control from few of them are given in the following sections.

6.6.3.1   VACUUM CLEANER

Vacuum cleaner is a very noisy and an irritating appliance in any house hold. Design improvements have been done to improve the sound quality of the radiated noise from such vacuum cleaner. A wet and dry domestic vacuum cleaner was used for noise control studies using jute derivatives. The physical dimensions and the weight of this dryer are 41.5 × 41.5 × 44.0 cm and 6 kg. The suction of the vacuum cleaner motor was 30 L/sec. From the sound intensity mapping using a two-microphone sound intensity probe the most noise producing component of the vacuum cleaner was found to be the exhaust pipe. A 2.5 cm lined dissipative muffler of 30 cm length consisting of jute fibers wrapped in jute textile was placed as shown in Fig. 6.18. An overall noise reduction of 8 dB was obtained by such a treatment with an improvement in its sound quality as well. The measured radiated sound power octave band spectrum of the vacuum cleaner with and without treatment is shown in Fig. 6.19. The overall radiated sound power of the vacuum cleaner was reduced to 57.1 dBA from 67.6 dBA with jute-based treatments[59]. The sound power measurements of the vacuum cleaner was done as per ISO 9614 standard[60]. It may be noted that the sound power of a sound radiating body is estimated by at first performing a normal sound intensity measurement over a surface area, and then summing up the product of the sound intensity times the surface area as given in Eq. (17). The sound intensity of the entire radiating area needs to be measured as a function of frequency. For a rectangular body, usually the intensity from the five radiating surfaces are measured, leaving aside the bottom surface, which may be placed against a sound radiating hard floor or an absorbing floor.

$$\text{Sound Power, } W = \Sigma\, I_i A_i \tag{17}$$

where I_i is the measured sound intensity and A_i is the corresponding normal surface area. Usually these areas are predefined and marked before the sound intensity measurements are done[61]. The sound power thus calculated is a function of frequency, and is usually represented in octave frequency bands. The sound intensity being a vector quantity is very much dependent on the direction of measurement. Thus enough care needs to be taken while traversing the two-microphone probe over a product during sound intensity measurements so that no directional error is made.

FIGURE 6.18   Vacuum cleaner with jute lined dissipative muffler.

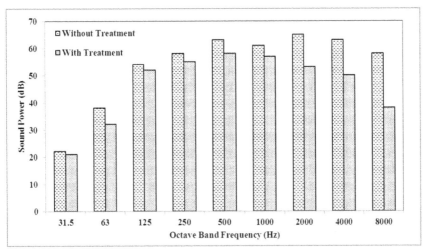

FIGURE 6.19   Radiated sound power level of the vacuum cleaner.

6.6.3.2   DOMESTIC CLOTHES DRYER

A domestic clothes dryer essentially consists of a rotating heated drum in which wet clothes are put. The drum is belt driven by an electric motor and hot air is forced onto the wet clothes through the drum and expelled out through a duct. The drum is enclosed in a sheet metal box. At the bottom corner of the box the drive electric motor is mounted. In order to reduce the structure borne noise, the motor is mounted at four locations through elastomeric mounts. The physical dimensions and mass of the dryer are 53 cm × 60 cm × 72 cm and 26 kg, respectively. Motor rating and Heater rating are 300 W and 1.8 kW, respectively. The drum used inside the dryer for placing the wet clothes is epoxy coated; a maximum of 5.5 kg of wet clothes can be dried in this dryer in an operation cycle. The dryer has a safety cut-off switches which limits the temperature inside the drum to 105°C. All measurements were done in the clothes dryer while running empty. Through sound intensity method noise mapping measurements it has been found that some of the major source of noise in a clothes dryer are the rectangular sheet metal shell, the motor and the blower exhaust. Various derivatives of jute are used for the noise control; jute felt is stuck to the inner walls of the rectangular shell by glue, jute felt is attached to the rear panel of the shell, the duct which carries the hot air out of the clothes drum is lined with jute fiber faced of 400 gsm jute textile. An overall noise reduction of 6 dB was obtained by such a treatment. Fig. 6.20 shows an opened view of the jute-lined shell of the clothes dryer. The octave band spectrum of the radiated sound power of the treated and untreated clothes dryer is shown in Fig. 6.21[62]. A similar type of treatment can be done for clothes washer, dishwasher and window type room air-conditioner.

FIGURE 6.20   Jute felt lined shell of domestic clothes dryer.

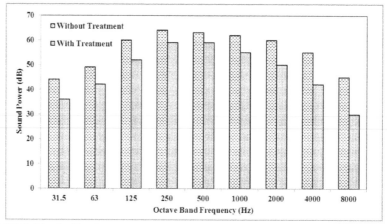

FIGURE 6.21   Radiated sound power level of the domestic clothes dryer.

6.6.3.3   REFRIGERATOR

Using the sound intensity method the noise radiated by a 3-door domestic frost free refrigerator was measured. The overall sound intensity contours of the five surfaces of the refrigerator in the frequency band from 20 Hz to 2000 Hz are shown in Fig. 6.22. This method helps to rank the noise source of the refrigerator. It is seen that the compressor located at the rear bottom of the refrigerator is the most noise producing part followed by the evaporator fan in the freezer compartment. A detailed measurement is done to determine the air-borne and structure borne noise path in the refregirator[63]. The radiated noise spectrum is rich in the harmonics of 50 Hz of the compressor operating speed, and the harmonics of the vane pass frequency of the evaporator fan. In order to reduce the noise radiated from the compressor apart from applying dampening materials on to the compressor shell, the sheet metal of the refrigerator body around the compressor are lined ure6.23 shows a view of the compressor compartment with the jute lining. Usually the compressor generates heat which is to be radiated while in operation, thus to ensure that the heat transfer from the compressor is not significantly affected, a temperature measurement on the compressor was done by monitoring round the clock using thermocouples. The temperature rise of the compressor with the jute treatment was 4°C with the maximum temperature reaching to 74°C. Further the jute-based temperature can withstand such increase in temperature with no adverse effects on the performance of the refrigerator.

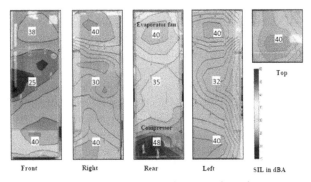

FIGURE 6.22   Sound intensity contour of the refrigerator for noise source ranking.

FIGURE 6.23   Jute lining in compressor compartment.

6.6.4 MACHINERY NOISE CONTROL

Jute can withstand high temperatures and are thus suitable for use in machinery enclosures placed around engines and compressors for noise control. In machinery enclosures they can be used in the inside lining of the walls. Usually such materials are faced with perforated sheets, so that most of the material is exposed to the incident sound field. They can also be used has lagging material around hot exhaust pipes and in HVAC ducts.

In noisy shop floors particularly in a punch press or a forging shop to reduce the reverberant field, usually sound absorber type panels are hung from the ceilings. Such sound absorbing panels can be made of the biocomposites like jute, coir, and cotton. The walls can be padded with biocomposite materials, curtains on factory windows and doors can be made up of such biocomposite textiles.

6.6.4.1   LINED DUCT

HVAC (Heating, ventilation and air-conditioning duct) are usually wrapped in fiber glass for thermal insulation. Fiber glass also have good sound absorbing properties and help in the breakout noise from the sheet metal duct. Jute-based composites have yet another application for noise control in ducts. The jute felt faced with a jute textile is applied to the inner walls of the duct. Since jute felt have good sound absorbing properties they help in the noise reduction. Since they are faced with jute textiles, the fibers do not separated due to the flow of air in the HVAC duct. It is recommended by ASHRAE (American Society of Heating, Refrigeration and Air-conditioning Engineers) that the maximum flow velocity of air at the exit of a duct from an AHU (Air Handling Unit) should be less than 7.5 m/s[64]. At these speeds in the duct the fibers do not separate from the felt. Figure 6.24 shows a sheet metal duct lined with 25 mm jute felt faced with jute textile of 400 gsm (grams per square meter). Fiber glass is not recommended to be used inside a duct, since if the fibers get carried by the air flowing inside the duct, they can be harmful to persons breathing the conditioned air, which is not so in the case of jute. Jute being hygroscopic can also withstand and moisture or condensate getting deposited in it in the inner of the duct walls. Jute felt can also be wrapped outside the duct to reduce the breakout noise.

The inside linings of AHU (Air Handling Units) of HVAC systems can also be lined with jute felt with a perforated facing backed with a 5 mm thick composite jute panel, for substantial noise reduction. Particularly in hospitals and classrooms where low noise levels are required, such a treatment can be done.

FIGURE 6.24   Jute lined duct.

6.6.4.2   MACHINERY ENCLOSURE

For noise control in machineries, many a times a lined enclosure is placed around a machine to reduce the radiated noise. Enclosures are used around IC engines, generator sets, compressors, blowers, punching press, etc. The walls of the enclosures are usually made out of thick steel sheets and are lined with sound absorbing materials of a certain thickness, which can withstand high temperature. Once enclosures are put around the machine, the temperature inside the enclosure increases which may affect the components of the machinery in particular the lubricants, which are used in bearings, etc. Thus usually enclosures for machines, which run continuously are provided with some sort of inlet air and exhaust airports. Since the ports on the enclosures have openings, it is recommended that an inline inlet and exhaust silencer is provided at the ports for bringing in fresh air and expelling out the warm air. Sometimes in few applications when the exhaust ports are not available cooling arrangement of the inlet air by a chiller is made. The inlet and exhaust silencers in such enclosure are of the dissipative type with jute-based lined ducts. Figure 6.25 shows a view of an IC engine-dynamometer setup covered by a perforated faced 50 mm jute fiber lined enclosure. The noise in the test cell reduced by around 10 dB by using such a jute-lined enclosure.

FIGURE 6.25   Perforated faced jute-lined enclosure for engine-dynamometer set

6.6.5   AUTOMOBILE NOISE CONTROL

The major sources of noise in an automobile are due to the power train, wind noise and tire road interaction noise. The power train noise is predominant at low engine speeds and in frequencies below 500 Hz whereas the wind noise and tire road interaction noise are predominant at high engine speeds at frequencies above 500 Hz. Automobile designers are always seeking for innovative technologies for reduced interior noise both for the driver and passengers as well as less pass-by noise for the bystander. Due to global competition among the manufacturers, designers are looking for greener materials, which are light weight, biodegradable with less car-bon foot print for noise control. In an automobile the predominant path of noise to the vehicle interior is either air-borne and structure borne. Conventionally structure borne noise path is impeded for the flow of energy by incorporating elastomeric vibration isolators. Whereas, airborne attenuation of noise is done by application of noise control materials at different parts of the vehicle.

Biocomposites made of jute, coir, cotton are used in automobile for reduction of the airborne noise[6,7]. The major path of airborne noise from the predominant sources of noise into an automobile interior passenger cavity are through the firewall (the wall below the dashboard separating the engine compartment from the passenger cavity), the floor of the vehicle, the roof of the vehicle, the parcel tray (area behind the rear seats above the rear bonnet), the engine hood or bonnet and the doors. By treatment with biocomposite materials in these locations the interior noise can be reduced. The biocomposite-based material can be broadly classified into sound ab-sorbing materials and sound blocking materials. In Table 6.13, a list of the major parts of an automobile is given with the type of treatment to be used for improvement in the interior noise of the automobile. The different parts of the vehicle are shown in Fig. 6.26. Researchers have used natural fiber like kenaf, jute, waste cotton and flax blended with polypropylene (PP) and polyester (PET) based nonwoven fibers for floor carpet in automobiles. In some applications a polyurethane foam is bonded below the floor carpet made of a blend of natural fibers for noise attenuation[65]. An-other advantage of using such biocomposites in automobiles is that after the end of the life of the vehicle when it is scrapped and the metal recycled, the biocomposite materials usually land up in landfills and would not harm the environment[65].

TABLE 6.13   Use of Biocomposites in Different Components of an Automobile for Interior Noise Reduction

Automobile Component	Sound Absorbing Material	Sound Blocking Material
Engine Bonnet Lining	X	
Firewall	X	X
Headliner	X	X
Door trim	X	
Floor	X	X
Parcel Tray	X	X
Rear trunk liner	X	X

Noise in the vehicle interior can be reduced by using these biocomposite materi-als in addition to improving the sound quality of the vehicle interior. After apply-ing these materials the efficacy of these materials can be evaluated by performing binaural sound recordings in an automobile using a human audio torso simulator as shown in Fig. 6.27. The improvements can be evaluated in terms of the sound quality metrics like sound pressure level, loudness level, sharpness, roughness and tonality ratio[58]. Such a facility exists at the Indian Institute of Technology Kharagpur to evaluate the sound quality of products.

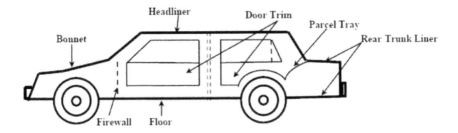

FIGURE 6.26   Different parts of a car where materials are used for noise control.

FIGURE 6.27   Sound quality measurements in automobile interior using human audio torso simulator.

In Table 6.13, it is noticed that at some places both the type of sound absorbing and sound blocking materials are used. This is for the fact that in a vehicle interior from the inside the sound has to be absorbed at the walls, and no sound from outside should enter into the vehicle interior space. Usually two layers of the materials are used in such situations. The sound-absorbing layer is facing the vehicle interior and the sound blocking is facing the vehicle exterior. These two materials are usually bonded at the interface with strong glue.

Though the exhaust system of an automobile is considered as a part of the powertrain system, here for a noise control purpose it is separately dealt. Usually a muffler or a silencer is used to reduce the automobile exhaust noise, which has a significant effect on the pass-by noise[66]. These mufflers are of two broad types, the reactive and the dissipative type. The reactive mufflers are usually designed based on the principles of impedance mismatch by introducing perforates, baffles and extended inlet/outlet pipes in an expansion chamber. Whereas the dissipative mufflers have high temperature withstanding capability sound absorbing material lined expansion chambers and tubes. Traditionally fiberglass has been used in such linings since they can withstand the very high exhaust gas temperature. The possibility of replacing these fiberglass materials with fire-retardant treated bio-composites exists. However other than the in-flow exhaust pipe noise, which is reduced by using the appropriate reactive or dissipative silencers, the muffler shell usually of sheet metal radiates noise, and this is known as the muffler breakout noise. By wrapping the muffler shell/chamber by a bio-composite made of jute felt the noise reduction can be brought. A tractor muffler with a jute felt jacket to reduced the breakout noise is shown in Fig. 6.28.

(a) without treatment   (b) with treatment

FIGURE 6.28   Tractor muffler with jute jacket for reduction of breakout noise (a) without treatment (b) with treatment.

6.7   CONCLUSIONS AND FUTURE DIRECTIONS

Human beings exposed to high levels of noise for a prolonged time can have permanent damage done to their ears. The human ear is nonlinear and the perception of sound heard by human being is to be clearly understood for implementing an effective noise control solution. Traditionally noise control has been done using expensive synthetic polymer based materials. However efforts are being made to use biocomposites for industrial noise control. Researchers around the globe are moving towards greener ecofriendly biocomposite materials for their use in various industrial noise control applications. Biocomposites based on ecofriendly materials like jute, coir, cotton, hemp and so forth are being used.

The three important elements in sound generation, propagation and reception are the source of sound, the path of sound and the receiver of sound. There can be an acoustical energy transfer between any of the three elements. The very basic step in noise control is to identify these three elements and then to rank and characterize the noise sources in terms of the amplitude of the sound and its frequency content.

Many experimental methods currently exist for the noise source identification like the sound pressure level mapping, the sound intensity mapping and acoustical holography.

In this chapter the application of jute a plant based fiber and its derivatives in the form of composites in industrial noise control has been discussed. The jute derivatives can be in the form of chopped pieces, fibers, felts, yarns, textile and composite panels. The physical, mechanical, thermal and acoustical properties of jute derivatives have been reported. Most of the values reported in this chapter are from actual measurements done at the various facilities of the Indian Institute of Technology Kharagpur as per international standards. For noise control, the jute derivatives can either be used as sound absorbers or sound barriers. The sound absorbers are quantified by the normal specific sound absorption coefficient and the sound barriers are quantified by the sound transmission loss properties. These values vary with frequency. Thus for effective noise control in a particular frequency the acoustical properties of the noise control material at that frequency need to be known.

Case studies from various applications made by the authors research group in using jute derivatives for noise control has been reported. Noise reduction using jute derivatives have been achieved in home appliances like domestic vacuum cleaner, clothes dryer and a refrigerator. Jute derivatives have been used for building acoustics to improve the reverberation time and in to reduce the transmission loss in office partitions. Jute has excellent fire retardant properties and has a high temperature stability, thus can be used in lining of machinery enclosures, HVAC ducts, breakout noise reduction in engine silencers. Jute derivatives in various forms can be used at different locations in an automobile for noise reduction, as well.

Research is progressing on the engineering of these materials for use in the transportation sectors of aviation and rail for noise reduction. In the materials front, materials researchers are working on improving the bonding strength of the fibers by suitable chemical pretreatment to the fibers. Research is being done on using the properties of these derivatives in the numerical prediction of the noise produced by different products where these materials are being used for noise control.

KEYWORDS

- **Acoustic Properties**
- **Biocomposites**
- **Environmental Friendly Materials**
- **Jute**
- **Natural Fibers**
- **Noise Control**
- **Sound Absorption**

REFERENCES

1. Ginn, K. B. (1978). Architectual Acoustics, 2nd ed. Bruel & Kjaer.
2. Hassall, J. R., & Zaveri, K. (1988). Acoustic Noise Measurements, 5th ed. Bruel & Kjaer.
3. Ramesh, M., Palanikumar, K., & Reddy, K. H. (2013). Mechanical Property Evaluation of Sisal- Jute Glass Fiber Reinforced Polyester Composites *Journal of Composites Part B*, 48, 1–9.
4. Gowda, T. M., Naidu, A. C. B., & Chhaya, R. (1999). Some Mechanical Properties of Untreated Jute Fabric-Reinforced Polyster Composites, *Composites part A: Applied Science and Manufacturing*, 30, 277–284.
5. Fatima, S., & Mohanty, A. R. (2013). Effect of Pre-Treatment Process on the Mechanical Strength and Transmission Loss Characteristics of Jute Fabric Composites, Proceedings of the 20th International Congress on Sound and Vibration, Bangkok, Thailand, July 7–11.
6. Koronis, G., Silva, A., & Fontul, M. (2013). Green Composites: A Review of Adequate Materials for Automotive Applications, *Composite: Part B*, 44, 120–127.
7. Alves, C., Ferrao, P. M. C., Silva, A. J., Reis, L. G., Freitas, M., Rodrigues, L. B., & Alves, D. E. (2010). Ecodesign of Automotive Components Making Use of Natural Jute Fiber Composites Journal *of Cleaner Production*, 18, 313–327.
8. ASTM D570–598 Standard Test Method for Water Absorption of Plastics.
9. Behera, A. K., Avancha, S., Basak, R. K., Sen, R., & Adhikari, B. (2012). Fabrication and Characterization of Biodegradable Jute Reinforced Soy Based Green Composite *Carbohydrate Polymer*, 329–335.
10. ASTM C518–610 Standard Test Method for Steady-State Thermal Transmission Properties by Means of the Heat Flow Meter Apparatus.
11. E1461–1511 Standard Test Method for Thermal Diffusivity by the Flash Method.
12. ASTM Standard D 2863–2906a Standard Test Method for Measuring the Minimum Oxygen Concentration to Support Candle-Like Combustion of Plastics (Oxygen Index).
13. FMVSS (1990). Standard 302 Flammability of Interior Materials Passenger Cars, Multipurpose Passenger Vehicles, Trucks and Buses.
14. ASTM Standard D 2843–2899. Standard Test Method for Density of Smoke from the Burning or Decomposition of Plastics.
15. Kinsler, L. E., Frey, A. R., Coppens, A. B., & Sanders, J. V. (2000). Fundamentals of Acoustics, 4th ed., John Wiley & Sons: New York.
16. Nortan, M. P. & Karcub, D. G. (2003). Fundamental of Noise and Vibration Analysis for Engineers, 2nd ed., Cambridge University Press.
17. Crocker, M. J. (2007). Handbook of Noise and Vibration Control, John Wiley & Sons, September.
18. Mohanty, A. R., Seybert, A. F., & Strong, W. F. (1991). Acoustical Property Determination by a Personal Computer System, Proceedings of Noise-Con, New York, 597–602.
19. ASTM Standard E 1050–1098 Standard Test Method for Impedance and Absorption of Acoustical Materials using a Tube, Two Microphones and A Digital Frequency Analysis System.
20. Mohanty, A. R., St. Pierre B. D., & Narayansami P. S. (2000). Reduction of Structure-Borne Noise in a Truck Cab Interior by Numerical Techniques *Applied Acoustics*, 59(3), 1–17.
21. Seddeq, H. S., Aly, N. M., Marwa, A. A., & Elshakankery, M. H. (2012). Investigation on Sound Absorption Properties for Recycled Fibrous Materials *Journal of Industrial Textiles*, 43(1), 56–73.
22. Seddeq, H. S. (2009). Factors Influencing Acoustic Performance of Sound Absorptive Materials *Australian Journal of Basic and Applied Sciences*, 3(4), 4610–4617.
23. Shenoda, F. B., Melik, R. W., & Shukry, N. (1987). Egyption Cotton and Flax Shives as Acoustical Materials, *Research and Industry*, 32(3), 183–190.

24. Satyanarayana, K., Mukherjee, P. S., Pavithran, C., & Pillai, S. G. K. (1990). Natural Fiber Polymer Composites *Cement & Concrete Composites*, 12, 117–136.

25. Shoshani, Y., & Rosenhouse, G. (1990). Noise Absorption by Woven Fabrics *Applied Acoustics*, 30, 321–333.

26. Hanna, Y. I., & Kandil, M. M. (1991). Sound Absorbing Double Curtains From Local Textile Materials *Applied Acoustics*, 34, 281–291.

27. Joshi, S. V., Drzal, L. T., Mohanty, A. K., & Arora, S. (2004). Are Natural Fiber Composites Environmentally Superior to Glass Fiber Reinforced Composites? *Composites: Part A*, 35, 371–376.

28. Zulkifli, R., Nor, M. J. M., Tahir, M. F. M., Ismail, A. R., & Nuawi, M. Z. (2008). Acoustical Properties of Multi-Layer Coir Fibers Sound Absorption Panel *European Journal of Scientific Research*, 8(20), 3709–3714.

29. Zulkifli, R., Nor, M. J. M., & Zulkarnain, (2010). Noise Control Using Coconut Coir Fiber Sound Absorber with Porous Layer Backing and Perforated Panel *American Journal of Applied Sciences*, 7, 260–264.

30. Sezgin, E., & Haluk, K. (2009). Investigation of Industrial Tea-Leaf-Fiber Waste Material for its Sound Absorption Properties *Applied Acoustics*, 70, 215–220.

31. Wang, Y., Zhang C., Ren, L., Ichchou, M., Galland, M. A., & Bareille O (2013). Influences of Rice Hull in Polyurethane Foam on its Sound Absorption Characteristics *Polymer Composite*, 1–9.

32. Ingard, U. (1994). Notes on Sound Absorption Technology, Noise Control Foundation Version, 94–90.

33. Fatima, S., & Mohanty, A. R. (2011). Acoustical and Fire Retardant Properties of Jute-based Composite Materials *Applied Acoustics*, 78, 108–114.

34. Attenborough, K. (1993). Models for the Acoustical Characteristics of Air Filled Granular Materials *Acta Acustica*, 64, 27–30.

35. Allard, J. F. (1993). Propagation of Sound in Porous Media, Elsevier Applied Science.

36. Mohanty, A. R. December (2000). Acoustical Material for Automotive NVH Reduction, Proceedings of the IUTAM Symposium on Designing for Quietness, India, 12–14.

37. SAE Standard J 1400, Laboratory Measurement of the Airborne Sound Barrier Performance of Automotive Materials and Assemblies.

38. ASTM E 90–109, Standard Test Method for Laboratory Measurement of Airborne Sound Transmission Loss of Building Partitions and Elements.

39. Fatima, S. (2010). Noise Control of Domestic Appliances by Jute-based Materials. Master's Thesis, Indian Institute of Technology Kharagpur, India.

40. Inman, D. J. (1994). Engineering Vibration, First ed. Prentice Hall.

41. Ewins, D. J. (1986). Modal Testing: Theory and Practice, Research Studies Press Ltd, England.

42. ASTM–E756, Standard Test Method for Measuring Vibration Damping Properties of Materials.

43. Chakraborty, S., Mukhopadhyay, M., & Mohanty, A. R. (2000). Free Vibration Responses of FRP Composite Plates: Expermental and Numerical Studies *Journal of Reinforced Plastics and Composites*, 19(7), 535–551.

44. Bujang, I. Z., Awang, M. K., & Ismail, A. E. (2007). Study on the Dynamic Characteristic of Coconut Fiber Reinforced Composites, Regional Conference on Engineering Mathematics, Mechanics, Manufacturing & Architecture, Kuala Lumpur, November 27–28.

45. Tajvidi, M., Falk, R. H., & Hermanson, J. C. (2006). Effect of Natural Fibers on Thermal and Mechanical Properties of Natural Fiber Polypropylene Composites Studied by Dynamic Mechanical Analysis *Journal of Applied Polymer Science*, 101, 4341–4349.

46. Petrone, G., Rao, S., Rosa, S. D., Mace, B. R., Franco, F., & Bhattacharya, D. (2013). Initial Experimental Investigations on Natural Fiber Reinforced Honeycomb Core Panels *Composites: Part B*, 55, 400–406.

47. Reddy, N., & Yang, Y. (2011) Completely Biodegradable Soyprotein-Jute Biocomposites Developed Using Water without Any Chemicals as Plasticizer *Industrial Crops and Products*, 33, 35–41.

48. Seki, Y. (2009). Innovative Multifunctional Siloxane Treatment of Jute Fiber Surface and its Effect on the Mechanical Properties of Jute/ Thermoset Composites *Materials Science and Engineering A*, 508, 247–252.

49. Reddy, N., & Yang, Y. (2011). Novel Green Composites Using Zein as Matrix and Jute Fibers as Reinforcement *Biomass and Bioenergy*, 35, 3496–3503.

50. ASTM D638–710, Standard Test Method for Tensile Properties of Plastics.

51. ASTM D790–810, Standard Test Methods for Flexural Properties of Unreinforced and Reinforced Plastics and Electrical Insulating Materials.

52. Egan, M. D. (1988). Architectural Acoustics, Mc-Graw-Hill, Inc.

53. Ashori, A. (2008). Wood Plastic Composites as Promising Green-Composites for Automotive Industries *Bioresource Technology*, 99, 4661–4667.

54. Thilagavathi, G., Pradeep, E., Kannaian, T., & Sasikala, L. (2010). Development of Natural Fiber Nonwovens for Application as Car Interiors for Noise Control *Journal of Industrial Textile*, 39(3), 267–278.

55. Fatima, S., & Mohanty, A. R. (2012). Noise Control of Home Appliances-The Green Way *Noise and Vibration Worldwide* 43(7), 26–34.

56. Green Noise Control with Indian Eco Material, Bruel & Kjaer, Denmark, Waves (2013), 2013–2022.

57. ISO 3382 (1975), Acoustics- Measurement of Reverberation time in Auditoria.

58. Lyon, R. H. (2000). Designing for Product Sound Quality, Marcel Dekker, Inc.

59. Fatima, S., & Mohanty, A. R. (2010). Jute as an Eco-Friendly Noise Control Material A Case Study *Journal of the Acoustical Society of India*, 36(1).

60. ISO 9614–9622: (1996). Determination of Sound Power Levels of Noise Sources Using Sound intensity-Part 2: Measurement by Scanning.

61. Fahy, F. J. (1989). Sound Intensity, Elsevier Applied Science.

62. Fatima, S., & Mohanty, A. R. (2010). Jute as an Acoustical Material for Noise Control of a Domestic Dryer, Proceedings of the 17th International Congress on Sound and Vibration, Cairo, July 18–22.

63. Fatima, S., Mohanty, A. R., Paradhe, S., & Tamizharasan, S. T. (2013). Vibro-Acoustic Source Path Characterization of a Domestic Refregirator, Proceedings of the 20th International Congress on Sound and Vibration, Bangkok, Thailand, July 7–11.

64. ASHRAE Handbook, HVAC applications, Atlanta, G. A. (2003).

65. Parikh, D. V., Chen, Y., & Sun, L. (2006). Reducing Automotive Interior Noise with Natural Fiber Nonwoven Floor Covering Systems *Textile Research Journal*, 76(11), 813–820.

66. Mohanty, A. R., & Fatima, S. (2013). An Overview of Automobile Noise and Vibration Control *Noise and Vibration Worldwide*, 44(6), 10–19.

CHAPTER 7

COMPLEX SHAPE FORMING OF FLAX FABRICS: ANALYSIS OF THE SOLUTIONS TO PREVENT DEFECTS

PIERRE OUAGNE and DAMIEN SOULAT

ABSTRACT

The possibility of manufacturing complex shape composite parts with a good production rate is crucial for the automotive industry. The sheet forming of woven reinforcements is particularly interesting as complex shapes with double or triple curvatures with low curvature radiuses can be obtained. To limit the impact of the part on the environment, the use of flax fiber based reinforcements may be considered for structural or semistructural parts. This study examines the possibility to develop composite parts with complex geometries such as a tetrahedron without defect by using flax based fabrics. An experimental approach is used to identify and quantify the defects that may take place during the sheet forming process of woven natural fiber reinforcements. Wrinkling, tow sliding, tow homogeneity defects and tow buckling are discussed. The origins of the defects are discussed, and solutions to prevent their appearance are proposed. Particularly, solutions to avoid tow buckling caused by the bending of tows during forming are developed. Specially designed flax based reinforcement architecture has been developed. However, if this fabric design has been successful for the tetrahedron shape, it may not be sufficient for other types of shapes and that is why the optimization of the process parameters to prevent occurrence of buckles from a wide range of commercial fabrics was also investigated with success.

7.1 INTRODUCTION

With the view to answer the weight reduction question, the replacement of metallic materials by composite materials exhibiting lower densities and higher stiffness, has been a great success in the aeronautic industry. Composite materials are an assembly of reinforcement materials, which confer the stiffness and a binder, (generally a polymeric resin) for the cohesion of the composite. For aeronautical parts, carbon or glass reinforcements are generally used. These carbon and glass fiber composites cannot be easily recycled even though processes such as mechanical grinding, pyrolysis, fluidized bed or solvolysis are studied at the laboratory scale [1,2]. In order to reduce the environmental impact of the composite part on the environment, the idea to replace the synthetic carbon and glass fibers by natural fibers such as flax or hemp has motivated numerous studies [3,7]. However, very few studies deal with the scale of the reinforcement and particularly of structural or semistructural reinforcements constituted of aligned tows woven for example according to a specifying fabric style. Few publications can be used to constitute a database reporting the mechanical properties of the natural fiber based reinforcements. At this scale, the choice of reinforcement structure (size of the tows, weaving style, etc.) is essential as it influences its mechanical characteristics[8,9]. Indeed, to manufacture high performance composite parts, it is necessary to organize and to align the fibers. As a consequence, aligned fibers architectures such as unidirectional sheets, noncrimped fabrics and woven fabrics (bidirectional) are usually used as reinforcement.

When dealing with weight reduction, it appears that the best gain can be obtained on complex shape parts. However, the possibility to realize these shapes in composite materials is still a problem to be solved. For example, only 25% of the Airbus A380 is constituted of composite materials. Several low scale manual manufacturing processes exists to realize these complex shape composite parts particularly for the military or the luxury car industries. The sheet forming of dry or comingled (reinforcement and matrix fibers mixed in a same tow) can be considered as a solution to manufacture at the industrial scale complex shape composite parts as this process shows a good production rate/cost ratio. Numerical approaches have been used to determine the process parameters to be used [10,12]. However, few of these studies dealt with complex shape parts for which specific defects such as tow buckles may appear [13]. The appearance of such defects may prevent the qualification of the part and indicate the limit of the reinforcement material behavior under a single or a combination of deformation modes. It is therefore important to quantify and understand the mechanisms controlling the appearance of defects so that the numerical tools developed in the literature for complex shape forming [14] can simulate them.

Natural fibers and particularly plant fibers have been explored as an alternative to synthetic fiber reinforcement for composite as they are characterized by lower density than glass fibers (1.5 for flax fibers; 2.6 for glass fibers), and because they potentially can be recycled or even degraded at the end of the composite life[14]. As these fibers are extracted from vegetal resources, a lot of studies deal with the properties of the natural fibers and particularly with their variability according to the place they are extracted in the plant, the climatic conditions during the growth of the plant, the treatments used to extract the fibers from the plant (retting, combing, etc.)[15,20]

If natural fibers show a lot of advantages such as biodegradability, nontoxicity, good insulation properties, low machine wear, etc.[21,22], the level of production of these fibers needs to be considered in such a way that food production is not affected. This also needs to be placed in parallel to socioecological impacts that may be encountered around the sand mining necessary for the production of glass fibers [23,25]. As a consequence, a large amount of studies has been devoted to investigate the behavior of individual fibers or group of few fibers of different types [26,29]. The studies globally showed that the tensile properties of the natural fibers can advantageously be compared to the ones of the glass fibers especially if one considers the specific tensile modulus and strength of flax fibers. As a consequence, the automotive industry is a candidate for the use of such fibers as this could lead for the same part performance to weight reduction of the composite [30].

All these studies, at the fiber scale, are justified by the fact that the natural fibers may show important variability in their mechanical properties and particularly when tensile strength and modulus are considered, because an apparent diameter is generally considered instead of a true cross-sectional area in the calculation of mechanical properties.[31] Review articles synthesize and compare at the fiber scale the performances of the natural fibers considered for technical application such as flax, hemp, jute, sisal, kenaf, etc.[32,34] The property variability is also discussed in these reviews as well as the disadvantages that may appear when considering the use of natural fiber in composites for large-scale production (the variability of the mechanical properties, the compatibility between matrix and natural fiber and the moisture absorption).

In order to avoid long considerations about the variability of the fiber properties, it may be interesting to consider for some manufacturing processes such as filament winding[35] or pultrusion[36] the scale of the tow or the scale of the yarn. The scale of the composite (natural fiber reinforcement combined to a polymeric resin) is also interesting if one wants to avoid considering the variability of the fiber properties[37,41]. The homogenized behavior at the composite scale depends on the reinforcement type (mat, woven fabric, noncrimped fabric), the resin used and the process chosen to manufacture the composite. The study of composite samples is also used to analyze the impact of the composite part all along its life cycle[42,43]. The energetic record to produce flax fibers for composite materials has been analyzed by Dissanayake et al.[44,45] They showed, in the case of traditional production of flax mats, with the use of synthetic fertilizers and pesticides associated to traditional fiber extraction such as dew retting and hackling, that the energy consumption linked to the production of a flax mat is comparable to the energy consumed during the production of a glass mat. They also showed that the spinning to produce yarns is an energy intensive operation and in that case, the glass woven fabric may show lower impact on the environment than an equivalent flax woven fabric if one considers an environmental energy viewpoint. As a consequence, it is recommended to use aligned fibers instead of spun yarns tows to produce natural fiber based woven fabrics.

Between the fiber and the composite scales, it may be interesting to study the behavior of natural fibers assemblies such as strands, tows, fabric with the view to optimize the composite manufacturing processes using these entry materials. As an example, studies for glass and carbon reinforcements showed that during the sheet forming process of dry (first step of the Resin Transfer Moulding (RTM)[46]) that the fabrics may be the subject to tension, shear and bending loads. Some of them may even be combined[47,50]. These loads may induce specific deformation states at the origin of forming defects. These defects may impact the quality of the composite part[51,53] and also modify the reinforcement permeability[54,58] that is a crucial parameter to control the impregnation of the porous reinforcements by liquid resin if Liquid Composite Molding (LCM) processes and particularly the RTM process are considered. The presence of defects also indicates the behavior limits of the fabric in a specific deformation mode. As an example the presence of wrinkles is generally associated to a limit in-plane shear behavior. These limits may be tested for each fabric on the different modes of deformation independently of the forming process[47,52,59,62]. For shear and tension, the strains taking place during the sheet forming process can be evaluated in-situ [62,63]. This has been performed for carbon and glass reinforcements, but which will be presented in this Chapter proposes to analyze the forming potentialities of a flax based fabric. After presenting the sheet-forming device used to shape the flax reinforcements, the work will introduce the different types of defects that may be encountered during sheet forming of textile fabrics, before concentrating on the feasibility of forming complex shapes without defects. Particularly, the solutions developed to reach this goal will be presented and discussed.

7.2   EXPERIMENTAL PROCEDURES

7.2.1   SHEET FORMING DEVICE FOR DRY TEXTILE REINFORCEMENT

A device presented in Fig. 7.1 was especially designed to analyze the possibility to form reinforcement fabrics. Particularly, the device was developed to examine the local deformations during the forming process[64]. The device is the assembly of a mechanical part and an optical part. The mechanical part consists of a punch/open die system coupled with a classical blank-holder system. The punch used in this study (Fig. 7.1.b) is a tetrahedron form with 265 mm sides. Its total height is 128 mm and the base height is 20 mm. The edges and vertices possess 10 mm radius for the punch and 20 mm for the die. As the punch possesses low edges radiuses, it is expected that large shear stains take place during forming. A triangular open die $(314\times314\times314$ mm$^3)$ is used to allow for the measurement of the local strains during the process with video cameras associated to a marks tracking technique[65]. A piloted electric jack is used to confer the motion of the punch. Generally, the punch velocity is 30 mm/min and its stroke 160 mm. The maximum depth of the punch is 160 mm. A classical multipart blank-holder system is used to prevent the appearance of wrinkling defects during the preforming tests by introducing tension on the fabric. It is composed of independent blank-holders actuated by pneumatic jacks that are able to impose and sense independently a variable pressure. The quality of the final preform may depend on several process parameters such as the dimensions, positions, and the pressure applied by each of the blank-holders can be easily changed to investigate their influence on the quality of the final preform[66]. Before starting the test, a square piece of fabric is positioned between the die and the blank-holders. The initial positioning of the fabric is of particular importance as it partly conditions the final tow orientations within the part. However, it is not possible to establish before the test their final position at the end of the forming and as a consequence their mechanical stiffness. So, for the tests presented below, it was chosen to align the warp or the weft tows with an edge of the tetrahedron (the opposite edge of Face C (Fig. 7.1.b). To avoid bending of the fabric under its own mass, a draw bead system is used to apply low tensions at the tow extremities. At the end of the performing test, the dry preform can be fixed by applying a spray of resin on its surface so that the preform can be removed from the tools and kept in its final state.

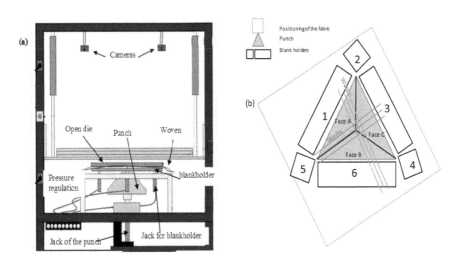

FIGURE 7.1 (a) The sheet forming device (b) Initial positioning of the fabric and position of the blank holders.

At the end of the preforming test, several analyzes at different scales can be performed. A first global analysis at the macroscopic scale concerning the final state of the preform before removing it from the tool can be performed. It consists in analyzing if the shape is obtained and if the shape shows defects. Another analysis, at the mesoscopic scale, consists in analyzing the evolution of the local strains (shear, tension) during forming.

Using this device, an experimental study to analyze with the tetrahedron shape, the generation of defects (wrinkles, tow buckles, tow sliding, vacancies, etc.) can be performed. The influence of the process parameter and particularly the blank-holder pressures on the generation and the magnitude of defects is also commented and analyzed.

7.2.2 TENSION AND SHEAR MEASUREMENT DEVICE

A biaxial tension device (Fig. 7.2) has also been used to characterize the tensile behavior of the reinforcements.

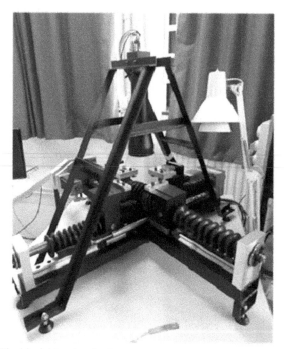

FIGURE 7.2   The biaxial tension device.

Biaxial and uniaxial tension tests as well as tensile test conducted on individual tows can be performed using this device. In-plane shear test can also be performed using the bias test samples. All these specific samples are shown in Fig. 7.3. For synthetic carbon or glass fabrics, the limit to failure is not reached during forming, and the tensile test are designed to analyze the possible nonlinearity of the stress-strain curves due to the 2D assembly of the woven textiles generally used. For natural fiber fabrics, the tensile limit of the fabric becomes particularly interesting as the tows used to elaborate the woven fabrics are manufactured from finite length fibers slightly entangled and held together by a natural binder are not expected to show comparable tensile resistance. The tensile strains for each considered tows are measured using a 2D version of the mark-tracking device described previously. The detailed description of the device as well as the procedure of the test may be found in Reference [67].

FIGURE 7.3 Mechanical characterization test of tows and woven fabrics: (a) tension of a tow; (b) uniaxial tension of a fabric; (c) in-plane shear behavior of a fabric; (d) biaxial tension of a fabric.

7.2.3 THE FLAX WOVEN FABRICS

The flax fabric (Fig. 7.4.a) used in this study, is a plain weave fabric with an areal weight of 280±19 g/m², manufactured by GroupeDepestele (France)[68]. The fabric is not balanced, as the space between the weft tows (1.59± 0.09 mm) is different to the one between the warp tows (0.26± 0.03 mm). The width of the warp and the weft tows are, respectively 2.53±0.12 mm and 3.25±0.04 mm. As a consequence, there are 360 warp tows and 206 weft tows per meter of fabric. The linear mass of the warp and the weft tows is the same and is equal to 494±17 g/km. The tows are constituted by globally aligned groups of fibers. The length of these groups of fibers varies between 40 to 600 mm with a maximum occurrence-taking place at 80 mm. This fabric is constituted of continuous tows (Fig. 4.b). Generally, when natural fibers are considered, twisted yarns are elaborated to increase its tensile properties. Indeed, as discussed by Goutianos et al.[69] sufficient tensile properties of the yarns are necessary for these ones to be considered for textile manufacturing or for processes such as pultrusion or filament winding. In this study, the flax tows used to elaborate the plain weave fabric are un-twisted and exhibit a rectangular shape. The fibers or groups of fibers are slightly entangled to provide a minimum rigidity to the tows. This geometry has been chosen as it generates low bending stiffness tows, therefore limiting the crimp effect in the fabric and therefore limiting empty zones between tows. It has also been chosen because fabric manufactured from highly twisted yarns exhibit low yarn permeability preventing or partially preventing the use of processes from the LCM (Liquid Composite Moulding) family. Un-twisted tows have also been chosen because manufactured composites display better mechanical properties than composites made with twisted yarns[70]. However, this reinforcement was originally developed to manufacture large panels with low curvature and was therefore not optimized for complex shape forming.

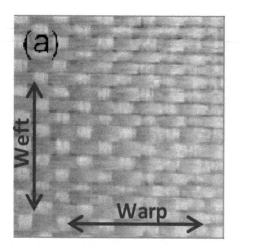 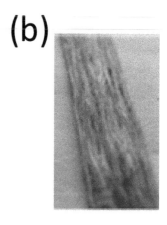

FIGURE 7.4   Reinforcement 1: (a) Flax fabric; (b) flax tow.

A second flax woven reinforcement has also been used. This reinforcement 2 is a 4/4 flax woven hopsack construction with an areal weight of 508 ± 11 g/m² manufactured by the Composites Evolution Company, UK[71]. The hopsack is presented in Fig. 7.5. The cylindrical yarns are manufactured from aligned fibers held together by a polyester yarn going in a spiral manner along the flax yarn. The linear density of the flax yarns is 250 ± 9 tex (g/km). The lineic mass of the yarn holding the flax fibers is 20 ± 3 tex. The reinforcement used in this work is not balanced. A difference of 20% in the number of flax yarns has been measured between warp and weft directions.

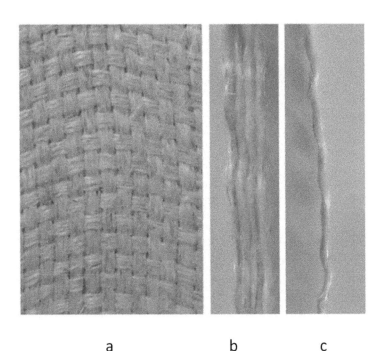

a   b   c

FIGURE 7.5   Reinforcement 2: (a) flax fabric; (b) 4 aligned yarns; (c) individual yarn.

7.3   RESULTS

7.3.1   GLOBAL SHAPE ANALYSIS

The first tests are carried out for reinforcement 1 with an orientation 0°. The blank holders' pressure is set to 1bar. The preform in its final state is presented on Figs. 7.6a and 7.6d. The final shape is in good agreement with the expected tetrahedron punch without large wrinkle or un-weaving on the useful zone. Only very small wrinkles can be observed on corners of the edges for both reinforcements 1 and 2. At the local scale, on faces and edges, the tow buckling defects (Figs. 7.6.b and 7.6.d) and misalignment of tows (Fig. 7.6.b) are identified. Tow buckling only takes place on the Face C and on the opposed edge (between Faces A and B) whereas misalignment of tows is observed on all faces.

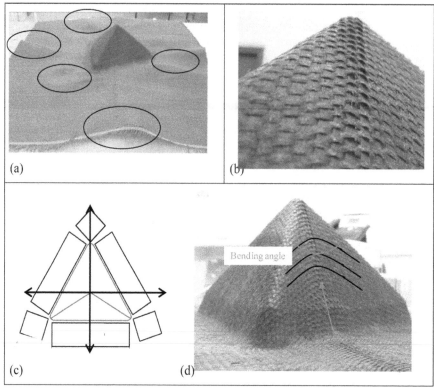

FIGURE 7.6   (a) Preform and Wrinkles around the useful zone. (b) Zoom on buckles. (c) Position of buckles. (d) tow is alignment and position of the buckles for orientation 0°.

Depending on the initial orientation of the fabric, tow buckling (Fig. 7.6.b) appears on faces and on one edge of the formed tetrahedron shape. These buckles zones converge from the bottom of the useful shape up to the triple point (top of the tetrahedron) (Fig. 7.6.d) Due to this defect the thickness of the preform is not homogeneous. The height of some of the buckles can reach 3 mm near the triple point. Due to this thickness in-homogeneity generated by these buckles, the preform could not be accepted for composite part manufacturing.

At the fabric scale, the buckles are the consequence of out of plane bending of the tows perpendicular to those passing by the triple point. The tows passing by the triple point (vertical ones, or weft tows for orientation 0°) are relatively tight. They seem to be much more stressed than the warp tows, perpendicular to the one passing by the triple point. It can be expected that the size of the buckles depends on those tows tension. In this zone, there is no homogeneity of the tensile deformation. This is illustrated by the orientation of the tows perpendicular to the one passing by the triple point on both sides of the buckle zone (drawn Fig. 7.2.d). These tows are curved instead of being straight, and this phenomenon is probably at the origin of the buckles. The tow misalignment is also observed for reinforcement 2 (Figs. 7.7.a and 7.7.b) globally at the same location as for reinforcement 1. However, tow buckles do not appear on the faces of the shape. Only very small buckles can be observed on the edge opposed to Face C in the case of a 0° orientation with a low blank holder pressure of 1 bar, as observed in Fig. 7.8.

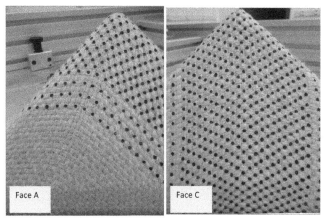

FIGURE 7.7   Tow misalignment in the faces of reinforcement 2.

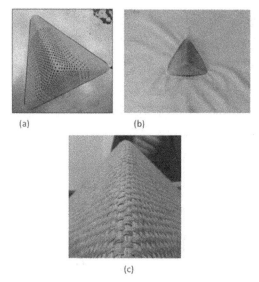

FIGURE 7.8   (a) global shape; (b) wrinkles outside the useful zone; small buckles on the edge at the opposite of Face C.

However, tow buckles do not appear on the faces of the shape. Only very small buckles can be observed on the edge opposed to Face C in the case of a 0° orientation with a low blank holder pressure of 1 bar, as observed in Fig. 7.8.

If tow buckles and small wrinkles can be observed on the shapes formed using the tetrahedron punch, the causes of the defects have not yet been discussed. It is therefore proposed to discuss the issues concerning the appearance of the defects that can be encountered during sheet forming of natural fiber based woven fabrics.

7.3.2 ANALYSIS OF THE FORMING DEFECTS

7.3.2.1 TOW BUCKLING

Tow buckles mainly appear during the forming of reinforcement 1 on faces and edges of the tetrahedron shape. The localization of the buckles seems to be influenced by the initial positioning of the fabric, and the size of the buckles probably depends on the tension state of the tows perpendicular to the ones passing by the triple point. As a consequence, two initial positioning of the fabric have been tested as shown in Fig. 7.9 in conjunction with different blank holder pressures.

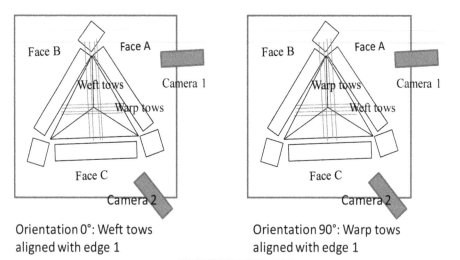

Face B
Face A
Weft tows
Camera 1
Warp tows
Face C
Camera 2
Orientation 0°: Weft tows aligned with edge 1
Face B
Face A
Warp tows
Camera 1
Weft tows
Face C
Camera 2
Orientation 90°: Warp tows aligned with edge 1

FIGURE 7.9 Initial positioning of the woven fabric.

Figure 7.10 shows that in the case of the orientation 0°, buckles only appear on edge 1 and on the middle of face 3. No buckles are observed on faces 1 and 2.

Fig. 7.10. Reinforcement 1: Localization of the buckle zone for initial fabric orientation of 0°

As the bending of the tows perpendicular to the ones passing by the triple point is the mechanisms supposed to be at the origin of the buckling defect, measurements of the bending angles on each faces has been carried out. Results are presented in Table 7.1:

TABLE 7.1 Bending Angle of the Horizontal Tows Measured on the Buckle Zone Orientation 0°

Face number	1	2	3
Bending angle (°)	138±5	136±4	146±4

Table 7.1 shows that the bending angles are globally situated in the same range of values. The bending angles on Faces A and B are slightly more pronounced than the one measured on Face C. Similar investigations were carried out for orientation 90°. Figure 7.11 shows that in this case of study that the buckles can be observed in Faces A and B only. For the orientation 90°, no buckles are observed on Face C and on edge 1 as it was the case for orientation 0°.

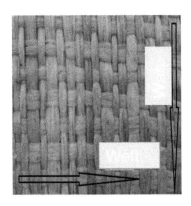

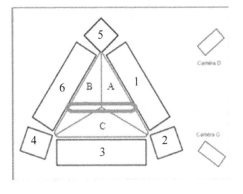

a (initial orientation of the fabric 90°)   (b) Localisation of the buckles

FIGURE 7.11   Localization of the buckle zone for initial fabric orientation of 90°

The bending angles of the tows exhibiting buckling on the 3 faces of the shape were also measured. The values are reported in Table 7.2.

TABLE 7.2 Bending Angle of the Horizontal Tows Measured on the Buckle Zone Orientation 90°

Face number	1	2	3
Bending angle (°)	138	141	143

For orientation 90° the measured bending angles are situated in the same range of values as the ones measured for the 3 faces for orientation 0°. As a consequence, the bending of the tows (Fig. 7.6d) is not responsible for the changes of the buckle zone location for the 2 tested orientations. The initial reinforcement orientation seems to be crucial. As a consequence, bending is not a sufficient criterion to predict the appearance of the buckles.

The reinforcement considered in this study is not balanced. The tows, used in the warp and the weft directions are similar. However, a space between the weft tows (about the width of a tow) is observed on the fabric whereas this space is not present between the warp tows. As buckles only appear on bending zones where the weft tows are vertical, (face 3 and edge 1 orientation 0° and face 1 and face 2 orientation 90°) one can conclude that the architecture of the reinforcement is a key parameter conditioning the appearance of the buckles. When the warp tows are vertical (without any space between them) the buckles do not appear even though the horizontal tows exhibit the same amount of bending. This suggests that the presence of the space between the weft tows is one of the parameter that controls the appearance of the buckles.

The process parameters may also play a role in the occurrence of the buckling defect. The tows showing the buckles are not tight, and the effect of increasing the blank holder pressure upon the occurrence of the tow-buckling defect has been performed. In a first extent, the bending angles are considered. Table 7.3 reports for the 0° orientation the values of the bending angles in the 3 faces of the tetrahedron for three uniform blank holder pressures (uniform pressure applied to the fabric around the shape).

TABLE 7.3   Evolution of the Bending Angle as a Function of the Uniform Blank Holder Pressure for Both Reinforcements

Reinforcement 1				Reinforcement 2			
Angle°/Face Pressure bar	A	B	C	Angle°/Face Pressure bar	A	B	C
1	138±5	136±4	146±4	2	139±5	143±4	141±5
1.5	137±5	137±4	142±5	4	144±6	144±5	149±6
2.5	135±4	139±4	138±4	5	137±4	147±6	147±5

For reinforcement 1, Table 7.3 shows that the bending angle measured on Face Cslowly decreases as a function of the increasing blank-holder pressure. This is probably due an increasing deformation of the tows passing by the triple point as these ones can drag in a larger extent to the top of the shape the perpendicular tows showing the buckles. The relative similar values observed on Faces A and B are probably due to measurement dispersion and to their relative inaccuracy. For reinforcement 2, no real tendency can be extracted from the bending angles values. It has to be noted that measurement were also performed for orientation 90° for both reinforcements and that similar conclusions can be emitted. As a consequence, the change of the blank holder pressure does not influence much the bending angle and therefore the tow orientation on the Faces and it is not really possible to control it.

However, the size, or height of the buckles may be influenced by an increasing tension of the tows. For reinforcement 2, small size buckles observed in Edge 1 disappear when a uniform blank holder pressure of 2 bar is applied. For reinforcement 1, the buckles remain. Table 7.4 shows the heights of the buckles, measured on edge 1 and face 3 for the 0° orientation and on faces 1 and 2 for the 90° orientation for different increasing uniform blank-holder pressures.

TABLE 7.4   Reinforcement 1: Size of the Buckles as a Function of the Uniform Blank Holder Pressure (orientation 0°)

Blank holder pressure(bar) Height of buckles (mm)	1	1.5	2	2.5
Edge 1	1.3±0.2	1.1±0.1	1±0.1	0.8±0.2
Face C	0.7±0.1	0.8±0.1	0.9±4	0.8±0.1

Table 7.4 shows that a relative reduction of the buckle size is observed on edge 1 while increasing the blank holder pressure. This reduction is probably due to a higher tension in the tows showing the buckles. In Face C, the size of the buckles can be considered as constant as the precision of the measurement is about ± 0.1 mm.

To locally increase the tension on the tows showing the buckles, differential blank holder pressures can be applied. The pressure of blank holders 1 and 6 was increased with the goal to raise the tension in the tows showing the buckles in edge 1 (opposite of Face C). The pressure in the other blank holders remains at 1 bar.

TABLE 7.5   Size of the Buckles as a Function of the Increasing Blank Holder 1 and 6 Pressures

Blank holder pressure(bar) Height of buckles(mm)	0.75	1.25	2
Edge 1	1.1±0.1	0.9±0.1	0.8±0.1
Face C	0.7±0.1	0.8±0.1	0.8±0.1

Table 7.5 shows that the size of the buckles decreases as it was expected by increasing the tension in blank holders 1 and 6 as the tension in the tows exhibiting the buckles is probably raised. In Face C, the size of the buckles remains constant as blank holders 1 and 6 do not influence their behavior. To reduce their size, the pressure of blank holders 2 and 4 was raised and the pressure in the other blank holders remains at 1 bar.

TABLE 7.6   Size of the Buckles as a Function of the Increasing Blank Holder 2 and 4 Pressures

Blank holder pressure (bar) Height of buckles (mm)	0.75	1.25	2
Edge 1	0.7±0.1	0.8±0.1	0.8±0.1
Face C	1.1±0.2	0.9±0.1	0.5±0.1

Table 7.6 shows that the size of the buckles this time decreases in Face C by increasing the tension in blank holder 2 and 4 as the tension in the tows showing the buckles is raised. However, the size of the buckles in Edge 1 remains constant.

The observations performed on the two previous tests indicate that it may be difficult to decrease simultaneously the size of the buckles and therefore to stop their occurrence by only working with the blank holder pressure and this for our test configuration.

It has to be noticed that the blank-holder pressure was not raised above values of 2.5 bar as another defect (sliding of tows within the membrane) appears in this case.

7.3.2.2   SLIDING OF TOWS

As indicated in Section 7.3.2.1, the raise of the blank holder pressure may have for consequence to also increase the tension in the fabric. Close to the basis of the shape, sliding of tows may appear when high blank holder pressures are applied therefore limiting the sliding of the fabric between the blank holder and the die. This phenomenon is shown in Fig. 7.12. The tow-sliding defect cannot obviously be accepted when manufacturing a composite part because the homogeneity of the reinforcement does not remain.

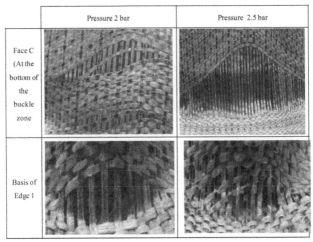

FIGURE 7.12 Reinforcement 1, orientation 0°: localization of the tow sliding phenomenon.

This type of defect also depends in a large extent of the complexity of the shape to be formed as well as the cohesion of the woven fabric. As an example, it is interesting to note that the tow sliding phenomenon appears for uniform blank holder pressures higher than 2 bar for reinforcement 1whereas it only appears for pressure higher than 5 bar for reinforcement 2.

The location of the phenomenon also depends on the orientation of the fabric. For reinforcement 1 at orientation 0°, tow sliding starts at places where tows are very tight (at the basis of Face C and at the bottom of Edge 1) as shown by Fig. 7.12.

For orientation 90°, the tow sliding phenomenon takes place on the basis of the faces A and B at the bottom of the buckle line where the tows seem to be the tightest (Fig. 7.13).

	Pressure 2 bar	Pressure 2.5 bar
Face A (Bottom of the buckle line)	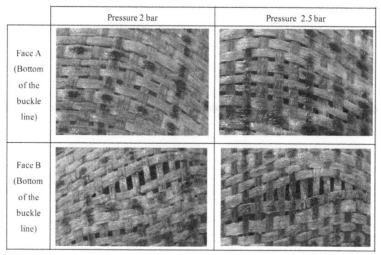	
Face B (Bottom of the buckle line)		

FIGURE 7.13   Reinforcement 1, orientation 90°: localization of the tow-sliding phenomenon.

For reinforcement 2, the tow-sliding phenomenon only appears when the blank holder pressure is higher than 5 bar. This therefore means that the cohesion of this reinforcement is higher than the one of reinforcement 1. However, the phenomenon also takes place for orientation 0° at the bottom of the buckle line as shown by Fig. 7.14

FIGURE 7.14   Reinforcement 2 orientation 0°: localization of the tow sliding phenomenon.

The appearance of the tow sliding phenomenon shows that it is particularly important to well control the blank holder pressure when a complex geometry is concerned and that it may not be possible to apply too high pressure if the cohesion of the reinforcement does not allow it. This may be a limitation of a fabric as increasing the blank holder pressure may be useful to get rid of tow buckles (reinforcement 2) or suppress small wrinkles by increasing the tension of the whole membrane.

7.3.2.3   WRINKLING DEFECT

Very small wrinkles may be observed in the corner basis of the tetrahedron as shown by Fig. 7.15.

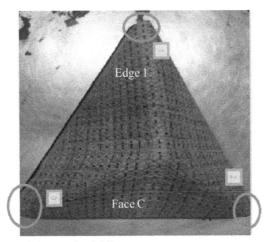

FIGURE 7.15   Potential zone of wrinkling.

The observed wrinkling defects have been observed for both reinforcement in the case of low blank holder pressures (lower than 1 bar). For higher pressure, the tension of the membrane become more important and a coupling effect[72]between in-plane shear and tension of the membrane already reported for carbon fabrics probably acts to suppress the wrinkle. However, it is important to apply pressures lower than the appearance limit of the tow sliding.

Shear angle measurements have been carried out using the mark tracking method to investigate the zones where the wrinkles may potentially appear, because it is commonly admitted that wrinkles appears when shear angles are too high. Figure 7.16 shows the shear angles measurements at different locations of Faces B (A to F) and A (G) for reinforcement 1 with orientation 0°.

	A	B	C	D	E	F	G
$\gamma°$	22.5	22.9	22.1	23.1	23	21.6	22.6

FIGURE 7.16   Reinforcement 1, orientation 0°: Shear angle at different locations of Faces B and A.

The results presented in Fig. 7.16 show relative shear angle homogeneity on the whole two faces of the tetrahedron. The magnitude of these angles is not very high and is probably lower than the locking angle (evaluated in Section 3.3) above which wrinkles may appear. Figure 7.17 for the reinforcement 1 shows the measurements of shear angles at the corner basis at the bottom of Edge 1. The measured angles are much higher than the ones recorded on Faces B and A. This therefore explains why wrinkles can be observed on that location and not on the Faces.

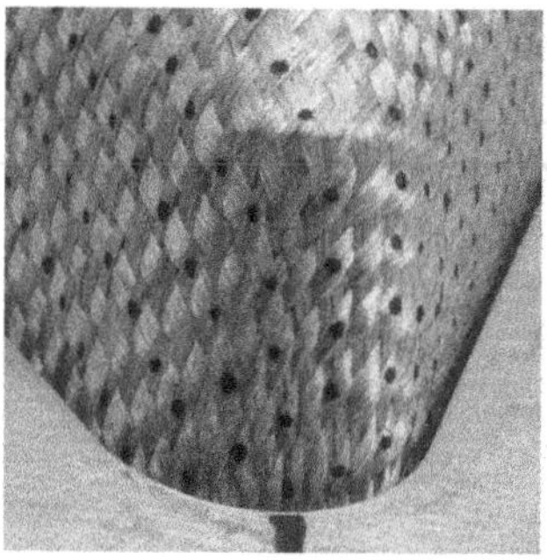

Blank holder pressure(bar)	1	1.5	2
Corner	Shear angle	Shear angle	Shear angle
1	32.5	40	45
2	45	45	50
3	50	50	55

FIGURE 7.17   Reinforcement 1, orientation 90°: Shear angles at the corner basis

Figure 7.18 shows the in-plane shear angles for corner 1 and 2 for orientation 0° for reinforcement 2 with a blank holder pressure of 5 bar. It shows that the shear angles have a tendency to rise as a function of the distance from the central line of the corner. The shear angles are also larger at the bottom of the corner. It is also interesting to note that the shear angles measured on reinforcements 1 and 2 are very close despite the fact that the fabric architectures are different. The shape therefore has a tendency to impose the shear angles in the corners.

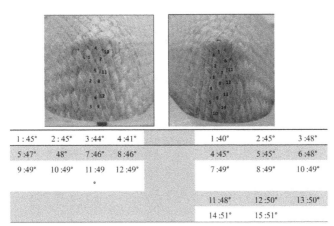

1 : 45°	2 : 45°	3 :44°	4 :41°		1 :40°	2 :45°	3 :48°
5 :47°	48°	7 :46°	8 :46°		4 :45°	5 :45°	6 :48°
9 :49°	10 :49°	11 :49	12 :49°		7 :49°	8 :49°	10 :49°
		°					
					11 :48°	12 :50°	13 :50°
					14 :51°	15 :51°	

FIGURE 7.18   Reinforcement 2, orientation 0°: shear angles at the corner basis.

7.3.2.4   TOW HOMOGENEITY DEFECT

Visually, some tows on the preform seem to be very tight. This is particularly the case of the vertical yarns passing by the triple point (top of the tetrahedron). A local analysis of the tensile strains using the mark tracking method is carried out on those tows to investigate if failure strain has not been overcome.

Figure 7.19 shows the position of the tested yarns and the place between which the strain was measured. Figure 7.20 shows the values of the strains measured on the five considered yarns for one face of the tetrahedron shape. The results show that the strain rises in a nonuniform way during the sheet forming process. At the beginning of the test, no strain is observed as the punch is not in contact with the fabric. Once in contact, the strain in the different tows rises in a regular manner up to values above which the strain increases with a lower slope. Figure 7.20 also shows that the strains measured on the tows passing by or close to the triple point (top of the tetrahedron) are higher than the other ones. Moreover, the tensile strain decreases as a function of the increasing distance from the triple point. A ratio of 1/2 is observed between the strain measured at the end of the test for tow 5 and the strain measured on tow 1.

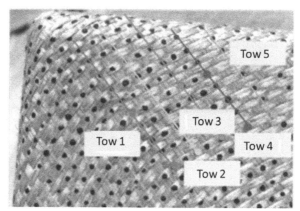

FIGURE 7.19   Position of the tested tows.

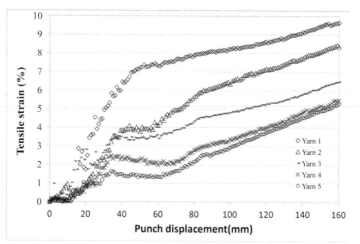

FIGURE 7.20   Reinforcement, orientation 0°: strain evolution in vertical tows of the Face C.

The maximum strain values measured at the end of the forming process indicated that these ones are all within the range 5–9.5%. These values seem to be relatively high in comparison to values evaluated for glass fiber fabric[73] and it is therefore important to investigate the mechanical behavior of the fabric independently of the process to find out if local failure in the tow took place during the forming test. Local failure in the tow causes local movements of the fibers within the tow, and this may lead locally to lower fiber density. This can certainly be a problem for the final composite part if the fiber density is not kept homogeneous as these places could be zones of weakness.

As reinforcement 1 is not balanced, the tensile strain has also been measured in the case of orientation 90° for the tow passing by the triple point of the Face C. Figure 7.21 shows the evolution of the tensile strains of the tows passing by the triple point for orientation 0 and 90° for Face C.

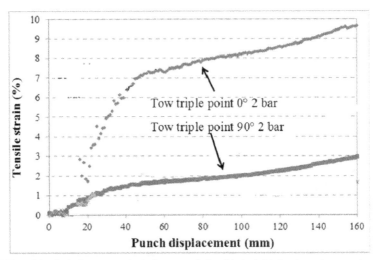

FIGURE 7.21   Reinforcement 1: influence of the fabric initial orientation on the tensile strain of tows in Face C.

For orientation 90°, the tensile strain values are much lower than the ones measured on the equivalent tow for orientation 0° and same process parameters. This may be explained by the fact that reinforcement 1 is not balanced. For orientation 0°, the vertical weft tows are more submitted to the crimp effect than the warp tows (vertical for orientation 90°). Because of the crimp (the fact that a tow passes above and then below the perpendicular tows) the tow is not completely tight when the fabric is not loaded. The phenomenon depends on the number of perpendicular tows met by the tow. In our case, the vertical weft tows for orientation 0° meet more perpendicular tows than the vertical warp tows for orientation 90°. A ratio of 1/3 is observed between the maximum tensile strains of the tows passing by the triple point for orientation 90° and 0°.

Figure 7.22 shows that for the reinforcement 2 and orientation 0° that the tensile strains are relatively lower than for equivalent tows of reinforcement 1. This may be due to the fact that the tows of reinforcement 2 are stiffer and therefore less strained than the tows of reinforcement 1. Figure 7.22 also shows that the tensile strains are symmetrical in both sides of the tows passing by the triple point.

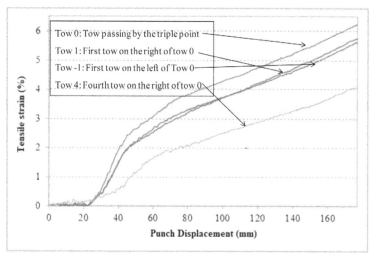

FIGURE 7.22   Reinforcement 2: strain evolution in vertical tows of the Face C; 2 bar.

Tensile strain measurements have also been carried out on tow −1 at different location from the top (close to the triple point) to the bottom of the tow. Figure 7.23 shows that the tensile strains raise when the measurement is performed close to the top of the shape.

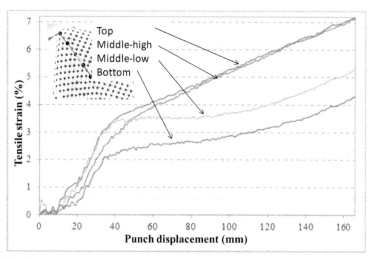

FIGURE 7.23   Reinforcement 2: Evolution of the strain alongside a tow; 2 bar.

This therefore means that the highest strains are recorded in the top zones of the shape close to the triple point. As a consequence, it may be expected that due to the high strains recorded in these tows that fiber movements within the tow takes place with some possible loss of fiber density. To confirm this hypothesis, it is important to investigate from which tensile strain these fiber movements may take place.

7.3.3 CHARACTERIZATION OF THE IN-PLANE SHEAR AND TENSILE BEHAVIOR

7.3.3.1 IN-PLANE SHEAR BEHAVIOR

Figure 7.24 shows a typical in plane shear curve for reinforcement 1.

The three characteristic zones usually observed during in plane analysis of woven fabrics are well defined in Fig. 7.24. In zone 1, the stiffness of the fabric is week, and the tows can easily rotate. The low stiffness is due to tow-tow frictions. In zone 2, lateral contact and compression between neighboring yarns takes place. A stiffness that can be nonlinear is observed up to defined angle called the locking angle. From that angle, the stiffness rises again, and out of plane bending of the fabric takes place due to its week bending stiffness. As a consequence, wrinkles may appear in the faces of the fabric. For the flax woven fabric studied in this work, Fig. 7.24 shows that the value of the locking angle can be estimated to be around 30°. This value confirms that the values of shear angle observed on the faces of the tetrahedron are lower than the locking angle. It is therefore normal that no wrinkles due to high shear angles appeared on the Faces of the preform. However, the locking angle is much lower than the shear angles measured in the corner of the tetrahedron. It is therefore normal that wrinkles were observed for low blank holder pressure (1 bar). When the blank holder pressure was raised the membrane became tighter and the wrinkles disappeared.

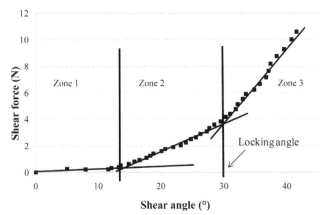

FIGURE 7.24 Reinforcement 1: Evolution of the shear force as a function of the shear angle from a BIAS test.

7.3.3.2 BIAXIAL TENSION BEHAVIOR

Biaxial tension tests were carried out to study the tensile behavior of the flax fabric. The results showing the biaxial behavior of the fabric in the weft direction are presented in Fig. 7.25 for different values of the parameter k_t. The parameter $kt = \varepsilon_{weft}/\varepsilon_{warp}$ is defined as the ratio between the strain in the weft direction over the one in the warp direction. The results show that an increasing value of k_t leads to higher strain in the weft tows. This means that the crimp effect decreases in this case. The crimp effect is the lowest in the case of the unidirectional test ($k_c = \varepsilon_{warp}/\varepsilon_{weft} = 0$).

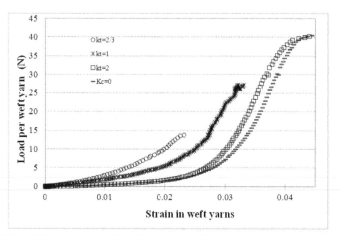

FIGURE 7.25   Reinforcement 1: Tensile behavior of the weft tows during a biaxial test.

Figure 7.25 indicates that the maximum failure strain is obtained in the case of the uniaxial test. This phenomenon is normal and is due to the crimp effect. The maximum strain to failure is observed for the case of the uniaxial test ($k_c = 0$). For all the considered cases, the failure strain is lower than 4.5%. These values are much lower than the ones measured on the tetrahedron face (Figs. 20, and 21). It is therefore probable that failure occurs in the tightest tows of the face during the process.

In a general way, the biaxial tensile behavior of the fabric is very similar to the ones observed on woven fabrics made from glass or carbon fibers.[67,73] Even if the tows are not constituted with continuous fibers, the entanglement between the fibers provides a sufficient continuity to the tow so that this one behaves like a homogeneous entity.

7.4   SOLUTIONS TO PREVENT DEFECTS

A Solution consisting in increasing the blank holder pressure has been already mentioned in Section 7.3.3.1 to get rid of wrinkles. However, low blank holder pressure should be preferred to avoid to homogeneity defects caused by too high strained tows. The use of high resistance tows could also be a solution to prevent this defect. Using low blank holder pressure would also delay the appearance of tow sliding described in Section 7.3.2.2 as this defect only appears when the blank holder pressure is increased. This therefore means that compromises need to be found already at this level to prevent the appearance of these defects.

To prevent the appearance of buckles different solutions may be developed. A first solution consists in designing specific fabric architecture as it was showed in Section 7.3.2.1 that the architecture of the fabric was a critical parameter. Indeed, for the two orientations the buckles appear on the warp tows (on the face C and its opposed edge for 0°orientation and on faces A and B for the 90°orientation).

Reinforcement 1 considered in this study is not balanced since an important space is observed between the weft tows whereas it is almost nonexistent between the warp tows as shown schematically in Fig. 7.26.a. This space controls the appearance of buckles. Its presence between the weft tows allows the warp tows to bend out of plane. Between the warp tows the lack of space prevents the movement of the weft tows. As a consequence, a balanced fabric with no space between two consecutive warp and weft tows could prevent tow buckling (see Fig. 7.26.b).

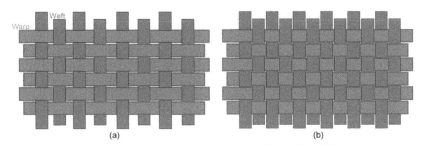

(a)   (b)

FIGURE 7.26   (a) Un-balanced fabric model, (b) Specially designed fabric model.

This new fabric (reinforcement 3) manufactured with the same un-twisted tows was manufactured by GroupeDepestele and tested under the same process conditions as the previous studied fabric. The results presented in Fig. 7.5 confirm the absence of buckles.

A second type of solution was investigated to prevent buckling on the final preform; it focuses on the optimization of the forming process parameters so that the local tensions in the preform can be changed in the defect zones. Nevertheless, the change of the local tensions with no modification of stresses in the rest of the fabric is not easy with the geometry of the bank holders used in this study (Fig. 7.1.b).

To reach this goal, new specially designed blank holders have been elaborated to apply minimum pressure to the tows passing by the triple point. New tests are conducted on reinforcement 1 for the 0°orientation. The final preform presented in Fig. 7.28 is obtained for a blank holders' pressure of 3bar, applied on the warp tows on which the buckles appeared previously. No buckle is observed on the Face C and its opposed edge, unlike with the previous blank holder system.

FIGURE 7.27   Reinforcement 3: tetrahedron shape without any tow-buckling defect.

FIGURE 7.28   Reinforcement 1, orientation 0°, Face C and Edge 1: Final preform obtained with specially designed blank holders.

7.5   CONCLUSIONS

The possibility of manufacturing complex shape composite parts with a good production rate is crucial for the automotive industry. The sheet forming of woven reinforcements is particularly interesting as complex shapes with double or triple curvatures with low curvature radiuses can be obtained. To limit the impact of the part on the environment, the use of flax fiber based reinforcements may be considered for structural or semistructural parts. This study examines the possibility to develop composite parts with complex geometries such as a tetrahedron without defect by using flax based fabrics. An experimental approach is used to identify and quantify the defects that may take place during the sheet forming process of woven natural fiber reinforcements. Wrinkling, tow sliding, tow homogeneity defects and tow buckling are discussed. The origins of the defects are discussed, and solutions to prevent their appearance are proposed. Particularly, solutions to avoid tow buckling caused by the bending of tows during forming are developed. Specially designed flax based reinforcement architecture has been developed. However, if this fabric design has been successful for the tetrahedron shape, it may not be sufficient for other types of shapes and that is why the optimization of the process parameters to prevent occurrence of buckles from a wide range of commercial fabrics was also investigated with success.

KEYWORDS

- **Automotive Industry**
- **Composite**
- **Flax Fabricis**
- **Mechanical Properties**
- **Natural Fibers**
- **Woven Reinforcements**

REFERENCES

1. Yang, Y., Boom, R., Irion B., Van Heerden, D. J., Kuiper, P., & DeWit, H. (2012). "Recycling of Composite Materials" Chemical Engineering and Processing, 51, 53–58.
2. Witik, R., Teuscher, R., Michaud, V., Ludwig, C., & Manson J. A. (2013). Carbon fiber reinforcement composite waste: An assessment of recycling, energy recovery and landfilling. Composite Part A. 49, 89–99.
3. Wambua, P., Ivens, J., & Verpoest, I. (2003). "Natural fibers can they replace glass in fiber reinforced plastics?" Composites Science and Technology 63, 1259–1264.
4. Baley, C. Fibers naturelles de renfort pour matériaux composites techniques de l'Ingénieur AM 5–130.
5. Saheb, D. N., & Jog, J. P. (1999). "Natural Fiber Polymer Composites: A Review." Advances in Polymer Technology, 18(4), 351–363.
6. Bodros, E., Pillin, I., Montrelay, N., & Baley, C. (2007). "Could biopolymers reinforced by randomly scattered flax fiber be used in structural applications?" Composites Science and Technology 67, 462–470.
7. Summerscales, J., Dissanayake, N., Virk, A., & Hall, W. (2010). "A Review of Bast Fibers and their Composites Part 2 composites"Composites Part A, doi: 10.1016/j.compositesa. 2010.05.020.
8. Cao, J. et al. (2008). "Characterization of mechanical behavior of woven fabrics: Experimental methods and benchmark results." Composites: Part A, 39, 1037–1053.
9. Sharma, S. B., Sutcliffe, M. P. F., & Chang, S. H. (2003). "Characterization of material properties for draping of dry woven composite material" Composite Part A. 34, 1167–1175.
10. Bickerton, S., Simacek, P., Guglielmi, S. E., & Advani, S. G. (1997). "Investigation of draping and its effects on the mold filling process during manufacturing of a compound curved composite part." Composite Part A, 28, (9–10), 801–816.
11. Hamila, N., & Boisse, P. (2008). "Simulations of textile composite reinforcement draping using a new semidiscrete three node finite element" Composite Part B 39, 999–1010.

12. Boisse, P., Hamila, N., Helenon, F., Hagege, B., & Cao, J. (2008). "Different approaches for woven composite reinforcement forming simulation." International journal of Material Forming DOI: 10.1007/s12289-008-0002-7.

13. Ouagne, P., Soulat, D., Hivet, G., Allaoui, S., & Duriatti D. (2011). "Analysis of defects during the performing of woven flax" Advanced Composites Letters, 20, 105–108.

14. Baley, C. (2002). "Analysis of the flax fibers tensile behavior and analysis of the tensile increase" Composites: Part A, 33, 2143–2145.

15. Shah, D. (2012). "Developing plant fiber composites for structural applications by optimizing composite parameters: a critical review" Journal of Materials Science. 48, 6083–6107.

16. Brutch, N., Soret-Morvan, O., Porokhovinova, E., Sharov, I., Morvan, C. (2007). "Characters of Fiber Quality in Lines of Flax Genetic Collection" Journal of Natural Fibers, 5, 95–126.

17. Coroller, G., Lefeuvre, A., Le Duigou, A., Bourmaud, A., Ausias, G., Gaudry, T., & Baley, C. (2013). "Effect of flax fiber individualization on tensile failure of flax/epoxy unidirectional composite" Composites Part A: 51, 62–70.

18. Placet, V. (2009). "Characterization of the thermo-mechanical behavior of hemp fibers intended for the manufacturing of high performance composites" Composites Part A. 40, 1111–1118.

19. Venkateshwaran, N., Elayaperumal, A., & Sathiya, G. (2012). "Prediction of tensile properties of hybrid-natural fiber composites" Composites part B, 43, 793–796.

20. Ku, H., Wang, N., & Pattarachaiyakoop, N., & Trada, M. (2011). "A review on the tensile properties of natural fiber reinforced polymer composites." Composites part B, 42, 856–873.

21. Ku, H., Wang, H., Pattarachaiyakoop, N., & Trada, M. (2011). "A review on the tensile properties of natural fiber reinforced composites" Composites: Part B, 42, 856–873.

22. Dittenber, D. B., & Ganga Rao, H. V. S. (2012). "Critical Review of Recent Publications on Use of Natural Composites in Infrastructure" Composites: PartA, 43, 1419–1429.

23. Kim, C. S., & Lim, H. S. (2009). Sediment dispersal and deposition due to sand mining in the coastal waters of Korea. Continental Shelf Research, 29, 194–204.

24. Thornton, E. B., Sallenger, A., Sesto, J. C., Egley, L., McGee, T., & Parsons, R. (2006). Sand mining impacts on long-term dune erosion in southern Monterey Bay. Marine Geology, 229, 45–58.

25. Ashraf, M. A., Maah, M. J., Yusoff, I., Wajid, A., & Mahmood, K. (2011). Sand mining effects, causes and concerns A case study from Bestari Jaya, Selangor, Peninsular Malaysia. Scientific Research and Essays, 6(6), 1216–1231.

26. Charlet, K., Jernot, J. P., Eve, S., Gomina, M., & Breard, J. (2010). "Multi-Scale Morphological Characterization of Flax from the Stem to the Fibrils" Carbohydrate Polymers, 82, 54–61.

27. Bodros, E., & Baley, C. (2008). "Study of the tensile properties of stinging nettle fibers." Materials Letter, 62, 2143–2145.

28. Pillin, I., Kervoelen, A., Bourmaud, A., Goimard, J., Montrelay, N., & Baley, C. (2011). "Could oleaginous flax fibers be used as reinforcement for polymers?" Industrial Crops and Products, 34, 1556–1563.

29. Ochi, S. (2008). "Mechanical Properties of Kenaf Fibers and Kenaf/PLA Composites" Mechanics of Materials, 40, 446–452.

30. Zini, E., & Scandola, M. (2011). "Green Composites an overview." Polymer composites, 1905–1915.

31. Virk, A. S., & Hall, W. (2010) Summerscales J. Failure strains as the key design criterion for fracture of natural fiber composites Composites Science and Technology, 70, 995–999.

32. La Mantia, F. P., & Morreale, M. (2011). "Green composites: A brief review." Composites: Part A, 42, 579–588.

33. Biagiotti, J., Puglia, D., & Kenny, J. M. (2004). "A Review on Natural Fiber-Based Composites-Part I" Journal of Natural Fibers, 1, 2, 37–68.

34. Puglia, D., Biagiotti, J., & Kenny, J. M. (2005). "A Review on Natural Fiber Based Composites Part II." *Journal of Natural Fibers*, 1, 3, 23–65.
35. Päivi Lehtiniemi, K. D., Tommi Berg Mikael Skrifvars, & Järvelä, P. (2011). Natural fiber-based reinforcements in epoxy composites processed by filament winding. Journal of Reinforced Plastics and Composites, 30(23), 1947–55.
36. Angelov, I., Wiedmer, S., Evstatiev, M., Friedrich, K., & Mennig. G. (2007). Pultrusion of a flax/polypropylene yarn. Composites Part A: Applied Science and Manufacturing, 38, 1431–1438.
37. Summerscales, J., Dissanayake, N., Virk, A., & Hall, W. (2010). "A review of bast fibers and their composites Part 2 Composite" *Composites: PartA*, 41, 1336–1344.
38. Satyanarayana, K. G., Arizaga, G. C., & Wypych, F. (2009). "Biodegradable composites based on lignocellulosic fibers. An overview" *Progress in Polymer Science*, 34, 982–1021.
39. Pandey, J. K., Ahn, S. H., Lee, C. S., Mohanty, A. K., & Misra, M. (2010). "Recent Advances in the Application of Natural Fiber Based Composites." *Macromolecular Materials and Engineering*, 295, 975–989.
40. Ouagne, P., Bizet, L., Baley, C., & Bréard, J. (2010). "Analysis of the Film stacking Processing Parameters for PLLA/Flax Fiber Biocomposites" *Journal of Composite Materials*, 44, 1201–1215.
41. Alawar, A., Hamed, A. M., & Al-Kaabi, K. (2009). "Characterization of treated data palm tree fiber as composite reinforcement." *Composites: Part B*, 40, 601–606.
42. Le Duigou, A., Davies, P., & Baley, C. (2011). "Environmental Impact Analysis of the Production of Flax Fibers to be used as Composite Material Reinforcement." *Journal of Biobased Materials and Bioenergy*, 5, 153–165.
43. Kim, S., Dale, B. E., Drzal, L. T., & Misra, M. (2008). "Life Cycle Assessment of Kenaf Fiber Reinforced Biocomposite." *Journal of Biobased Materials and Bioenergy*, 2, 85–93.
44. Dissanayake, N., Summerscales, J., Grove, S., & Singh, M. (2009). "Life Cycle Impact Assessment of Flax Fiber for the Reinforcement of Composites" *Journal of Biobased Materials and Bioenergy*, 3, 245–248.
45. Dissanayake, N., Summerscales, J., Grove, S., & Singh, M. (2009). "Energy use in the production flax fiber for the reinforcement of composites" *Journal of Natural Fibers*, 331–346.
46. Rudd, C. D., & Long, A. C. (1997). Liquid Molding Technologies Woodhead Publishing Limited.
47. Buet-Gautier, K., & Boisse, P. (2001). "Experimental Analysis and Modelling of Biaxial Mechanical Behavior of Woven Composite Reinforcements" *Experimental Mechanics,* 41(3), 260–269.
48. Launay, J., Hivet, G., Duong, A. V., & Boisse, P. (2008). "Experimental analysis of the influence of tensions on in plane shear behavior of woven composites reinforcements." *Composite Science and Technology,* 68, 506–515.
49. Cao, J., et al. (2008). "Characterization of mechanical behavior of woven fabrics: Experimental methods and benchmark results." *Composites: Part A*, 39, 1037–1053.
50. DeBilbao, E., Soulat, D., Hivet, G., & Gasser, A. (2010). "Experimental Study of Bending Behavior of Reinforcements" *Experimental Mechanics,* 50(3), 333–351.
51. Prodromou, A. G., & Chen J. (1997). "On the relationship between shear angle and wrinkling of textile composite performs" *Composites Part,* 28, 491–503.
52. Lomov, S. V., Boisse, Ph., Deluycker, E., Morestin, F., Vanclooster, K., Vandepitte, D., Verpoest, I., & Willems, A. (2008). "Full field strain measurements in textile deformability studies" *Composites: PartA*, 39, 1218–1231.
53. Skordos, A. A., Aceves, C. M., & Sutcliffe, M. P. F. (2007). "A simplified rate dependent model of forming and wrinkling of preimpregnated woven composites" *Composites: Part A,* 38, 1318–1330.

54. Arbter, R., Beraud, J. M., Binetruy, C., Bizet, L., Bréard, J., Comas-Cardona, S., Demaria, C., Endruweit, A., Ermanni, P., Gommer, F., Hasanovic, S., Henrat, P., Klunker, F., Laine, B., Lavanchy, S., Lomov, S. V., Long, A., Michaud, V., Morren, G., Ruiz, E., Sol, H., Trochu, F., Verleye, B., Wietgrefe, M., Wu, W., & Ziegmann, G. (2011). "Experimental Determination of the Permeability of Textiles: A Benchmark Exercise." *Composites: Part A*, 42, 1157–1168.

55. Heardmann, E., Lekakou, C., & Bader, M. G. (2001). "In plane permeability of sheared fabrics" *Composites: Part A.*, 32, 933–940.

56. Long, A. C. (2001). "Process modeling for liquid molding of braided preform" *Composites: Part A*, 32, 941–953.

57. Ouagne, P., & Bréard, J. (2010). "Continuous Transverse Permeability of Fibrous Media" *Composites Part A*, 41, 22–28.

58. Ouagne, P., Ouahbi, T., Park, C. H., Bréard, J., & Saouab, A. (2012). Continuous measurement of fiber reinforcement permeability in the thickness direction: Experimental technique and validation. *Composites: Part B*, 45, 609–618.

59. Robitaille, F., & Gauvin, R. (1998). "Compaction of textile reinforcements for composites manufacturing I: Review of experimental results." *Polymer Composites*, 19, 198–216.

60. Kelly, P. A., Umer, R., & Bickerton, S. (2006). "Viscoelastic response of dry and wet fibrous materials during infusion processes." *Composites: Part A*, 37, 868–873.

61. Ouagne, P., Bréard, J., Ouahbi, T., Saouab, A., & Park, C. H. (2010). "Hydro Mechanical Loading and Compressibility of Fibrous Media for Resin Infusion Processes" *International Journal of Materials Forming*, 3, 1287–1294.

62. Hivet, G., & Duong, A. V. (2011). "A Contribution to the Analysis of the Intrinsic Shear Behavior of Fabrics" *Journal of Composite Materials*, 45, 695–717.

63. Willems, A., Lomov, S. V., Verpoest, I., & Vandepitte, D. (2008). "Optical strain fields in shear and tensile testing of textile reinforcements." *Composite Science and Technology*, 68, 807–819.

64. Soulat, D., Allaoui, S., & Chatel, S. (2009). Experimental device for the optimization step of the RTM process. *Int J Mat Form*, 2, 181–184.

65. Bretagne, N., Valle, V., & Dupré, J. C. (2005). "Development of the Marks Traking Technique for Strain Field and Volume Variation Measurements," *NDT&E International*, 38, 290–298.

66. Zhu, B., Yu, T. X., Zhang, H., & Tao, X. M. (2011). Experimental Investigation of Formability of Commingled Woven Composite Preform in Stamping Operation, Composites: Part B, 42, 289–295.

67. Boisse, P., Gasser, A., & Hivet, G. (2001). "Analyzes of fabric tensile behavior: determination of the biaxial tension-strain surfaces and their use in forming simulation." *Composites: Part A*, 32, 1395–1414.

68. GroupeDepestele: www.groupedepestele.com

69. Goutianos, S., & Peijs, T. (2003). "The optimization of flax fiber yarns for the development of high-performance natural fiber composites" *Advanced Composite Letters*, 12, 237–241.

70. Goutianos, S., Peijs, T., Nystrom, B. & Skrifvars, M. (2006). "Development of flax based textile reinforcements for composite applications" *Applied Composites Materials*, 13, 199–215.

71. Composite Evolution ltd: http://www.compositesevolution.com/

72. Allaoui, S., Boisse, P., Chatel, S., Hamila, N., Hivet, G., Soulat, D., & Vidal Salle, E. (2011). Experimental and Numerical Analyzes of Textile Reinforcement Forming of a Tetrahedral Shape, Composite Part A, 42, 612–622.

73. Duong Anh, Vu. (2008). Étude expérimental du comportement mécanique de reinforts composites tissues lors de la mise en forme sur géométries non développables PhD thesis, University of Orléans.

CHAPTER 8

INJECTION MOLDING OF NATURAL FIBER REINFORCED COMPOSITES

INDERDEEP SINGH and SAURABH CHAITANYA

ABSTRACT

With increasing awareness about environmental concerns, Natural fiber composites (NFCs) have emerged as a potential replacement to traditional polymer composites, which are derived from nonrenewable resources and are nonbiodegradable. NFCs can be classified into green and partially green composites based on the polymer matrix used. NFCs have several advantages over traditional polymer composites leading to worldwide increase in their demand. In order to meet this demand, primary processing techniques should be developed exclusively for fabrication of NFCs. Injection molding process is the most extensively used process in the industry for the production of polymer composites. In this chapter, injection molding process parameters and the various issues and challenges during fabrication of NFCs using injection-molding process has been discussed.

8.1 INTRODUCTION

The continuous demand of materials having high strength and stiffness, light weight and low cost has led to the development of fiber reinforced composites to replace metals in structural applications. The polymer matrix composites (PMCs) based on petroleum derived polymers and synthetic fibers have been developed and used as structural material for engineering applications for decades, owing to their high strength and stiffness and light weight.[1]PMCs have been used for numerous applications ranging from aerospace, automobiles, electrical, sports goods, construction and household items. Due to superior mechanical properties and wide applications, the use of PMCs has increased in the last few decades. Traditionally used PMCs consists of synthetic fibers like glass, aramid and carbon fibers, resulting in composites that can be easily used for structural applications. These PMCs are made up of nonrenewable and nonbiodegradable materials and the resulting composites are also nonrecyclable. These drawbacks have led to the disturbance in the ecological balance. Limited petroleum resources, rapidly depleting land fill space and strict environmental rules and regulations have forced the researchers and plastic industries worldwide to look for alternate matrix and reinforcement materials to overcome the drawbacks as well as to meet the performance of traditional PMC.

The possibility of reinforcing natural fibers into the polymer matrix has been explored by many researchers and industries worldwide. Natural fiber composites (NFCs) have been seen as a potential replacement to the traditional PMCs in many structural and nonstructural applications. Natural fibers are the largest and fastest growing renewable resource of fibers available that can serve as a potential replacement to synthetic fibers. NFCs consisting of natural fibers reinforced polymer matrix are being developed and studied as a potential material for the replacement of traditional PMCs. Natural fibers have distinct advantages over traditionally used glass fibers as natural fibers are renewable, biodegradable and have a low density ($1.2–1.6$ g/cm^3) as compared to glass fibers (2.5 g/cm^3) which result in the fabrication of lighter composites. Natural fiber composites can also be recycled and cause no abrasion to tools used for their processing. Natural fibers can be broadly classified into three types as plant fibers, animal fibers and mineral fibers (see Fig. 8.1).

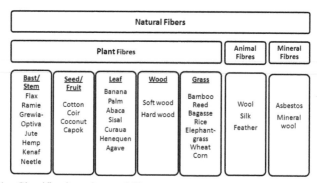

FIGURE 8.1   Classification of natural fibers.

The incorporation of natural fibers in the petroleum derived polymer matrix results in PMCs called Partially Green Composites (PGCs). PGCs have been studied by many researchers and it has been found that they have a distinct advantage of being recyclable and have comparable properties to traditional PMCs. However, PGCs still remain nonbiodegradable. To overcome the drawbacks of traditional PMCs a material called biocomposite was innovatively developed by reinforcing natural fibers into a biopolymer matrix (fully biodegradable polymers) by the DLR Institute of Structural Mechanics, in 1989.[2] Biocomposites are still being developed and have attracted the attention of researchers worldwide in the last decade. Biocomposites consists of a renewable and biodegradable matrix (cellulose, starch, lactic acid, etc. derived) like polylactic acid (PLA), poly hydroxyl alkanoates (PHA), polyhydroxy butyrate covalerate (PHBV), etc. and natural fibers (plant, animal and mineral based) like sisal, hemp, flax, etc. The classification of polymers is shown in Fig. 8.2.

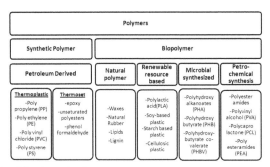

FIGURE 8.2   Classification of polymers.

Biocomposites derived from renewable natural resources are termed as Green Composites. The cost and availability of biopolymers is restricting their wide spread use. The incorporation of natural fibers into the biopolymer not only reduces its cost (cheaper comparable to glass fibers) and weight but at the same time provides more strength and stiffness compared to base bio polymer.

FIGURE 8.3   Natural fiber composites.

Due to the increasing awareness among society about environmental concerns, the demand for NFCs has drastically increased during the last decade. According to a market forecasting report by BCC research (leading market forecasting agency), the global market for applications of wood-plastic composites, cellulosic plastics, plastic lumber and NFCs during the 5 year period (2011–2016) is estimated to grow at a compound annual growth rate (CAGR) of 13.8%. The global market for building products and automotive application is expected to experience a growth at a CAGR of 12.4% and 17.1%, respectively.[3] Also the global use of bioplastics is expected to increase up to 3.7 million metric tons by 2016, at a CAGR of 34.3%.[4] With the increase in demand, the need to identify and develop primary processing techniques especially for the fabrication of NFCs arises. As other composites, NFCs can also be tailor made according to the specific properties required for specific applications by careful selection of the polymer matrix, natural fiber and a suitable manufacturing process. NFCs can be manufactured in a similar way as traditional PMCs by compression molding, resin transfer molding, hot pressing, direct extrusion and injection molding. The NFCs are being manufactured using the processes designed for the manufacturing of traditional PMCs for a controllable output. However, the properties of NFCs are highly variable compared to synthetic fibers. The properties of the natural fibers vary in terms of mechanical, thermal and structural properties. Also, there are several problems in fabrication of NFCs such as distribution of fibers in the matrix, fiber attrition during mechanical mixing, interfacial bonding between hydrophobic matrix and hydrophilic fibers and thermal degradation of the fibers during processing. Hence, there is a need for identification and development of processing technologies for NFCs.

Although compression molded parts exhibit better mechanical properties than other processes, but the compression molding process can only be used for small to medium parts with simple geometries. Short fiber reinforced polymers are used extensively as structural material as they provide superior mechanical properties and can be easily processed by the rapid, low-cost injection molding process.[5]Injection molding process is used for the fabrication of small to medium parts with complex geometries. The parts which require precision, dimensional accuracy and excellent surface finish can be easily processed by injection molding. Also the polymers and the fibers are exposed to higher temperatures for a very short period preventing their degradation due to exposure to higher temperature for a long time. Apart from these advantages, there are several issues and challenges related to the fabrication of NFCs by injection molding process. This chapter focuses on the identification and selection of the processing parameters to successfully overcome the issues and challenges faced by the industry in fabrication of NFC by injection molding.

8.2   SELECTION OF MANUFACTURING PROCESS FOR NFCS

NFCs can be processed by various processing techniques like compression molding, hot pressing, resin transfer molding and injection molding. The ideal manufacturing process should be able to transform the raw materials to the desired shape without any defects. The selection of the manufacturing process for the fabrication of NFCs is based on:
1.   Desired properties of NFCs.
2.   Geometry of resultant composites.
3.   Processing limitations of matrix and fibers.
4.   Production output desired.
5.   Manufacturing cost.

NFCs can be tailor made according to the desired properties of the composite by varying the percent fraction of reinforcement and additives into the polymer matrix. The manufacturing processes might limit the amount and size of fibers to be used as in case of injection molding process. The manufacturing process should be carefully chosen while designing the composite. The geometry of the desired composites plays an important role in determining the appropriate fabrication process. Large components are usually manufactured by open mold process where as for small to medium size products compression and injection molding process is preferred. The complexity of geometry of a component also plays an important role in determining the manufacturing process. Usually, complex components which require precision are made by injection molding process. The shape of the component to be made is the replica of the mold cavity. During the molding process the polymer matrix and natural fibers are subjected to heat and both the matrix and natural fibers have a tendency to degrade at higher temperatures. Hence, suitable raw materials and appropriate processing technology should be selected based on the desired performance of the composite. The selection of the manufacturing process for NFCs involves various considerations to be taken into account. The aspect (length/diameter) ratio, interfacial bonding, orientation, and amount of percentage of natural fibers highly influence the properties of the end product.

8.3   INJECTION MOLDING PROCESS

Injection molding process is the most extensively used molding method in the industry used for the production of polymer composites due to its simplicity and fast processing cycle. Injection molding machine mixes and injects a measured amount of matrix and fiber mixture into the mold resulting in the desired product. It consists of three major sections: the injection unit, mold, and ejection and clamping unit (see Fig. 8.4).

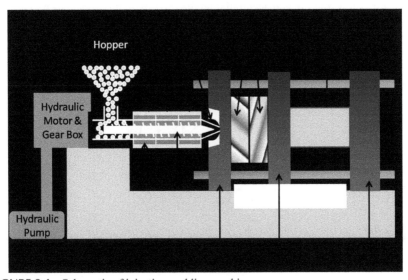

FIGURE 8.4   Schematic of injection molding machine.

The injection unit consists of a heated screw barrel having a compression screw, which can rotate as well as reciprocate. The function of the heated barrel is to provide heat to the polymer matrix to melt before injection. The function of the reciprocating screw is to carry and compress the pellets from the hopper into the heated barrel, mix the polymer matrix and fiber, provide heat to the matrix by viscous shearing and inject the mixture into the closed mold by acting as a piston. In other words injection unit consists of an extruder with an added function of reciprocating screw that injects the mixture into the mold. The cavities in the mold are the replica of the desired geometry of the product. The molds consist of cooling and/or heating coils to regulate the mold temperature. The mold temperature determines the cooling rate of the product. The clamping unit clamps the mold tightly against the injection pressure to prevent burr formation and the ejector unit actuates the ejectors in the mold to eject the part when the cycle completes.

A typical injection molding process cycle is shown in Fig. 8.5. Generally the injection molding cycle is assumed to start from mold close position. After the mold closes it is tightly clamped against the injection pressure by the clamping unit. In the mean time the screw is retracted to its back position and then it injects the molten mixture with the desired injection pressure and speed into the mold cavity. The injected mixture undergoes shrinkage during solidification and to compensate that the screw is kept forward by the desired holding pressure for some time. After this point the screw starts to retract and plasticize the mixture while the part is being cooled in the mold. The part is allowed to cool sufficiently to be able to bear the ejector force and meet the desired dimensions. In the mean time the screw is being pushed backward as it is rotating and accumulating the mixture in the front. The part is then ejected and the cycle repeats itself.

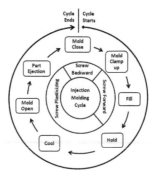

FIGURE 8.5   Injection molding cycle.

8.3.1 INJECTION MOLDING PROCESS PARAMETERS

The typical injection-molding machine is designed for the fabrication of thermoplastics and the same is used for fabrication of NFCs without any major change in the machine tool. As the properties of the natural fibers used as reinforcement are different from that of synthetic fibers used in traditional PMCs, an understanding of the various process parameters of injection molding machine and their effects is necessary to achieve the desired flawless composites. In case of natural fiber reinforced composites the polymer pellets and the fibers are first dried in the hot air circulation dryer, hopper dryer, or dehumidification dryer depending upon the material predrying recommendations to remove excessive moisture. The fibers in the chopped form and matrix in pellet form are either precompounded or directly fed into the injection-molding hopper. The common precompounding device used prior to injection molding is twin-screw extruder or a melt mixer, which uniformly blends the matrix and fibers prior to injection molding resulting in uniform distribution of natural fibers in the composite. There are various process parameters to be taken care of while injection molding of NFCs. Optimization of these process parameters leads to reduction in the cycle time and as a result a reduction in the operating cost and increased productivity. The important process parameters are identified as screw barrel temperature, screw speed, injection speed, injection time, injection pressure, mold temperature and back pressure.

8.3.1.1 SCREW BARREL TEMPERATURE

The screw barrel temperature is an important process parameter as it directly influences the properties of the end product. The melting of the polymer matrix takes place in the screw barrel. It is divided into various heating zones from feeding zone to the nozzle. The temperature of these zones depends upon the processing temperature of polymer matrix used. The temperature of the heaters has to be adjusted carefully as heating of the mixture in the barrel is not only done by the heaters alone. The shearing action of the screw also provides some heat during plasticizing. The viscosity of the fiber matrix mixture reduces as the temperature is increased but at the same time higher temperatures can lead to degradation of matrix as well as natural fibers used.

8.3.1.2 SCREW SPEED

The screw speed during plasticizing of the mixture is responsible for the homogeneity of the NFCs. Higher screw speed leads to the high shear forces leading to heat generation resulting in the reduction in viscosity of the mixture. At the same time high screw speed leads to fiber attrition (fiber breakage) and in extreme cases causes the polymer properties to degrade (due to high heat generation). A proper balance between the screw speed and the backpressure should be maintained to allow mixing of the fibers and matrix during the entire cooling cycle.

8.3.1.3   INJECTION SPEED

Injection speed is an important process parameter during injection molding of thermoplastics. Injection speed is dependent upon the melt viscosity of the polymer and the fiber load. The injection speed has to be increased with the rising viscosity of the melt as it becomes difficult for the viscous mixture to flow into thin cavities. High injection speed causes high shear rates while passing through the mold, which reduces the viscosity of the melt. The same has been reported in a study where, when the fiber content of natural fiber was increased above 30% the injection speed was increased in order to overcome the increased viscosity and fill the mold cavities completely.[6]Injection speed should be faster in case of precision parts having thin cross-section or in case of multi cavity mold. Injection speed is kept comparatively slower for thick parts. Injection molding machines now come equipped with programmable injection speed control, which can vary the injection speed during injection process.

8.3.1.4   INJECTION TIME

Injection time directly depends upon the gate solidification time. The gate solidification time is the time when the material stops flowing due to its solidification at the cavity gate. Injection time consists of filling time and holding time. The gate sealing time is determined by measuring the weight of the product at regular intervals while increasing the injection time till the weight of the product becomes constant. If the injection time is less, means the holding pressure is removed before the material is solidified the material in the mold cavities would flow back due to high pressure in the mold cavity leading to various defects like voids, sink, war page and dimensional mismatch. Therefore in order to optimize the injection time, holding time should be reduced gradually till the optimum hold time is achieved (till gate solidification).

8.3.1.5   INJECTION PRESSURE

Injection pressure consists of the filling pressure and the holding pressure. The filling pressure during injection molding is usually more than the holding pressure. It should be able to inject 90 to 95% of the shot volume into the mold cavity. Ideally the filling pressure should be the maximum pressure that can be achieved without flashing but high pressure would lead to increased stresses in the molded part. Holding pressure is important to overcome the shrinkage during low temperature solidification of the crystalline polymer. The holding pressure should be set according to the material shrinkage properties. High holding pressure also reduces the void and sink defects, but it should be regulated carefully as high holding pressure also leads to burr formation.

8.3.1.6   MOLD TEMPERATURE

Mold temperature is also an important process parameter in case of injection molding process. Mold temperature can be regulated by the use of heaters and the cooling channels provided in the mold. Low temperatures of the mold can be attained using chilled water channels, which continuously take the heat from the mold, but low temperature of the mold leads to rapid cooling of the part resulting in the development of residual stresses in the part. High mold temperature is desired to obtain glossy parts. To avoid warpage the molded parts should be cooled below the heat deflection temperature.

8.3.1.7   BACK PRESSURE

Back pressure is the pressure exerted on the screw by the melt while screw is recovering to back position after the injection stroke. While returning the screw feeds the pellets into the heated screw barrel for plasticizing, this material after melting accumulates in front of the screw pushing the screw backwards. It is used for the uniform mixing of the fibers in the polymer melt. It also removes the entrapped air and ensures consistent shot density. However, high back pressure might also decrease the plasticization ability.

8.3.2   PROCESSING OF NF COMPOSITES BY INJECTION MOLDING

Due to the ease of fabrication, high production rate and high process variability, the injection molding has been widely used in the industry. Many researchers have used injection molding to study the mechanical behavior of NFCs.[7,12] The researchers working on NFCs have mainly focused their studies on PP and PLA based polymers. This is also due to the limitations of natural fibers as they limit the use of polymers, which have processing temperatures above $210°$ C.[13] Natural fiber reinforced PLA is being studied as a potential replacement to various engineering plastics. A comparison of different compounding processes (mixer-injection molding, mixer-compression molding and direct compression molding process) on the mechanical properties of PP/(30 wt.%) abaca composites with different fiber lengths (5, 25 and 40 mm) was done and it was found that the parts produced by mixer-injection molding process has better mechanical performance compared to other processes.[14] The mechanical and thermal properties of silk worm fiber reinforced PLA fabricated by injection molding process were studied and a good interfacial interaction between PLA and silk fibers was reported, showing a good wettability of PLA during extrusion and injection process.[15] A study on the extruded and kneaded PP/Flax fibers found that the length of the fibers after both the processes is significantly reduced. However, the extruded samples showed a slightly higher fiber length and thickness.[16] The compounded pellets for injection molding of flax reinforced PLA were prepared by using an extruder having a speed and flow rate of 500 rpm and 5 Kg/h, respectively. The temperature zone near the feeder was maintained at 150°C and the die end at 190°C. The output was in the form of strands, which were cut into pellets using a granulator. The injection molded PLA/flax test specimens showed good fiber dispersion without any large fiber clusters.[17] A comparative study between short sisal fiber (5 mm and 1–3 wt.%) reinforced PP and PLA fabricated by injection molding process was done. The precompounding of the matrix and the fibers was done using an extruder at 190°C. A uniform distribution of the fibers after the processing was reported and the length of fibers was found to be in the range of 2 to 5 mm. Also scanning electron microscopy (SEM) micrographs showed an intimate contact and better interfacial interaction between the fibers and PLA.[11] A new precompounding processing method for fabrication of NFCs was explored in which hemp fibers were blend spun with a small amount of PLA fibers to form compounded fiber pellets followed by the traditional extrusion and injection molding process to develop composites. The use of this method resulted in the feed control and uniform dispersion of fibers in the PLA matrix. Also this fabrication method resulted in the significant improvement in the mechanical properties of the developed composites.[18] A significant improvement in the mechanical properties of PP/(30 wt.%) wheat straw fibers fabricated by melt mixing followed by injection molding was reported.[19] The compounding techniques namely twin-screw extruder, high-speed mixer and two-roll mill were compared to study the PP/wood fiber composites. The twin screw extruded composites resulted in better mechanical properties as compared to the fibers compounded by other processes.[20] The extrusion (twin-screw extruder) followed by injection molding process was used to fabricate microcrystalline cellulose (MCC) reinforced PLA composites. The injection molding of the composites was done at a temperature of 200°C with an injection speed of 60 mm/s. The mold temperature, pack pressure and cooling time was kept as 50°C, 400 bars and 15 s, respectively.[21,22] In another study, composites with PP, polyethylene, high impact polystyrene, and PLA reinforced with man made cellulose filament yarn (rayon tire cord yarn) were fabricated by injection molding process in which modified pultrusion process and an extruder (twin screw) was used for precompounding.[23] The recycled newspaper cellulose fiber (RNCF) and chopped glass fibers reinforced PLA composites were fabricated using a full size twin screw extruder followed by an injection molder. A uniform temperature profile in all the zones of the extruder was maintained at 183°C and screw speed was set at 100 rpm. Prior to injection molding, the pelletized composites thus formed were dried in a convection oven for 2 h at 80°C. The injection molding was carried out on 85-ton capacity injection molding machine at 183°C with a nozzle temperature of 185°C. The cooling time, hold, pack and fill pressures were maintained at 50 s, 6.89, 8.96, and 6.89 MPa, respectively.[10] In another study the wood fibers and the PLA pellets were mixed mechanically in a kitchen mixer prior to extrusion at 100 rpm and a temperature of 183°C followed by the transfer of extruded material to injection molder through a preheated cylinder at 183°C having a mold temperature of 40°C.[24]

8.4   ISSUES AND CHALLENGES IN INJECTION MOLDING OF NFCS

The NFCs have been studied by several researchers, using injection molding as a fabrication process. Several issues and challenges regarding the fabrication of NFCs by injection molding have been observed.

8.4.1   *DISTRIBUTION AND ORIENTATION OF NATURAL FIBERS*

The distribution and orientation of the fibers in the composite plays an important role in determining the mechanical properties of the composites. In injection molded composites, the orientation of the reinforced fibers is a critical factor to control. The fiber orientation of short fibers vary with respect to the thickness as well as the in plane direction. During the injection molding of composites, the fibers are oriented according to the complex molten polymer flow generated during the process. Convergent flow of the fiber matrix mixture results in fiber alignment along the flow axis while the divergent flow causes the fibers to align perpendicular to the flow direction.[25]The fibers near the surface are generally aligned parallel to the direction of flow and are aligned perpendicular to the direction of flow at the center. A similar orientation of fibers was also observed, in a study where distribution of flax fibers reinforced PP was studied. The fibers close to the surface and sides of the molded part were well aligned with the direction of flow, whereas, near the mid plane the fibers were randomly distributed.[26]This shows the dependence of fibers to align in the direction of shearing and stretching. During injection of the matrix-fiber mixture into the mold, the shear flow near the mold walls aligns the fibers in the direction of flow. This outer layer in contact with the mold surface is called as skin. Below this layer, the mixture in the molten state continues to experience shear and fibers are aligned along the flow direction. Then a core layer is formed in the center where bulk deformation of the flow occurs causing the material to stretch in and out of the paper direction aligning the fibers.[25] The fiber distribution in PLA/Jute composites was studied using long fiber pellets (LFT) (fabricated via pultrusion process) and recompounded pellets (RP) (extruded LFT) and it was found that the fiber dispersion and separation in RP was better than that of LFT, resulting in a better interfacial interaction and improved mechanical properties, although severe fiber attrition was observed in case of recompounding process.[27] Hence, it shows that the orientation and distribution of the fibers not only depends on the processing and flow conditions but also on the precompounding of the fibers and the matrix. A balance between the amount of compounding and the distribution of the fibers is required to keep the fiber attrition to minimum. Also, along with the processing parameters, the rheological properties of the matrix should be taken into account while designing the mold for injection molding to ensure a better flow and less fiber attrition.

8.4.2   FIBER BREAKAGE/ATTRITION

Fiber attrition is a major problem while dealing with natural fiber reinforced composites during the melt mixing process.[28] The strength of the natural fiber reinforced composites depends on the amount of applied load transmitted to the fibers. The extent of load transmitted depends upon the length of the fiber and the fiber matrix interfacial bonding. The load sharing capacity of the fiber in short fiber reinforced composites (as in case of injection molded composites) depends upon the critical fiber length (CFL). If the fiber length is less than the CFL, debonding of matrix and fiber takes place resulting in failure at low load. If the fiber length is more than the CFL, a failure due to breaking of fibers takes place indicating high composite strength. In case of injection molding process, a significant amount of fiber attrition is found in the molded part due to high shear rates during plasticizing, injection and passage through narrow gates and openings of the mold.[26] Fiber attrition also takes place during precompounding process employing a melt mixer, kneader or twin screw extruder. Reduction of the fiber length below the CFL would lead to degradation of the composite properties as the short fibers would not be able to bear the load for which the composites are designed. Hence, the determination of the CFL of fibers is important prior to injection molding of fiber-reinforced composites. Although, increasing the fiber content would lead to better mechanical properties but the injection molding process limits the amount of fibers that can be injected due to increased viscosity of the mixture and narrow gate and sprue of the mold. To overcome this problem of fiber attrition during processing of NFCs by injection molding, the process parameters and mold dimensions should be adjusted according to the fiber load and viscosity of the melt. The process parameters should be adjusted to cause minimum shear rate during processing of NFCs. Also the gate and sprue dimensions should be increased in order to accommodate the fibers and reduce the shear rate. Better precompounding techniques should be developed to reduce the fiber attrition during precompounding of fiber and polymer. In a study regarding incorporation of a counter rotating extruder for compounding of biopolymer-wood composites, it was reported that a higher aspect ratio of the fibers was achieved due to the use of counter rotating extruder which acts more likely as a refiner and the fiber breakage, compared to other mixing techniques was reduced.[6]

8.4.3   RESIDUAL STRESSES

Residual stresses are the internal stresses that exist inside the molded part in the absence of the external load. The residual stresses in case of injection molding are flow induced and thermal induced stresses. The flow-induced stresses in polymer chains depend on orientation and packing pressure while thermal stresses are induced as a result of non uniform cooling of the molded part.[29] In other words, residual stresses in case of injection molding are a result of temperature variations, high pressure generated and polymer chains orientation after cooling.[30] The characteristic residual stress distribution in an injection molded part shows tensile stresses at the core and surface, and compressive stresses at intermediate region.[25,31] The major causes of residual stresses in case of fiber reinforced composites have been listed as high pressure gradient, orientation of the polymer chains, non uniform temperature profile and the difference in the thermal expansion coefficient of fibers and the matrix. These stresses are introduced during injection molding process stages of filling, packing, and cooling. Residual stresses in the molded composite cause an early fracture. The stress distribution in injection molded part depends upon the pressure history to which the melt mixture is subjected from start to filling up of the mold cavity.[25] The residual stresses in case of injection molded parts may cause defects like stress cracking, warpage and long-term deformation.[32] The residual stresses can be reduced by the gradual cooling of the molded part. The gradual cooling of the injection-molded part is achieved by setting higher mold temperature. Heat treatment of the injection-molded parts is another way of relieving the residual stresses developed due to non uniform heating. The residual stresses can also be reduced by carefully adjusting the process parameters like screw speed, injection pressure, melt and mold temperature as well as taking the rheological properties of the polymer matrix into consideration. The appropriate mold design helps in the reduction of residual stresses. The mold designing includes the injection gate size and location, cavity shape and vents for air to escape.

8.5   CONCLUSION

The driving forces behind the use of NFCs are environmental as well as economic considerations. The use of NFCs is rapidly increasing in aerospace, automobile, house hold, electric, sports goods and biomedical applications. This has made NFCs as a potential candidate for research efforts. The literature available on NFCs has shown that NFCs have the potential to replace the traditional PMCs in many applications. As the demand is increasing; fast, easy and economical processing techniques are required for fabrication of NFCs. Injection molding of NFCs has been reported by many researchers but still much work has not been presented on the selection and optimization of the process parameters. In other words, there is no set of thumb rules available for the processing of NFCs. In this chapter different aspects regarding fabrication of injection molded NFCs have been discussed. Injection molding process parameters like screw barrel temperature, screw speed, injection speed, injection time, injection pressure, mold temperature and back pressure have been discussed. The study of the effect of these parameters on the injection molding of NFCs is important to ensure a defect free injection molded composite. Distribution and orientation of fibers, fiber attrition and residual stress generation are the main issues that have to be taken care of during injection molding of NFCs. These issues can be minimized by careful selection of process parameters and by optimal mold design. The rheological properties of the polymer matrix should also be taken into consideration while selecting the parameters and designing of the mold.

Finally, it can be concluded that injection-molding process has a huge potential for processing of natural fiber reinforced composites. A judicious selection of process parameters and an optimal mold design can further enhance the application areas of injection molding process in context of the natural fiber reinforced composites.

KEYWORDS

- **Composites**
- **Injection Molding**
- **Mechanical Characterization**
- **Natural Fibers**
- **Polymers**
- **Structural Applications**

REFERENCES

1. Bajpai, P. K., Singh, I., & Madaan, J. (2012). Development and characterization of PLA based green composites: a review. *Journal of Thermoplastic Composite Materials*, DOI: 10.1177/0892705712439571.
2. Mohanty, A. K., Misra, M., & Hinrichsen, G. (2000). Biofibers, biodegradable polymers and biocomposites an overview. *Macromolecular Materials and Engineering*, 276, 1–24.
3. Smock, D. A. (2011). Wood-plastic composites: technologies and global markets, PLS034B, BCC Research.
4. Smock, D. A. (2012). Global markets and technologies for bioplastics: PLS050B, BCC Research.

5. Fu, S., & Lauke, B. (1996). Effects of fiber length and fiber orientation distributions on the tensile strength of short-fiber-reinforced polymers. *Composites Science and Technology*, 56, 1179–1190.

6. Sykacek, E., Hrabalova, M., Frech, H., & Mundigler, N. (2009). Extrusion of five biopolymers reinforced with increasing wood flour concentration on a production machine, injection molding and mechanical performance. *Composites Part A: Applied Science and Manufacturing*, 40, 1272–1282.

7. Adam, J., Korneliusz, B. A., & Agnieszka, M. (2013). Dynamic mechanical thermal analysis of biocomposites based on PLA and PHBV-A comparative study to PP counterparts. *Journal of Applied Polymer Science*, 130, 3175–3183.

8. Bledzki, A. K. (2007). Abaca fiber reinforced PP composites and comparison with jute and flax fiber PP composites. *eXPRESS Polymer Letters*, 1, 755–762.

9. Akos, N. I., Wahit, M. U., Mohamed, R., & Yussuf, A. A. (2013). Comparative studies of mechanical properties of poly(ε-caprolactone) and poly(lactic acid) blends reinforced with natural fibers. *Composite Interfaces*, 20, 459–467.

10. Huda, M., Drzal, L., Mohanty, A, & Misra, M. (2006). Chopped glass and recycled newspaper as reinforcement fibers in injection molded poly(lactic acid) (PLA) composites: a comparative study. *Composites Science and Technology*, 66, 1813–1824.

11. Mofokeng, J. P., Luyt, A. S., Tabi, T., & Kovacs, J. (2011). Comparison of injection molded, natural fiber-reinforced composites with PP and PLA as matrices. *Journal of Thermoplastic Composite Materials*, 25, 927–948.

12. Bax, B., & Müssig (2008). Impact and tensile properties of PLA/cordenka and PLA/flax composites. *Composites Science and Technology*, 68, 1601–1607.

13. Le, A., Pillin, I., Bourmaud, A., Davies, P., & Baley, C. (2008). Effect of recycling on mechanical behavior of biocompostable flax / poly (L-Lactide) composites. *Composites Part A: Applied Science and Manufacturing*, 39, 1471–1478.

14. Bledzki, A. K., Faruk, O., & Mamun, A. A. (2008). Influence of compounding processes and fiber length on the mechanical properties of abaca fiber-polypropylene composites. *Polimery*, 53, 120–125.

15. Cheung, H. Y., Lau, K. T., Tao, X. M., & Hui, D. (2008). A potential material for tissue engineering: silkworm silk/PLA biocomposite. *Composites Part B: Engineering*, 39, 1026–1033.

16. Bos, H. L., Müssig, J., & van den Oever, M. J. A. (2006). Mechanical properties of short-flax-fiber reinforced compounds. *Composites Part A: Applied Science and Manufacturing*, 37, 1591–1604.

17. Rozite, L., Varna, J., Joffe, R., & Pupurs, A. (2011). Nonlinear behavior of PLA and lignin-based flax composites subjected to tensile loading. *Journal of Thermoplastic Composite Materials*, 26, 476–496.

18. Song, Y., Liu, J., Chen, S., Zheng, Y., Ruan, S., & Bin, Y. (2013). Mechanical properties of poly (lactic acid)/hemp fiber composites prepared with a novel method. *Journal of Polymers and the Environment*. doi: 10.1007/s10924-013-0569-z.

19. Panthapulakkal, S., Zereshkian, A., & Sain, M. (2006). Preparation and characterization of wheat straw fibers for reinforcing application in injection molded thermoplastic composites. *Bioresource technology*, 97, 265–272.

20. Bledzki, A. K., Letman, M., Viksne, A., & Rence, L. (2005). A comparison of compounding processes and wood type for wood fiber PP composites." *Composites Part A: Applied Science and Manufacturing*, 36, 789–797.

21. Mathew, A., Oksman, K., & Sain, M. (2005). Mechanical properties of biodegradable composites from poly lactic acid (PLA) and microcrystalline cellulose (MCC). *Journal of applied polymer*, 97, 2014–2025.

22. Mathew, A. P., Oksman, K., & Sain, M. (2006). The effect of morphology and chemical characteristics of cellulose reinforcements on the crystallinity of polylactic acid. *Journal of Applied Polymer Science*, 101, 300–310.

23. Ganster, J., Fink, H. P., & Pinnow, M. (2006). High-tenacity man-made cellulose fiber reinforced thermoplastics–injection molding compounds with polypropylene and alternative matrices. *Composites Part A: Applied Science and Manufacturing*, 37, 1796–1804.

24. Huda, M. S., Drzal, L. T., Misra, M., & Mohanty, A. K. (2006). Wood-fiber-reinforced poly(lactic acid) composites: evaluation of the physicomechanical and morphological properties. *Journal of Applied Polymer Science*, 102, 4856–4869.

25. Ho, M., Wang, H., Lee, J. H., Ho, C., Lau, K., Leng, J., & Hui, D. (2012). Critical factors on manufacturing processes of natural fiber composites. *Composites Part B: Engineering*, 43, 3549–3562.

26. Nyström, B. (2007). Natural fiber composites: optimization of microstructure and processing parameters. PhD Dissertation, Luleå University of Technology.

27. Wang, C., Uawongsuwan, P., Yang, Y., & Hamada, H. (2013). Effect of molding condition and pellets material on the weld property of injection molded jute polylactic acid. *Polymer Engineering and Science*, doi. 10.1002/pen.23411.

28. Joseph, P. V., Joseph, K., & Thomas, S. (1999). Effect of processing variables on the mechanical properties of sisal-fiber-reinforced polypropylene composites. *Composite Science and Technology*, 59, 1625–1640.

29. Wang, T. H., & Young, W. B. (2005). Study on residual stresses of thin-walled injection molding. *European Polymer Journal*, 41, 2511–2517.

30. Azaman, M. D., Sapuan, S. M., Sulaiman, S., Zainudin, E. S., & Abdan, K. (2013). An investigation of the processability of natural fiber reinforced polymer composites on shallow and flat thin-walled parts by injection molding process. *Materials and Design*, 50, 451–456.

31. Kim, S. K., Lee, S. W., & Youn, J. R. (2002). Measurement of residual stresses in injection molded short fiber composites considering anisotropy and modulus variation. *Korea-Australia Rheology Journal*, 14, 107–114.

32. Kim, C. H., & Youn, J. R. (2007). Determination of residual stresses in injection-molded flat plate: simulation and experiments. *Polymer Testing*, 26, 862–868.

CHAPTER 9

DEVELOPMENT AND PROPERTIES OF SUGAR PALM FIBER REINFORCED POLYMER COMPOSITES

S. M. SAPUAN, L. SANYANG, and J. SAHARI

ABSTRACT

Natural fibers have recently become attractive as an alternative reinforcement for fiber reinforced polymer composites. They are gaining more attention due to their low cost, easy availability, less health hazards, fairly good mechanical properties, high specific strength, nonabrasive, ecofriendly and bio-degradability characteristics. Polymers from renewable resources have attracted tremendous amount of attention to researchers and engineers over two decades. The increasing appreciation for biopolymers is mainly due to environmental concerns, and the rapid petroleum resources depletion. Sugar palm fiber (SPF) reinforcement of a novel biodegradable sugar palm starch (SPS) has been studied in this chapter. The result shows that the mechanical properties of plasticized SPS improved with the incorporation of fibers. Fiber loading also increased the thermal stability of the biocomposite. Water uptake and moisture content of SPF/SPS biocomposites decreased with the incorporation of fibers, which is due to better interfacial bonding between the matrix and fibers as well as the hindrance to absorption caused by the fibers. It can be seen that tensile strength and impact strength of biocomposites increase with increasing fiber content. This enhancement indicates the effectiveness of the SPF act as reinforcement. SPF reinforcement of epoxy and high impact polystyrene (HIPS) have also been looked into. Overall, SPF treatments enhanced the mechanical properties of both polymers (epoxy and high impact polystyrene). Thus, indicating that SPF has a promising potential to be used as reinforcement in polymer composites.

9.1 INTRODUCTION: BIOCOMPOSITES

Composites are attractive materials that consist of two (or more) distinct constituents, which when coupled together provides a material with completely different properties from those of the individual original components.[1] Composite materials are unique because they combine material properties in the manner not existing in nature. Such combination often results in lightweight materials with high stiffness and tailored properties for specific applications.[2] Glass fiber is the most dominant fiber and is generally used in 95% of cases to reinforce thermoplastic and thermoset composites. Recent research developments manifested that in certain composite applications, natural fibers demonstrate competitive performance to glass fibers.[2] The most widely used composite in industries today is glass fiber-reinforced composite. Although glass fiber composite has numerous advantages, it also has demerits. Glass fiber can cause irritation to the skin, eyes and upper respiratory tract. These inherent health hazards of glass fiber has fuel the extensive search for safer, cheaper and maybe better fiber than glass fiber.[3,7] Natural fibers are highly potential alternatives for glass fibers. Natural fibers are less abrasive to tooling and not causing as many respiratory problems for workers or consumers. Moreover, they are low cost and have loading bearing potential. Therefore, the use of natural fiber based composites has extended to various sectors, such as aircraft, construction, automotive, etc. (Fig. 9.1).[8]

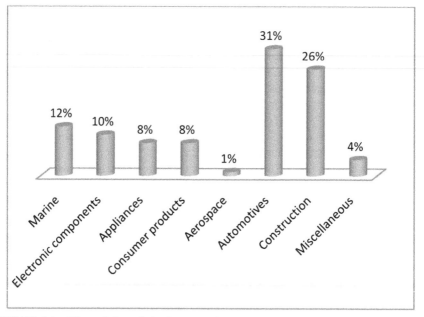

FIGURE 9.1   Fiber reinforced plastic composites used in 2002 (Adapted from John, M. J., Thomas, S., Biofibers and biocomposites. Carbohydrate Polymers 2008, 71, 343–364. With permission.).

Biodegradable composite types are not strange materials to human civilization. Their use dates to antiquity, such as the Great Wall of China whose construction started initially in 121 B.C. as earth works were connected and made strong by clay bricks made of local materials initially using red willow reeds and twigs with gravel during the Han dynasty (209 B.C.). The Wall was later built with clay, stone, willow branches, reeds, and sand during the Qin dynasty (221–206 B.C.).[8,10] However, biocomposite materials have transited through significant developments in terms of using different raw materials, processes and even applications. The history of fiber-reinforced plastics began since 1908 with cellulose fiber in phenolics, later extend to urea and melamine and reaching commodity status with glass fiber reinforced plastics. The fiber reinforced composites industry is now a multibillion-dollar business.[11]

Biocomposites are broadly defined as composite materials made from natural fiber and petroleum derived nonbiodegradable polymers (i.e., PP, PE) or biodegradable polymers (PLA, PHA). The latter category of biocomposites which are derived from natural fiber and biobased polymers (bioplastic/biopolymer) are likely to be more environmental friendly and such composites are termed as green composites (Fig. 9.2) (John and Thomas, 2008).[11]

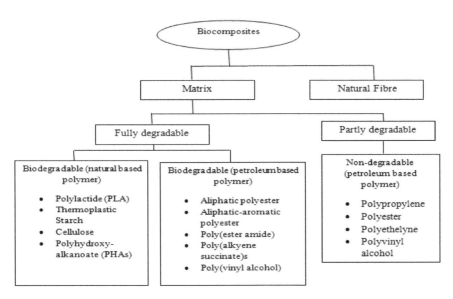

FIGURE 9.2 Classifications of biocomposites (Adapted from Mohanty, A. K., Misra, M., Drzal, L. T., Selke, S. E., Harte, B. R. and Hinrichsen, G. (2005a). Natural fibers, biopolymers, and biocomposites: An introduction. In *Natural Fibers, Biopolymers and Biocomposites,* Boca Raton, FL: CRC Press [Awaiting permission approval]).

9.2.1 MAIN FACTORS CONTRIBUTING TO THE GROWTH OF BIOCOMPOSITE MATERIALS

The swift growth of new materials based on natural fibers and biopolymers is primarily accelerated by factors such as greater environmental awareness and the depletion of petrochemical resources.[11]

9.2.1.1 ENVIRONMENTAL CONCERN

The disposal of composite derived from glass fibers and petroleum based polymer matrices, after their intended life span is becoming a problem and very expensive. Their recycling and reuse is critical because they are made from two different materials. The two most implemented disposal alternatives are land filling and incineration. Landfill space is drastically decreasing due to the heavy ongoing waste disposal.[2] The increasing environmental pollution due to the use of abundant plastics and emissions during incineration is greatly affecting the food we eat, water we drink, and even the air we breathe.[2] Composite consisting of biofibers which are biodegradable and conventional thermoplastics or thermosets which are nonbiodegradable would partially resolve these mentioned problems. However, if the matrix resin/ polymer is biodegradable such biofiber reinforced biopolymer composites would become completely biodegradable. Biopolymer or synthetic polymers reinforced with natural fibers (known as biocomposites) are a feasible alternative to glass fiber composites. Figure 9.3 shows the life cycle of natural fiber reinforced composites.

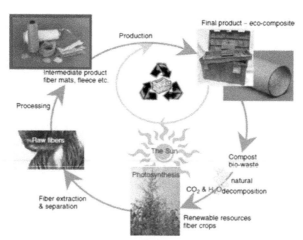

FIGURE 9.3 The life cycle of fiber reinforced composites. (Adapted from Fakten and Trends 2002 Zur Situation der Landwitschaft, Eggenfelden, 2002, p.193 [Awaiting permission approval]).

9.2.1.2 PETROLEUM RESOURCE DEPLETION

The large use of petroleum based polymers and polymer-based synthetic fiber composites in different sectors led to disposal problems which also triggers environmental issues. Most of the polymers used in composites are based on nonrenewable petroleum, whose price is increasing and unstable. Furthermore, the production of fossil-based materials like petroleum from biomass takes approximately 1 million years, compared to 1–10 years for the conversion of many chemical into CO_2 (see Fig. 9.4). For the production of 100% renewable and biodegradable composites, both the polymeric matrix and the reinforcement must be from renewable resources which are often derived from plants within few years.[8,12] Hence, to overcome the dependence on petroleum-based polymers and synthetic fibers, attempts have been made to use 100% renewable and biodegradable biopolymer and natural fiber in the production of biocomposites.

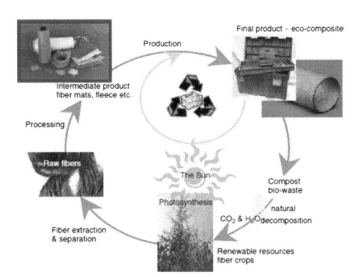

FIGURE 9.4 Global carbon cycle-sustainability driver (Adapted from Satyanarayana, K. G.; Arizaga, G. G. C.; Wypych, F., Biodegradable composites based on Lignocellulosic fibers–An overview. *Progress in Polymer Science 2009*, 34, 982–1021. With permission.).

9.2.2 BIODEGRADABLE POLYMER/BIOPOLYMER MATRICES

Biodegradable polymers can originate from biomass or petroleum-based and can be classified as green polymeric matrices.[14] It is essential to differentiate biodegradable polymers from biopolymers because the origin of the raw material can vary. For instance, poly(caprolactone) (PCL) is a petroleum-based polymer but is completely biodegradable and thus can be referred to as green matrix. These polymers can be degraded by anaerobic and aerobic biological processes, which mainly produce carbon dioxide, water, and biomass.[13] Biopolymers (bio-based polymers) are considered biodegradable but not all biodegradable polymers are biopolymers. Reinforcement of the biodegradable materials with natural fibers give improved material properties, desired in various applications without compromising biodegradability. Figure 9.5 shows the life cycle of biodegradable polymers, which includes conservation of fossil resources, water and CO_2 production. The rate of biodegradation depends on humidity, temperature (50–70°C), amount and type of microbes.

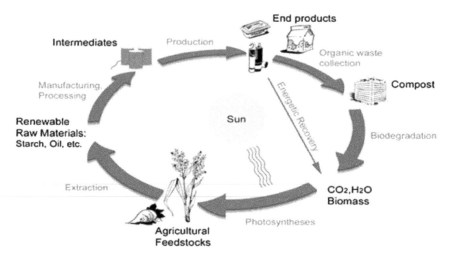

FIGURE 9.5 The life cycle of biodegradable polymers (Siracusa, V.; Rocculi, P.; Romani, S.; Rosa, M. D., Biodegradable polymers for food packaging: a review. Trend in Food Science and Technology 2008, 19, 634–643. With permission.).

Polymers from renewable resources have attracted tremendous amount of attention over two decades. The increasing appreciation for biopolymers is predominantly due to: environmental concerns, and the rapid petroleum resources depletion.[14] Figure 9.6 presents the classification of polymers in four different groups.[11,15] All the three groups of polymer are derived from renewable resources except the last group, which is from petroleum products.[11] Generally, biopolymers can be classified into three groups: (1) natural polymers, like starch, protein and cellulose; (2) synthetic polymers from natural monomers, such as polylactic acid (PLA); and (3) polymers from microbial fermentation, such as polyhydroxybutyrate (PHB).[14]

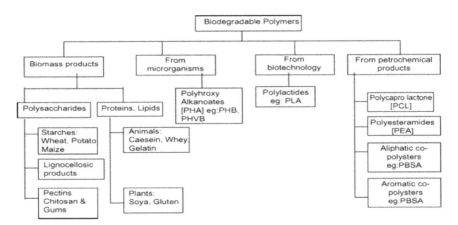

FIGURE 9.6   A classification of biodegradable polymers based on their origin (Adapted from John, M. J., Thomas, S., Biofibers and biocomposites. Carbohydrate Polymers 2008, 71, 343–364. With permission.).

Numerous new polymer materials were developed from renewable resources, such as starch, which is a natural polymer. Other biopolymers are poly lactic acid from fermentable sugar and polyhydroxyalkanoate (PHAs) from vegetable oils next to other bio-based feedstocks.[16]

9.3   STARCH

Nowadays, many investigations are conducted regarding the development and characterization of biopolymers since conventional synthetic plastic materials are resistant to microbial attack and biodegradation.[17,18] Among all biopolymers, starch has been considered as one of the most promising one due to its easy availability, biodegradability, lower cost and renewability.[19] Starches are commonly used biopolymers. Starches are the major form of stored carbohydrate in plants such as corn, wheat, rice, and potatoes. They are hydrophilic polymers that natively exist in the form of discrete and partially crystalline microscopic granules which are held together by an extended micellar network of associated molecules.[8] Starches are composed of both linear and branched polysaccharides well known as amylose and amylopectin, respectively (Fig. 9.7). Native starches contain about 70–85% amylopectin and 15–30% amylose. The ratio of amylose and amylopectin in starches varies with the botanical origin.[20] The Amylopectin is mainly responsible for the crystallinity of the starch granules. Starch granules exhibit hydrophilic properties and strong intermolecular connection through hydrogen bonding formed by hydroxyl groups on the granule surface.[21]

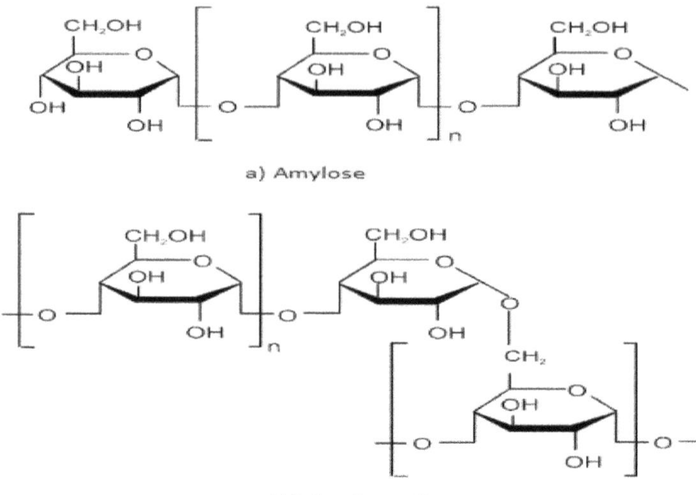

a) Amylose

(b) Amylopectin

FIGURE 9.7   Unit structure of Amylose and amylopectin.

Amylose molecules consist of 200–20,000 glucose units, which form a helix as a result of the bond angles between the glucose units. Amylopectin is a highly branched polymer containing short side chains of 30 glucose units attached to every 20–30 glucose units along the chain. Amylopectin molecules may contain up to 2 million glucose units.[22] Starches from various sources are chemically similar and their granules are heterogeneous with respect to their size, shape, and molecular constituents. Proportion of the polysaccharides amylose and amylopectin become the most critical criteria that determine starch behavior.[23,24] Most amylose molecules (molecular weight ~10^5–10^6 Da) are consisted of (1→4) linked α-D-glucopyranosyl units and formed in linear chain. But, few molecules are branched to some extent by (1→6) α-linkages.[25,26] Amylose molecules can vary in their molecular weight distribution and in their degree of polymerization (DP) which will affect to their solution viscosity during processing, and their retrogradation/recrystallization behavior, which is important for product performance. Meanwhile, amylopectin is the highly branched polysaccharide component of starch that consists of hundreds of short chains formed of α-D-glucopyranosyl residues with (1→4) linkages. These are interlinked by (1→6)-α-linkages, from 5 to 6% of which occur at the branch points. As a result, the amylopectin shows the high molecular weight (10^7–10^9 Da) and its intrinsic viscosity is very low (120–190 mL/g) because of its extensively branched molecular structure.[25,26]

Starches are highly potential candidates for developing sustainable materials, for it is simply generated from carbon dioxide and water by photosynthesis in plants.[21,27,29] However, they display poor melt processability and are highly water soluble and hence, they are difficult to process and are brittle. The presence of numerous intermolecular hydrogen bonds affects processability of starches.[20,30] Thus, for application purposes, starches need a plasticizer to render them processable. Plasticizers such as water, glycerol and sorbitol assist in increasing the starch flow and also decrease the glass transition temperature and melting point of starch.[8,20,31] Starch can be transformed into thermoplastic-like material, when the molecular interactions are disrupted by using plasticizers under specific conditions. The heating of starch granules in the presence of plasticizers yield a nonirreversible transition and swelling of amorphous areas.[20] The process of disrupting starch molecular structure is known as gelatinizing and the plasticized starch is called thermoplastic starch (TPS). It is vital to note that starches are not real thermoplastic but they act as synthetic plastic in the presence of plasticizers (water, glycerol, sorbitol, etc.) at high temperature. The various properties of thermoplastic starch product such as mechanical strength, water solubility and water absorption can be prepared by altering the moisture/plasticizer content, amylose/amylopectin ratio of raw material and the temperature and pressure in the extruder (Mohanty et al., 2000).[32] Plasticizers are the most important material to increase the flexibility and processibility of TPS. There are large number of researches that were performed on the plasticization of TPS using glycerol [33], sorbitol [34], urea and formamide[35], dimethyl sulfoxide[36]and low molecular weight sugars [37]. The properties of TPS also depend a lot on moisture. As water has a plasticizing power, the materials behavior changes according to the relative humidity of the air through a sorption-desorption mechanism.[38]The factors that greatly influence the final morphology of TPS are composition, mixing time, temperature, shear and elongation rate of the operation.[20]

TPS alone can be used for the production of useful products but the moisture sensitivity of starch limits its usage in many commercial applications. To enhance the properties of starch, various physical or chemical modifications of starch such as blending, derivation and graft copolymerization are employed.[21] Blending of TPS with PHA, PLA, and PCL produce 100% biodegradable materials. The aim of blending low cost starch with completely degradable polyester is to lower the cost of the latter while maintaining other significant properties at an acceptable level.[21,39,40]

9.3.1 POLY(LACTIC ACID) (PLA)

PLA is one of the most promising and important biodegradable polyesters with many desirable properties. It is obtained from the fermentation of sugar feedstock (i.e., corn, sugarcane, etc.), which are renewable resources and are readily biodegradable.[41,42] PLA is a versatile polymer, recyclable and compostable, with high transparency, high molecular weight, and good water solubility resistance.[41] In addition, it also has good processability as well as high strength and modulus.[21]

PLA is generated either by ring opening polymerization (ROP) of lactic or by direct polycondensation of lactic acid.[20,43]A cyclic dimer of lactic acid is used in ROP.

The monomer lactic acid is a chiral molecule and occurs as D- and L-Lactic acids. The purification of lactic acid at the time of its production determines the molecular weight of PLA and ensures the production of PLA with consistent properties.[20] The main drawback with the production of PLA via direct polycondensation reaction is the problem of trace water removal in the last stages of polymerization. This remaining water in the polymer can reduce the molecular weight of the final product. Thus, most commercial PLA productions are performed by ROP of lactides.[20]

PLA is among the most widely used biobased polymers in varies applications, due to its ability to be stress crystallized, thermally crystallized, impact modified, filled, copolymerized, and processable in most processing equipment.[2] PLA can rapidly degrade in the environment and the by-products (CO_2 and water) are of less toxicity to living organisms and have no significant environmental impact (see Fig. 9.8). Although PLA is an attractive biodegradable polymer, it has inferior properties such as thermal stability and impact resistance as compared to conventional polymers used for thermoplastics. However, the stiffness, permeability, crystallinity and thermal stability of PLA can be improved via PLA modification, copolymerization with other monomers, and composites.[20]

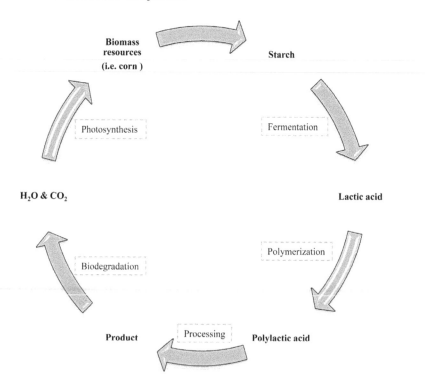

FIGURE 9.8   The life cycle of PLA biopolymer.

9.3.2 POLY(HYDROXYALKANOATE) (PHA)

PHA is a member of biopolyesters that is obtained by a broad variety of microorganisms. These microorganisms consume PHA as carbon and energy sources.[8,20,44] More than 159 different types of PHAs, that is, homopolymers and copolymers can be produced by using various bacterial species and growth conditions.[20] The most popular polymers of the PHA family are polyhydroxybutyrate (PHB) and poly(hydroxybutyrate-cohydroxyvalerate) (PHBV). These polymers are synthesized when bacteria are exposed to carbon source while all other necessary nutrients are limited.[20,45] The properties of PHBV alter by varying the valerate content. PHAs are renewable, biodegradable and biocompatible. However, they possess a narrow processing window and are sensitive to processing conditions as well high temperature and shears. Therefore, additives, blends, and composites are potential techniques that can be employed to resolve these shortcomings of PHA.[20]

9.3.3 NATURAL FIBERS AS ALTERNATIVE FOR SYNTHETIC FIBERS

Natural fibers have accompanied human society since the start of our life. In early history, humans collected the fiber from the plant for rope and textile. Natural fibers are also used as a paper sheets, fish nets and old rags. According to Mohanty et al.,[46] natural fibers are derived from plants and animals as shown in Fig. 9.9. Production of materials from renewable resources has risen, due to the increasing awareness that nonrenewable resources are becoming scarce. This century can be considered as the cellulosic century because numerous renewable plant resources for products are being discovered. The increasing attention to natural fibers is primarily due to their economical production with few requirements for equipment and low specific weight. Such attributes of natural fibers result in higher specific strength and stiffness as compared to glass reinforced composites.[11] Natural fibers also provide safer handling and working conditions. They are nonabrasive to mixing and molding equipment, which can help in cost reductions.[11] The most significant virtue of natural fibers is their positive environmental impact. They are carbon dioxide neutral, meaning they do not emit excess carbon dioxide into the atmosphere as they are composted or combusted. In this way, natural fibers contribute to the mitigation of global warming. Working with natural fibers reduces dermal and respiratory irritation owing to their friendly processing atmosphere with better working conditions.[11] The abundant availability of natural fibers provides an added advantage over the use of synthetic fibers.

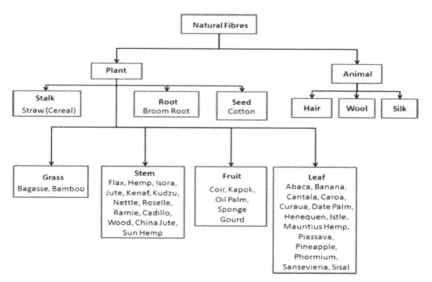

FIGURE 9.9 Natural fibers based on origin (Adapted from Azwa, Z. N.; Yousif, B. F.; Manalo, A. C.; Karunasena, W., A review on the degradability of polymeric composites based on natural fibers. Materials and Design 2013, 47, 424–442; John, M. J., Thomas, S., Biofibers and biocomposites. Carbohydrate Polymers 2008, 71, 343–364. With permission.).

Among the natural fibers, plant fibers are the main sources of fibers that can be found in a large quantity. Figure 9.10 shows the two main categories of natural fibers from plants. The nonwood natural fibers are classified into bast fibers, leaf fibers, seed fibers, stem fibers, grass and straw fibers.[16,47] Natural fibers from plant resources can also be categories as primary and secondary depending on their utilization. Primary plants are those that are grown solely for their fiber content (i.e., jute, hemp, kenaf, and sisal), whereas secondary plants are the ones in which the fiber are produced as a by-product (i.e., oil palm, pineapple, date palm and coir).

The most widely used natural fibers among the list showed in Table 9.1; flax, jute, hemp, sisal, ramie, and kenaf fibers were extensively researched and used in numerous applications.[16] Nowadays, sugar palm and oil palm fibers which are abundantly available in tropical regions are gaining more interest and significance in research due to their specific properties.

One of the key parameters of the plant fibers is to understand their chemical composition, as the chemical compositions of natural fibers have strong relationship to its performance for application in composite materials. Plant fiber also referred as lignocellulosics, consist mainly of cellulose, hemicelluloses and lignin. It also consists of minor amounts of free sugars, starch, proteins and other organic compounds.[48] Cellulose, hemi-cellulose, and lignin are the three main constituents of any plant fibers and the proportion of these components in a fiber depends on the age, source of the fibers and the extraction conditions used to obtain the fibers.[49] Table 9.1 shows the chemical composition of different types of fibers.

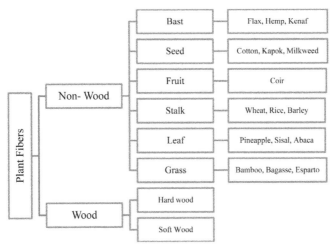

FIGURE 9.10   Classification of plant fibers.

Natural fibers are multicellular in nature, consisting of a continuous numbers, mostly in a form of cylindrical honeycomb cells which have different sizes, shapes, and arrangements for different types of fibers.[50] Different types of fibers provide different properties as shown in Table 9.2.

TABLE 9.1   Chemical Composition of Different Types of Natural Fibers

Fiber	Cellulose (wt.%)	Hemicellulose (wt.%)	Lignin (wt.%)	Waxes (wt.%)
Bagasse	55.2	16.8	25.3	–
Bamboo	26–43	30	21–31	–
Flax	71	18.6–20.6	2.2	1.5
Kenaf	72	20.3	9	–
Jute	61–71	14–20	12–13	0.5
Hemp	68	15	10	0.8
Ramie	68.6–76.2	13–16	0.6–0.7	0.3
Abaca	56–63	20–25	7–9	3
Sisal	65	12	9.9	2
Coir	32–43	0.15–0.25	40–45	–
Oil palm	65	–	29	–
Pineapple	81	–	12.7	–
Curaua	73.6	9.9	7.5	–
Wheat straw	38–45	15–31	12–20	–
Rice husk	35–45	19–25	20	14–17
Rice straw	41–57	33	8–19	8–38

Source: Faruk, O.; Bledzki, A. K.; Fink, H-P; Sain, M., Biocomposites reinforced with natural fibers: 2000–2010. *Progress in Polymer Science* 2012, 37, 1552–1596. With permission.

TABLE 9.2   Physico-Mechanical Properties of Different Natural Fibers

Fiber	Tensile strength (MPa)	Young's modulus (GPa)	Elongation at break (%)	Density [g/cm³]
Abaca	400	12	3–10	1.5
Bagasse	290	17	–	1.25
Bamboo	140–230	11–17	–	0.6–1.1
Flax	345–1035	27.6	2.7–3.2	1.5
Hemp	690	70	1.6	1.48
Jute	393–773	26.5	1.5–1.8	1.3
Kenaf	930	53	1.6	–
Sisal	511–635	9.4–22	2.0–2.5	1.5
Ramie	560	24.5	2.5	1.5
Oil palm	248	3.2	25	0.7–1.55
Pineapple	400–627	1.44	14.5	0.8–1.6
Coir	175	4–6	30	1.2
Curaua	500–1150	11.8	3.7–4.3	1.4

Source: Faruk, O.; Bledzki, A. K.; Fink, H-P; Sain, M., Biocomposites reinforced with natural fibers: 2000–2010. *Progress in Polymer Science* 2012, 37, 1552–1596. With Permission.

9.3.4   MALAYSIA: SUGAR PALM TREE

Sugar palm tree is a member of the *Palmae* family and naturally a forest species.[51] It belongs to the subfamily *Arecoideae* and tribe *Caryoteae*.[52,53] Hyene[54] reported that sugar palm have approximately around 150 local names indicating its multiple uses by the villagers. The names includes, *Arengapinnata*, Areng palm, Black fiber palm, Gomuti palm, Aren, Irok, Bagot and Kaong. In Malaysia, it is known as either *enau* or *kabung*. Sugar palm plant was originally from Assam, India and Burma. It originates from an area covering South East Asia up to Irian Jaya in the east of Indonesia. Sugar palm tree are widespread to Malaysia, Indonesia and other South East Asian countries. It is one of the most diverse multipurpose tree species in culture.

Malaysia as a tropical country has ample resources of natural fibers. One of those abundant natural fibers found in Malaysia but has not been widely used in the field of reinforcement is the sugar palm fiber.[3] This fiber is traditionally used by the local people to make brooms, brushes, septic tank base filter, door mats, carpet, chair/ sofa cushion, and rope. Although the fiber is popular among locals to have high strength and stiffness, little research has been conducted up to date on the full potential of sugar palm fibers and their composites.[3,55,58] Another attractive potential of sugar palms is their ability to produce biopolymers (i.e., starch). The starch obtained from the trunks of sugar palm trees can be use to make biodegradable polymer which in turn can be reinforced with natural fibers to make green composites. This composite possesses the advantage of being renewable, biodegradable, abundantly available (especially in tropical countries like Malaysia) and inexpensive as such they have a promising future in the field of biocomposite materials. Figure 9.11 shows the image of sugar palm fiber and sugar palm starch.

FIGURE 9.11   Sugar palm fiber extracted from sugar palm tree.

9.3.5   SUGAR PALM FIBERS REINFORCED SUGAR PALM STARCH BIOCOMPOSITES

Of late, biocomposites are yet to be seen in high magnitude. The depletion of petroleum resources coupled with awareness of global environmental problem provides the alternatives for a new green material in which the products are compatible with the environment and independent of petroleum based resources. Biocomposite with starch used as a matrix is one of the most popular biodegradable biocomposite and is highly investigated by researchers.[59,60] Biodegradable matrices are reinforced with natural fibers to improve the composites properties and these composites provide positive environmental advantages, good mechanical properties and light weight.[61,63]

There is a wide variety of studies that have been reported on the performance of incorporation sugar palm fiber in petroleum based matrix. Bachtiar et al.[64] carried out a study on properties of sugar palm fiber reinforced High Impact Polystyrene (HIPS) while Ishak et al.[65] and Sahari et al.[66,67] deals with sugar palm fiber reinforced unsaturated polyester. In the meantime, other research groups have conducted investigation on sugar palm fiber combined with epoxy composites.[3,68,69] However no previous research has been done on sugar palm fiber reinforced polymer matrix derived from natural resources. It is important to note that, we can extract the biopolymer which acts as matrix from sugar palm tree itself. So, the interesting part of these study is we investigate the properties of an environmentally friendly composite where the matrix (sugar palm starch) and fiber (sugar palm fiber) are derived from one source, that is, sugar palm tree., that is, sugar palm tree as shown in Fig. 9.12.

FIGURE 9.12   Sugar palm fiber (SPF) and sugar palm starch (SPS).

9.3.6   SUGAR PALM FIBERS

Sugar palm fiber is black in color, with diameter up to 0.50 mm.[70] According to Siregar,[68] sugar palm fiber has heat resistant of up to 150°C and the flash point is around 200°C. It has been reported that the fiber length of sugar palm fiber is up to 1.19 m and density is 1.26 kg/m^3.[71,72] Traditionally, sugar palm fiber was used as ropes, filters, broom, roof and handicraft application such as for making '*kopiah.*'[72,73] Tomlinson[74] reported that the ropes made from sugar palm fiber have better performance than the ropes made from rattan fiber (*Calamus sp.*). The main advantages of sugar palm fiber are durable and good resistant to seawater. It is also not affected by heat and moisture compared to coir fiber. Unlike other natural fibers, sugar palm fiber can directly be obtained from the trees which do not need secondary processes to yield the fibers.[3] Due to these advantages, sugar palm fiber should be a good material in the development of new 'green' materials.

Sugar palm fiber locally known as *ijuk* is one of the most popular fiber among the researcher over the last decade. The fiber is originally wrapped along the sugar palm trunk.[75] The tree can grow up to 12.3 m tall and has a thick, black/brown hairy fibrous trunk, with the dense crown of leaves, which are white on the outside. The tree begins to produce black sugar palm fiber after about 5 years, before flowering and the type of its fibers are depending to age and altitude of sugar palm tree.[71] The fibers that are taken after flowering will produce fiber approximately 1.4 m long. It can yield about 15 kg for each tree and around 3 kg is very best and stiffest. In Malaysia, black sugar palm fiber started used since 1416 during Malacca Sultanate History. In 1800, the sugar palm tree was planted by British East India Company in *Penang* to yield its high durability of rope made from black sugar palm fiber.

Previously, the characterization (tensile and chemical properties) of single fibers from different morphological parts of sugar palm tree, that is, sugar palm frond (SPF), sugar palm bunch (SPB), *ijuk* and sugar palm trunk (SPT).[76] From the investigation, it was found that the tensile strength of *ijuk* was 276.64 MPa and the tensile modulus was5.86 GPa. The elongation at break of *ijuk* was 22.3% which was approximately the same with oil palm and coir fibers in term of physical and mechanical properties because they were from same palmae family.[77] For the chemical analysis, it was shown that *ijuk* has a high cellulose content which is 52.29%. This proved that mechanical properties of sugar palm fiber are strongly influenced by the cellulose content.[78] Cellulose was the main structural component that provides strength and stability to the plant cell walls and the fibers.[49]Generally, *ijuk* can be used as reinforcement in composites due to higher tensile strength and cellulose content in comparison with other established natural fibers such as kenaf, pineapple leaf, coir and oil palm bunch.

9.3.7 SUGAR PALM STARCH

Sugar palm tree is one of multipurpose trees grown in Malaysia. The inner part of sugar palm stem contains starch. It was used as raw material for starch and glue substances.[79] For the production of one ton of starch, ten to 20 trees are needed which suggests that one tree can produce 50 to 100 kg of starch.[76] This accumulated starch is harvested from the trunk of matured palms (after 25 years) and can be applied as 'green' material. Starch will act as biopolymer in the presence of a plasticizer such as water, glycerol and sorbitol at high temperature. Previously, the characterization of sugar palm starch as a biopolymer have been done by using the glycerol as a plasticizers.[76] From the research, it was found that the tensile strength of SPS/G30 showed the highest value 2.42 MPa compared to the other concentration of the plasticizer. The higher the concentration of the plasticizers, the higher the tensile strength of plasticized SPS and optimum concentration was 30 wt.%. The tensile strength decrease to 0.5 MPa when concentrations of plasticizer was 40 wt.%. As the plasticizer content increased to 40%, not enough SPS to be well bonded with glycerol and thus poor adhesion occurred which reduce the mechanical properties of SPS/G40. This result well agreed with the finding of Laohakunjit and Noomhorm[80] who claimed that the films outside this range are either too brittle (< 20 wt.%) or too tacky (> 45 wt.%).

Generally, as the plasticizer increase, the tensile strength and elongation of plasticized SPS increase, while the tensile modulus decreased. This phenomena indicates that the plasticized SPS is more flexible when subjected to tension or mechanical stress. The results from this study prove the finding done by Beerler and Finney[81] whereby they reported that the plasticizers such as glycerol will interfere the arrangement of the polymer chains and the hydrogen bonding. It is also most likely affect the crystallinity of starch by decreasing the polymer interaction and cohesiveness. Thus, this make the plasticized SPS become more flexible with the increasing of glycerol.

For the thermal properties, the Tg of dry SPS reaches 242.14 °C and decreased with addition of glycerol. This value was higher than Indica rice starch where Tg values was237°C.[82] Meanwhile, Myllärinen et al.[83] claimed that for dry starch the Tg reaches 227 °C. The sample with high glycerol concentrations showed lower Tg values and Tg of starch without plasticizer were higher than those of samples with glycerol (Table 9.3). This behavior was also observed by Mali et al.[84] for yam starch and by Forssell et al.[33], for films based on barley starch, in both cases with glycerol as plasticizer. According to Guilbert and Gontard[85], plasticization decreases the intermolecular forces between polymer chains, consequently change the overall cohesion, leading to the reduction of Tg. Mitrus[86] has been claimed that the plasticizer decreased Tg because it facilitates chain mobility. Brittleness is one of the major problems connected with starchy material due to its high Tg.[87] In the absence of plasticizers, starch are brittle. The addition of plasticizers overcomes starch brittleness and improves its flexibility and extensibility of the polymers.

TABLE 9.3   Glass Transition Temperature (Tg) of Plasticized SPS

Sample	Glass transition temperature, Tg (onset)	Glass transition temperature, Tg (midpoint)
Native SPS	237.91 °C	242.14 °C
SPS/G15	225.68 °C	229.26 °C
SPS/G20	206.44 °C	217.90 °C
SPS/G30	189.57 °C	187.65 °C
SPS/G40	176.71 °C	177.03 °C

9.4   PREPARATION OF MATERIALS

The sugar palm fiber (SPF) was collected at Jempol, Negeri Sembilan in Malaysia. All of the fibers were grind and screened using Fritsch pulverisette mill to obtain 2 mm fiber size. For the extraction of sugar palm starch (SPS), firstly, the woody fibers and starch powder was obtained from the interior part of the trunk (see Fig. 9.13). Then, this mixture (woody fibers and starch powder) was carried out for washing process to obtain the starch. Then starch was kept in an open air for a moment and dried in an air circulating oven at 120°C for 24 hrs.

FIGURE 9.13   The process of preparing sugar palm starch from sugar palm tree.

9.4.1   FABRICATION OF SPF/SPS BIOCOMPOSITES

Sugar palm starch (SPS) 70 wt.% and glycerol 30 wt.% were mixed using mechanical stirrer for 30 min. Subsequently, the sugar palm fibers were added and stirred for 20 min. The mixture was cast in an iron die and was kept for precuring at room temperature (28°C) for 24 h. Finally, the sample was cured by hot pressing in a Carver hydraulic hot press at 130°C for 30 min under the load of 10 ton (see Fig. 9.14). The final products were in the form of plates with dimensions 150 cm × 150 cm × 0.3 cm. By changing the content of sugar palm fibers of 0, 10, 20 and 30 wt.%, a series of SPF/SPS biocomposites was prepared with different content of SPF and coded as SPS, SPF10, SPF20 and SPF30, respectively.

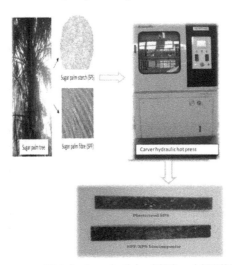

FIGURE 9.14   The process of fabricating plasticized SPS and SPF/SPS biocomposites.

9.4.2   MECHANICAL PROPERTIES

The tensile strength and modulus of SPF/SPS biocomposites showed increasing trend with increasing SPF loading, while the addition of the SPF made the elongation fall from 8.03% to 3.32%. A considerable increase of tensile strength with the increase of SPF indicated that the natural cellulose fiber has a great impact and contributes to the formation of good bonding between SPS and SPF. This was due to the remarkable intrinsic adhesion of the fiber–matrix interface caused by the chemical similarity of starch and the cellulose fiber.[35] SPF is believed to work as a carrier of load in the SPS matrix. The stronger bonding between the matrix and fiber leads to enhanced interfacial adhesion between them and therefore a greater transfer of stress from the matrix to the fibers during tensile testing.[88] Increasing the fiber loading resulted in increase of tensile modulus. The SPF/SPS biocomposite consists of low stiffness matrix and high stiffness fibers, thus by increasing the weight percentage of fibers increases the stiffness of the biocomposites. Meanwhile, elongation showed decreasing trend when increasing fiber loading. This is also a normal consequence of the increase of fibers wt.%, which is having a low strain, compared to rubbery starch materials, SPS.[15,89]

The flexural strength and modulus had an increasing trend with increasing fiber loading. It is normal to have increase in flexural strength and modulus when introducing more fibers to SPS matrix. Similar trend also was obtained for impact strength of SPF/SPS biocomposites where the impact strength increases with increasing fiber loading. Like other mechanical properties, the impact strength of biocomposites is highly affected by the fiber. Overall, the mechanical property results observed in the present study agree with those reported by other researchers for lignocellulosic fibers and/or cellulose nanofibers incorporated in thermoplastic starches.[90,92]

9.4.3   THERMAL PROPERTIES

It was found that the fiber loading increased thermal stability of the biocomposite. The first mass loss, in the range 31–100°C, is due to evaporation of moisture. Then the mass loss between 150°C and 380°C for SPF/SPS biocomposites, are due to the decomposition of the three major constituents of the natural fibers; hemicellulose, cellulose and lignin.[94] In general, the thermal decomposition of these fibers consists of four phases. The first phase was decomposition of hemicelluloses, followed by cellulose, lignin and lastly their ash.[95,96] Yang et al.[97] reported that hemicelluloses decompose at 220°C and substantially completed at 315°C. As soon as hemicelluloses had completely decomposed, the decomposition of cellulose will take place as the second phase of decomposition. Because of its highly crystalline nature of their cellulose chain than amorphous, cellulose has relatively thermally stable. It does not start to decompose until hemicelluloses had completely decomposed which normally starts at higher temperature at about 315°C.[97] This is supported by Kim et al.[98] where they reported that the critical temperature of decomposition of crystalline cellulose is 320°C. The third phase was the decomposition of lignin. It is the most difficult to decompose compared to hemicelluloses and cellulose. Although the decomposition of lignin had started as early as 160°C, it decomposes slowly and extends its temperature as high as 900°C to complete its decomposition.[97] This is contributed by lignin which is very tough component and known as the compound that gives rigidity to the plant materials. Finally, when the lignin had completely decomposed, the component that is left is inorganic material in the fibers which can be assumed as ash content. This is due to the presence of inorganic materials such as silica (silicon dioxide, SiO_2) in the fiber which can only be decomposed at a very high temperature of 1723°C. The increase of fibers increased the lignin and ash content, compared to pure plasticized SPS. It was found that the lignin and ash content of SPF was 31.5% and 4%, respectively.[99] The large degradation at 310 °C appears for plasticized SPS are due to the elimination of the polyhyroxyl groups, accompanied by depolymerization and decomposition in starch.[100]

9.4.4 WATER ABSORPTION

The effect of fiber loading on water content of SPF/SPS biocomposites was investigated. It can be noted that the water uptake or absorption of the SPF/SPS biocomposite decreases with increasing of fiber loading. Similar trends were also found for moisture content results. This could be attributed to the hydrophobic behavior of SPF compared to SPS. Ma et al.[101] claimed that starch is a multihydroxyl polymer with three hydroxyl groups per monomer. The remarkable decreases of the water content of SPF/SPS biocomposites compared to SPS indicated that SPF was a better water resistant material compared to plasticized SPS. From the present study it be generally said that the greater the fiber content, the lower the water absorption in the biocomposite. This phenomenon also can be attributed to better interfacial bonding between the matrix and fibers as well as the hindrance to absorption caused by the fibers. Moisture absorption of the composites was also studied by leaving in open air the untreated jute-starch composites for 72 days. The humidity absorbed was very low and they claimed that the increase in the percentage of reinforcement did not affect substantially the moisture absorption.[102]

The tensile strength of SPF/SPS biocomposites increased with the increase of fiber content, that is, 2.42 MPa, 3.12 MPa, 4.01 MPa and 5.31 MPa, respectively. This enhancement indicates the effectiveness of the SPF act as reinforcement. But after 72 h exposed in 75% RH it is clear that the tensile strength for all SPF/SPS biocomposites decrease drastically to 0.42 MPa, 0.87 MPa, 1.24 MPa and 1.73 MPa, respectively. Similar finding were also observed to the impact strength of SPF/SPS biocomposites where the impact strength increase with increasing of fiber content and decrease accordingly after exposed to aqueous environmental. This phenomenon could be attributed to SPS, which tends to absorb high amount of moisture after exposed to aqueous environmental and leads to higher degradation rate. This tendency can be also related to the weak SPF–SPS interface due to water absorption after exposed to 75% RH.[103,105] The decrease in mechanical properties with increasing in moisture content could be attributed to the formation of hydrogen bonding between the water molecules and SPS. The hydroxyl groups in SPS form an intramolecular hydrogen bonds inside the molecule and intermolecular hydrogen bonds with other starch molecules as well as hydroxyl groups of moist air. With the presence of high hydroxyl group, starch tends to show low moisture resistance. This may lead to dimensional variation and poor interfacial bonding of SPS. The rigidity of the starch structure is destroyed by the water molecules. The mechanical deterioration of SPF/SPS biocomposite after water immersion was due to formation of hydrogen bonding between the water molecules and SPS matrix. This leads to the degradation of fiber–matrix interface region creating poor stress transfer efficiencies resulting in a reduction of dimensional and the mechanical properties of SPF/SPS biocomposites.

9.4.5 SUGAR PALM FIBER REINFORCED EPOXY COMPOSITE

Epoxy is a large family of resins which represent some of the high-performance resins.[106] It is a thermosetting polymer formed from the reaction of an epoxide resin with polyamine hardener.[107] Epoxy is one of the best matrix materials for industrial application, due to its excellent properties such as good adhesion, mechanical properties, chemical and heat resistance, low moisture content, little shrinkage, and processing ease.[16,107] The outstanding performance of epoxy in terms of adhesive and mechanical properties, and resistance to environmental degradation lead to its wide usage in the aircraft and boat building industry.[106] Over the years, there were many studies conducted on SPF-epoxy composites.[108,114]

9.4.5.1 EFFECT OF AGING ON SUGAR PALM FIBER REINFORCED EPOXY COMPOSITE

Tensile and impact tests were conducted by Ali et al.[16] on original and aged SPF/epoxy composite to compare their mechanical properties. An accelerated aged composite was obtained by using the standard material age acceleration under ASTM F1980 with 10% error. The parameters used in their investigation were; (a) time of real time aging in a natural environment (70 days), (b) ambient temperature (25 °C), (c) accelerated aging temperature (70 °C), and (d) accelerated aging factor (2.0). Equation 1 and 2 were used to determine the accelerated aging rate and accelerated aging time duration.

Where AAR was accelerated aging rate, AAT was accelerated aging temperature, AT was ambient temperature, Q_{10} was accelerated aging factor, AATD was accelerated aging time duration and DRTA was the desires real time aging.[16] When the parameters were substituted into the equations, the accelerated aging time (AAT) was 74 hrs. and 10 mins which was equivalent to 70 days of aging in a natural environment.

Aged SPF/epoxy composite manifested superior tensile strength compared to those of the original composite. The tensile strength of the aged composite increased by 50.4%, which indicates that the aging process rendered the composite matrix more compact, yielding a harder and denser composite material. This may happen due to slight increase in specific volume of the epoxy polymer chains and energy was stored within the chains during the heating process. The polymer chains eventually release the stored energy during the cooling stage (after heating the composite), causing a decrease in the composite specific volume. This phenomenon causes the epoxy polymer chains to become denser and further strengthen the composite bond.[108] However, lower tensile modulus of the aged composite was obtained from the tensile test as compared to the original composite. It was evident from their findings that the aged SPF/epoxy composites had very low ductility properties. The resistance towards deformation decreased for the aged composite due to the dense epoxy matrix polymer chains. The impact strength of both composites explicitly shows that SPF/epoxy composites are brittle and that aging has a minimal effect on the impact properties of the composite.[108]

9.4.5.2   EFFECT OF FIBER TREATMENT

Leman et al.[3] conducted series of tensile tests to determine the maximum stress of SPF reinforced epoxy composites. The composites were fabricated using SPF treated with seawater and freshwater for different time periods (6, 12, 18, 24, and 30 days). The aim of their study was to examine whether SPF can be effectively used in the marine sector as a potential substitute for the conventional glass fiber for manufacturing fishing boats. The results proved that SPF treated with seawater for 30 days had the highest stress value (23042.48 kPa), followed by freshwater treated SPF for 30days (21266.5 kPa). The SPF/epoxy composite strength improved by 67.26 and 54.37% for seawater and freshwater treated SPF for 30 days, respectively, as compared to the untreated SPF. The SPFs that were treated for 30 days with seawater or freshwater had the smoothest surfaces. This was attributed to the removal of their first surface layers, which consisted of hemicelluloses and pectin. Therefore, the adhesion between SPF and epoxy polymer matrix significantly improved, leading to higher tensile strength.

The effect of alkaline treatment on tensile and impact properties of SPF/epoxy composites was reported by Bachtiar et al.[112] The treatment was conducted using two different sodium hydroxide solution concentrations (0.25 and 0.5 M) and three different soaking times (1, 4 and 8 h). The alkali treatment in all conditions (different alkali concentration and soaking time) significantly enhanced the tensile properties of SPF/epoxy composite particularly for tensile modulus. SPF treatment with alkali removed the hemicellulose and lignin content, making the fiber relatively ductile and provided rougher fiber surface than the untreated SPF. This created better interlocking mechanism between the fiber and matrix surface. However, the treatment had a negative effect on the tensile strength of SPF/epoxy at higher alkali concentration. Very high alkali solution for fiber treatment certainly cause damage to the fiber and consequently decrease the tensile strength of fiber and also their composites.[110] On the contrary, higher alkali concentration provided better impact performance for the SPF/epoxy composite. At strong alkali concentration treatment, the lignin and hemicellulose were washed out enabling better exposure of the fiber to epoxy matrix, leading to better bonding between their surfaces.[107,112,115]

9.4.5.3 EFFECT OF FIBER LOADING AND ORIENTATION

SPF/epoxy composite tensile properties in different fiber orientations (long random, chopped random, and woven roving) and contents (10, 15, and 20%) was investigated.[70] The results of different orientations proved that the 10 wt.% woven roving of SPF had the highest tensile strength and tensile modulus values, that is, 51.725 MPa and 1255.825 MPa. The high tensile properties of the woven roving SPF orientation were ascribed to the longitude and transversal orientation of woven roving fiber formation. Consequently, the bonding between their fibers are stronger compared to the long random SPF and chopped random SPF. Meanwhile, increasing the fiber loading from 10 to 15 wt.% for both long random and chopped random SPF orientations improved the tensile strength and tensile modulus of their respective composites. Further addition of long random and chopped random SPFs up to 20 wt.% in their composites decreased their tensile strength and tensile modulus. This might be due to the inadequate wetting of the fiber by the matrix resin as well as the easy crack initiation and propagation at the fiber ends.

9.4.6 SUGAR PALM FIBER REINFORCED HIGH IMPACT POLYSTYRENE (HIPS) COMPOSITES

Polystyrene is an aromatic polymer made from an aromatic monomer (styrene) which is commercially manufactured from petroleum.[107] High Impact Polystyrene (HIPS) is a type of polystyrene that is versatile, economical and easy to fabricate. HIPS is mostly used for low strength structural applications where impact resistance, machinability and low cost are needed. Polybutadiene is added during the polymerization of HIPS. This polybutadiene is generally known as rubber, which provides HIPS with the required toughness and impact resistance.[57]

9.4.7   MECHANICAL PROPERTIES

9.4.7.1   EFFECT OF FIBER LOADING

A study on the effect of SPF loading on the tensile properties of SPF/HIPS composite was reported by Sapuan and Bachtiar[57]. Fiber contents of 0, 10, 20, 30, 40 and 50% by weight were incorporated in HIPS matrices. The tensile strength of short SPF/HIPS composites slightly decreased compared to the neat HIPS (0% loading) as the fiber content in them increased from 10 to 30%. The average tensile strength of the neat HIPS was 29.92 Mpa, which was approximately 12.4, 22.5 and 35.5% higher than SPF/HIPS composites with 10, 20 and 30% fiber contents, respectively (see Fig. 9.15). The decrease of tensile strength of the composites with fiber loadings from 10 to 30% by weight emanated from the weak fiber-matrix interface as a result of differing polarities of the hydrophilic SPF and hydrophobic HIPS matrix. On the other hand, considerable increment in the tensile strength of the composites was realized as the fiber loadings increased from 40 to 50%. This may be attributed to the better interaction and improved dispersion of SPF in the HIPS matrix, which enhanced the interfacial bonding between fibers and matrix. Generally, good tensile strength depends more on effective and uniform stress distribution.[96] The high tensile strength of composites at fiber content of 40 to 50% indicates the fibers were effectively capable of transfer the load to one another. This illustrates that the fibers (40 to 50% weight fraction) effectively participated in the stress transfer causing the effect of crack inhibition more dominant over the effect of crack initiators.[96]

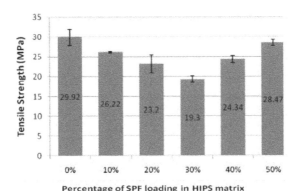

FIGURE 9.15   Tensile strength of SPF-HIP composite for different fiber loading (Adapted from Sapuan, S. M.; Bachtiar, D., Mechanical Properties of Sugar Palm Fiber Reinforced High Impact Polystyrene Composites. *Procedia Chemistry* 2012, 4, 101–106. With permission.).

The tensile moduli of short SPF/HIPS composites increased from 1516 to 1706 MPa with increase fiber loading from 10 to 30% by weight (see Fig. 9.16). The maximum tensile modulus value obtained at fiber content of 30% in HIPS matrix was 25.6% higher than neat HIPS (0% fiber loading). However, the addition of short SPF by 40 to 50% inflicted slight decrease on the tensile modulus of the composites. The decrease of tensile modulus beyond SPF loading of 30% was due to the decrease of bond quality between the fibers and HIPS matrix. At high fiber contents, the extent of fiber wetting reduces because it is difficult to achieve good consolidation of the composites during fabrication process.[107]

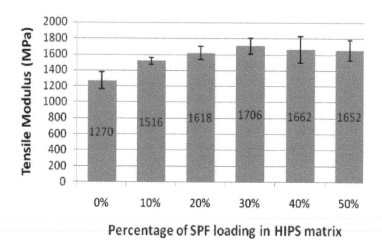

FIGURE 9.16   Tensile modulus of SPF-HIP composite for different fiber loading (Adapted from Sapuan, S. M.; Bachtiar, D., Mechanical Properties of Sugar Palm Fiber Reinforced High Impact Polystyrene Composites. *Procedia Chemistry* 2012, 4, 101–106. With permission.).

9.4.7.2   EFFECT OF FIBER TREATMENTS

The effect of polystyrene-block-poly(ethylene-ran-butylene)-block-poly(strene-graft-maleic-anhydride) as compatibilizing agent (2 and 4%) and alkali treatment (4 and 6%) of short SPF on the flexural strength and flexural modulus of SPF/HIPS composites were studied by Bachtiar et al.[116] using 40 wt.% of fiber content. SPF alkali treatment using 6% NaOH solution improved the flexural strength, flexural modulus and impact strength of the composites as compared to the untreated composites by 12%, 19% and 34%, respectively. On the contrary, the SPF/HIPS composites treated with compatibilizing agent indicated no improvement in flexural strength and flexural modulus. However, significant improvements of impact strength of the alkali and compatibilizing agent treated composites were obtained. The impact strength of the 4% alkali and 3% compatibilizing agent treated composites were about 16% higher than the untreated SPF/HIPS composites. The enhancement of the impact strength of alkali treated SPF/HIPS composites were due to: (1) development of rough surface fibers which offers good fiber-matrix adhesion; and

(2) removal of hemicellulose and lignin parts of the SPF fibers, whereas the strong cellulose components on the fibers remained. The compatibilizing agent also enhanced the impact strength of the composites due to chemical reaction of hydroxyl groups of SPF fibers with the anhydride groups of the copolymers which resulted into good interface adhesion between SPF fiber and HIPS matrix.[116]

9.5 CONCLUSIONS

The rapid advancement in the development of greener materials based on natural fibers and biopolymers is gaining more attention due to increase environmental awareness coupled with depletion of petroleum resources. Utilization of fibers and polymers that are biodegradable and obtained from renewable resources will help to preserve our environment. Hence, sugar palm tree is a potential 'green resource' for natural fibers and biocomposites.

Sugar palm tree is one of the multipurpose trees grown in tropical regions. Of recent, sugar palm fiber with its desirable properties has manifested high potential to be used as reinforcement in polymer composites. Sugar palm fibers can be used as reinforcement for bio-based polymers to produce 100% biodegradable composites. The use of sugar palm fiber and bio-based polymers or even petroleum-based polymers to develop greener composites helps in: (1) reducing the negative environmental impact of synthetic polymers and fibers; (2) decreasing the pressure for the dependence on petroleum products; and (3) developing sugar palm trees as new crop in the future for tropical countries most especially in Malaysia. When biocomposite materials from sugar palm tree are increasingly used for industrial applications, such would boost Malaysia's status as global promoter, developer and manufacturer of green composites. This would lead to increasing revenues and create more jobs. The successful development of green composites from sugar palm tree would provide opportunities to improve the standard of living of the sugar palm tree farmers in Malaysia. These would generate nonfood source of economic development for farming and rural areas in Malaysia.

9.6 ACKNOWLEDGEMENTS

The authors wish to acknowledge the financial support from Ministry of Education Malaysia with Exploratory Research Grant Scheme (ERGS) project vote number 5527190. Thanks also due to Universiti Putra Malaysia for granting sabbatical leave to S. M. Sapuan in 2013–2014 and to Ministry of Education Malaysia for scholarship award (MyPhD) to J. Sahari. The assistance of Dr. Nukman Yusoff of Universiti Malaya is highly appreciated.

KEYWORDS

- **Biodegradable Polymers**
- **Composites**
- **Mechanical Properties**
- **Polymer Matrix**
- **Starch**
- **Sugar Palm Fibre**
- **Thermal Properties**

REFERENCES

1. Fowler, P. A., Hughes, J. M., & Elias, R. M. (2006). Biocomposite: technology. Environmental credentials and market forces. *Journal of the Science of Food and Agriculture*, 86, 1781–1789.
2. Mohanty, A. K., Misra, M., Drzal, L. T., Selke, S. E., Harte, B. R. & Hinrichsen, G. (2005). Natural fibers, biopolymers, and biocomposites: An introduction. In *Natural Fibers, Biopolymers and Biocomposites*. Boca Raton, FL: CRC Press.
3. Leman, Z., Sapuan, S. M., Azwan, M., Ahmad, M. M. H. M., & Maleque, M. A. (2008a) The effect of environmental treatments on fiber surface properties and tensile strength of sugar palm fiber-reinforced epoxy composites. *Polymer-Plastics Technology and Engineering*, 47(6), 606-612.
4. Leman, Z., Sastra, H. Y., Sapuan, S. M., Hamdan, M. M. & Maleque, M. A. (2005). Study on impact properties of arenga pinnata fiber reinforced epoxy composites. *Journal of Applied Technology*, 3(1), 14–19.
5. Maleque, M. A., Saiful, A., & Sapuan, S. M. (2005). Effect of volume fraction of pseudostem banana fiber on the properties of epoxy composite. *Proceedings of the International Conference on Mechanical Engineering* 2005 (ICME2005), Dhaka, Bangladesh, December, 28–30, 471–485.
6. Sapuan, S. M., Zan, M. N. M., Zainudin, E. S., & Arora, P. R. (2005). Tensile and flexural strengths of coconut spa the-fiber reinforced epoxy composites. *Journal of Tropical Agriculture*, 43, 67–69.
7. Sapuan, S. M., Harimi, M., & Maleque, M. A. (2003). Mechanical properties of epoxy/coconut shell filler particle composites. *Arabian Journal for Science and Engineering*, 28, 171–181.
8. Satyanarayana, K. G., Arizaga, G. G. C., & Wypych, F. (2009). Biodegradable composites based on Lignocellulosic fibers an overview. *Progress in Polymer Science*, 34, 982–1021.
9. Encyclopedia Britannica online. Great Wall of China [www.britannica.com, accessed October 2013].
10. Construction of the Great Wall [http://library.thinkquest.org/c004471/tep/en/ cultures/chinese. html, accessed in October 2013].
11. John, M. J., & Thomas, S. (2008). Biofibers and biocomposites. *Carbohydrate Polymers*, 71, 343–364.
12. Narayan, R. (2006). Rationale, drivers, standards and technology for biobased materials. In Graziani, M., Fornasiero, P., editors. Renewable resources and renewable energy a global challenge. Boca Raton, USA: CRC Press.

13. Vilaplana, F., Strömberg, E., & Karlsson, S. (2010). Environmental and resource aspects of sustainable biocomposites. *Polymer Degradation and stability*, 95, 2147–2161.
14. Yu, L, Dean, K, & Li, L. (2006). Polymer blends and composites from renewable resources. Progress in Polymer Science, 31, 576–602.
15. Averous, L., & Boquillon, N. (2004). Biocomposites based on plasticized starch: Thermal and mechanical behaviors. *Carbohydrate Polymers*, 56, 111–122.
16. Faruk, O., Bledzki, A. K., Fink, H. P., & Sain, M. (2012). Biocomposites reinforced with natural fibers: 2000–2010. *Progress in Polymer Science*, 37, 1552–1596.
17. Arvanitoyannis, I. *(1999)*. Totally and-partially biodegradable polymer blends based on natural and synthetic macromolecules: preparation and physical properties and potential as food packaging materials. *Journal of Macromolecular Science,* 39(2), 205–271.
18. Fang, J., Fawler, P., Eserig, C., González, R., Costa, J., & Chamudis, L. (2005). Development of biodegradable laminate films derived from naturally occurring carbohydrate polymers. *Carbohydrate Polymers,* 60(1), 39–42.
19. Xu, Y., Kim, K., Hanna, M., & Nag, D. *(2005)*. Chitosan-starch composite film: preparation and characterization. *Industrial Crops and Products*, 21(2), 185–192.
20. Reddy, M. M., Vivekanandhan, S., Misra, M., Bhatia, S. K., & Mohanty, A. K. (2013). Bio-based plastics and bionanocomposites: current status and future opportunities. *Progress in Polymer Science*, 38, 1653–1689.
21. Lu, D. R., Xiao, C. M., & Xu, S. J. (2009). Starch-based completely biodegradable polymer materials. *EXPRESS Polymer Letters*, 3(6), 366–375.
22. Ray, S. S., & Bousmina, M. (2005). Biodegradable polymers and their silicate nanocomposites. In greening the twenty-first century materials world. *Progress in Materials Science,* 50(8), 962–1079.
23. Ellis, R. P. *(1998)*. Starch production and industrial use. *Journal* of the *Science* of *Food* and *Agriculture,* 77(3), 289–311.
24. Zobel, H. F. (1988). Molecules to granules: a comprehensive starch review. *Starch*, 40(2), 44–50
25. Buleon, A. *(1998)*. Starch granules: structure and biosynthesis. *International Journal of Biological Macromolecules*, 23(2), 85–112.
26. Whistler, R. L. & Daniel, J. R. (1984). Molecular structure of starch. In Whistler, R. L., BeMiller, J. N., Paschall, E. F. (Eds.,) *Starch: Chemistry and Technology,* Orlando, FL: Academic Press.
27. Araújo, M. A., Cunha, A., & Mota, M. (2004). Enzymatic degradation of starch-based thermoplastic compounds used in prostheses: Identification of the degradation products in solution. *Biomaterials*, 25, 2687–2693.
28. Zhang, J. F., & Sun, X. Z. (2004). Mechanical properties of PLA/starch composites compatibilized by maleic anhydride. *Biomacromolecules*, 5, 1446–1451.
29. Teramoto, N., Motoyama, T., Yosomiya, R., & Shibata, M. (2003). Synthesis, thermal properties, and biodegradability of propyl-etherified starch. *European Polymer Journal*, 39, 255–261.
30. Halley, P. J. (2005). Thermoplastic starch biodegradable polymers. In: Smith, R., editor. Biodegradable polymers for industrial applications. Cambridge UK CRC Press/Woodhead, 140–162.
31. Coffin, D. R., & Fishman, M. L. (1994). Physical and mechanical properties of highly plasticized pectin/starch films. *Journal of Applied Polymer Science*, 54, 1311–1320.
32. Mohanty, A. K., Misra, M., & Hinrichsen, G. (2000). Biofibers, biodegradable polymers and biocomposites: an overview. *Macromolecular Materials and Engineering*, 276/277, 1–24.

33. Forssell, P. M., Mikkia, J. M., Moates, G. K., & Parker, R. (1997). Phase and glass transition behavior of concentrated barley starch-glycerol-water mixtures, a model for thermoplastic starch. *Carbohydrate Polymers*, 34, 275–282.

34. Gaudin, S., Lourdin, D., Le Botlan, D., Ilari, J. L., & Colonna, P. *(1999)*. Plasticization and mobility in starch-sorbitol film. *Journal of Cereal Science*, 29, 273–284.

35. Ma, X., Yu, J., & Jin, F. *(2004)*. Urea and formamide as a mixed plasticizer for thermoplastic starch. *Polymer International*, 53, 1780–1785.

36. Nakamura, S., & Tobolsky, A. V. (1967). Viscoelastic properties of plasticized amylose films. *Journal of Applied Polymer Science,* 11, 1371–1381.

37. Kalichevsky, M. T., Jaroszkiewicz, E. M., & Blanshard, J. M. V. (1993). A study of the glass transition of amylopectin-sugar mixtures, *Polymer,* 34, 346–358.

38. Vilpoux, O., & Avérous, L. (2004). Starch-based plastics. In M. P. Cereda and O. Vilpoux (Eds.), *Technology, use and potentialities of Latin American starchy tubers*. Sao Paolo: NGO Raízes and Cargill Foundation.

39. Mani, R., & Bhattacharya, M. (2001). Properties of injection molded blends of starch and modified biodegradable polyesters. *European Polymer Journal*, 37, 515–526.

40. Ratto, J. A., Stenhouse, P. J., Auerbach, M., Mitchell, J., & Farrell, R. (1999). Processing, performance and biodegradability of a thermoplastic aliphatic polyester/starch system. *Polymer*, 40, 6777–6788.

41. Siracusa, V., Rocculi, P., Romani, S., & Rosa, M. D. (2008). Biodegradable polymers for food packaging: a review. *Trend in Food Science and Technology*, 19, 634–643.

42. Cabedo, L., Feijoo, J. L., Villanueva, M. P., Lagaron, J. M., & Gimenez, E. (2006). Optimization of biodegradable nanocomposites based application on a PLA/PCL blends for food packaging application. *Macromolecular Symposium*, 233, 191–197.

43. Jem, K., Van der pol, J., & de vos, S. (2010). Microbial lactic acid. Its polymer poly (lactic acid) and their industrial applications. In: Chen, G. Q., editor. Plastics from Bacteria-Natural Functions Applications. Heidelberg: Springer-verlag, 323–346.

44. Scholz, C., & Gross, R. A. (2000). Poly (hydroxyalkanoates) as potential biomedical materials: an overview. In: Polymers from renewable resources. Washington: American Chemical Society, 328–334.

45. Doi, Y. (1990). Microbial Polyesters. New York: Wiley VCH 166 pp.

46. Mohanty, A. K., Tummala, P., Liu, W., Misra, M., Mulukutla, P. V. & Drzal, L. T. (2005b). Injection molded biocomposites from soy protein based bioplastic and short industrial hemp fiber. *Journal of Polymer and the Environment,* 13(3), 279–285.

47. Abdul Khalil, H. P. S., Bhat, I. U. H., Jawaid, M., Zaidon, A., Hermawan, D., & Hadi, Y. S. (2012). Bamboo fiber reinforced biocomposites: A review. *Materials and Design*, 42, 353–368.

48. Rowell, R. M. (1998). Property enhanced natural fiber composite material based on chemical modification. In P. N., Prasad, J. E., Mark, S. H., Kandil, Z. H., Kafafi, (Eds.), *Science and Technology of Polymers and Advanced Materials*. New York: Plenum Press.

49. Reddy, N., & Yang, Y. (2005). Biofibers from agricultural by products for industrial applications. *Trends in Biotechnology,* 23(11), 22–27.

50. Satyanarama, K. G., Pai, B. C., Sukumaran, K., & Pillai, S. G. K. (1990). Fabrication and properties of lignocellulosic fiber incorporated polyester composites. In Cheremisioff, N. P. (Eds.), *Handbook of Ceramic and Composites.* New York A Wiley Interscience Publication.

51. Siregar, J. P. (2005). *Tensile and flexural properties of arenga pinnata filament (Ijuk Filament) reinforced epoxy composites*, MSc Thesis, Universiti Putra Malaysia.

52. Moore, H. E. (1960). A new subfamily of palms. The Gryotoideae. *Principes,* 4, 102–117.

53. Dransfield, J. & Uhl, N. W. (1986). An outline of a classification of palms. *Principes,* 30, 3–11.

54. Hyene, K., & Sago. (1950). De uttige planten van Indonesie (3e Ed.). Netherlands/Indonesia: Van Hoeve, S. Gravenhage/Bandung, 330–339.

55. Ishak, M. R., Sapuan, S. M., Leman, Z., Rahman, M. Z. A., Anwar, U. M. K., & Siregar, J. P. (2013). Sugar palm (*Arenga pinnata*): Its fibers, polymers and composites*Carbohydrate Polymers*, 91, 699–710.

56. Sahari, J., Sapuan, S. M., Zainudin, E. S., & Maleque, M. A. (2013a). Thermo-mechanical behaviors of thermopolymer starch derived from sugar palm tree (*Arenga pinnata*). *Carbohydrate Polymers*, 92, 1711–1716.

57. Sapuan, S. M., & Bachtiar, D. (2012). Mechanical Properties of Sugar Palm Fiber Reinforced High Impact Polystyrene Composites. *Procedia Chemistry*, 4, 101–106.

58. Bachtiar, D., Sapuan, S. M., & Hamdan, M. (2008). Mechanical properties of sugar palm fiber reinforced epoxy composites. *In book of abstract of international conference on advances in polymer science and technology*, New Delhi. India.

59. Hermann, A. S., Nickel, J., & Riedel, U. (1998). Construction materials based upon biologically renewable resources from components to finished parts. *Polymer Degradation and Stability*, 59, 251–261.

60. Bastioli, C. (1998). Biodegradable materials present situation and future perspectives. *Macromolecular Symposia*, 130, 379–391.

61. Narayan, R. *(1992)*. Biomass (renewable) resources for production of materials, chemicals and fuels a paradigm shift. *ACS Symposium Series*, 476, 1–9.

62. Rana, A. K., & Jayachandran, K. *(2000)*. Jute fiber for reinforced composites and its prospects. *Molecular Crystals and Liquid Crystals*, 353, 35–45.

63. Mohanty, A. K., & Misra, M. *(1995)*. Studies on jute composites a literature review. *Polymer Plastics Technology and Engineering*, 34(5), 729–792.

64. Bachtiar, D., Sapuan, S. M., Zainudin, E. S., Abdan, K., & Dahlan, K. Z. M. (2011). Effects of alkaline treatment and a compatibilizing agent on tensile properties of sugar palm fiber-reinforced high impact polystyrene composites. *Bioresources*, 6, 4815–4823.

65. Ishak, M. R., Leman, Z., Sapuan, S. M., Rahman, M. Z. A., & Anwar, U. M. K. (2011). *Effects of impregnation pressure on physical and tensile properties of impregnated* sugar palm (*Arenga pinnata*) fibers. *Key Engineering Materials*, 471–472, 1153–1158.

66. Sahari, J., Sapuan, S. M., Ismarrubie, Z. N. & Rahman, M. Z. A. (2011a). Comparative study on physical properties of different part of sugar palm fiber reinforced unsaturated polyester composites. *Key Engineering Material*, 471–472, 455–460.

67. Sahari, J., Sapuan, S. M., Ismarrubie, Z. N. & Rahman, M. Z. A. (2011b). Investigation on Bending Strength and Stiffness of Sugar Palm Fiber from Different Parts Reinforced Unsaturated Polyester Composites. *Key Engineering Material*, 471–472, 502–506.

68. Siregar, J. P. (2005). *Tensile and flexural properties of arenga pinnata filament* (*Ijuk Filament*) *reinforced epoxy composites*, MSc Thesis, Universiti Putra Malaysia.

69. Suriani, M. J., Hamdan, M. M. H. M., Sastra, H. Y., & Sapuan, S. M. (2006). Study of interfacial adhesion of tensile specimens of *Arenga pinnata* fiber reinforced composites. *Multidiscipline Modelling in Material and Structure*, 3(2), 213–224.

70. Sastra, H. Y., Siregar, J. P., Sapuan, S. M., Leman, Z., & Hamdan, M. M. (2006). Tensile properties of *ArengaPinnata* fiber-reinforced epoxy composites. *Polymer-Plastic Technology and Engineering*, 45, 1–8.

71. Harpini, B. (1987). Quality improvement, product diversification and developing the potentials of sugar palm. *Annual report for 1986/1987 of the Coconut Research Institute in Manado, Sulawesi (Indonesia)* Manado: BALITKA.

72. Bachtiar, D., Sapuan, S. M., Ahmad, M. H. M. & Sastra, H. Y. (2006). Chemical composition of ijuk (*ArengaPinnata*) fiber as reinforcement for polymer matrix composites. *Journal Teknologi Terpakai*, 4, 1–7.

73. Ishak, M. R., Leman, Z., Sapuan, S. M., Edeerozey, A. M. M., & Othman, I. S. (2010). Mechanical properties of kenafbast and core fiber reinforced unsaturated polyester composites. *IOP Conf. Series: Materials Science and Engineering*, 11, 012006.
74. Tomlinson, P. B. (1962). The leaf bases in palms its morphology and mechanical biology. *Journal of the Arnold Arboretum*, 43, 23–45.
75. Ismail, J. (1994). *Kajian Percambahan dan Kultur in-vitro Enau (Arenga pinnata)*, MSc Thesis. Universiti Putra Malaysia.
76. Sahari, J. (2011). *Physio-Chemical and Mechanical Properties of Different morphological Parts of Sugar Palm Fiber Reinforced Polyester Composites*. MS Thesis, Universiti Putra Malaysia.
77. Gu, H. (2009). Tensile behaviors of the coir fiber and related composites after NaOH treatment. *Materials and Design*, 30, 3931–3934.
78. Habibi, Y., El-Zawawy, W., Ibrahim, M. M. & Dufresne, A. (2008). Processing and characterization of reinforced polyethylene composites made with lignocellulosic fibers from Egyptian agro-industrial residues. *Composites Science and Technology*, 68, 1877–1885.
79. Haris, T. C. N. (1994). *Development and Germination Studies of the Sugar Palm (Arenga Pinnata Merc.) Seed*, PhD Thesis, Universiti Putra Malaysia.
80. Laohakunjit, N. & Noomhorm, A. (2004). Effect of plasticizers on mechanical and barrier properties of rice starch film. *Starch*, 56, 348–356.
81. Beerler, A. D. & Finney, D. C. (1983). Plasticizers, in *Modern Plastics Encyclopedia*, New York McGraw-Hill.
82. Vasudeva, S., Hiroshi, O., Hidechika, T., Seiichiro, I., & Ken'ichi, O. (2000). Thermal and physicochemical properties of rice grain, flour and starch. *Journal of Agricultural and Food Chemistry*, 48, 2639–2647.
83. Myllärinen, P., Partanen, R., Seppälä, J., & Forsella, P. (2002). Effect of glycerol on behavior of amylose and amylopectinfilms. *Carbohydrate Polymers*, 50, 355–361.
84. Mali, S., Grossmann, M. V. E., Garcia, M. A. Martino, M. N., & Zaritzky, N. E. (2002). Microstructural characterization of yam starch films. *Carbohydrate Polymers*, 50, 379–386
85. Guilbert, S. & Gontard, N. (1995). Technology and applications of edible protective films. VII Biotechnology and food research. *New shelf-life technologies and safety assessments*, Helsinki, Finland.
86. Mitrus, M. (2005). Glass transition temperature of thermoplastic starches. *International Agrophysics*, 19, 237–241.
87. De Graaf, R. A., Karman, A. P., & Janssen, L. P. B. M. (2003). Material properties and glass transition temperatures of differential thermoplastic starches after extrusion processing. *Starch*, 55, 80–86.
88. Herrera Franco, P. J, & Valdez-Gonzalez, A. (2005). Fiber–matrix adhesion in natural fiber composites. In Mohanty, A. K., Misra, A., Drzal, L. T. (Eds.). *Natural fibers, biopolymers and biocomposites,* Boca Raton: CRC Press.
89. Maria Guadalupe Lomeli Ramirez, Kestur, G., Satyanarayana, Setsuo Iwakiric, Graciela Bolzon de Muniz, Valcineide Tanobe, Thais Sydenstricker Flores-Sahagun. (2011). Study of the properties of biocomposites. Part I. Cassava starch-green coir fibers from Brazil *Carbohydrate Polymers*, 86, 1712–1722.
90. Curvelo, A. A. S., Carvalho, A. J. F., & Agnelli, J. A. M. (2001). Thermoplastic starch- cellulosic fibers composites. *Carbohydrate Polymers*, 45, 183–188.
91. Gaspar, M., Benko, Z., Dogossy, G., Reczey, K., & Czigany, T. (2005). Reducing waterabsorption in compostable starch-based plastics. *Polymer Degradation and Stability*, 90, 563–569.
92. Guimaraes, J. L., Wypych, F., Saul, C. K., Ramos, L. P., & Satyanarayana, K. G. (2010). Studies of the processing and characterization of corn starch and its compositeswith banana and sugarcane fibers from Brazil. *Carbohydrate Polymers*, 80, 130–138.

93. Mueller, D. H. & Krobjilowski, A. (2003). New discovery in the properties of composites reinforced with natural fibers. *Industrial Textiles,* 33(2), 111–130.
94. Mohanty, S., Verma, S. K., & Nayak, S. K. *(2006).* Dynamic mechanical and thermal propertiesof MAPE treated jute/HDPE composites. *Composites Science and Technology,* 66, 538–547.
95. Ishak, M. R., Sapuan, S. M., Leman, Z., Rahman, M. Z. A., & Anwar, U. M. K. (2012). *Characterization of* sugar palm (*Arenga pinnata*) fibers: tensile and thermal properties. *Journal of Thermal Analysis and Calorimetry,* 109, 981–989.
96. El-Shekeil, Y. A., Sapuan, S. M., Khalina, A., Zainudin, E. S., & Al-Shuja'a, O. M. (2012). Effect of alkali treatment on mechanical and thermal properties of kenaf fiber-reinforced thermoplastic polyurethane composite. *Journal of Thermal Analysis and Calorimetry,* 109, 1435–1443.
97. Yang, H., Yan, R., Chen, H., Lee, D. H., & Zheng, C. (2007). Characteristics of hemicelluloses, cellulose and lignin pyrolisis. *Fuel,* 86, 1781–1788.
98. Kim, D. Y., Nishiyama, Y., Wada, M., Kuga, S., & Okano T. (2001). Thermal decomposition of cellulose crystallites in wood. *Holzforschung,* 55(5), 521–524.
99. Sahari, J. & Sapuan, S. M. (2012). Natural Fiber Reinforced Biodegradable Polymer Composites. *Reviews on Advanced Materials Science,* 30, 166–174.
100. Aggarwal, P., & Dollimore, D. (1998). A thermal analysis investigation of partially hydrolyzed starch. *Thermochimica Acta,* 319, 17–25.
101. Ma, X., Yu, J., & John, F. K. (2005). Studies on the properties of natural fibers-reinforced thermoplastic starch composites. *Carbohydrate Polymers,* 62, 19–24.
102. Vilaseca, F., Mendez, J. A., Pe`lach, A., Llop, M., Can~igueral, N., Girone`s, J., Turon, X., & Mutje, P. (2007). Composite materials derived from biodegradable starch polymer and jute strands. *Process Biochemistry,* 42, 329–334.
103. Dhakal, H. N., Zhang, Z. Y., & Richardson, M. O. W. *(2007).* Effect of water absorption on the mechanical properties of hemp fiber reinforced unsaturated polyestercomposites. *Composites Science and Technology,* 67, 1674–1683.
104. Alam, M. K., & Khan, M. A. (2006). Comparative study of water absorption behavior in Biopol and jute-reinforced Biopol composite using neutron radiography technique *Journal ofReinforcedPlastics and Composites,* 25, 1179–1187.
105. Hazizan, Md. Akil., Leong Wei Cheng, Z. A., Mohd Ishak, A., Abu Bakar, M. A., & Abd Rahman. *(2009).* Water absorption study on pultruded jute fiber reinforced unsaturated polyester composites. *Composites Science andTechnology,* 69, 1942–1948.
106. Ray, D., & Rout, J. (2005). Thermoset Biocomposites. In: Mohanty, A. K., Misra, A., Drzal, L. T. (Eds.). *Natural fibers, biopolymers and biocomposites,* Boca Raton, CRC Press.
107. Shinoj, S., Visvanathan, S., Panigrahi, M., & Kochubabu, M. (2011). Oil palm fiber (OPF) and its composites: A review. *Industrial Crops and Products,* 33, 7–22.
108. Ali, A., Sanuddin, A. B., & Ezzeddin, S. (2010). The effect of aging on Arenga pinnata fiber-reinforced epoxy composite. *Materials and Design,* 31, 3550–3554.
109. Bachtiar, D. (2008). *Mechanical properties of alkali-treated sugar palm (Arenga Pinnata) fiber-reinforced epoxy composites,* MSc Thesis, Universiti Putra Malaysia.
110. Bachtiar, D., Sapuan, S. M., & Hamdan, M. M. (2008a). The effect of surface alkali treatment on impact behavior of sugar palm fiber reinforced epoxy composites. *In proceedings of postgraduate seminar on natural fiber composites,* Serdang, Malaysia.
111. Bachtiar, D., Sapuan, S. M., & Hamdan, M. M. (2008b). The effect of alkaline treatment on tensile properties of sugar palm fiber reinforced epoxy composites. *Material and Design,* 29, 1285–1290.

112. Bachtiar, D., Sapuan, S. M., & Hamdan, M. M. (2009). The influence of alkaline surface treatment on the impact properties of sugar palm fiber reinforced epoxy xomposites. *Polymer-Plastics Technology and Engineering*, 48, 379–383.
113. Leman, Z., Sapuan, S. M., Ishak, M. R., & Ahmad, M. M. H. M. (2010). Pre-treatment by water retting to improve the interfacial bonding strength of sugar palm fiber reinforcedepoxy composite. *Polymers from Renewable Resources*, 1, 1–12.
114. Leman, Z. (2009). Mechanical properties of sugar palm fiber-reinforced epoxy composites. PhD thesis Universiti Putra Malaysia.
115. Yousif, B. F., & Tayeb, E. N. S. M. (2008). High-stress three-body abrasive wear of treated and untreated oil palm fiber-reinforced polyester composites. *Proceedings for the Institution for Mechanical Engineers Part J: Journal of Engineering Tribology*, 222, 637–646.
116. Bachtiar, D., Sapuan, S. M., Khalina, A., Zainudin, E. S., & Dahlan, K. Z. M. (2012). Flexural and Impact Properties of Chemically Treated Sugar Palm Fiber Reinforced High Impact Polystyrene Composites. *Fiber and Polymers* 13(7), 894–898.

CHAPTER 10

BIOCOMPOSITES BASED ON NATURAL FIBERS AND POLYMER MATRIX—FROM THEORY TO INDUSTRIAL PRODUCTS

ANITA GROZDANOV, IGOR JORDANOV, MARIA E. ERRICO, GENNARO GENTILE, and MAURIZIO AVELLA

ABSTRACT

Over the last decades, due to the increased environmental awareness, numerous studies for production of biocomposites based on natural fibers have been published and many comprehensive reviews have been published. Compared with conventional reinforcements, such as glass and carbon fibers, natural fibers which are renewable resources, offer several other advantages including a wide availability (based on different vegetable species), recyclability, low density and low costs, low abrasion and preserving mechanical properties. Application of cellulose fibers in composites is not only beneficial from an ecological point of view, lowering the environmental impact of the final product within the production, usage and disposal period, but it offers further technical and economical benefits. Especially, natural fiber-reinforced biocomposites have the potential to replace current materials used for automotive industrial applications. In order to obtain composites with the best mechanical properties, most of the research activities in the last decades have been concentrated on the surface physical and chemical modifications of the fibers mainly to optimize their interfacial behavior.

In the first part of this work, besides the overview of the state-of-the-art description regarding biocomposites, we will also present characterization results of the lab-scaled flax fiber reinforced biopolymer matrices (PLA) as well as compared with the same-industrially produced composites.

In the second part of the paper, biocomposites based on "self-reinforced cellulose" or "all-cellulose" composites prepared from cotton textile fabrics by partial fiber surface dissolution in lithium chloride dissolved in N, N-dimethylacetamide will be presented. Two different parameters have been studied: (i) surface treatment medium (alkaline/enzyme/bleaching) and (ii) cotton textile preforms (knits, woven).

10.1   INTRODUCTION

Due to the excellent characteristics, such as lower weight/higher strength, fiber-reinforced composite materials have found wider technical application.[1,2] Composite materials maximize weight reduction (as they are typically 20% lighter than aluminum) and are known to be more reliable than other traditional metallic materials, leading to reduced aircraft maintenance costs, and a lower number of inspections during service. Additional benefits of composite technologies include added strength, noncorrosive materials with superior durability for a longer lifespan. However, because of the remarkable increased environmental consciousness, the substitution of traditional synthetic polymer based composites reinforced with glass and carbon fibers with new biocomposites was considered of fundamental importance.

Compared to traditional composites, biocomposites are materials based on natural fibers of different preforms and fabrics and biocompatible polymer matrices. The interest for using biocomposites has increased because they are lightweight, nontoxic, nonabrasive during processing, have low cost and are easy to recycle. Actually, the first natural fiber composites were used more than 100 years ago. In 1896, for example, airplane seats and fuel tanks were made of natural fibers with a small content of polymeric binders. However, these attempts were without recognition of the composite principles and the importance of fibers as the reinforcing part of composites as well as with the importance of having biodegradable matrix. The use of natural fibers was suspended due to low cost and growing performance of technical plastics and, moreover, synthetic fibers. A renaissance in the use of natural fibers as reinforcements in technical applications began during the late twentieth century.

The automotive and space market is growing in terms of quantity, quality and product variety. The key factor for future growth is fuel efficiency. A 25% reduction in vehicle weight is equivalent to a saving of 250 million barrels of crude oil and a reduction in CO_2 emissions of 220 billion pounds per year.[2,3] Over the last several years European, American and Japanese recycling regulations have encourage the use of biomass in automotive materials. Additionally, European Union legislation implemented in 2006 has mandated that 85% of a vehicle must be reused or recycled by 2015. Japan requires 95% of a vehicle to be recovered (which includes incineration of some components) by 2015.[4]

So, in the last 2 to 3 decades, scientists and engineers have worked together to improve and to enhance the performance of natural fiber based biocomposites, as well as to find some other possible application for them. The aim of this review is to give the overview of the state-of-the-art potential of biocomposites and in particular, biocomposites realized with Flax fibers and PLA matrix, and cotton-based "All-cellulose," composites from lab-scale up to the industrial products.

10.2 BIOCOMPOSITES BASED ON NATURAL FIBERS AND POLYMER MATRIX

Extensive research has been underway to study the potential of different natural fibers as reinforcement for biodegradable (a synthetic or renewable) polymer matrices in order to develop the components for different body parts of automobiles.[5,6] Natural fibers are incorporated into door panel trims, package trays, trunk trims and other interior parts.[5]

The interest in using biocomposites based on natural fibers and biocompatible polymer matrices has grown because they are lightweight, biodegradable, nontoxic, nonabrasive during processing, have low cost and are easy to recycle. Natural fiber reinforced materials offer environmental advantages such as reduced dependence on nonrenewable energy/material sources, lower pollutant and greenhouse emissions. Lacarin et al. have compared the environmental impacts of the biocomposites and the glass/PP composite for the different steps of the life cycle.[7] The energy consumption to produce a flax-fiber mat (9.55 MJ/kg/1), including cultivation, harvesting and fiber separation, amounts to approximately 17% of the energy needed to produce a glass-fiber mat (54.7 MJ/kg/1).[2] Environmental aspects reveal that natural fibers display an increase of about 15% of the performance of the composites, while focusing on economical aspects they cost about seven times less than glass fibers (Table 10.1).

TABLE 10.1   Natural Fibers as Reinforcing Material

Fiber	Economic	Weight reduction
Glass fib.	~US $2/kg	2.5–2.8 g/cm^3
Natural fib.	~US $0.44–0.55/kg	1.2–1.5 g/cm^3

Depending on their performance when they are included in the polymer matrix, lignocellulosic fibers can be classified into three categories: 1) Wood flour particulate which increase the tensile and flexural modulus of the composites, 2) Fibers of higher aspect ratio that contribute to improve the composites modulus and strength when suitable additives are used to regulate the stress transfer between the matrix and the fibers, 3) Long natural fibers with the highest efficiency among the lignocellulosic reinforcements.

Natural fibers may be classified by their origin as either cellulosic (from plants), protein (from animals) or mineral. Plant fibers may be further categorized as: seed hairs (e.g., cotton), bast or stem fibers (e.g., linen from the flax plant), hard (leaf) fibers (e.g., sisal), or husk fibers (e.g., coconut).

Cellulose is one of the most abundant renewable and biodegradable biopolymer resource with high mechanical performance. It is a hydrophilic glucan polymer consisting of a linear chain of β-1,4–bonded anhydroglucose units that contains alcoholic hydroxyl groups. Cellulose represents the main structural component of plant cell walls. These hydroxyl groups form intra and intermolecular hydrogen bonds inside the macromolecule and among other cellulose macromolecules, respectively, as well as with hydroxyl groups from the surrounding air and polymer matrices. In terms of primary walls, cellulose fibrils have been found to be preferentially deposited perpendicular to the axis of cells during their initial state of growing. Due to the great stiffness and strength of cellulose fibrils, it is much easier to expand the cell wall perpendicular to the orientation of cellulose. The secondary cell walls consist of different layers that are deposited on the primary cell wall in a characteristic manner (strictly parallel). The interaction between the stiff cellulose fibrils and the plant matrix polymers in the cell wall is one of the key issues to elucidating the mechanical performance of plants.[8] The most efficient natural fibers considered include threads with a high cellulose content coupled with a low microfibril angle, resulting in high filament mechanical properties. Due to their hollow and cellular nature, natural fiber preforms have up to 40% lower density. They act as acoustic and thermal insulators, and exhibit reduced bulk density.

The absolute mechanical data of natural fibers are inferior relative to E-glass and Carbon fibers. But when they are used in composites, their mechanical properties are even higher than E-glass reinforced composites (Table 10.2).[9]

TABLE 10.2 Comparison of Mechanical Properties of Natural and Conventional Fiber Reinforcements

Fiber	Tensile strength [GPa]	Tensile modulus [GPa]	Specific strength [GPa/g.cm³]	Specific modulus [GPa/g.cm³]
Flax	2.00	85	1.60	71
Hemp	0.7	35	0.5	25
E-Glass	3.50	72	1.35	28
Carbon (standard)	3.00	235	1.71	134

So, natural fibers have lower densities and they can be found to be cheaper than glass fibers, although their strength is usually significantly less. Because of their good specific modulus values, natural fibers can be preferable to glass fibers in applications where stiffness and weight are primary concerns. Theoretically, tensile and flexural moduli of composites are strongly dependent on the modulus of the components and display slight sensitivity to interfacial adhesions. In natural fiber reinforced biocomposites, the inclusion of a rigid phase such as cellulose fibers, contribute to increase the polymer matrix stiffness.

In fact, not only the modulus, but also the tensile and flexural strengths are sensitive of the fiber/matrix interfacial adhesion, and interface is a determent factor in transferring the stress from the matrix to the fibrous phase. So, in order to create a good and strong fiber/matrix interfacial adhesion between fibers (highly polar) and common polymer matrices (nonpolar), a proper strategy to improve fiber/matrix compatibility is required. Today, for optimization of a strong fiber/matrix interfacial adhesion, generally two approaches are considered as effective: the fiber surface modification and the use of an appropriate compatibilizing agent. Generally, the mechanical properties of natural fiber reinforced biocomposites have been improved by using surface modification treatment of the fibers such as dewaxing, mercerization, bleaching, cyanoethylation, silane treatment, benzoylation, peroxide treatment, acylation, acetylation, latex coating, and steam-explosion.[9,10]

Poly(lactic acid) (PLA) is currently the most popular polymer derived from renewable resources, which is fermented to lactic acid. The lactic acid is then, via a cyclic dilactone, lactide, ring opening polymerized to the desired polylactic acid. This polymer is modified by certain means, which enhance the temperature stability of the polymer and reduce the residual monomer content. The resulting polylactic acid can be processed similarly as polyolefins and other thermoplastics although the thermal stability could be enhanced. The polylactide is fully biodegradable. According to our current understanding, the degradation occurs by hydrolysis to lactic acid, which is metabolized by microbes to water and carbon monoxide. By composting together with other biomasses, the biodegradation occurs within two weeks, and the material fully disappears within 3–4 weeks. PLA is a thermoplastic, aliphatic polyester, which is useful in the packaging-, electrical- and automotive industry, for example, applications where biodegradable materials started competing with cheaper synthetic plastics.

It was widely reported that tensile and flexural modulus of PLA could be improved by increasing the cellulose content or cellulose based reinforcements in PLA based composites [11,12]. Regarding the impact properties, it was shown that toughness results were impaired for PLA composites reinforced with cellulose fibers[12], while small improvement were obtained with the addition of cotton or kenaf fibers[6]. Different natural fibers have been employed in order to modify the properties of PLA. Up to now, the most studied natural fiber reinforcements for PLA have been kenaf[6,11,12], flax[13,14], hemp[15], bamboo[16], jute[17], wood fibers [18]. Besides conventional natural fibers, recently reed fibers have been tested in appropriate PLA composites in order to improve the tensile modulus and strengths[19]. Other innovative methods over the last few years to improve the mechanical properties of PLA based composites have been utilization of continuous hybrid fiber reinforced composite yarn obtained by the microbraiding technique[20]. Naturally derived microbraided-yarn was fabricated by using thermoplastic biodegradable PLA resin fiber as the resin fiber and jute spun yarn as the reinforcement. Using jute spun yarn/PLA microbraided-yarn, continuous natural fibers reinforced biodegradable resin composite plates was molded by hot press molding with various molding conditions.

10.3   FLAX /PLA BASED BIOCOMPOSITES

Among all the natural fiber reinforced biocomposites, the flax based composite shows the best properties when compared to other natural composites, including glass-reinforced traditional composites. Flax fibers offer higher reinforcing properties than hemp and kenaf natural fibers. Namely, comparison of the mechanical properties of the natural fiber reinforced composites has shown that composites based on flax fibers exhibited higher tensile strength relative to those based on hemp or kenaf fibers. Flax fibers exhibit a higher fineness and more unique distribution compared to hemp or kenaf. According to current theories, a higher fiber fineness should results in better fiber embedment during compression molding and consequently higher mechanical properties. Generally, mechanical properties of natural fibers are determined by the cellulose content and microfibrillar angle. The cells of the flax fibers consist mostly of pure cellulose cemented by means of noncellulosic incrusting such as lignin, hemicellulose, pectin or mineral substances, resins, tannins and small amount of waxes and fats. Flax cell wall consists of about 70–75% cellulose, 15% hemicellulose and pectin materials. The Young's modulus of the natural fibers decreases with the increase of diameter. The mechanical properties of the natural fibers are also closely related to the degree of polymerization of the cellulose in the fiber[21]. Basic physicochemical properties and cellulose content for flax fibers versus other natural reinforcing fibers are shown in Table 10.3.

TABLE 10.3   Physicochemical Properties and Cellulose Content of Natural Fibers

Fiber	Density [g/cm³]	Young modulus [GPa]	Fracture stress [MPa]	Elongation [%]	Cellulose [%]	Hemicellulose [%]	Lignin [%]	Pectin [%]	Wax [%]	Microfi. / Spiral angle [°]
Flax	1.4–1.5	10–80	345–500	1.2–3.3	62–72	18.6–20.6	2–5	2.3	1.5–1.7	5–10
Hemp	1.48	20–70	270–900	1.0–3.5	68–74	15–22.4	3.7–10	0.9	0.8	2–6.2
Kenaf	1.4	14.5–53	220–930	1.5–2.7	31–72	20.3–21.5	8–19	3–5	/	/
Sisal	1.2–1.5	3.0–98	510–700	2.0–2.5	60–78	10–14.2	8–14	10	2.0	10–22
Cotton	1.5	5.5–12.6	287–597	7.0–8.0	82–90	5.7	<2	0–1	0.6	/

Oksman et al.[22] have studied the mechanical properties of PLA/Flax composites versus PP/Flax. The addition of flax fibers increase the modulus, but the higher fiber content has not improved the modulus in PLA composites as it has been observed for PP composites due to the fiber orientation in the polymer matrix. The test composites were compression molded and the fibers could be orientated differently from one sample to another. Because of the brittle nature of PLA, triacetin was used to plasticize the pure PLA and for the PLA/Flax composites. The addition of triacetin has shown a positive effect on the elongation to break for pure PLA and PLA/Flax composites, which was expected because of the softening effect. The highest triacetin addition (15%) clearly shows a negative effect for PLA/Flax composites, both the stress and stiffness were strongly decreased. As expected, it was shown that the addition of triacetin did not affect the impact properties of the PLA/Flax composites. The addition of 5% triacetin in PLA has shown the best results on impact strength. The authors also reported that thermal properties of PLA were increased with the incorporation of flax fibers. The softening temperature was increased from about 50°C for pure PLA to 60°C with flax fibers, and it is further increased if the composite is crystallized.

10.3.1   FLAX MODIFICATION FOR BIO-COMPOSITE PRODUCTION

The interest in using natural flax fibers as reinforcement in biocomposites has increased dramatically, and in the same time, it also represents one of the most important uses. However, flax fibers are hygroscopic in nature; moisture absorption can result in swelling of the fibers, which may lead to microcracking of the composite and degradation of mechanical properties. This problem can be resolved by treating these fibers with suitable chemicals to decrease the hydroxyl groups; these groups may be involved in the hydrogen bonding within the cellulose molecules or by different type of pretreatments.[23,24,25] Chemical treatments may activate these groups or can introduce new moieties that can effectively interlock with the matrix.

Kozlowski et al. modified flax fiber with plasma, boiling and bleaching for PLA/Flax composite production. Modified flax gives composites with better mechanical properties that are more resistant to flame than unmodified ones[25].

As mentioned, the type of pretreatment has a strong influence on the natural composite properties. Susheel Kalia et al.[26], have reported numerous treatments of natural fibers. Mercerization, bleaching, grafting, coupling, treatment with silane, benzoylation as well as plasma treatment to fibers improve most of the usable properties of natural fibers composites.

B. Wang et al. presented results of the modification of short flax fibers, which were derived from Saskat-chewan-grown flax straws use in fiber-reinforced composites, performed by mercerization, silane treatment, benzoylation and peroxide treatment[27]. SEM analysis has shown that physical microstructure changes occurred at the fiber surface. Silane treatment provided surface coating to the fibers. Benzoylation treatment produced a smooth fiber surface, while after dicumyl peroxide treatment the fibrillar structure of the individual ultimate fibers was observed. This may be due to the leaching out of waxes and pectic substances. Micropores, particles adhering to the surface, groove like portions and protruding structures made the fiber surface very rough. Application of the coupling agents has been shown to be effective improving the surface properties of flax fibers, forming a mechanically interlocked coating on its surface. The results have shown that silane and peroxide treatment on flax fiber bundles lead to a higher tensile strength than that of the untreated fiber bundles. Comparatively lower tensile strength was observed in benzoylation treated fibers[27].

Recently, introduction of enzymes in numerous wet processing of natural fibers, such as desizing, bio-scouring, bio-polishing, etc., are powerful methods for surface modification. Enzymes alone, or with combination with mercerization gives cottons that have less hydrophilic surface relative to traditionally treated fibers, but with increased numbers and types of functional groups.[28,29]

A. Grozdanov et al. have worked on comparison of Flax/PLA biocomposites versus Kenaf/PLA biocomposites.[30,31] Natural fibers as a nonwoven preforms, flax and kenaf were kindly supplied by KEFI-Italy. Surface modification was carried out as follows: (a) dewaxing: the fibers were treated with 1:2 mixture of ethanol/benzene for 72 h at 50 °C, followed by washing with distilled water and air drying to get defatted fibers; (b) vinyl monomer grafting: acrylonitrile (ACN) graft copolymerization onto dewatted fibers was carried out using 0.01 M Ce^{4+}/0.1 M HNO_3 as initiator at temperature of 50°C; (c) alkali treatment: the defatted fibers were treated with 2 and 10% NaOH solution for 1 h at 30 °C; (d) acetylation: dewatted fibers were placed in 100 mL flask and covered with the appropriate amount of acetic anhydride for 0.5 h at 20 °C, followed by Soxhlet extraction and drying. Biocomposites based on PLA matrix reinforced with flax and kenaf nonwoven preforms (20% wt. fiber content) have been prepared by compression molding at 170°C under the pressure of 50 bars. The flexural strength and the flexural modulus were measured in three-points bending mode using an Instron machine (model 5564), at a cross-head speed of 1 mm/min and at room temperature. The test span was 48.0 mm. For each sample 10 specimens were tested and the average values of the flexural strength and modulus were calculated. Thermogravimetrtic analysis (TGA) was performed in the range of 25 to 800°C, which had heating rate of 20 K/min (under nitrogen), using the Perkin Elmer DIAMOND system. The morphology of chemically treated and dried (12 h in vacuum) flax and kenaf fibers, as well as of their composites was examined by SEM (JEOL, model JSM-T20 (U_w = 20 kV).

Characteristic TGA data for different treated flax and kenaf nonwoven preforms are presented in Table 10.4. Comparison of the decomposition temperatures of flax and kenaf nonwoven preforms has shown that higher thermal stability (about 70°C) exhibited flax nonwoven forms probably as a result of higher crystallinity as well as higher cellulose and hemicellulose content.

TABLE 10.4 Decomposition Temperature at Different Weight Loss Levels of Treatedflax and Kenaf Fibers

Differently treated fibers	T_d flax fib. (~50%)	[°C] kenaf fib. (~50%)	T_d flax fib. (~90%)	[°C] kenaf fib. (~90%)
As received	448	384	492	415
Alkali treated	450	360	494	416

Due to the chemical treatments, overall morphology of the flax and kenaf fibers has been changed. Characteristic SEM photos of alkali treated flax and kenaf fibers are shown in Fig. 10.1 (kenaf fib.) and 10.2 (flax fib.). The untreated fibers represent the bundles with relatively smooth surface (Figs. 10.1a and 10.2a), although small particles attached to the surface are also seen. The alkali treated fibers have a rough surface topography with significant defibrillation of individual fibers (Figs. 10.1b and 10.2b).

a) Kenaf-untreated (x200)   b) Kenaf-alkali treated (x500)

FIGURE 10.1 SEM microphotographs of fibers (a) and alkali (b) treated kenaf fibers.

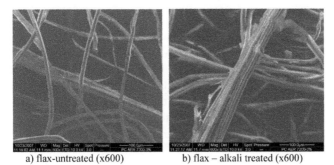

a) flax-untreated (x600)   b) flax – alkali treated (x600)

FIGURE 10.2 SEM microphotographs of untreated fibers (a) and alkali (b) treated flax fibers.

The flexural modules and flexural strength for the obtained PLA/Flax biocomposites with various treated Flax preforms were shown in Table 10.5. It is evident that flax fibers improved the flexural modulus and flexural strength of the polymer matrix compared to neat PLA.

TABLE 10.5   The Flexural Modulus and Flexural Strength of Biocomposites with Various Treated Flax Preforms

Biocomposite	Flexural modulus [MPa]	Flexural strength [MPa]
PLA neat	3550	30,9
PLA/Flax (80/20%wt)	3721	60,7
PLA/Flax-Alkali tret. (80/20%wt)	3595	65,6
PLA/Flax-Acetil. (80/20%wt)	4210	58,1

Mechanical properties of the obtained biocomposites reinforced with flax and kenaf nonwoven preforms have been shown in Table 10.6. Evidently, flax and kenaf nonwoven preforms increased the mechanical properties of the PLA neat polymer. Higher flexural modulus was obtained for the composites reinforced with kenaf fibers, while higher flexural strength was measured for PLA/Flax biocomposites.

TABLE 10.6   Flexural Data for PLA Based Eco Composites Reinforced With Flax and Kenaf Nonwoven Performs

Sample	Flexural modulus [MPa]	Flexural strength [MPa]
PLA neat	3550 ± 50	30.9 ± 0.2
PLA/Kenaf (80/20%wt)	4630 ± 40	32.7 ± 0.4
PLA/flax (80/20%wt)	4400 ± 40	36.6 ± 1.8

Characteristic morphology in the obtained PLA/Flax biocomposites have shown that flax fibers were well covered with the polymer matrix, which resulted in good stress transfer between the flax fibers and PLA polymer matrix. SEM photographs of the studied PLA/Flax biocomposites with various treated flax performs are shown in Fig. 10.3. Surface treatments have resulted in defibrillation of the flax fibers and increased the possibility for mechanical interlocking of flax fibers and polymer matrix, which consequently increased the fiber/matrix interfacial strength.

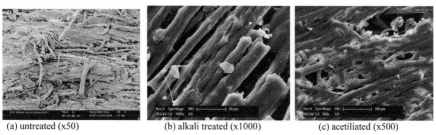

| (a) untreated (x50) | (b) alkali treated (x1000) | (c) acetiliated (x500) |

FIGURE 10.3 SEM microphotographs of PLA/Flax biocomposites with various treated flax fiber performs: (a) untreated, (b) alkali treated, (c) acetylated flax fiber preforms

10.4 ALL-CELLULOSE BIOCOMPOSITES

Recently, development of biodegradable, ecofriendly polymer composite materials has been focused towards monocomponent, all-cellulose composites[32]. As it was previously mentioned, cellulose is one of the most abundant renewable and biodegradable biopolymer resources with high mechanical performance. In their cell walls, the spirally oriented cellulose plays the role of reinforcements in a soft hemi-cellulose and lignin matrix. All-cellulose composites were produced based on the original concept of self-reinforced composites, a composite with a matrix and reinforcement from the same polymer, which has been primary developed for thermoplastic high-density polyethylene[33]. In this new type of so-called self-reinforced composites, the interfacial bonding problems are circumvented by the use of cellulose for both the reinforcement and the matrix. These composites have exhibited significant prospects as bio-based and biodegradable materials that have excellent mechanical properties. Because cellulose does not exhibit a melting point, it must be dissolved in order to aid in processing. Fiber surface selective dissolution of aligned cellulose fibers has been employed in the solvent by controlling the immersion time. Since the cell wall of the natural fibers is build of several layers, the surface layer of the fibers can be partially dissolved and transformed into the composite matrix phase. The remaining fiber cell cores maintain their original structure and impact a reinforcing effect to the composite. Due to the fact that both the fiber and matrix phases of this cellulose composite are from the same origin, and they are chemically identical, a strong interfacial adhesion could be expected between them[33].

In this procedure, activated fibers are immersed in lithium chloride/N,N-dimethylacetamide (LiCl/DMAc) for a specified immersion times. The fibers are then removed from the solvent and the partially dissolved fibers start to gel. Finally, this fiber-incorporated gel are coagulated in a nonsolvent system to extract the DMAc and LiCl, and then dried under vacuum. The cell wall of a cellulose fibers is constituted by a number of layers; therefore, the surface layer of the fibers can be partially dissolved and transformed into the matrix phase of the composites, whereas the undissolved part of the fibers, preserve the original structure, thus imparting the reinforcing effect to the composite.

Currently, several kinds of solvent systems have been used to dissolve cellulose, such as lithium chloride/N,N-dimethylacetamide (LiCl/DMAc), dimethyl sulfoxide (DMSO)/tetrabutylammonium fluoride, NH_3/NH_4SCN, NaOH/urea, ionic liquids, PEG/NaOH, etc.[34] All-cellulose composites have been prepared by dissolving pretreated cellulose pulp and then impregnation of the cellulose solution into the aligned fibers followed by coagulation in methanol and drying. Examples of starting materials are pulp,[35] filter paper[36,] and long fibers[37]. Nishino et al. have prepared all-cellulose composites from pure cellulose and ramie fibers in LiCl/DMAc[32]. Duchemin et al. have studied the effect of dissolution time and cellulose concentration on the crystallography of precipitated cellulose, using microcrystalline cellulose (MCC) as a model material[38]. The results of their work have contributed to a further understanding of the phase transformations that occur during the formation of all-cellulose composites by partial dissolution.

All-cellulose composites have been obtained, as an example, from aligned ramie fibers[32]. Due to the high fiber volume fraction in these composites (up to 80%), the tensile strength of these uniaxially reinforced all-cellulose composites has been found as high as 480–540 MPa. A similar approach has been also used to prepare random all-cellulose composites from filter paper[36]. As concerning the preparation methodology, by increasing the immersion time of cellulose fibers, larger fractions of the fiber skin are dissolved to form a matrix phase. Therefore, an improvement of interfacial adhesion has been observed by increasing the dissolution time. In the case of aligned ramie fibers, longitudinal tensile tests have shown that an immersion time of 2 h is the optimum processing condition to produce all-cellulose composites. In these conditions, it has been found that the amount of fiber surface selectively dissolved to form the matrix phase. This is adequate to provide sufficient interfacial adhesion to the composite, whereas the undissolved fiber cores retain their original structure and strength.

Lu et al.[39] have published the results of their work on all-cellulose composites prepared by molding slightly benzylated sisal fibers. In contrast to plant fiber/synthetic polymer composites, water resistance of the current composites was greatly increased as characterized by the insignificant variation in the mechanical properties of the composites before and after being aged in water. They have found that sisal/cyanoethylated wood sawdust and sisal/benzylated wood sawdust all-composites exhibit mechanical properties similar to those of glass fiber reinforced composites[39]. Physical heterogeneity in these all-plant fiber composites was favorable for the interfacial interaction. Biodegradability of the self-reinforced sisal composites was also followed. They found that in the case of enzymolysis aided by cellulose, the degradation rate of the composites gradually slowed down due to the hindrance of the lignin, which cannot be hydrolyzed by cellulose[39].

A. Grozdanov et al. have worked on all-cellulose composites based on cotton textile fabrics, prepared by partial fiber surface dissolution in lithium chloride dissolved in N,N-dimethylacetamide (LiCl/DMAc)[40]. Two different parameters have been studied: (i) the influence on type of scouring (alkaline or enzymatic) and (ii) the cotton textile preforms (knits, woven). In order to improve the interface and protect against fiber degradation for the all-cellulose composites, alkaline scouring was performed by using 4% NaOH for 60 min treatment at 100°C. Enzymatic scouring was done with two conditions: alkaline pectinase–BioPrep3000 L at 55°C for 30 min and acid pectinase–NS 29048 at 45°C for 30 min[28,29]. All-cellulose composites with ~90–95% fiber volume fraction were successfully prepared by using solutions of 3 (wt./v) cellulose concentrations in 8% (wt./v) LiCl/DMAc for impregnation of cotton textile preforms. Characterization protocols of the obtained all-cellulose composites have included FTIR, SEM, TGA/DTA, ^{13}C-NMR, mechanical tests and Dp-determination.

It was found that a dissolution time of 24 h lead to bio-based materials with the best overall mechanical performance, since this time allowed dissolution of a sufficient amount of the fiber surface to obtain good interfacial bonding between fibers, while keeping a considerable amount of remaining fiber cores that provide a strong reinforcement to the composite. Characteristic mechanical curves of the studied all-cellulose composites based on various treated cotton woven fabrics are presented in Fig. 10.4. The measured data for the maximum load and deformations are presented in Table 10.7. Comparison of the mechanical performances of all-cellulose composites based on alkali- and enzyme-treated cotton-woven preforms has shown that the treatments can effectively improve the mechanical strength of the composites. The higher values for the mechanical strength were obtained for the all-cellulose composites based on enzyme-treated cotton-woven preforms. Han et al.[34] have confirmed the same effect. Moreover, they found that with an increase of the immersion time from 1 to 3h, the values of the mechanical strength sharply increased.

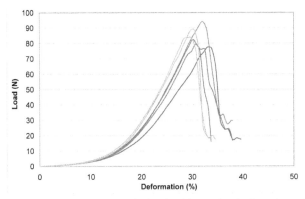

FIGURE 10.4   Characteristic curves obtained with mechanical testing of all-cellulose composites.

TABLE 10.7   Maximum Load and Deformations for the All-Cellulose Composites based on Cotton Woven Pre-Forms

Sample	Maximum load [N]	SD [N]	Deformation [%]	SD [%]
Control sample	78.0	3.4	42	1.5
Alkali treated + bleached	84.3	6.2	31.0	1.4
Enzyme treated (alkaline pectin-ase) + bleached	86.1	4.1	31.0	1.0
Enzyme treated (acid pect.) + bleached	87.2	3.9	30.0	1.2

(all-cellulose composites based on alkali treated and bleached woven preform).

Although the biocomposites based on enzyme- and bleach-treated preforms have shown the best mechanical properties, alkali-treated cotton-woven preforms have shown higher lateral crystalline indices in the obtained composites compared to enzyme treated ones. Crystalline indices for the studied all-cellulose composites, obtained as a ratio of the FTIR bands at 1430 cm^{-1} (CH$_2$ symmetric band) and 898 cm^{-1} (Group C1 frequency: –CH2=C–R) are presented in Table 10.8. The results obtained for the crystalline indices confirmed there were not significant changes in the crystalline structure of the cotton based composites.

TABLE 10.8   Crystalline Index of the Studied All-Cellulose Composites Based on Cotton Woven Pre-Forms

Sample	CrI (A_{1430}/A_{898})
Control sample	3.7
Alkali treated + H$_2$O$_2$	3.8
Enzyme treated + H$_2$O$_2$ (alkali pectinase)	3.1
Enzyme treated + H$_2$O$_2$ (acid pectinase)	3.4

The crystalline structure was studied also by ^{13}C-NMR spectroscopy. The obtained results are shown in Table 10.9. The crystalline fraction X$_c$ was calculated by deconvolution of the spectra in the 80–90 ppm region, according to the following equation:

TABLE 10.9 Crystalline Fraction Obtained by ^{13}C-NMR of the Cotton Based All-Cellulose Composites

Sample Treatment	X_c (%)
Control sample	78.8
Alkaline + H_2O_2	79.4
Enzyme (alkaline pectin) + H_2O_2	79.7
Enzyme (acid pectin) + H_2O_2	79.1

$I_{88.5}$ and $I_{83.5}$ are the intensity of the peaks assigned to the crystalline and amorphous fraction, respectively.

No significant changes in cellulose structure were evidenced, as all the spectra are very similar (almost identical) (Fig. 10.5). For all of the studied samples, it is evident that the fiber core was not damaged and the fibers kept their structural and strength performance.

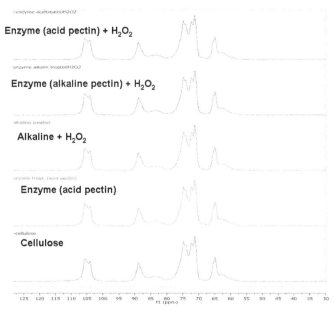

FIGURE 10.5 C-NMR spectra of the all-cellulose composites based on various treated cotton woven preforms.

Characteristic SEM microphotographs of the morphology in the obtained all-cellulose composites are shown in Figs. 10.6 and 10.7. Figure 10.6 show morphology of alkali-treated cotton all-cellulose composites, while in Fig. 10.7, morphology of enzyme (alkali pectinase) treated cotton performs are shown. SEM images provide direct information regarding the interfacial bonding of the studied all-cellulose composites based on alkali and enzyme treated cotton performs confirming where good fiber-fiber adhesion was registered in both types of cotton performs. For alkali scoured performs, progressive build-up of covering thermoplastic films around the fibers were found, while for enzymatic scoured performs bonding bridges were registered between two fibers.

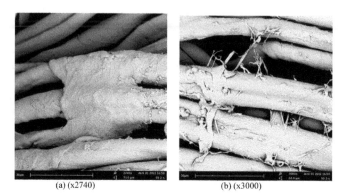

(a) (x2740)   (b) (x3000)

FIGURE 10.6   Morphology of all-cellulose composites based on alkali treated and bleached cotton preforms.

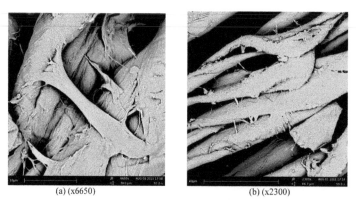

(a) (x6650)   (b) (x2300)

FIGURE 10.7   Morphology of all-cellulose composites based on enzyme-treated (alkali-pectinase) cotton preforms.

Therefore, it can be affirmed that all-cellulose composites based on various treated cotton fiber performs show very interesting mechanical properties and represent a new class of bio-derived and biodegradable composite materials with interesting possibilities of future development.

10.5   INDUSTRIAL APPLICATION OF BIOCOMPOSITES

Today, natural fiber reinforced biocomposites are mainly applied in the automotive industry. From 1996 to 2000 year, the use of natural fibers in the European automotive sector has increased from 5000 to 28,000 tons, respectively. It is evident that flax, hemp and kenaf fibers are among the most applied types of natural fibers. According to automotive industry reports, about 5 to 10 kg of natural fibers are incorporated into every European car[5]. This figure includes flax fibers, hemp, jute, sisal and kenaf, which all are used in composite production[41]. The use of flax was reported by the suppliers to be circa 1.6 k ton in 1999, and is expected to rise to 15 to 20 k ton in the near future. The German and Austrian car industry alone employed 8.5 to 9 k ton of flax fibers in 2001[42]. The introduction of every new car model increases the demand, depending on the model, by 0.5 to 3 k ton per year.

The automotive industry gives a long list of presumed benefits of natural fiber composites, which includes the general reasons for the application of natural fibers as discussed briefly in the previous sections:
- Low density, which may lead to a weight reduction of 10 to 30%.
- Acceptable mechanical properties, good acoustic properties.
- Favorable processing properties, for instance low wear on tools.
- Options for new production technologies and materials.
- Favorable accident performance, high stability, less splintering.
- Favorable ecobalance for part production.
- Favorable ecobalance during vehicle operation due to weight savings.
- Occupational health benefits compared to glass fibers during production.
- No off-gassing of toxic compounds (in contrast to phenol resin bonded wood and recycled cotton fiber parts).
- Relatively easy recycling (it is not clear whether they mean thermal recycling here).
- Price advantages both for the fibers and the applied technologies.

Obviously the production and application of natural fiber reinforced parts also brings along some difficulties:
- For the production of nonwovens: presence of shives, dust, very short fibers.
- Uneven length distribution and uneven decortications of the fibers (especially for nonwovens).
- Irreproducible fiber quality combined with availability.
- Variations in nonwoven quality and uniformity due to fiber quality variation.
- Moisture sensitivity, both during processing and during application.
- Limited heat resistance of the fibers.
- Specific smell of the parts.
- Limited fire retardancy.
- Variations in quality and uniformity of produced parts.
- Possible molding and rotting.

TABLE 10.10   Application of Natural Fibers in Automotive Parts

Manufacturer Model Application (dependent on model)
Audi A3, A4, A4 Avant, A6, A8, Roadster, Coupe
Seat back, side and back door panel, boot lining, hat rack, spare tire lining
BMW 3, 5 and 7 Series and others
Door panels, headliner panel, boot lining, seat back
Daimler/ A-Series, C-Series, E-Series, S-Series
Chrysler Door panels, windshields/dashboard, business table, pillar cover panel
Fiat Punto, Brava, Marea, Alfa Romeo 146, 156
Ford Mondeo CD 162, Focus
Door panels, B-pillar, boot liner
Opel Astra, Vectra, Zafira
Headliner panel, door panels, pillar cover panel, instrument panel
Peugeot New model 406
Renault Clio
Rover 2000 and others
Insulation, rear storage shelf/panel
Saab Door panels
SEAT Door panels, seat back
Volkswagen Golf A4, Passat Variant, Bora
Door panel, seat back, boot lid finish panel, boot liner
Volvo C70, V70

The higher volume fraction of lower density natural fibers in natural fiber composites also reduces the weight of the final component. Joshi et al. have reported that natural fiber composite components based on hemp fibers applied in Audi-A3 car resulted in 20–30% reduction in weight[43]. In fact, natural fiber composites are becoming popular in automotive applications because of this weight reduction. Lower weight components improve fuel efficiency and in turn significantly lower emissions during the use phase of the component life cycle. It was estimated that the coefficient for reduction in fuel consumption on gasoline powered vehicles ranges from 0.34 to 0.48 L/(100 kg × 100 km) in the New European Driving Cycle, while the saving on diesel vehicles ranges from 0.29 to 0.33 L/(100 kg × 100 km). In other words, over the lifetime travel of 175,000 km an automobile, a kilogram of weight reduction can result in fuel savings of 5.95–8.4 L of gasoline or 5.1–5.8 L of diesel, and corresponding avoided emissions from production and burning of these fuels.

Mueller and Krobjilowski have studied application of nonwoven composite fabrics in automotive interior components. They have compared carded and nee-dle-punched nonwoven fabrics of 750 g/m² produced from 50/50 flax/PP, hemp/PP and kenaf/PP[5]. They concluded that fine fibers improve the mechanical properties on natural composites. For that purpose cotton fiber composites were prepared as acoustic materials and compared with synthetic based composites. The type of res-ins slightly influence on the acoustic properties, while composite thickness depends from the type of synthetic composite should be replaced.

Besides the automobile industry, a growing number of the nonautomotive ap-plications and products are being presented for natural fiber biocomposites. Some of these applications are in the field of energy and impact absorption, such as floor cov-erings that use excellent acoustic properties of the natural fibers, bicycle helmets, security helmets for the construction area, and monitor housings for the computers[5].

The construction sector and infrastructure are the second large sector with huge potential for increased applications. This application field is particularly interesting for rural and agri-cultural societies as well as for the countries where natural fiber production is very high.

Other uses of natural fiber based composites are for various furniture elements such as, deck surface boards, and picnic tables[42].

10.6   CONCLUSIONS

It is very clear and evident that natural fiber reinforced biocomposites offer a huge potential for future applications not just in the automotive industry, but also in other sectors such as construction, infrastructure and furniture production. The main chal-lenges related to the lower moisture absorption, higher fire resistance, better me-chanical properties, durability, variability, and manufacturing/processing of natural fiber reinforced biocomposites are being addressed by many recent research efforts. Moisture absorption can be reduced through surface modifications of fibers and/or by special coatings. Fire resistance can be improved by the use of in tumescent coat-ings, which eventually may also be made from renewable resources. Mechanical properties and durability are the main areas of research into natural fiber reinforced biocomposites, and many proposed solutions have been found to improve the fiber/ matrix interface. Fiber variability is itself largely uncontrollable, but the develop-ment of quality assurance protocols and diversification of fiber growing sources can address the issue before the fibers reach composite manufacturers. Natural fi-ber reinforced biocomposites have been successfully adapted to nearly every major manufacturing process currently used with synthetic composites, usually with few or no modifications to the processes themselves.

New types of all-cellulose composites were successfully prepared by a surface selective or partial dissolution method of cotton woven textile fabrics. Two different media were used for the fiber surface treatment: i) alkaline scouring with bleaching and ii) enzymatic scouring with acid and alkali pectinases combined with bleaching.

Therefore, future research in the field of natural fiber reinforced biocomposites for infrastructure applications would be most beneficial if directed at one of the highlighted challenging areas, particularly focused on continuing to improve mechanical properties, moisture resistance, and durability.

KEYWORDS

- **Biocomposites**
- **Cellulose**
- **Mechanical Properties**
- **Natural Fibers**
- **PLA**
- **Polymer Matrix**
- **Renewable Resources**

REFERENCES

1. Avella, M., Buzarovska, A., Errico, M. E., Gentile, G., & Grozdanov, A. (2009). Ecochallenges of bio-based polymer composites, *Materials*, 2, 911–925.
2. Pandey, J. K., Ahn, S. H. Lee, C. S., Mohanty, A. K., & Misra, M. (2010). Recent Advances in the Application of Natural Fiber Based Composites, *Macromolecular Materials and Engineering*, 295, 975–989.
3. Mair, R. L. (2000). Tomorrow's Plastic Cars, *ATSE Focus* No. 113, July/August.
4. "Directive 2000/53/EC of the European Parliament and of the Council of 18 September 2000, on End of Life Vehicles,' in: Official Journal of the European Communities, 21 October 2000.
5. Mueller, D. H., & Krobjilowski, A. (2003). New Discovery in the Properties of Composites Reinforced with Natural Fibers, *Journal of Industrial Textiles*, 33, 111.
6. Avella, M., Bogoeva-Gaceva, G., Buzarovska, A., Errico, M., Gentile, G., & Grozdanov, A. (2008). Poly (lactic acid) based biocomposites reinforced with kenaf fibers, *J. Appl. Polym. Sci.*, 108, 6, 3542–3551.
7. Lacarin, M., Perwuelz, A., Pesnel, S., Rault, F., & Vroman, P. (2012). Environmental study of biocomposites intended for passenger cars, *Proceedings 2nd LCA Conference*, Lille, France, 6–7 November.
8. Dittenber, D. B., & GangaRao, H. V. S. (2012). *Composites P. A.*, 43, 1419–1429.
9. Bogoeva-Gaceva, G., Avella, M., Malinconico, M., Buzarovska, A., Grozdanov, A., gentile, G., & Errico, M. E. (2007). Natural fiiber Eco Composites, *Polymer composites*, 28(1), 98–107.
10. Lee, S. G., Choi, S. S., Park, W. H., & Cho, D. (2003). *Macromolec Symp.*, 197, 89.
11. Huda, M. S., Drzal, L. T., Mohanty, A. K., & Misra, M. (2008). Effect of fiber surface-treatments on the properties of laminated biocomposites from poly (lactic acid) (PLA) and kenaf fibers, *Comp. Sci. and Technol.*, 68(2), 424–432.

12. Graupner, N. (2008). Application of lignin as a natural adhesion promoter in cotton fiber rein-forced poly (lactic acid) (PLA) composites, *J. Mater. Sci.*, 43, 5222–5229.

13. Bax, B., & Müssig, J. (2008). Impact and Tensile Properties of PLA/Cordenka and PLA/Flax Composites. *Composites Science and Technology*, 68(7–8), 1601–1607.

14. Bodros, E., Pillin, I., Montrelay, N., & Baley, C. (2007). Could biopolymers reinforced by ran-domly scattered flax fibers be used in structural applications, *Comp. Sci. & Technol.*, 67(3–4), 462–470.

15. Hu, R., & Lim, J. K. (2007). Fabrication and mechanical properties of completely biodegrad-able hemp reinforced PLA composites, *J. Compos. Mater*, 41(13), 1655–1669.

16. Tokoro, R., Vu, D. M., Okubo, K., Tanaka, T., Fujii, T., & Fujiura, T. (2008). How to improve mechanical properties of polylactic acid with bamboo fibers. *J. Mater. Sci.*, 43(2), 775–787.

17. Shikamoto, N., Ohtani, A., Leong, Y. W., & Nakai, A. (2007). Fabrication and mechanical properties of jute/PLA composites, *Proceedings of the American Society for Composites, Technical Conference* 22nd 151/1-151/10. Publisher: DEStech Publications, Inc.

18. Huda, M. S., Drzal, L. T., Mohanty, A. K., & Misra, M. (2005). Wood fiber reinforced Poly (lactic acid) composites, *5th Annual SPE Automotive Composites Conference*, Sept. 12–14, Troy, Michigan.

19. Bourmaud, A., & Pimbert, S. (2008). Investigations on mechanical properties of poly (pro-pylene) and poly (lactic acid) reinforced by miscanthus fibers, *Composites PartA: Applied Science and Manufacturing*, 39(9), 1444–1454.

20. Shikamoto, N., Ohtani, A., Leong, Y. W., & Nakai, A. (2007). Fabrication and mechanical properties of jute/PLA composites, *Proceedings of the American Society for Composites, Technical Conference* 22nd 151/1-151/10. Publisher DEStech Publications, Inc.

21. Williams, G. I., & Wool, R. P. (2000). *Appl. Compos. Mater*, 7, 421.

22. Oksmana, K. M., & Skrifvars, J. F. (2003). Selinc, Natural fibers as reinforcement in polylac-tic acid (PLA) composites, *Composites Science and Technology*, 63, 1317–1324.

23. Walter, P. (2000). Beurteilung der Qualitaë t von Flachsfasern in Tuë rinnenverkleidungen mittels pflanzenanatomischer Methoden, In: Proceedings, 3. *International Wood and Natural Fiber Composites Symposium*, Kassel/Germany, September 19/20th.

24. Singha, A. S., & Vijay Kumar Thakur, (2009). Chemical Resistance, Mechanical and Physical Properties of Biofibers-Based Polymer Composites, *Polymer-Plastics Technology and Engi-neering*, 48, 736–744.

25. Marek Kozlowski, et al. (2005). Influence of Fibers Modification on Biocomposite Properties, FP6-BIOCOMP (project no. NMP2-CT-515769).

26. Susheel Kalia, B., Kaith, S., & Inderjeet Kaur. Pretreatments of Natural Fibers and their Ap-plication as Reinforcing Material in Polymer Composites. A Review, *Polymer Engineering and Science*, 1253–1272.

27. Wang, B., Panigrahi, S., Tabil, L., Crerar, W., & Sokansanj, S. (2003). Modification of flax fibers by chemical treatment, *CSAE/SCGR* June 6–9, Meeting Montréal, Québec.

28. Jordanov, I., Mangovska, B., Simoncic, B., & Forte-Tavcer, P. (2010). Changes in the Non-Cellulosic Components of Cotton Surface after Mercerization and Scouring, *AATCC Review*, 10, 6, 65–72.

29. Jordanov, I., & Mangovska, B. (2009). Characterization on Surface of Mercerized and En-zymatic Scoured Cotton after Different Temperature of Drying, *The Open Textile Journal*, 2, 39–47.

30. Grozdanov, A., Buzarovska, A., Avella, M., Prendjova, M., Gentile, G. & Errico, M. E. (2008). Application of nonwoven performs based on natural fibers as reinforcements in ecocompos-ites, *ECCM 13–13th European Conference on Composite Materials, ID 0215*, June 2-5, Stock holm Sweden.

31. Buzarovska, A., Grozdanov, A., Gentile, G., Errico, M. E., & Avella, M. (2009). Morphology and mechanical properties of fully bio-degradable flax reinforced Poly (Lactic Acid) composites, *European Polymer Congress 2009, EPF-2009*, Graz, Austra, July 12–17.

32. Nishino, T., Matsuda, I., & Hirao, K. (2004). All Cellulose Composites, *Macromolecules* 37, 7683–7687.

33. Cabrera, N., Alock, B., Loos, J., & Peijs, T. (2004). Processing of all PP composites for ultimate recyclability Proc. Inst. Mech. Eng. Part. L: *J. Mater. Design Appl.*, 37, 7683–7687.

34. Han, D. L., & Yan, L. (2010). Preparation of all Cellulose Composite by Selective Dissolving of Cellulose Surface in PEG/NaOH Aqueous Solution *Carbohydrate Polymers*, 79, 614–619.

35. Gindl, W., Schoberl, T., & Keckes, J. (2006). Structure and Properties of a pulp fiber reinforced composites with regenerated cellulose matrix, *Appl. Phys. A.*, 83(1), 19–22.

36. Nishino, T., & Arimoto, N. (2007). All cellulose composites prepared by selective dissolving of fiber surface, *Biomacromolecules*, 8(9), 2712–2716.

37. Soykeabkaew, N., Arimoto, N., Nishino, T., Peijs, T. (2008). All cellulose composites prepared by selective dissolution of aligned lingo-cellulosic fibers, *Compos. Sci Tecnol* 68(10–11), 2201–2207.

38. Duchemin, B., Newman, R., & Staiger, M. (2007). Phase transformations in microcrystalline cellulose due to partial dissolution, *Cellulose*, 14(4), 311–320.

39. Lu, X., Zhang, M. Q., Rong, M. Z., Yue, D. L., & Yang, G. C. (2004). The preparation of self-reinforced sisal fiber composites, *Polym & Polym compos.*, 12(4), 297–307.

40. Grozdanov, A., Jordanov, I., Parizova, A., Gentile, G., Errico, M. E., & Avella, M. (2012). Surface treatment effects of the cotton textile preforms on the properties of All-cellulose composites," Mo3. 14(5). *ECCM 15–15th European Conference on Composite Materials*, June 24–28 Venice-Italy.

41. Karus, M., Kaup, M., & Lohmeyer, D. (2000). Study on Markets and Prices for Natural Fibers (Germany and EU), FNR–FKZ: 99NR 163, nova Institute, available at www.nachwachsende-rohstoffe.info.

42. Kaup, M., Karus, M., & Ortman, S. (2002). Auswertung der Markterhebung: Naturfasereinsatz in Verbundwerkstoffen in der deutschen und österreichischen Automobilindustrie, Status 2002, Analyze und Trends, nova Institut 2003, available at www.nachwachsenderohstoffe.com

43. Joshia, S. V., Drzal, L. T., Mohanty, A. K., & Arorac, S. (2004). Are natural fiber composites environmentally superior to glass fiber reinforced composites, *Composites: Part A*, 35, 371–376.

CHAPTER 11

FIRE RESISTANCE CELLULOSIC FIBERS FOR BIOCOMPOSITES

MINH-TAN TON-THAT, TRI-DUNG NGO, and BOUCHERVILLE

ABSTRACT

The incorporation of renewable resources in composite materials is a viable means to reduce environmental impact and support sustainability development in the composites industry. Cellulosic fiber polymer composites have received very much attraction for different industrial applications because of its low density and its renewable ability. However, the uses of cellulosic fibers in the composite are limited in many applications that require fire resistance due to their flammability and their low thermal resistance.

This chapter reports an innovative and sustainable treatment approach to retard the burning of cellulosic fibers for composite production in which a minimum amount of nontoxic and low cost inorganic chemicals have been used. Different types of reacting minerals and different treatment parameters have been investigated in order to determine the most cost-effective treatment solution. The cellulosic fibers obtained from this approach become self-extinguished while there is no negative effect on fiber strength. The composite with the treated cellulosic fibers also shows their good fire resistance with minor effect on the mechanical properties. Thus this solution will open the door for the use of the cellulosic fibers in composites for applications where fire resistance is an important issue, particularly in aerospace, transportation, and construction.

"Fire Resistance Cellulosic Fibers for Biocomposites" by Minh Tan Ton-That and Tri Dung Ngo was originally published with the National Research Council of Canada (© by the authors).

11.1   INTRODUCTION

Cellulosic materials (natural and synthetic) in different forms (fiber, film, powder, particle, pellet, chip, etc.) at different sizes (nano, micro or macro) are often flammable and have low thermal resistance. They can be burned and also can spread the fire in the presence of oxygen. Thus, their use either in direct or nondirect form is limited in applications requiring fire resistance. Due to their flammability, the use of cellulosic materials in polymer composites is also limited in certain applications.

Cellulosic materials are treated with different flame retardants depending on the application, for example in furniture, textiles or composites. The most commonly used flame retardants are based on halogen, phosphorous, boron, ammonium, graphite, alkaline-earth metallic compounds or mixtures thereof. To improve fire resistance of organic polymer composites, the incorporation of flame retardants based on halogen, phosphorous, metallic hydroxide (magnesium hydroxide, aluminum hydroxide, calcium hydroxide, layer double hydroxide), metallic oxide (antimony oxide, boron oxide), silicate (clay, talc), etc., in the polymer matrix has been widely used.

Among the compounds listed above, halogen based flame retardants are well known to be the most efficient as they can be used at a low concentration in the final composition thus limiting their impact on other properties of the product. However, halogen compounds are considered to be harmful to the environment. Boron compounds are supposed to be efficient, however, they tend to be washed off due to their good solubility in water. Less harmful flame retardants based on phosphorous, graphite or alkaline-earth metallic compounds are much less efficient, thus a large amount of those additives must be used in the formulation. The use of flame retardant incorporated in a polymer matrix alone does not satisfactorily resolve the flammability problem in cellulose-polymer composites, especially when the concentration of cellulose is quite significant in the formulation of the composite.

It is generally known that metal hydroxides, including barium hydroxide, can be used as a flame retardant for cellulosic materials[1–4] and for polymer materials.[5] Further, Herndon[6] used a flame retardant composition for cellulosic material comprising sodium hydroxide and a metal salt of boron among other ingredients. The metal salt of boron is defines as borax, which is a sodium tetraborate. De Lissa[7] suggested a flame-proofing composition comprising potassium hydroxide and/or potassium carbonate and possible a small amount of sodium hydroxide and/or sodium carbonate and may include another potassium salt. Musselman[8] proposed inorganic additives to impart flame resistance to polymers. The additives include hydroxides and metal salts that evolve gas. One such metal salt is barium chloride dihydrate. The use of a mixture of a polycondensate of a halogenated phenol and an alkaline earth metal halide in a flame retarding composition has also been suggested.[9]

Flame retardant compositions in which ancillary flame retardant additives may be used alone or in combination, such as metal hydroxides and metal salts, including alkaline earth metal salts, has also been reported[10].

Fukuba[11] discloses the use of "alkali compounds" for use in flame resistant plaster board. The "alkali compounds" are defined as at least one of an alkali metal hydroxide, alkali metal salt, alkaline earth metal hydroxide or alkaline earth metal salt. It is preferred to use a mixture of alkali metal salts and alkaline earth metal salts, for example a mixture of sodium and calcium formate.

Yan demonstrate the use of a flame retardant composition which initially involves the step of making magnesium hydroxide from the reaction of magnesium sulfate and sodium hydroxide.[12]

It is known that treatment of cellulosic materials with alkaline earth metal carbonates (e.g., barium carbonate) imparts fire resistance to the cellulosic material[13]. Here, the alkaline earth metal carbonate is applied to the cellulosic material by first coating the cellulosic material with an alkaline earth metal chloride and then treating the so-coated material with sodium carbonate. It is also known to use both a clay and a metal hydroxide in a fire retarding composition comprising a polymer material.[14,15]

However, there is no disclosure treating a cellulosic material with an aqueous reaction mixture of an alkali metal hydroxide and alkaline earth metal salt simultaneously with or shortly after mixing the alkali metal hydroxide with alkaline earth metal salts.

There remains a need for an environmentally friendlier, effective approach to producing fire-resistant cellulosic materials. This chapter presents an innovative method for improving fire resistance of cellulosic materials, especially when the cellulosic material is to be used in polymer composites, which is simple, cost-effective and environmentally friendly.

11.2 PRINCIPLE CONCEPT

In this method cellulosic material is treated with an aqueous mixture of alkali metal or ammonium hydroxide and alkaline-earth or aluminum metal salt simultaneously with or within a short period of time of preparing the mixture. The treated cellulosic material becomes self-extinguishing and may also have improved thermal stability, improved interfacial thermal resistance, improved resistance to damage by oxidants and other chemical agents, improved resistance to damage by ultra-violet light and/ or reduced negative impact on fiber strength and/or modulus. The fire-resistant cellulosic material may also be treated with a layered nanoparticulate material either simultaneously with, subsequent to or prior to treatment with the aqueous mixture of alkali metal or ammonium hydroxide and alkaline-earth or aluminum metal salt to impart further fire resistance to the cellulosic material.

In principle, the mentioned chemicals attached on the cellulose surface to form a nonflammable layer that can protect cellulose effectively from fire. Single or double or multiple layer can deposit on the cellulose surface as desired. These layers can be based on the same or different chemical compositions as desired. Figure 11.1 describes the principle of the cellulosic fiber after the treatment.

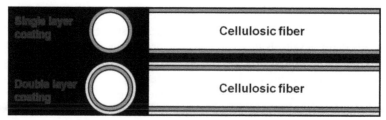

FIGURE 11.1   Description of the principle structure of the treated cellulose fiber.

The treatment is very simple and easy to scale-up and it consists of the soaking of the cellulose in the aqueous chemical solution bath and drying as illustrated in Fig. 11.2.

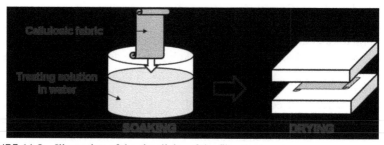

FIGURE 11.2   Illustration of the simplicity of the fiber treatment process.

Polymer composites produced from cellulosic material treated according to this method have significantly improved fire resistance with minimum negative impact on the mechanical performance, and may have the added benefit of one or more of improved thermal stability and improved interfacial thermal resistance.

11.3   EXPERIMENTAL

11.3.1   MATERIALS

Different cellulosic materials tested are shown in Table 11.1. The chemical products used in this work are summarized in Table 11.2.

TABLE 11.1   Description of Cellulosic Fibers

Sample	Fiber	Weight (g/m²)
C1	Flax fabric C 20 M-2/2 twill from Moss Composites, Belgium, received in 2008	149

TABLE 11.1 *(Continued)*

C2	Flax fabric C 20 M-2/2 twill from Moss Composites, Belgium, received in 2010	149
C3	Flax fabric C10 M-8H satin from Moss Composites, Belgium, received in 2010	258
C4	Canada woven flax fabric from JB Matin, Canada, received in 2010	240
C5	Hemp mat supplied by Composite Innovation Centre, Canada	350

TABLE 11.2 Description of Chemicals

Chemicals	Company	Information
$CaCl_2.2H_2O$	Fisher	
$Ca(NO_3)_2.4H_2O$	Aldrich	
$MgCl_2$	Sigma Life Science	
$MgSO_4$	Sigma-Aldrich	
$Mg(NO_3)_2.6H_2O$	Fluka	
$Mg(OH)_2$	Alfa Aesar	
$NaOH$	Aldrich	
KOH	Sigma-Aldrich	
$Al(OH)_3$	Aldrich	
$AlCl_3$	Sigma-Aldrich	
NH_4OH	Sigma-Aldrich	
$BaCl_2$	Fisher	
$Ba(OH)_2$	Aldrich	
Clay MMT	Southern Clay Products, Inc.	Montmorillonite Cloisite Na™
Clay LDH	AkzoNobel	Layered double hydroxides (LDH)

11.3.2 SOLUTION PREPARATION

Different aqueous solutions, which can be single, or bicomponents were used in this study. In a single component solution, only one chemical is dispersed in demineralized water. In a bi-component solution or a bi-component suspension, separate solutions or suspensions of each of the two chemicals were prepared in an equal amount of demineralized water and then they were mixed together.

11.3.3   FIBER TREATMENT PROCESSES

Prior treated with the chemical the celluloses were cleaned with the detergent at 80°C for 2 h to remove the impurities and contaminants as much as possible then rinsed three times with demineralized water.

Two different fiber treatment processes were used. In one-step treatment processes (P1), cellulosic fiber was soaked in a prepared solution for a period of time. The fibers were then dried in air for 6 h and then in an oven at 120°C for 2 h prior to testing.

In the two-step treatment processes (P2) cellulosic fiber was soaked in a first solution for 5 to 300 seconds. The fibers were then removed from the treating medium and allowed to dry in air for 6 h, and then dried in an oven at 120°C for 2 h. The dried fibers were then soaked in a second solution for 5 to 300 seconds. Finally the fibers were dried in air for 6 h and then in an oven at 120°C for 2 h prior to testing.

11.3.4   COMPOSITE FABRICATION

For phenolic (PF) composites, phenolic resin was then wetted on the fibers and dried in an oven to remove solvent from the resin and to let the resin transfer to stage B before compression. Wabash PC 100–2418–2TM compression was used to fabricate the composites under 100 psi pressure at 150°C. The amounts of resin and fiber in the final product were about 60 wt.% and 40 wt.%, respectively. The thickness of the composite plaque was about 3 mm.

Laminate epoxy composites were prepared by compression molding similar to the fabrication of phenolic composites but at 80°C. The amounts of resin and fiber in the final product were about 60 wt.% and 40 wt.%, respectively.

Laminate unsaturated polyester (UPE) composites were prepared by compression molding similar to the fabrication of phenolic composites but at 50°C. The amounts of resin and fiber used were about 70 wt.% and 30 wt.%, respectively. The UPE resin contains 20 wt.% alumina trihydrate Hubert SB332.

11.3.5   FIBER SURFACE OBSERVATION

JEOL JSM–6100 SEM at a voltage of 10 kV was used to observe the surface of fibers before and after treatment. This SEM was also used to observe the fracture surface of the composites after tensile test.

11.3.6   GENERAL PROCEDURE FOR BURNING TESTS

A Govmark UL94 chamber was used to conduct burning tests. For each example, results are provided using numbers and the terms "NB" and "G." The term "NB" means "no burning" and is an indication that there was no flame and no glow after removing the flame. "NB" represents excellent fire resistance as the sample did not continue to burn after the external flame source was removed, thus the sample was self-extinguishing. The term "G" means "glow" and is an indication that the sample continued to glow after removal of the flame. The numbers are the time in seconds that the sample continued to glow after removal of the flame.

11.3.7 HORIZONTAL BURNING TEST (HB)

A minimum of five specimens of each fiber sample having width × length (W × L) of 0.5 × 6.0 inch (12.7 × 152.4 mm) were cut from bulk fiber. Specimens were held at one end in a horizontal position and tilted at 45° with marks at 1, 1.5, 2.0, 2.5, 3.0, 3.5, 4.0, 4.5, 5.0 inch from the free end. A flame was applied to the free end of the specimen for 30 seconds or until the flame front reached the 1 inch mark. If combustion continued, the duration was timed between each 0.5-inch mark. A thin metallic wire was inserted to support the specimen.

For the composite samples, five specimens having W×L of 0.5×6.0 inch (12.7×152.4 mm) were cut from the 3 mm thick composite plaque. Specimens were held at one end in a horizontal position and tilted at 45° with marks at 1, 2.0, 3.0, 4.0, 5.0 inch from the free end. A flame was applied to the free end of the specimen for 30 seconds or until the flame front reached the 1 inch mark. If combustion continued, the duration was timed between each 1.0-inch mark.

11.3.8 VERTICAL BURNING TEST (VC-2)

A Govmark VC-2 chamber was used to conduct burning tests for some composites. This chamber is widely cited through out the USA and internationally to measure the ignition resistant properties of aircraft and transportation materials, tents and protective clothing.

For phenol formaldehyde composite samples, three specimens having width × length (W×L) of 3×12 inch (76.2×304.8 mm) were cut from the 3 mm thick composite plaque. Specimens were held at one end in the vertical position. The flame was applied for 60 seconds and then removed until flaming stopped. The combustion time and burning length was recorded. If the specimen has burning length and burning time less than 8 inch and 15 seconds, respectively, it is considered to be passed the standard (self-extinguished) Each separate set of specimens prepared for testing will consist of at least three specimens (multiple places).

11.3.9 TENSILE TEST

Tensile tests on fibers were conducted on a tow (strand) disassembled from the fabric. The tows in the longitudinal direction in the fabric were separated from the ones in the orthogonal direction. Tests were carried out for both series separately. The tensile properties of the fiber tow were determined at room temperature and 50% relative humidity on an Instron 5548 microtester machine, with crosshead distance of 50 mm and speeds of 120 mm/min. The maximum load at break was recorded for each specimen. A minimum 10 specimens were tested for each type of sample.

The tensile properties of the composites were evaluated at room temperature and 50% relative humidity on an Instron 5500R machine, with crosshead speeds of 5 mm/min according to ASTM 3039-00. A minimum 5 specimens were tested for each type of sample.

11.4   RESULTS AND DISCUSSIONS

11.4.1   FIBER TREATMENT

11.4.1.1   FIBERS TREATED WITH SINGLE COMPONENT SOLUTIONS

Flax fiber C1 and C2 as described in Table 11.2 was treated with different single component solutions as indicated in Table 11.3 for 120 s using the process P1. Burning tests were conducted in accordance with the general procedure described above and the results from the burning tests are also shown in Table 11.3. It is evident from Table 11.3 that all of the C1 fibers treated with various single component systems are not self-extinguishing, although these treatments slowed down flame propagation. Fibers treated with NaOH or KOH did not continue to burn but did continue to glow. Fibers treated with NaOH and then washed with water did continue to burn, demonstrating that any fire resistant effect afforded by an alkali metal hydroxide alone is easily removed if the fibers get wet. Collectively, Table 11.3 demonstrates that single component systems of metal hydroxides, metal salts or clays do not impart self-extinguishing properties on fibers treated with the systems.

TABLE 11.3   C1 Fibers Treated with Single Component Solutions Using P1

Name	Description	Burning characteristics
C1	Untreated	Burned
C1–1	Clay MMT2%	Burned
C1–2	Clay MMT4%	Burned
C1–3	Clay LDH2%	Burned
C1–4	Clay LDH4%	Burned
C1–5	$(BaCl_2)2\%$	Burned
C1–6	$(Ba(OH)_2)2\%$	Burned
C1–7	$(BaCl_2)2\%$ then washed with water	Burned

TABLE 11.3 *(Continued)*

Name	Description	Burning characteristics
C1–8	$(Ba(OH)_2)2\%$ then washed with water	Burned
C2	Untreated	Burned
C2–1	$Ba(OH)_2$	Burned
C2–2	$BaCl_2$	Burned
C2–3	$BaCl_2$ twice	Burned
C2–4	$MgNO_3$	Burned
C2–5	$MgCl_2$	Burned
C2–6	$MgSO_4$	Burned
C2–7	$Mg(OH)_2$	Burned
C2–8	$Ca(NO_3)_2$	Burned
C2–9	$CaCl_2$	Burned
C2–10	KOH	Glowed
C2–11	NaOH	Glowed
C2–12	NaOH twice	Glowed
C2–13	NaOH then washed with water	Burned
C2–14	$AlCl_3$	Burned
C2–15	$Al(OH)_3$	Burned

The difference in surface structure between the untreated and treated flax fibers are illustrated in Fig. 11.3. In general the single component systems do not provide a good coating on the flax fiber surface. Among them LDH and MMT provide better coverage but the can be peeled off easily during handling the fibers. These can be the reason for their poor fire retardant performance.

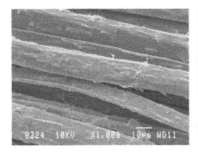 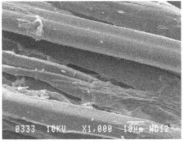

FIGURE 11.3 *(Continued)*

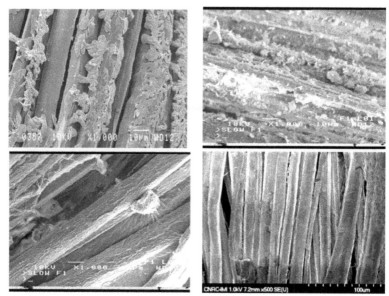

FIGURE 11.3   SEM image of the flax fibers: treated with a) NaOH, b) BaCl$_2$, c) Ba(OH)$_2$, d) LDH, e) MMT and f) untreated.

11.4.1.2   FIBERS TREATED WITH BI-COMPONENT SOLUTIONS

Flax fiber C1 and C2 were treated with different bi-component systems as indicated in Table 11.4. It is evident that all of the C1 fibers treated with bi-component systems involving the mixture of barium chloride and sodium hydroxide are self-extinguishing. Fibers treated with barium chloride alone then with clay or barium hydroxide alone then with clay are not self-extinguishing. Thus, single component systems are not self-extinguishing, even with the subsequent addition of clay. A mixture of both the alkaline metal salt and the alkali metal hydroxide is needed to make the fibers self-extinguishing. It is further clear that washing the fibers after treatment with a bi-component system does not remove the self-extinguishing properties imparted by the treatment. Further, the order in which clay is introduced into the bi-component does not affect the self-extinguishing properties of the fibers after treatment.

For the C2 series fibers treated with (MgCl$_2$+NaOH) and with (CaCl$_2$+NaOH) are self-extinguishing. Fibers treated with (Mg(NO$_3$)$_2$+NaOH) and with (Ca(NO$_3$)$_2$+NaOH) did not burn but continued to glow. Fibers treated with (MgSO$_4$+NaOH) continued to burn, but at a slower rate than untreated fibers. The efficiency of the (MgCl$_2$+NaOH) system is greater than the (Mg(NO$_3$)$_2$+NaOH) system, which is greater than the (MgSO$_4$+NaOH) system. This is also similar for the calcium-containing systems where the efficiency of the (CaCl$_2$+NaOH) system is greater than the $(Ca(NO_3)_2+NaOH)$ system. Thus, chloride is the most preferred counter anion for the alkaline earth metal cation.

TABLE 11.4   C1 and C2 Fibers Treated with a Solution of Barium-Containing Bi-component Systems

Name	Description	Burning characteristics
C1	Untreated	Burned
C1–9/P2	$BaCl_2$ then + clay MMT	Burned
C1–10/P2	$Ba(OH)_2$ then + clay LDH	Burned
C1–11/P1	$BaCl_2$ + NaOH	Self-extinguished
C1–12/P1	$BaCl_2$ + NaOH then washed	Self-extinguished
C1–13/P2	$BaCl_2$ + NaOH then + clay MMT	Self-extinguished
C1–14/P2	$BaCl_2$ + NaOH then + clay LDH	Self-extinguished
C1–15/P2	Clay MMT then + $BaCl_2$ + NaOH	Self-extinguished
C1–16/P2	Clay LDH then + $BaCl_2$ + NaOH	Self-extinguished
C2	Untreated	Burned
C2–16/P1	$MgCl_2$ + NaOH	Self-extinguished
C2–17/P1	$Mg(NO_3)_2$ + NaOH	Glowed
C2–18/P1	$MgSO_4$ + NaOH	Burned
C2–19/P1	$CaCl_2$ + NaOH	Self-extinguished
C2–20/P1	$Ca(NO_3)_2$ + NaOH	Glowed
C2–21/P1	$AlCl_3+NH_4OH$	Self-extinguished
C2–22/P2	$AlCl_3$ + NH_4OH then clay MMT	Self-extinguished

Figure 11.4 illustrates the fibers treated with the bi-component systems providing better coating and adhesion of the chemical on the fiber surface thus preventing the treated fiber from burning.

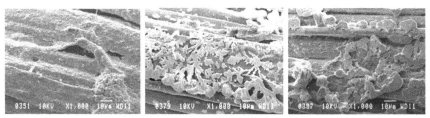

FIGURE 11.4   SEM image of the flax fibers treated with a) $NaOH+MgCl_2$, b) $NaOH+BaCl_2$ and c) $NaOH+BaCl_2+MMT$.

Figure 11.5 illustrates the remains of the flax fiber after burning test. The nontreated flax burned completely to form the gray ash while the flax treated with $BaCl_2$ formed the black char and the fibers treated with ($NaOH+BaCl_2$) or ($NaOH+BaCl_2+MMT$) become self extinguishing. This indicates this treatment method is very effective depending on the selective chemical combination and event the treatment can be performed directly in the fabric and not necessary at individual filament level.

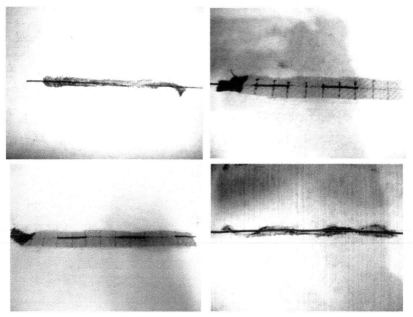

FIGURE 11.5   Photo of the flax fibers after burning test: treated with a) $BaCl_2$, b) $NaOH+BaCl_2$ c) $NaOH+BaCl_2+MMT$, and d) untreated.

11.4.1.3   DIFFERENT FIBERS TREATED WITH A MAGNESIUM-CONTAINING BI-COMPONENT SYSTEMS

Various fiber samples C2, C3, C4, C5 and C6 were treated with a magnesium-containing bi-component systems as indicated in Table 11.5. It is evident that all of the fibers were self-extinguishing after treatment with the ($MgCl_2+NaOH$) system. It shows that the treatments are very effective across a range of cellulose materials and their forms.

TABLE 11.5 Different Fibers Treated with a Magnesium-containing Bi-component Systems Using P1

Name	Description	Burning characteristics
C2	Untreated C2	Burned
C2–21	C2 + MgCl$_2$ + NaOH	Self-extinguished
C3	Untreated C3	Burned
C3–1	C3 + MgCl$_2$ + NaOH	Self-extinguished
C4	Untreated C4	Burned
C4–1	C4 + MgCl$_2$ + NaOH	Self-extinguished
C5	Untreated C5	Burned
C5–1	C5 + MgCl$_2$ + NaOH	Self-extinguished

11.4.1.4   TENSILE PROPERTIES OF FIBER TOWS

It is very difficult to test the tensile properties of individual fibers thus the test was performed on the fiber tow removed from the fabric tested instead in accordance with the procedure described above. Table 11.6 lists the fiber tows come from different treated and untreated fibers that were tested as well as their tensile properties. The tows in the longitudinal direction in the fabric are denoted as parallel, whereas the ones in the orthogonal direction are denoted as perpendicular.

It is evident from Table 11.6 that the tensile properties of the fiber tow did not change much for most of the systems indicating that the treatment did not generally have a detrimental effect on tensile properties except for fibers treated with alkali metal hydroxide alone (e.g., KOH and NaOH). It is clear, therefore, that cellulose materials treated with both alkaline earth metal salt and alkali metal hydroxide are advantageously fire retardant, often self-extinguishing, while retaining good tensile properties, in contrast to fibers treated only with alkali metal hydroxide or treated with another metal salt.

TABLE 11.6 Tensile Force of Tows of Treated C2 Fibers

Fiber	Description	Max load pounds force (N)	
		Parallel	Perpendicular
C2	Untreated C2	4.6 (20.4)	5.4 (23.8)
C2–1/P1	Ba(OH)$_2$	4.9 (21.7)	5.4 (24.1)
C2–2/P1	BaCl$_2$	4.7 (21.1)	5.6 (25.1)
C2–4/P1	Mg(NO$_3$)$_2$	5.4 (23.8)	5.7 (25.5)
C2–7/P1	Mg((OH)$_2$	4.3 (19.2)	5.3 (23.6)
C2–10/P1	KOH	3.6 (15.8)	4.5 (20.2)

TABLE 11.6   *(Continued)*

| Fiber | Description | Max load pounds force (N) | |
		Parallel	Perpendicular
C2–11/P1	NaOH	3.6 (15.8)	4.3 (19.0)
C2–13/P1	NaOH then washed	3.0 (13.2)	4.1 (18.4)
C2–15/P1	$Al(OH)_3$	5.1 (22.8)	5.3 (23.6)
C2–17/P1	$MgCl_2$ + NaOH	4.8 (21.3)	5.3 (23.7)
C2–18/P1	$MgSO_4$ + NaOH	5.2 (23.0)	5.8 (25.8)
C2–19/P1	$CaCl_2$ + NaOH	4.4 (19.5)	5.6 (24.9)
C2–20/P1	$Ca(NO_3)_2$ + NaOH	5.5 (24.4)	5.6 (24.8)

11.4.2   BIOCOMPOSITES

11.4.2.1   PF/FLAX FIBER COMPOSITES

Vertical VC–2 burning test was conducted on the PF composites and the results are shown in Table 11.7. Sample PF–C2 is a comparative example of a PF composite containing untreated flax fiber. PF–C2–21/P2 is PF composite containing flax fibers treated with a bi-component aluminum-containing system. Burning time or burning length are the time or the length it takes for the fire stop after the external ignition flame was removed. Thus, a shorter burning time and length indicate a more fire resistant material. PF is a thermoset resin which itself has considerable resistance to fire. Thus, in the PF composites flax fiber mainly contributes to burning the composite specimen. As is evident from Table 11.8, fire-resistant flax fibers of this approach provide a tremendously significant greater resistance to burning in the PF matrix than untreated flax fibers that allows the obtained composite with the treated flax fibers to be classified as self-extinguished (Fig. 11.6).

TABLE 11.7   Burning Test Results of PF/Flax Fiber Composites

Sample	Composition	Max flame time (s)	Max burn length (inches)	Glow	Pass VC–2 test
PF–C2	Phenol formaldehyde matrix	35.8 ± 8.2	1.3 ± 0.1	No	No
	Flax fabric C2 (untreated)				
PF– C2–21/P1	Phenol formaldehyde matrix Flax fabric C2–34/P1 (C2+(AlCl3+NH4OH))	7.5±5.6	0.5 ± 0.1	No	Yes

FIGURE 11.6   Photos after burning of PF/flax fiber composites made with a) untreated and b) treated flax fibers.

Figure 11.6 illustrates the remains of the PF composite flax fabric after VC–2 burning test. The PF composite with nontreated flax fabric has burned very much before the fire stops. However, the PF composite with treated flax behaves very differently, the fire stops very shortly after the torque was removed.

The SEM observation of the fracture composite specimens demonstrates a good fiber-matrix in both the untreated and treated flax composites. Figure 11.7 illustrates the resin sticks on the fracture fibers in both, untreated and treated fiber after tensile test. Thus, the treatment does not cause any harm to the fiber-matrix interface, which determines the mechanical properties.

FIGURE 11.7   SEM image of fractured PF/flax fiber composites made with a) untreated and b) treated flax fibers.

Flexural properties of the PF/flax fiber composite samples are shown in Table 11.8. It is evident that PF composites containing fibers treated with a bi-component aluminum-containing system have comparable flexural properties with the reference if standard deviation is taken into account. It is in coherent with the SEM observation.

TABLE 11.8   Flexural Properties of PF/Flax Fiber Composites

Sample	Composition	Flexural strength (MPa)	Flexural modulus (GPa)
PF–C2	Phenol formaldehyde matrix Flax fabric C2 (untreated)	129.8 ± 8.2	10.2 ± 0.3
PF–C2–21/P1–2%	Phenol formaldehyde matrix Flax fabric C2–21/P1 (C2+(AlCl3+NH4OH))	118.1± 4.2	10.1 ± 0.8

11.4.2.3   EPOXY/FLAX FIBER COMPOSITES

The glass transition temperature of the epoxy/flax fiber composite samples determined by differential scanning calorimetry is about 80–82°C and there is no significant difference between them, indicating that the treatment has no effect on the curing of the epoxy matrix.

Horizontal burning tests were conducted in the composites and the results are shown in Table 11.9. It is clearly seen that the conventional epoxy/flax fiber composite is flammable while the treated flax fibers have stopped the composites from burning.

TABLE 11.9   Burning Test Results of Epoxy/Flax Composites

Sample	Burning characteristics
Epoxy–C2	Burned
Epoxy–C2–21/P1	Self-extinguished
Epoxy–C2–22/P2	Self-extinguished

The SEM observation of the fracture composite specimens demonstrates the resin sticks on the fracture fibers, illustrating a good fiber-matrix in both the untreated and treated flax composites (Fig. 11.8).

FIGURE 11.8 SEM image of fractured epoxy/flax fiber composites made with a) untreated and b) treated flax fibers.

Tensile properties of the epoxy/flax fiber composite samples are shown in Table 11.10. The epoxy composites containing treated flax fibers have very slightly reduction in tensile strength and modulus while improving the energy to break which represents the composite toughness. It indicates that the treatment did not destroy the composite mechanical properties as for other cases often observed in the literature.

TABLE 11.10 Tensile Properties of Epoxy/Flax Fiber Composites

Sample	Tensile stress (MPa)	Tensile modulus (GPa)	Energy to break (J)
Epo–C2	117.7 ± 4.0	9.8±0.6	33.7±2.0
Epo–C2–21/P1	106.4 ± 1.0	7.2±0.3	36.7±2.6
Epo–C2–22/P2	103.7 ± 4.2	8.4±0.2	36.7±2.6

11.4.2.4 UPE/FLAX FIBER COMPOSITES

Sample UPE–C2 is a comparative example of an UPE composite containing untreated flax fiber. UPE–C2–21/P1 is a composite containing flax fibers treated with a bi–component aluminum-containing system. Horizontal burning test was conducted (Table 11.11) and the results demonstrate that the fire-resistant treated flax fibers have made the composites to be self-extinguish which is not the case for the untreated fibers. The flexural properties of the treated flax composite are comparable with those of the untreated flax composites (Table 11.12) confirming that the treatment does not affected the composite mechanical performance.

TABLE 11.11   Burning Tests on UPE/Flax Fiber Composites

Sample	Burning length (inches)					
	0.0	1.0	2.0	3.0	4.0	5.0
	Burning Time (seconds)					
UPE–C2	0	114	-	-	421	522
UPE–C2–21/P1	0	NB	NB	NB	NB	NB

TABLE 11.12   Flexural Properties of UPE/Flax Fiber Composites

Sample	Flexural stress (MPa)	Flexural modulus (GPa)
UPE–C2	95.6 ± 4.2	4.2±0.6
UPE–C2–21/P1	100.2 ± 9.5	3.4±0.4

11.5   CONCLUSIONS

Coating of a layer of effective chemicals on the cellulosic fibers significantly improves its fire resistance. The treated cellulosic material becomes self-extinguishing and has no negative impact on fiber strength and/or modulus. The fire-resistant cellulosic material may also be treated with a layered nano-particulate material either simultaneously with, subsequent to or prior to treatment with the effective chemicals to impart further fire resistance to the cellulosic material. Fire-resistant flax fibers provide a tremendously significant fire resistance to polymer matrix than untreated flax fibers while they provide the composites equivalent mechanical properties as of untreated flax fiber. This will allow the production of green composites from cellulosic fibers with improved fire resistance.

KEYWORDS

- **Biocomposites**
- **Cellulose Fibers**
- **Fire Resistance**
- **Mechanical Properties**
- **Polymers**

REFERENCES

1. Rock, M. (December 3, 2009). Flame Retardant Fabrics U.S. Patent Publication, 2009 298370.
2. Gordon, I. (April 9, 1901) Composition for Fire proofing Paper U.S. Patent, 671, 548.
3. Chen, Y., Frendi, A., Tewari, S., & Sibulkin, M. (1991). Combustion Properties of Pure and Fire-Retarded Cellulose, *Comb. Flame*, 84, 121–140.

4. Mostashari, S. M., Kamali, N. Y., & Fayyaz, F. (2008). Thermogravimetry of Deposited Caustic Soda Used as a Flame-Retardant for Cotton Fabric. *J. Therm. Anal. Cal.*, 91(1), 237–241.

5. Ohkoshi, M., Okazaki, H., Hoshio, T., & Yasuno, M. (April 8, 2008). Photopolymerizable Composition and Flame-Retardant Resin-Formed Article U. S. Patent, 7, 354, 958.

6. Herndon, J. F., & Morgan, D. J. (March 29, 1994). Flame Retardant Composition and Method for Treating Wood. Canadian Patent Publication, 2,079,302.

7. De Lissa, R. C. F., & Schwarze, W. G. (1976). Flame-Proof Cellulosic Product, U.S. Patent 3,973,074, August 3.

8. Musselman, L. L., & Greene, H. L. (January 2, 1996). Materials for Use as Fire Retardant Additives, U. S. Patent, 5,480,587.

9. Nishibori, S., Komori, H., Saeki, S., & Kinoshita, H. (January 28, 1986). Flame Retarder for Organic High Molecular Compounds Prepared from Polycondensates of Halogenated Phenols, U.S. Patent 4,567,242.

10. Yoshifumi, N., Tadao, Y., Yuji, T., & Yoichi N. (November 27, 2003). Cross linked Phenoxyphosphazene Compounds, Flame Retardants, Flame-Retardant Resin Compositions, and Moldings of Flame-Retardant Resins. U.S. Patent Publication 2003-0220515.

11. Fukuba, K., & Miyazaki, M. (December 20, 1977). Flame Resistant Plaster Board and its Manufacture, U. S. Patent, 4,064,317.

12. Yan, X. B. (November 29, 2006). Preparation Method of Hydrophobic Ultrafine Nanometer Fire Retardant Magnesium Hydroxide Abstract of Chinese Patent Publication 1869154.

13. Mostashari, S. M. (2004). The Impartation of Flame-Retardancy to Cotton Fabric by the Application of Selected Carbonates of Group II. *J. Appl. Fire Sci.*, 13(1), 1–8.

14. Ebrahimian, S., & Jozokos, M. A. March 27, (2002). Zero Halogen Polyolefin Composition, Great Britain Patent Publication 2367064.

15. Seietsu, K. (June 26, 2002). Method for Producing Flame-Retardant of Nonfusible Fiber Japanese Patent Publication, 2002-180374.

CHAPTER 12

REINFORCING FILLERS AND COUPLING AGENTS' EFFECTS FOR PERFORMING WOOD POLYMER COMPOSITES

DIÈNE NDIAYE, MAMADOU GUEYE, COUMBA THIANDOUME, ANSOU MALANG BADJI, and ADAMS TIDJANI

ABSTRACT

Wood polymer composites (WPC) have attracted a lot of researchers, mainly due to their low densities, low cost, high filling levels, renewable and none toxic organic fibers to the glass or carbon, biodegradability and above all their availability from renewable sources. They present considerable commercial interest due to these potential opportunities. WPCs are used in a variety of innovative applications, such as the automotive sector, construction products or packaging industries. The product has the esthetic appearance of wood and the processing capability of thermoplastics and its performance in mechanical properties.

12.1 INTRODUCTION

The incorporation of various types of fillers into polymer matrices is an interesting route to produce polymer composites with different properties. Considerable researches have been focused extensively in the use of natural fibers as reinforcement material in a thermoplastic matrix. The utilization of vegetable fibers is driven by growing market trends in terms of environmental impact. The most common composites with natural fibers are made with polyolefin (polyethylene PE and polypropylene PP) and polyvinyl chloride (PVC) matrices and wood fibers as reinforcement.

In last few years, the utilization of fibers and powders derived from agricultural sources has attracted attention of many researchers mainly due to their low densities, low cost, none abrasiveness, high filling levels, renewable and none toxic organic fibers to the glass or carbon, biodegradability, and above all the availability from renewable sources.[1–5]

The acronym 'WPC' covers an extremely wide range of composite materials that use plastics ranging from PP to PVC and binders/fillers ranging from wood flour to natural fibers (e.g., flax)[6]. Wood polymer composites (WPC) are experiencing a growing market demand; hence it is logical to study ways to enhance the performance attributes of WPCs.

The use of various reinforcing fillers in the composites and their effect has to be typified. This will augment the industry with better understanding of properties and consequently delivering better products. Extensive research and product development has been done to reinforce polyolefin and other none biodegradable plastics, but research on reinforcing biodegradable polymers is limited because of the incompatibility between the two entities. The application of biodegradable polymers has primarily focused on the medical, agricultural, and consumer packaging industries[7,8]. The lignocelluloses' fibers used in polyolefins include cellulose fibers, wood fiber, flax, cannabis sativa (hemp), jute fiber, pine, sisal, rice husk, sawdust, wheat straw paper, mud, coir, kenaf, cotton, pineapple leaf fiber, bamboo fiber and palm tree[9].

Reinforcements such as wood have been successfully used to improve the mechanical properties of thermoplastic composites. Extensive efforts are being made to develop biodegradable composites using renewable resources in an attempt to replace the nonbiodegradable synthetic polymers used for composites[10]. Composites made from blends of thermoplastics and natural fibers have gained popularity in a variety of applications because they combine the desirable durability of plastics with the cost effectiveness of natural fibers as fillers or reinforcing agents[11]. The product has the esthetic appearance of wood and the processing capability of thermoplastics and its performance in humid area.

Another attraction is the fact that these materials are obtained easily from natural wastes. Because of these attributes, WPCs are used in a variety of innovative applications, such as the automotive (door panels or trims, door trims, trunk liners), construction products (decking, fencing, siding, windows, door frames, interior paneling or decorative trim), or packaging industries. Natural fibers possess excellent sound absorbing efficiency and are more shatter resistant and have better energy management characteristics than glass fiber reinforced composites. The incorporation of natural fibers into polycaprolactone (PCL) has been shown to enhance the biodegradability of the resulting composites[12]. However, the biodegradability of the resulting product is limited if all polymers are not biodegradable[13].

Properties of WPCs depend on the characteristics of matrix and fillers, chemical interaction between wood fibers and polymer, humidity absorption and processing condition. However, the use of wood-fibers shows some drawbacks such as fiber-polymer incompatibility and their low temperature of thermal degradation due to the presence of cellulose and hemicelluloses. The presence of hemicellulose, lignin and other impurities in these organic reinforcements causes a lack of adhesion between fibers and polymers. The low thermal degradation limits the allowed processing temperature to less than 200 °C. The compatibility between the wood fibers and polymeric matrix constitutes one important factor in the production of WPCs with improved mechanical properties[14,15,16].

These disadvantages of wood led some researchers to use other materials as reinforcements instead of wood. In resents years, natural fillers such as jute, kenaf, hemp, sisal, pineapple, rice husk, have been successfully used to improve the mechanical properties of thermoplastic composites[17]. The hydrophobic nature of PP poses a potential problem in achieving good fiber/matrix adhesion in these systems however, as cellulose is an inherently hydrophilic polymer because of the numerous hydroxyl groups contained within it. To alleviate this obstacle, chemical compatibilizers or coupling agents have been developed which alter the surface of the hydrophilic cellulose fiber in order to improve the dispersion and interfacial adhesion between the fiber and matrix. The most popular coupling agent that is being used by many researchers is maleic anhydride grafted polyolefin, such as polyethylene (MAPE) and polypropylene (MAPP)[18,19]. Many in-depth studies have elucidated the mechanisms of adhesion between MAPP treated wood fibers and the PP matrix that cause the improvement by the formation of linkages between the OH groups of wood and maleic anhydride[20,18]. It was found that coupling agent can form chemical bonds on the surface of wood and the interface between wood and polymer and it can well infiltrate the surface of wood, which finally result in lower surface tension of wood material[21,22,23]. One such compatibilizer is maleic anhydride grafted polypropylene (MAPP), a waxy polymer system that has been proven useful in the processing and production of cellulose reinforced PP composites[24,25]. MAPP is formed by reacting maleic anhydride (MA) with PP in the presence of an initiator to produce PP chains with pendant MA groups (Fig. 12.1).

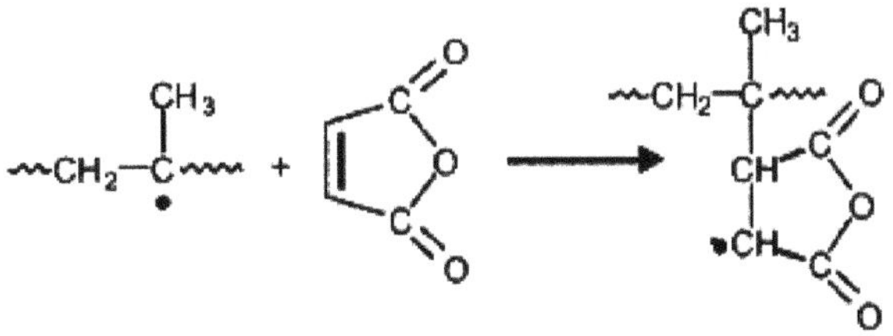

FIGURE 12.1 Schematic of the modification of PP with maleic anhydride group.

The PP portion of MAPP can entangle and co crystallize with the unmodified PP, while the maleic anhydride groups can bond to the hydroxyl (–OH) groups on the fibers. When mixed with cellulose, the hydroxyl group of the cellulose breaks one of the C-O bonds in the MA group and forms a new bond between one of the carbons from the MAPP group and the oxygen from the cellulose (Fig. 12.2).

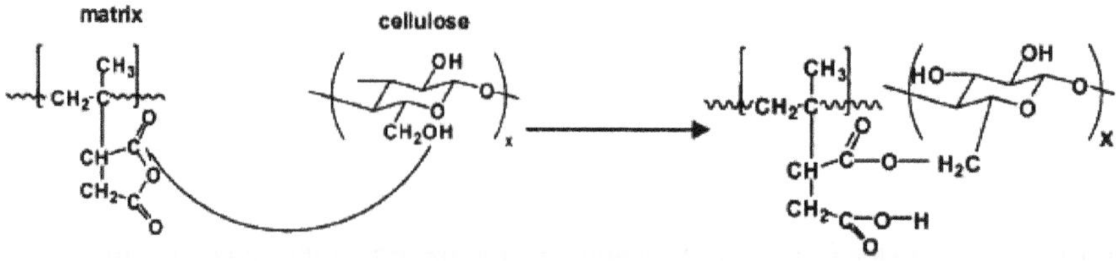

FIGURE 12.2   Schematic of the MA group attaching to a hydroxyl group on cellulose

The resulting chemical bond between the oxygen of MAPP[26,27] is also able to compensate for insufficient breakup forces during processing, such as low shear stress, by reducing the interfacial tension between PP and the cellulose, which leads to finer dispersion of the fiber throughout the system[28]. Addition of MAPP also reduced the degree of water absorption (>20%), making these materials more suitable for using in damp environments.

There are some studies on using other coupling agents such as silane[29,30] and isocyanates[31,32]. In all cases, with the coupling agents used, there is usually a significant improvement of the mechanical properties of the final composites.

Our study is inspired by the principle of ecological replacement of inorganic polluting substances by agricultural products (wood fibers, rice straw and Dried Distillers Grain with Soluble (DDGS) as reinforcements in polymer matrices. DDGS is a coproduct of the dry grind corn process that is used to produce fuel ethanol from corn. Considerable efforts have been made to use the coproducts obtained during the processing of corn, wheat and soybeans, wheat gluten and soy proteins, respectively for composite applications[33].

Usually Rice husk is a coating or protective layer formed during the growth of grains of rice. Removed during the refining of rice, these shells have low commercial value, because the SiO_2 and the fibers contained have poor nutritional value and are used in few quantities in animal ration. Rice Husk (RH) has been a problem for rice farmers due to its resistance to decomposition in the ground, difficult digestion and low nutritional value for animals[34]. Therefore, the development of new polymer composites filled with RH turns out to be a very interesting approach.

Another problem in our country, ethanol industry has grown exponentially in recent years, and the supply of distillers dried grains with solubles (DDGS) has subsequently increased dramatically. Currently, the ethanol industry's only outlet for DDGS is animal feed ingredients[35,36].

The voluminous production rate of DDGS is exceeding its consumption rate as animal feed. Most research publications about DDGS focus on its feed application[37]. Successfully making the separated DDGS fiber-based WPC would benefit the dry grind plants, wood composites manufacturers and the rural economies in which these production facilities are largely located by increasing their revenues. As technology develops towards the utilization of natural by products, classical wood fibers have shown strong potential as reinforcement in polymer matrix composites.

Composites were prepared by extruding DDGS with polypropylene and phenolic resin[38]. The need for materials with environmentally friendly characteristics has increased due to limited natural resources and increasing environmental regulation[39,40,41]. It is interesting to note that natural fibers such as jute, rice husk, DDGS, banana, sisal, etc., are abundantly available in developing countries in Africa but are not optimally used.

In this study, polypropylene was reinforced with pine wood, rice husk and distillers dried grains with soluble, byproducts of the ethanol process. The objectives of this study were to develop composites with PP matrix and three different reinforcing fillers: pinewood, rice husk and DDGS and to explore the performances and limitations of these reinforcing fillers on the mechanical, thermal and morphological properties of the composites. The results obtained in these tests, are discussed. We also discuss how to provide competitive alternative materials to natural wood, which becomes increasingly expensive and is diminishing in supply.

12.2   EXPERIMENTAL

12.2.1   MATERIALS

The basic materials used in this study are listed below.

Polypropylene (PP) is selected as one of the most popular candidates as a matrix material due to its versatility to accept numerous types of fillers and reinforcements. PP in the form of pellets, donated by Eastman Chemical Co. (Kingsport, TN) was used as the matrix. It had a melt flow index of 5.2 g/10 min (at 190 °C and a 2.16 kg load) and a density of 0.910 g/cm³. In order to increase the interface adhesion, polypropylene grafted with maleic anhydride was added as coupling agent to all the composites studied. MAPP (G-2010) was supplied by Eastman Chemical Co.

The wood fillers in particles of approximately 425 mm (40-mesh size) were kindly donated by American Wood fibers (Schofield, WI) and are constituted predominantly with ponderosa pine, maple, oak and spruce (hard wood).

Rice-husk flour (RHF) was used as natural fiber reinforcement. The fibers were hand chopped to an average length of 30 mm. After, the fibrous material was ground into flour form with particle size of 60-mesh size using a Thomas-Wiley mill. Rice husk was first treated with dilute hydrochloric acid (10% for 60 mn) to hydrolyze and remove the low molecular weight hemicellulose present in rice husk. After this step, rice husk was alkali treated by 0.5 NaOH solution (alkali treatment is described below) to remove portion of silica and lignin from rice husk and also some minor quantities of low molecular weight cellulose may also become water soluble. NaOH is known to be an effective reagent for reducing the protein content of organic fiber. Then, rice husk flour was oven dried at 100 °C for 24 h to adjust the moisture content to less than 2 wt.% and then stored over desiccant before compounding.

DDGS was obtained as the commercial animal feed pellet product. DDGS were milled with a Thomas-Wiley mill grinder and the particles exited through a 2 mm diameter stainless screen and were collected into a 1.81–L Mason jar. It is well known that the presence of oil in the DDGS acts as a plasticizer to lubricate relative molecular motion, hence lowering the modulus. We have extracted DDGs with hexane (to remove oils) then with dichloromethane (to remove polar extractible) employing a Soxhlet extractor. This treatment removes a certain amount of oils and other impurities, and the surface becomes rougher. The residue protein components in the separated DDGS fibers degraded at the melting temperature of the thermoplastic polymers.

12.2.1.1 ALKALI TREATMENTS

The rice husk and DDGS were soaked in a 0.5 NaOH solution at room temperature maintaining a ratio of (500 mL alkali solution/50 g reinforcing filler) reinforcement was kept immersed in the alkali solution for 2 h. The fibers were then washed several times with distilled water to remove any NaOH on the fiber surfaces, neutralized with dilute acetic acid, and again washed with distilled water. Below, in Fig. 12.3 are exposed photos of different reinforcements used in our study. Alkali treatment is one of the well-known processes to increase mechanical properties. The process alters the chemical content in crude fiber by removing lignin, pectin, hemi-cellulose, and changing the state of the materials from hydrophilic to hydrophobic. The large amount of hemi-cellulose lost made the fibers lose their cementing capacity and caused them to separate out from each other, making them finer[42,43,44].

FIGURE 12.3   Photos different reinforcements: (a) Pinewood, (b) Rice Husk, (c) DDGS.

12.2.2 COMPOUNDING AND PROCESSING

Before compounding, wood flour, rice husk and DDGS were dried in an oven for at least 48 h at 105 °C to expel moisture before blending with PP and then stored in polyethylene bags. First, the PP was put in the high-intensity mixer (Papenmeier, TGAHK20, Germany), and the reinforcement was added after the PP had reached its melting temperature. The mixing process took 10 min on average. After blending, the compounded materials were stored in a sealed plastic container. Several formulations were produced with various contents of PP, wood flour, rice husk, DDGS and 5% of MAPP in all the samples (Table 12.1). For the extraction of volatile and harmful gases, the hood was open.

TABLE 12.1 Composition and code of the Reinforcement/polypropylene composites (percentage is in weight).

Sample	PP (%)	Reinforcement (50%)	MAPP (%)
PP	100	-------	-------
PM	95	-------	5
PWM	45	Wood	5
PRM	45	Rice	5
PDM	45	DDGS	5

For the mechanical property experiments, test specimens were molded in a 33-Cincinnati Milacron reciprocating screw-injection molder (Batavia, OH). The nozzle temperature was set to 204 °C. The extrudate, in the form of strands, was cooled in the air and pelletized. The resulting pellets were dried at 105 °C for 24 h before they were injection-molded into the ASTM test specimens for tensile (Type I, ASTM D 638) and Izod impact strength testing. The dimensions of the specimens for the tests were $120 \times 3 \times 12$ mm^3 (Length × Thickness × Width).

12.2.3 PROPERTY EVALUATION

12.2.3.1 SCANNING ELECTRON MICROSCOPY (SEM)

The state of dispersion of the wood flour in the polymeric matrix was analyzed with SEM. A FEI Quanta 400 microscope (NE Dawson Creek Drive, Hillsboro, Oregon) working at 30 kV was used to obtain microphotographs of the fractured surfaces of the composites. Samples were cut in liquid nitrogen to avoid any deformation of the surfaces.

12.2.3.2 DIFFERENTIAL SCANNING CALORIMETRY (DSC)

DSC is widely used to characterize the thermal properties of WPCs. DSC can measure important thermoplastic properties, including the melting temperature (Tm), heat of melting, degree of crystallinity $\chi(\%)$, crystallization, and presence of recyclates, nucleating agents, plasticizers, and polymer blends (the presence, composition, and compatibility). Thermal analysis of the WPC samples was carried out on a differential scanning calorimeter (Perkin Elmer Instruments, Pyris Diamond DSC,

Shelton, Connecticut) with the temperature calibrated with indium. All DSC measurements were performed with samples of about (9.5 ± 0.1) mg under a nitrogen atmosphere with a flow rate of 20 mL/min. Three replicates were run for each specimen. All samples were subjected to the same thermal history with the following thermal protocol, which was slightly modified from the one reported by Valentini et al.[45].

1. First, the samples were heated from 40 to 180 °C at a heating rate of 20 °C/min to remove any previous thermal history.
2. Second, the samples were cooled from 180 to 40.00 °C at a cooling rate of 10 °C/min to detect the crystallization temperature (Tc).
3. Finally, the samples were heated from 40 to 180 °C at a heating rate of 10 °C/min to determine Tm. Tm and the heat of fusion (ΔH_m) were calculated from the thermograms obtained during the second heating. The heats of fusion were normalized on the basis of the weight fraction of PP present in the sample. The values of ΔH_m were used to estimate χ, which was adjusted for each sample in χ_{cor} (%) based on the percentage of polypropylene in the composite. Crystallinity (χ_{cor}) was estimated according to the following equation:

$$\chi_{cor} (\%) = \frac{\Delta H_m}{\Delta H_0 \cdot X_P} \qquad (1)$$

ΔH_m and ΔH_0 are, respectively heats (J/g) of melting of composite and 100% crystalline PP, taken as 207.1 J/g,[46] and X_{pp} is the PP fraction in the composite.

12.2.4   MECHANICAL PROPERTIES

The tensile tests like tensile strength (TS) and flexural modulus (FM) were carried out on an Instron 5585H testing machine (Norwood, MA) with crosshead rates of 12.5 and 1.35 mm/min according to the procedures outlined in ASTM standards D 638 and D 790, respectively. The measured flexural modulus of samples was obtained from three-point bending test. Eight replicates were run to obtain an average value for each formulation. Before each test, the films were conditioned in a 50% relative humidity chamber at 23 °C for 48h. Notched Izod impact resistance was tested according to ASTM D 256 standard by ZWICK 5101 testing pendulum at room temperature. Each mean value represented an average of eight tests. The impact strength (IS) is defined as the ability of a material to resist the fracture under stress applied at a high speed. The impact properties of composite materials are directly related to their overall toughness. In the Izod standard test, the only measured variable is the total energy required to break a notched sample.

12.3   RESULTS AND DISCUSSIONS

12.3.1   CHARACTERIZATION OF THE COMPOSITES' MORPHOLOGY

Analyzing the SEM micrographs provides helpful information about distribution and compatibility of the different phases in the composite. Figure 12.4 shows typical state of adhesion between the reinforcing fillers and the PP matrix.

Figure 12.4 shows the SEM micrographs (magnification of $100\,\mu m$) of the composites. As seen from Fig. 12.4b–d, PP matrix had melted and had a nonuniform surface, the reinforcement embedded in the polymer matrix suggesting that polymer has been plasticized. Fiber pullout was observed on all of PP composite with organic reinforcing fillers fracture surfaces examined using SEM. These results indicate a lack of complete adhesion between PP and reinforcing filler. However, the micrograph of Fig. 12.1a (PP with MAPP) exhibits finer morphology and a smoother appearance. The higher miscibility (PP with MAPP) made the morphology of the polymer blend appeared the cocontinuous structure. However, in all the composites with reinforcement, the micrographs taken from the fractured surface of the specimens showed different organization of the fibers in the composites (Figs. 12.4b, 12.4c and 12.4d). Reinforcement particles were not uniformly spread throughout the polymer matrix but could occur randomly or even in clumps. Reinforcement particles are of various sizes and irregular shapes, the incorporation of filler into the polymer matrix disrupted the homogeneity of the matrix. However, more rice clumps were observed on the fractured surfaces of the PRM (Fig. 12.4c) with rice reinforcement than in the blend of PM and PDM. It is visible that the external surface of the PRM is very rough; showing many aligned lumps. This observation suggests that rice was less favorable for better adherence with polymer matrix. Further, when large aggregates of particles of rice appear, it follows large cracks between the reinforcement and the polymer matrix (Fig.4c). In Fig.4d, the surface of the sample has characteristic even-sized rectangular nodules, creating a crocodile skin effect. A more detailed analysis of the micrograph at higher magnifications revealed a very rough surface formed by rounded nodules, which would facilitate adhesion between the particles and the matrix[47,48]. The dark voids visible in Fig. 12.4b, 4c and 4d are due to the pullout of reinforcement agglomerates during the deformation, which indicate weak filler–matrix adhesion. Large plastic deformation and fibrillation can clearly be seen in these figures indicating the ductile mode fracture of the former composites. Poor fiber-matrix adhesion is confirmed by the presence of gaps between the fibers and the matrix and fiber pullouts. The coupled composite displayed a rough morphology with the presence of voids between the filler particles and the polymer matrix, clearly indicating weak interaction between them. Higher void contents usually mean lower fatigue resistance, greater susceptibility to water penetration and weathering, and increased variation or scatter in strength properties. In Fig. 12.4a, there are any voids; the surface is well homogeneous confirming the effect on promoting adhesion played by MAPP in the interfacial region in the composite. We can notice the necessity to use coupling agents or additives to provide good binding between PP matrix and the fibers leading to composites with good properties.

FIGURE 12.4 SEM micrographs of fractured surfaces of composites: (a) PM, (b) PWM, (c) PRM, (d) PDM.

The SEM micrographs show that the level of dispersion of the reinforcement in the matrix is not as good as MAPP with PP. The SEM morphological study shows that the composites of PP with reinforcement have less fiber-matrix adhesion and wettability than PP with MAPP. Besides, aggregates and net porosities of the reinforcing particles are clearly observed even in the presence of MAPP

12.3.2 THERMAL PROPERTIES OF THE COMPOSITES

The thermal properties of the composite blends containing different reinforcement (rice, wood and DDGS) measured by DSC are shown in Table 12.2.

TABLE 12.2 Thermal Properties of Composites of PP Matrix with Different Reinforcements

Sample	T_m (°C)	ΔH_m (J/g)	T_c (°C)	χ_{cor} (%)
PP	160.8	79.5	120.5	38.9
PM	160.5	82.8	121.0	39.0
PWM	164.7	84.7	125.1	42.5
PRM	163.6	83.9	124.0	40.7
PDM	163.2	84.0	124.3	41.0

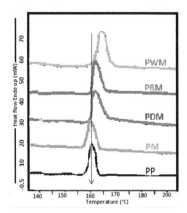

FIGURE 12.5 DSC curves (melting) of neat PP and composites with different reinforcements (the curves are vertically moved for clarity). The codes and compositions of the samples are described in Table 12.1.

Figure 12.5 presents the thermograms of the second heating of the PP/reinforcement flour blends subjected to the same rate flow. Only one endothermic peak corresponding to PP can be observed in these figures. All the composites regardless have invariably exhibited a slightly higher T_m compared to the T_m of neat PP. Figure 12.5 shows DSC curves corresponding to the cooling scan for PP and its composites. All curves show exothermic peaks corresponding to the crystallization of the polymeric matrix. A shift of T_c towards higher temperatures in the presence of MAPP and reinforcement was observed in the composites. This indicated that the phenomenon of crystallization during the cooling occurred more rapidly in composites containing MAPP than in the pure PP. The effect of rising crystallization rates was clear for all of the composites containing MAPP. The results imply that MAPP acted as a precursor and increased crystallization. The presence of reinforcement (wood, rice or DDGS) decreases the thermal stability, heat in turn causes scissions of chains and all these phenomena generate an early fusion and it is more pronounced whit samples with reinforcement agent.

Figure 12.6 shows DSC curves corresponding to the cooling scan for PP and its composites. All curves show exothermic peaks corresponding to the crystallization of the polymeric matrix. Every bio-filler affects the thermal properties of the composite differently[49,50]. The cooling characteristics have shown very interesting behavior. The temperatures corresponding to onset of crystallization and peak crystallization have increased due to presence of filler reinforcement. These temperatures have further increased due to chemical treatment of rice husk and DDGS. Thus it seems that addition of reinforcement is causing early crystallization of PP. This indicates that reinforcement is influencing the degree of super cooling of PP. Hattotuwa et al.[51] also have reported similar results. The presence of MAPP does not significantly modify the crystallization temperature but leads to an increase in the degree of crystallinity. It is recognized that wood and MAPP act as nucleating agents[52,16]. The presence of these two elements generates the formation of more crystals. T_m, T_c, ΔH_m and χ_{cor} are reported in Table 12.2 for composites of PP matrix and different reinforcements of wood, rice and DDGS. An increase in T_m was observed when reinforcement was loaded into the polymer matrix. The addition of reinforcement had the effect of shifting T_m to higher values. This increase was accompanied by an increase of the composites' degree of crystallinity χ (%) which was corrected as χ_{cor} (%) by taking into account there enforcement concentration[53,54]. These results suggest that crystallization occurred earlier with the incorporation of reinforcement, which played the role of a nucleating agent. Reinforcement provided sites for heterogeneous nucleation; this induced crystallization of the polymeric matrix. This was ascribed to the poor thermal conductivity of reinforcement. In the composite, reinforcement acted as an insulating material, hindering the heat conductivity. As a result, the composites compounds needed more heat to melt. Similar findings were previously reported by Matuana and Kim[55] for PVC based wood–plastic composites. They found that the addition of wood flour to the PVC resin caused significant increases in the temperature and energy at which fusion between the particles started. The delayed fusion time observed in rigid PVC/wood flour composites was attributed to the poor thermal conductivity of the wood flour; this decreased the transfer of heat and shear throughout the PVC grains. These phenomena were consistent with the results of this study. For a composite, the impact strength depends on the composition and structure as well as the testing method. Adding reinforcement in all cases was shown to increase both the crystallization temperature and extent of crystallization of polymer matrix in WPC systems as compared to controls.

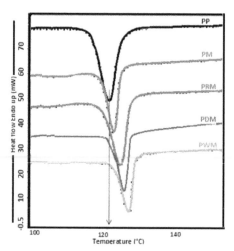

FIGURE 12.6   DSC curves (cooling) of neat PP and composites with different reinforcements.

The thermal stability of the composites was investigated using DSC analysis under nitrogen atmosphere. The results have shown a spectacular improvement of thermal stability of the composites and an increase of the degree of crystallinity. Although the properties of some blends are acceptable for some applications, further improvement will be necessary, mainly by optimizing fiber-polymer.

12.3.3 MECHANICAL PROPERTIES OF THE COMPOSITES

The effect of the reinforcement agent on the notched Izod impact energy for the composite is also listed in the following table.

The average for five test specimens and their significant standard error is given for each property. Figures 12.7 and 12.8 graphically summarize the data listed in Table 12.3.

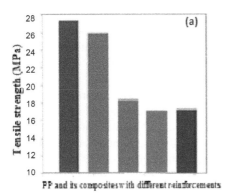 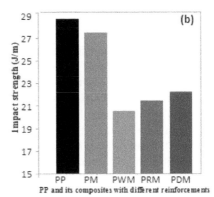

FIGURE 12.7 Curves of mechanical properties of PP and its composites with different reinforcements: (a) Tensile strength, (b) Impact strength of notched samples.

TABLE 12.3 Tensile Strength (TS), Flexural Modulus (FM) and Notched Izod Impact Strength (IS) of PP and Its Composites with Different Reinforcements

Sample	TS (MPa)	FM (GPa)	IS (J/m)
PP	27.5	1.40	29.0
PM	29.5	1.85	27.5
PWM	15.4	4.85	20.6
PRM	17.1	4.87	21.5
PDM	17.5	4.88	22.3

Figure 12.7(a) shows the behavior of tensile strengths of both PP and its composites, where it becomes obvious that the addition of reinforcement weakens the

matrix. This behavior is observed for all matrix reinforcement combinations, although the rate in reduction of the tensile strength and Izod impact strength varied from case to case, depending on the reinforcement. Most of all plant fibers are hydrophilic in nature with a moisture content enough high due to the presence of cellulose in cell structure. All these organic reinforcements generally have high aspect ratio, so, the efficiency of transmitting stress from matrix to these types of agents is quite poor. On the other hand, lignin increased the hard segments of composite the films, making the films less elastic and more brittle, which led to the impact strength decreasing. This explains why the impact strength decreased as the filler content reached 40% in weight. Poor interfacial bonding causes partially separated micro spaces between the filler and the matrix polymer, which obstructs stress propagation, when tensile stress is applied, and induces decreased strength and increased brittleness but compatibilizing agent can solve partially this problem. This is what justifies the use of a coupling agent such as MAPP in all formulations. As the reinforcement loading (50%) is higher, filler–filler agglomeration occurs and degree of weak interface regions between reinforcement particles and matrix become more and leads deterioration in tensile strength. As the degree of agglomeration increases, the filler– matrix interaction becomes poor, leading to the decrement in the tensile strength. Incorporation of both types of fillers (rice, wood and DDGS), generates a reduction in tensile strength of PP and its composites.

Figure. 12.7(b) shows that notched impact strength decreases in all the composites. This may be explained by the fact that the presence of reinforcing fillers ends within the body of the composite can cause crack initiation and subsequent failure. The reason is that the ends of reinforcing fillers act as notches and generate considerable stress concentrations, which could initiate micro cracks in the ductile PP matrix.

The impact test machine used in this study did not provide enough energy to break the neat PP because of the high flexibility of the PP matrix. By contrast, all specimens broke completely into two pieces. Introducing the reinforcement in the composites led to an increased stress concentration because of the poor bonding between the reinforcement (wood, rice or DDGS) and the polymer. As impact wave met different phases such as fiber, polymer, and voids in the cross machine direction, it would lose its energy as dissipation energy. Although crack propagation became difficult in the polymeric matrix reinforced with filler, the decrease in the impact energy observed was ascribed to fiber ends, at which micro cracks formed and fibers debonded from the matrix. These micro cracks were a potential point of composite fractures. Another reason for the decreased impact strength may have been the stiffening of polymer chains due to the bonding between the wood fibers and the matrix. For high-impact properties, in fact, a slightly weaker adhesion between the fiber and polymer is desirable, as it results in a higher degradation of impact energy and supports the so-called fiber pullout[56]. In composites, the effect of the reinforcement is to increase the tendency to agglomerate, which generates a low interfacial adhesion leading to the weakening of the interfacial regions. These agglomerates then act as sites for crack initiation. Poor interfacial bonding has been indicated in the literature as the major reason for the loss in strength and elastic modulus[57].

Adding fillers also resulted in an increment of void content, which contributes to stress concentration, thus reducing strength. This behavior is consistent with what is observed in the impact tests that revealed a decrease in composites samples. The presence of numerous cavities is clearly visible in Fig. 12.4b (PWM) which has the lowest impact strength, this indicates that the level of interfacial bonding between the fibers and the matrix is weak and when stress is applied it causes the fibers to be pulled out from the matrix easily leaving behind gaping holes. These two properties are indicators of the plasticity of the material, and showed that the PP has a tendency for the occurrence of fracture with loading reinforcement. The flexural modulus of composites is influenced mainly by the adhesion between the matrix and dispersion of reinforcing fillers inside it. The results for this mechanical property also supported the existence of a certain degree of miscibility in the composite plastics (Fig. 12.8).

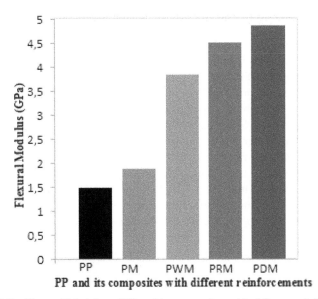

FIGURE 12.8 Flexural Modulus of PP and its composites with different reinforcements.

From Fig. 12.8, all the compositions showed a flexural modulus higher than the pure PP. This increase in flexural properties was expected due to the improved adhesion between components in the blends. For some authors[58], this is due to the restriction of the mobility and deformability of the matrix with the introduction of mechanical restraint. Many researchers[59,60,61] have observed that the inclusion of wood fibers or lignocellulosic fibers into thermoplastics such as, polyethylene, or PP generally results in a decrease in tensile strength and elongation at break but an increase in Young's modulus. This increase of flexural modulus can be attributed to the increase in volume fractions of high-modulus fibers in plastic composites[62]. When increasing the reinforcement, tensile and compression strengths constantly decreased. The presence of the wood or other reinforcement in the polymeric matrix augmented the polymer's rigidity, increasing the value of the modulus in relation to the pure polymer. This phenomenon has also been reported by other researchers who studied the effect of wood flour on mixtures of recycled polystyrene and polyethylene[63,64].

and virgin polystyrene[65]. The Flexural modulus of composite with DDGS is significantly higher than all the other ones, may be, this is likely due to the oils being removed in the DDGS material. So, we can notice that the initial chemical treatments on DDGS certainly has a positive effect on the improvement of mechanical properties (FM and IS) that are higher than those of untreated wood. These results are in agreement with those of Julson et al[61]. The improved dispersion obtained from the composite with treated DDGS was also responsible for the highest flexural modulus. For none coupled composite (wood and PP only), the filler particles began to form aggregates. Direct physical bonds between filler particles are weak and, thus, easily broken during tensile loading, which explains the decrease in the flexural modulus (WPM). Compatibilizers can change the molecular morphology of the polymer chains near the fiber–polymer interphase. Yin et al.[66] reported that the addition of coupling agent (MAPP) even at low levels (1–2%) increases the nucleation capacity of wood-fibers for polypropylene, and dramatically alters the crystal morphology of polypropylene around the fiber. When MAPP is added, surface crystallization dominates over bulk crystallization and a transcrystalline layer can be formed around the wood–fibers. Crystallites have much higher moduli as compared to the amorphous regions and can increase the modulus contribution of the polymer matrix to the composite modulus[67]. The flexural modulus of the composites can be correlated with the morphology of these ones. Composites whose surfaces are smoother and more homogeneous exhibit the greatest flexural modulus. The resultant increase in flexural modulus properties (Fig. 12.8) can be explained on the basis of improved wettability (compatibility) of the reinforcement fibers with the polymer matrix. The increased compatibility is obtained by reducing the polarity of the wood fiber surface nearer to the polymer matrix.

The mechanical results of this study show that loading of PP with these natural fibers leads to a decrease in tensile and impact strength of the pure polymer. On the other hand, the flexural modulus increases due to the higher stiffness of the fibers. The significant improvements in flexural properties of the blends composites made with MAPP and reinforcing fillers were further supported by SEM micrographs.

12.4 CONCLUSION

Wood fibers, rice husk and DDGS, which originate from renewable resources, are an interesting alternative to mineral fibers. All these samples with reinforcement exhibited a markedly heterogeneous and highly rough fracture surface with large voids, or cavities, around the filler particles due to the accumulation of stresses in the particle–matrix interface zone. This produced an adverse effect on mechanical properties such as tensile strength and impact resistance.

The SEM micrographs reveal that interfacial bonding between the treated filler and the matrix has significantly improved, suggesting that better dispersion of the filler into the matrix was achieved upon treatment of rice husk and DDGS. The thermal properties revealed the strong nucleation ability of the reinforcement flour and MAPP on PP crystallization. Crystallization of all the composites with the coupling agent MAPP only or with reinforcement began earlier compared to that of pure PP. This suggested that MAPP and organic reinforcement acted as nucleation agents and were responsible of the shift of crystallinity towards higher temperatures.

Tensile and impact strength exhibited a marked downward tendency as reinforcement was loaded. This is due to the weak interfacial adhesion and low compatibility between matrix and filler. The weak bonding between the hydrophilic lignocellulosic agent and the hydrophobic matrix polymer obstructs the stress propagation and causes the decrease of these properties. These properties were not significantly affected by changing rice husk by DDGS. However, the flexural modulus for all composites with organic reinforcement is higher than the value for neat PP, as a consequence of the high modulus of cellulosic agent.

In summary, the use of organic reinforcements as fillers to polymer matrix composites proved to be a viable alternative. Reductions in tensile and impact strength properties reported with the addition of fillers may be tolerable to some applications. Increment in flexural modulus is achieved in all cases. The development of alternatives for recycling rice husks and DDGS as reinforcements in polymer matrix composites is an important step to provide a good destination for these wastes and opens an opportunity to produce a new value added product. This development can permit reduction of production costs and less use of wood in composites based on polymer matrix, especially with the scarcity of wood across the world. This novel application of rice husk and DDGS for bio composites has significantly higher economic value than its traditional use as a feed stuff. The distillers' dried grains with soluble (DDGS) from corn ethanol industry; the rice husk products show immense opportunities in engineering new green composites when integrated with thermoplastics. The properties of PP composites can be adjusted by mixing different reinforcements species for filler blend.

12.5 FUNDING

This research received no specific grant from any funding agency in the public, commercial or not-for-profit sectors.

12.6 CONFLICTS OF INTEREST

None declared.

KEYWORDS

- **Composite**
- **Morphology**
- **Alkali Treatment**
- **Mechanical Properties**
- **Thermal Properties**

REFERENCES

1. George, J., Bhagawan, S. S., Prabhakaran, N. & Thomas, S. (1995). Short Pineapple leaf fiber reinforced Low density Polyethylene Composites. *J Appl Polym Sci.,* 57, 843–854.
2. Manikandan, N. K. C., Diwan, S. M., & Thomas, S. (1996). Tensile Properties of Short Sisal Fiber Reinforced Polystyrene Composites. *J Appl Polym Sci,* 60, 1483–1496.
3. Uma Devi, L., Bhagawan, S. S. & Thomas, S. (1997). Mechanical Properties of Pineapple Leaf Fiber-Reinforced Polyester Composites. *J Appl Polym Sci.,* 64 1739–1748.
4. Oksman, K., & Clemons, C. (1998). Mechanical Properties and Morphology of Impact Modi-fied Polypropylene Wood Flour Composites *J Appl Polym Sci,* 67, 1503–1513.
5. Rozman, H. D., Tan, K. W., Kumar, R. N., & Abubakar, A. (2001). The effect of hexa methy-lene di isocyanate modified ALCELL lignin as a coupling agent on the flexural properties of oil palm empty fruit bunch–polypropylene composites. *Polym Int,* 50, 561–567.
6. Rails, T., Wolcott, M. P., & Nassar, J. M. (2001). Interfacial contributions in lignocellulosic fiber reinforced plastic composites. *J Appl Polym Sc,* 80, 546–555.
7. Saleh, O. S. B. (2012). Enhancement of Polyolefins Compatibility with Natural Fibers through Chemical Modification *Am J Polym Sci,* 2(5), 102–108.
8. Chandra, R., & Rustgi, R. *(1998).* Biodegradable Polymers Prog *Polym Sci.,* 23, 1273–1335.
9. Malkapuram, R., Vivek, K., & Negi, Y. S. (2009). Recent Development in Natural Fiber Rein-forced Polypropylene Composites. *J Appl Polym Sci,* 28 1169–1189.
10. Mohanty, A. K., Khan, M. A., Sahoo, S., & Hinrichsen, G. (2000). Effect of Chemical Modi-fication on the Performance of Biodegradable Jute Yarn Biopol Composites *J Mater Sci.,* 35:2589–2595.
11. Pilarski, J. M., & Matuana, L. M. (2005). Durability of Wood Flour Plastic Composites Ex-posed to Accelerate Freeze-Thaw Cycling Part I: Rigid PVC Matrix. *J Vinyl & Add Technol,* 11(1), 1–8.

12. Di, Franco. C. R., Cyras, V. P., Busalmen, J. P., Rueseckaite, R. A., & Vazquez, A. (2004). Degradation of Polycaprolactone/Starch blends and Composites with Sisal Fiber *Polym Degrad Stabil*, 86(1), 95–103.

13. Tilstra, L., & Johnsonbaugh, D. (1993). The biodegradation of blends of polycaprolactone and polyethylene exposed to a defined consortium of fungi. *J. Environmen. Polym. Degrad*, 1(4), 257–267.

14. Woodhams, R. T., Thomas, G., & Rodgers, D. K. (1984). Wood Fibers as Reinforcing Fillers for Polyolefins *Polym Eng. Sci.* 24, 1166–1171.

15. Kazayawoko, M., Balatinecz, J. J., & Matuana, L. M. (1999). "Surface Modification and Adhesion Mechanisms in Woodfiber Polypropylene Composites" *J Mater Sci*, 34(24), 6189–6199.

16. Ndiaye, D., Matuana, L. M., Morlat-Therias, S., Vidal, V., Tidjani, A., & Gardette, J. L. (2011). Thermal and Mechanical Properties of Polypropylene/Wood Flour Composites *J. Appl Polym Sci*, 119 3321–3328.

17. Bullions, T. A., Hoffman, D., Gillespie, R. A., Price-O'Brien, J., & Loos, A. C. (2006). Contributions of Feather Fibers and Various Cellulose Fibers to the Mechanical Properties of Polypropylene Matrix Composites *Comps Sci. Technol*, 66(1), 102–114.

18. Kazayawoko, M., Balatinecz, J. J., & Matuana, L. M. (1999). Surface Modification and Adhesion Mechanism in Wood Fiber-Polypropylene Composites *J Mater Sci*, 34(24), 6189–6199.

19. Lai, S. M., Yeh, F. C., Wang, Y., Chan, H. C., & Shen, H. F. (2003). Comparative Study of Maleated Polyolefins as Compatibilizers for Polyethylene/Wood Flour Composites *J Appl Polym Sci*, 87, 487–496.

20. Woodhams, R. T., Thomas, G., & Rodgers, D. K. (1984). Wood fibers as reinforcing fillers for polyolefins. *Polym Eng Sci*, 24, 1166–1171.

21. *Bledzki, A. K., & Gassan, J. (1999). Composites reinforced with cellulose based fibers. Prog. Polym Sci, 24, 221–274.*

22. Li, Q., & Matuana, L. M. (2003). Effectiveness of Maleated and Acrylic-Acid functionalized Polyolefin Coupling Agents for HDPE/Wood-Flour Composites. *J Thermopl Compos Mater*, 16(6), 551–564.

23. Chotirat, L., Chaochanchaikul, K., & Sombatsompop, N. (2007). On Adhesion Mechanisms and Interfacial Strength in Acrylonitrile Butadiene-Styrene/Wood Sawdust Composites *Int J Adhes*, 27, 669–678.

24. Chotirat, L., Chaochanchaikul, K., & Sombatsompop, N. (2007). On Adhesion Mechanisms and Interfacial Strength in Acrylonitrile-Butadiene-Styrene/Wood Sawdust Composites *Int J Adhes*, 27, 669–678.

25. Qui, W., Zhang, F., Endo, T., & Hirotsu, T. (2003). Preparation and characteristics of composites of high-crystalline cellulose with polypropylene: effects of maleated polypropylene and cellulose content. *J Appl Polym Sci*, 87(2), 337–345.

26. Panaitescu, D. M., Donescu, D., Bercu, B., Vuluga, D. M., Iorga, M., & Ghiurea, M. (2007). Polymer Composites with Cellulose Micro Fibrils *Polym Eng Sci*, 47, 1228–1234.

27. Trejo-O'Reilly, J. A., Cavaille, J. C., Paillet, M., Gandini, A. et al. (2000). Interfacial properties of regenerated cellulose fiber/polystyrene composite materials Effect of the Coupling Agent's Structure on the Micromechanical Behavior. *Polym Compos*, 21, 65–71.

28. Bullions, T. A., Gillespie, R. A., Price-O'Brien, J. & Loos, A.C. (2004). The effect of maleic anhydride modified polypropylene on the mechanical properties of feather fiber, kraft pulp, polypropylene composites. *J Appl Polym Sci*, 92(6), 3771–3783.

29. Kyu-Nam Kim, Hyungsu Kimm, Jae-Wook Lee. (2001). Effect of Interlayer Structure, Matrix Viscosity and Composition of a Functionalized Polymer on the Phase Structure of Polypropylene-Montmorillonite Nanocomposites *Polym Eng Sci*, 41, 1963–1969.

30. Keener, T. J., Stuart, R. K., & Brown, T. K. (2004). Maleated Coupling Agent for Natural Fiber *Compos PartA Appl Sc*, 35, 357–362.
31. Bengsson, M., & Oksman, K. (2006). Silanecrosslinked wood plastic composite: Processing and properties. *Compos Sci Technol*, 66, 2177–2186.
32. Raj, R. G., Kokta, B. V., Maldas, D., & Daneault, C. (1989). Use of wood fibers in thermoplastics VII the effect of coupling agents in polyethylene-wood fiber composites. *J ApplPolymSci*, 37, 1089–1103.
33. Kokta, B. V., Raj, R. G., & Daneault, C. (1989). Use of wood flour as filler in polypropylene: Studies on mechanical properties. *Polym Plast Technol Eng*, 28, 247–259.
34. Yang, H. S., Kim, H. J., Park, H. J., et al. (2004). Rice husk Filled Polypropylene Composites, Mechanical and Morphological Study. *Compos Struct*, 63, 305–312.
35. Piva, A. M., Steudner, S. H., & Wiebeck, H. (2004). *Proceedings of the 5ᵗʰ International Symposium on Natural Polymer and Composites, São Pedro, Brazil*.
36. Wu, Q., & Mohanty, A. K. (2007). Renewable Resource Based Biocomposites from Coproduct of Dry Milling Corn Ethanol Industry and Castor Oil Based Biopolyurethanes. *J Biobased Mater Bioenerg*, 1, 257–269.
37. Cheesbrough, V., Rosentrater, K., & Visser, A. (2008). Properties of Distillers Grains Composites: A Preliminary Investigation. *J Polym Environ*, 16, 40–50.
38. Lee, S. Y., Yang, H. S., Kim, H. J., Jeong, C. S., Lim, B. S., & Lee, J. N. (2004). Creep Behavior and Manufacturing Parameters of Wood Flour Filled Polypropylene Composites. *Compos Struct*, 65, 459–469.
39. Tatara, R. A., Rosentrater, K. A., & Suraparaju, S. (2009). Design properties for molded, corn-based DDGS-filled phenolic resin. *Ind Crops Prod.*, 29, 9–15.
40. Ismail, H., & Jaffri, R. M. (1999). Physicomechanical Properties of Oil Palm Wood Flour Filled Natural Rubber Composites *Polym Testing*, 18, 381–388.
41. Cheesborough, V., Rosentrater, K., & Visser, J. (2008). Properties of distilers grains composites: a preliminary investigation. J Polym Environ, 16, 40–50.
42. Chevanan, N., Rosentrater, K. A., & Muthukumarappan, K. (2008). Effect of DDGS, Moisture Content, and Screw Speed on Physical Properties of Extrudates in Single-Screw Extrusion *Cereal Chem*, 85(2), 132–139.
43. Mohanty, A. K., Khan, M. A., & Hinrichsen, G. (2000). Surface Modification of Jute and Its Influence on Performance of Biodegradable Jute-Fabric/Biopol Composites *Compos Sci Technol*, 60, 1115–1124.
44. Dipa Ray &. Sarkar, B. K. (2001). Characterization of Alkali-Treated Jute Fibers for Physical and Mechanical Properties *J Appl Polym Sci*, 80, 1013–1020.
45. Abdelmouleh, M., Boufi, S., Belgacem, M. N., & Dufresne, A. (2007). Short natural-fiber reinforced PP and natural rubber composites: effect of silane coupling agents and fibers loading. *Compos Sc Technol*, 67, 1627–1639.
46. Valentini, L., Biagiotti, J., Kenny, J. M et al. (2003). Morphological Characterization of Single-Walled Carbonnanotubes-PP Composites *Compos Sci Technol*, 63, 1149–1153.
47. Wunderlich, B. (1990). Thermal Analysis of Polymeric Materials, *New York: Academic Press.*
48. Pramanick, A. & Sain, M. (2006). Nonlinear Viscoelastic Creep Prediction of HDPE Agro Fiber Composites *J Compos Mater*, 40(5), 417–431.
49. Yao, X. F., Zhao, H. P. & Yeh, H. Y. (2006). Micro/Nanoscopic Characterizations of Epoxy Silica Nano Composites *J Reinf Plast Compos*, 25(2), 189–196.
50. Kalia, S., Kaith, B. S., & Kaur, I. (2009). Pretreatments of Natural Fibers and Their Applications as Reinforcement Materials in Polymer Composites A review *Polym Eng Sci*, 49, 1253–1272.

51. Nongnuch, S., Bryan, R., Brajendra, K., Sharma, et al. (2011). Physical Properties and Fatty Acid Profiles of Oils from Black, Kidney, Great Northern, and Pinto Beans: *J Am Oil Chemists Soc*, 88(2), 193–200.

52. Premalal, H. G. B., Ismail, H., & Baharin, A. (2003). Effect of processing time on the tensile, morpholical and thermal properties of rice husk powder-filled polypropylene composites *Polym Plast Tech Eng.*, 42, 827–851.

53. Peter Zugenmaier, (2006). Materials of Cellulose Derivatives and Fiber-Reinforced Cellulose Polypropylene Composites Characterization and Application *Pure Appl Chem*, 78(10), 1843–1855.

54. Mandelkern, L. (1964). Crystallization of Polymers, *Series in Advanced Chemistry, McGraw-Hill: New York, Chapter 5.*

55. Norma, E. M., & Villar, M. A. (2003). Thermal and Mechanical Characterization of Linear Low Density PE/Wood Composites *J Appl Polym Sci*, 90(10), 2775–2784.

56. Matuana, L. M. & Kim, J., & Fusion, W. (2007). Characteristics of Wood-Flour Filled Rigid PVC by Torque Rheometry *J Vinyl Add Technol*, 13(1), 7–13.

57. Bengtsson, M., Baillif, M. L., & Oksman, K. (2007). Extrusion and Mechanical Properties of Highly Filled Cellulose Fiber-Polypropylene Composites *Compos: Part A*, 38, 1922–1931.

58. DeRosa, R., Telfeyan, E., Gaustad, G., & Mayes, S. (2005). Strength and Microscopic Investigation Ofunsaturated Polyester BMC Reinforced with SMC-Recyclate *J Thermoplast Compos Mater* 18, 333–349.

59. Ismail, H., Nasaruddin, M. N., & Rozman, H. D. (1999). The Effect of Multifunctional Additive in White Rice Husk Ash Filled Natural Rubber Compounds. *Eur Polym J*, 35(8), 1429–1437.

60. Stark, N. M., & Berger, M. J. (1997). Effect of Particle Size on Properties of Wood Flour Reinforced Polypropylene Composites 4th Intern. *Conf. Wood fiberplastic Comp Forest Products Soc. Madison*, WI May 12–14, 134–143.

61. Julson, J. L., Subbarao, G., Stokke, D. D., Gieselman, H. H. et al. (2004). Mechanical Properties of Biorenewable Fiber/Plastic Composites *J Appl Polym Sci*, 93, 2484–2493.

62. Febrianto, F., Setyawati, D., Karina, M., Bakar, E. S. et al. (2006). Influence of wood flour and modifier contents on the physical and mechanical properties of wood flour-recycled polypropylene composites. *J. Biol. Sci.*, 6(2), 337–343.

63. Sanadi, A., Caulfield, D. F., & Rowell, R. M. (1998). Lignocellulosics/Plastic Composites, the Fibril Angle, Cellulose, Paper, and Textile Division *Am Chem Soc: Washington, DC*, 8–12.

64. Simonsen, J., & Rials, T. J. (1996). Morphology and properties of wood-fiber reinforced blends of recycled polystyrene and polyethylene. *J Therm Compos Mater*, 9(3), 292–302.

65. Maldas, D., & Kokta, B. V. (1990). Effect of Recycling on the Mechanical Properties of Wood Fiber Polystyrene Composites Part I: Chemi thermo mechanical pulp as reinforcing filler. *Polym Compos*, 11(2), 77–83.

66. Yin, S., Rials, T. G. & Wolcott, M. P. (1999). Crystallization Behavior of Polypropylene and Its Effect on Woodfiber Composite Properties *Fifth Internat Conf Wood Fiber Plast Compos. Madison, WI, Forest Products Society*, May 26–2, 139–146.

67. Sinha, R. S., Maiti, P., Okamoto M., et al. (2002). New Polylactide/Layered Silicate Nanocomposites 1 Preparation, Characterization, and Properties *Macromol*, 35, 3104–3110.

CHAPTER 13

PROPERTIES OF DRIED DISTILLERS GRAINS WITH SOLUBLES, PAULOWNIA WOOD, AND PINE WOOD REINFORCED HIGH DENSITY POLYETHYLENE COMPOSITES: EFFECT OF MALEATION, CHEMICAL MODIFICATION, AND THE MIXING OF FILLERS

BRENT TISSERAT, LOUIS REIFSCHNEIDER, DAVID GREWELL, GOWRISHANKER SRINIVASAN, and ROGERS HARRY O'KURU

There is a need to identify usable lignocellulosic materials that can be blended with thermoplastic resins to produced commercial lignocellulosic plastic composites (LPC) at lower costs with improved performance. The core objectives of this study are to: i) evaluate the use of dried distillers grain with solubles (DDGS) and Paulownia wood (PW) flour in high density polyethylene-composites (LPC); ii) assess the benefit of chemically modifying DDGS and PW flour through chemical extraction and modification (acetylation/malation); and iii) to evaluate the benefit of mixing DDGS with Pine wood (PINEW) in a hybrid LPC. Injection molded test specimens were evaluated for their tensile, flexural, impact, environmental durability (soaking responses), and thermal properties. All mechanical results from composites are compared to neat high-density polyethylene (HDPE) to determine their relative merits and drawbacks. HDPE composites composed of various percentage weights of fillers and either 0% or 5% by weight of maleate polyethylene (MAPE) were produced by twin screw compounding and injection molding. Chemical modification by acetylation and malation of DDGS and PW fillers prior to compounding was done to evaluate their potential in making an improved lignocellulosic material. Composite-DDGS/PINEW mixture blends composed of a majority of PINEW were superior to composites containing DDGS only. Composites containing MAPE had significantly improved tensile and flexural moduli compared to neat HDPE. Impact strength of all composites were significantly lower than neat HDPE. Chemical modification substantially improved the tensile, flexural, water absorbance, and thermal properties of the resultant composites compared to untreated composites. Differential scanning calorimeter and thermogravimetric analysis were conducted on the HDPE composites to evaluate their thermal properties as this may indicate processing limitations with conventional plastics processing equipment due to the exposure of the bio-material to elevated temperatures. Finally, because exposure to the moisture in the environment can affect the physical and color properties of wood, changes in the size and color of test specimens after prolonged soaking were evaluated.

Contact information: a) Functional Foods Research Unit, National Center for Agricultural Utilization Research, Agricultural Research Service, United States Department of Agriculture, 1815 N. University St., Peoria IL 61604, USA; b) Department of Technology, College of Applied Science and Technology, Illinois State University, Normal IL 61790, USA; c) Polymer Composites Research Group: Agricultural and Bio systems Engineering, College of Agriculture and Life Science, Iowa State University, Ames, IA, 50011, USA; d) Bio-oils Research Unit, National Center for Agricultural Utilization Research, Agricultural Research Service, United States Department of Agriculture, Peoria, IL 61604, USA; *Corresponding author: Brent.Tisserat@ars.usda.gov

13.1   INTRODUCTION

The U.S. wood plastic composite (WPC) industry is projected to increase 13% a year to amount to $5.3 billion by 2015 and is likely thereafter to continue to increase at a similar rate in the foreseeable future.[1,3] There is an ever increasing need to improve the quality of lignocellulosic plastic composite and WPC in order to obtain more useful, reliable and inexpensive commercial products.[4,10] The most common type of LPC is WPC, which uses wood flour fillers derived from wood waste materials such as shavings and sawdust generated from lumber processing.[6,10,12] WPC thermoplastics typically include polyethylene (PE), polypropylene (PP), and polystyrene and are mixed with up to 50% wood flour (w/w) depending on the desired mechanical and physical properties and industrial acceptance.[6,11,12] Cost is the most important consideration in the commercialization of any LPC/WPC products. Generally, wood flour fillers are used without elaborate chemical preparations; however, the fillers are sized (sieved) and dried to enhance their processing. The price of LPC is dictated by the price of petroleum and the cost of wood/lignocellulosic fillers. Currently, PE and PP sell for ≈$1.85 to $2.27/kg ($0.91 to $1.12/lb.) and ≈$2.23 to $2.47/kg ($1.10 to $1.22/lb.), respectively.[2,13] Commercial hardwood flour blends are derived from lumber milling byproducts is composed of various tree species (e.g., maple, birch, ash) and sells for ≈$0.18 to $0.48/kg ($0.08 to $0.22/lb.).[11] Wood waste material prices fluctuate on the basis of availability (housing demand) and the demand for their utilization.[14] For example, in 2006-2008, when the US housing market contracted, sawdust prices quadrupled due to a lack of supply.[14] Biomass energy usage competes with LPC/WPC filler availability and price. Currently, 85% of wood waste is consumed for energy production (fuel pellets and direct combustion).[15] The Energy Independence and Security Act of 2007 mandates that 36 billion gallons of biofuels be produced by 2022 and woody biomass materials will be increasingly used to achieve this goal.[16] A number of government subsidy programs are diverting the woody biomass into bio-energy facilities from their traditional markets. Changes in the cost, availability, and utilization of the biomass and wood waste markets are in flux.[16] As previously noted, since the demand for wood flour needed by the WPC industry will also increase and its cost will undoubtedly increase due to the bio-energy mandates, new sources of woody biomass are clearly needed.

Alternative woody biomass sources to provide wood flour are being developed.[7,17,19] Small-diameter trees obtained from forest under-stories or brush conditions offers a source of woody materials to satisfy both the bio-energy as well as wood flour for WPC.[17,18] Short-rotational woody crops using "fast-growing trees" grown in coppicing plantations are another option to obtain woody materials.[20] Marginal land utilization has been suggested as the potential site for planting large acreages of bio-energy woody tree crops.[5,20,21]

Paulownia elongate S. Y. Hu, family Paulowniaceae, a native to China, is an extremely fast-growing coppicing hardwood that is cultivated in plantations in China and Japan. Paulownia wood (PW) is highly valued in the construction and furniture industries.[22,23] There are several attributes of Paulownia wood that favor using it as a feedstock for WPC: a Paulownia plot containing 2000 trees per hectare can yield up to 150 to 300 tons of wood within 5 to 7 years, growth rate of heights up to 3.7–4.6 m and diameters of 3 to 5 cm a year are common, Paulownia trees are amenable to being established on marginal lands and have deep taproots, which make them drought resistant, PW is light weight, insect resistance, highly durable, and heat resistance. Paulownia species such as *P. elongate*, *P. kawakamii*, and *P. tomentosa*, are currently being grown and evaluated in the United States for their commercial wood properties.[23,24] For example, recent studies conducted at Fort Valley State University, Fort Valley, GA show that two to four-year-old trees can grow to a diameter of 16.5 cm and achieved a height of 10 m.[24] In addition, Paulownia could serve as a short-rotational woody crop that could be harvested frequently over a 10 year period. Therefore, in this study the utilization of juvenile wood materials harvested from 3 year old trees were used as a reinforcement materials with thermoplastic resins.

In many cases, lignocellulosic flour cost is less than wood flour; sometimes costing only a few cents a pound, thereby making it a very economically attractive material to be developed as a filler for LPC.[25,26] Ag-waste materials generated from processing seeds have not been vigorously exploited as possible fillers in bio-composites. In the U.S. Midwest, dried distillers grains and soluble (DDGS) offers an abundant, available and inexpensive lignocellulosic flour for biocomposites.[27,31] DDGS are processed corn seeds left over after the distillation of alcohol to generate the bio-based ethanol fuel.[31,32] Approximately, 25 million metric tons of DDGSs are produced annually in North America with this figure expected to increase further in the next few years.[29,31,33] Currently, DDGS is used almost entirely as an animal feed although other uses have been sought.[26,27,29] DDGS sells for about $0.06 to $0.10/kg ($0.03 to $0.05/lb.) which makes it an attractive bio-filler to blend with thermoplastic resins. However, to date, studies employing DDGS as a filler with thermoplastic resins have produced composites that have poor mechanical properties compared to the neat thermoplastic resin.[30,33,35] Further research is required to produce a DDGS material that has improved mechanical properties in order to become an acceptable filler material.

In addition, there are also numerous other seed residues (seed meals or press cakes) generated from seeds after their oil processing (soybean, cottonseed, pennycress). Roughly half of the oil seed's harvest mass remains as a press cake after oil extraction by pressing.[36] In 2012, 472 million tons of oil seeds were harvested globally to provide for culinary (e.g., edible oils and food additives) and industrial (e.g., soaps, cosmetics and biodiesel) products.[37] Many press cakes are used as an animal feed or fertilizer, when appropriate.[38] However, there is much interest in finding higher value uses for press cakes. Also, "new" oil energy-crops, such as jatropha (*Jatropha curcas* L.) and pennycress (*Thlaspi arvense* L.) containing even higher oil compositions than current oil seeds crops and are being developed to address the world's fuel needs.[36,39,40] In the U.S. Midwest, pennycress has a promising future as a bio-diesel crop. It contains more oil than soybeans and is unique in that it is a winter annual that can be grown on the same land used for soybeans without competition since their planting and harvesting dates do not coincide.[36,39,40] However, pennycress press cake cannot be used as animal feed since it contains high levels of toxic glucosinolates.[40] Therefore, alternative uses for pennycress press cakes are sought in order to maximize the utilization of this oil seed crop.[41] In the tropical and subtropical regions, jatropha is becoming a prominent bio-diesel crop. Likewise, its seed meal is also toxic due to the presence of phorbol esters and is not available to be used as an animal feed or fertilizer. Employment of these press cakes as a filler in LPC could be an ideal utilization. Press cakes price between a range of $0.09 to $0.55/kg ($0.04 to $0.25/lb.) depending on the species and extent of their preparation, which makes press cakes an attractive bio-filler to be blended with thermoplastic resins.[42] However, press cakes composition differs substantially from other lignocellulosic flour fillers because they contain high concentrations of extractives which includes residual vegetable oil (» 8–15%) and protein (» 20–35%) while having a low cellulose (» 11–25%) and lignin concentration (» 3–15%).[36] DDGS is composed of 25–33% protein, 39–60% carbohydrates, 5–12% oils and 2–9% ash.[33] In contrast, PW flour contains: water and solvent extractives (» 3–12%), protein content (» 1–2%), cellulose (45–50%), hemicellulose (22–25%) and lignin (20–25%).[4,43,44] Few published reports have dealt with using press cakes as a lignocellulosic flour filler.[45] Attempts to employ press cakes has resulted in composites with relatively poor mechanical properties when compared to neat thermoplastic resins.[45]

Chemical modification (acetylation and malation) of lignocellulosic and wood flour fillers is a common method to improve their physical and mechanical properties.[6,8,11,12,33,46,53] Chemical modification of a lignocellulosic material is defined as a chemical reaction between a reactive portion of the lignocellulosic material (hydroxyl group) and a chemical reagent, with or without a catalyst, to create an ester group.[6,8,11,12,33,46,52] Acetylation is the most common method to chemically modify lignocellulosic materials.[8,51,54,55] Acetylation offers a number of benefits to WPC/ LPC compared to nonacetylated WPC/LPC including superior weathering resistance,[8,53] greater thermal stability,[51] and enhanced mechanical properties.[56] Because there is no accepted method to administer acetylation and/or chemical modification treatments to lignocellulosic and wood flours there are a myriad of acetylation/malation techniques presented in the literature.[8,47,48,50,51,53,55,68] Generally, however, chemical medication by acetylation involves the treatment of lignocellulosic and wood flour particles by soaking or coating with a coupling agents (e.g., acetic anhydride) in order to reduce the presence of hydroxyl groups in exchange for esterification linkages (i.e., covalent bonds between the wood and the reagent).[8,47,53] In this study, chemical modifications were made on both Paulonia wood flour and DDGS prior to their blending with HDPE to determine if a chemical modification techniques could improve the mechanical properties of these lignocellulosic materials in the resulting composites.

There are three core objectives of this study. The first is to perform an assessment of the mechanical properties of thermoplastic composites made with DDGS. The methods developed to produce a usable DDGS composite that exhibits high mechanical properties can be transferred to the development of composites containing seed press cake residues from various species. This is a reasonable assumption due to the chemical compositional similarity between the DDGS and press cakes. Lignocellulosic materials are polar (hydrophilic) due to the occurrence of hydroxyl groups and are not compatible with thermoplastic resin polymers, which are nonpolar (hydrophobic). In order to obtain a LPC/WPC with superior physical and mechanical properties a coupling agent is often employed to aid in the binding of the lignocellulosic materials to thermoplastic resins.[50] A variety of different types of couplings agents are blended with LPC/WPC but maleate polyolefins are the most common due to their cost, performance and acceptability.[6,11,12,33,46,49,50,52] Since inclusion of a coupling agent is typical in WPC,[6,11,12] the effects of employing a commercial maleate polyethylene (MAPE) on the mechanical properties of HDPE-DDGS composites is included in this study. Residual oils in DDGS may adversely affect the performance of DDGS composites due to their lubricating effect. Therefore, a solvent extracted DDGS material was tested to assess the benefit of oil extraction.

In addition, DDGS was subjected to chemical modification treatments in order to obtain an improved DDGS composite.

The second core objective was to evaluate the mechanical, physical, and thermal properties of WPC obtained from blending Paulownia wood flour with high density polyethylene because there have been relatively few studies of the use of Paulownia wood (PW) as a fiber reinforcement for thermoplastics.[10,43,44,69] There is interest in using Paulownia wood flour derived from juvenile trees since small diameter short-rotation woody crop trees are likely to be a source of woody biomass needed by the US in the future. This study used PW flour derived from juvenile tree biomass (i.e., 36-month-old). The use of a maleate PE was employed as part of the scope of the project. Further, because chemical modification through acetylation or malation of filler materials may affect the performance of reinforcement, the mechanical and flexural properties of WPC derived from PW that had been acetylated and acetylated/maleate was examined.

The third core objective was an evaluation of the physical and mechanical properties of "mixed" composites composed of DDGS mixed with pine wood (PINEW) was conducted due to their relatively unique chemical make-up. The mechanical and flexural properties of these composites were benchmarked to formulations containing just PINEW or DDGS as well as to neat HDPE. The mechanical property outcomes were normalized to the control HDPE for ease of assessing the benefit of various filler treatments.

Because bio-composites are subject to degradation by water, water immersion tests were administered on tensile bars composites to evaluate their environmental durability. Weights, thickness, and mechanical properties were measured before and after the immersion tests. Finally, because bio-fiber materials are sensitive to heat exposure during processing, differential scanning calorimetry and thermogravimetric analysis were conducted on DDGS, PW, and PINEW composites to evaluate their thermal properties to assess any implications of processing on these materials.

13.2   EXPERIMENTAL

13.2.1   MATERIALS

The high-density polyethylene (HDPE) employed as the matrix material was Paxon BA50-120 (ExxonMobil Chemical Company, Houston, TX). It had a melt-flow index of 12 g/10 min, a density of 0.951 g/cm, and a melt temperature of 204°C. The binding agent was a polyethylene-graft-maleic anhydride, or maleate polyethylene (MAPE), supplied by Equistar Chemicals LP (product code NE542013). The MAPE had a melting point of 104–138°C with approximately 1% maleic anhydride by weight grafted on the polyethylene.

DDGS (corn-based meal) was obtained as the commercial animal feed pellet product (Archers Daniel Midland Co., Decatur, IL). White Pine (*Pinus strobus* L.)

wood (PINEW) material was obtained from packaged bedding shaving materials (American Wood Fiber, Schofield, WI). *Paulownia elongate* wood material was obtained from 36-mo-old trees grown in Fort Valley, GA. PW and PINEW shavings were milled successively through 4-, 2-, and then 1-mm screens with a Thomas-Wiley mill grinder, (Model 4, Thomas Scientific, Swedesboro, NJ). Particles were then sized through a Ro-Tap™ Shaker (Model RX-29, Tyler, Mentor OH) employing 203 mm diameter stainless steel screens. Screens employed were #10, #30, and #40 US Standards (Cole-Parmer/ ThermoFisher Scientific, Waltham, MA). PW consisted of ≤590 µm particles obtained from particles passing through the #30 mesh sieve and particles collected by the successive sieves were designated as #40 mesh and finer (≥#40), thereafter. DDGS materials were treated in the same manner as wood materials to produce a ≥#40 filler.

To examine what effect the oils in the DDGS may have on the mechanical properties of the composites, the vegetable oils were extracted from DDGS using hexane and then other polar extractables were extracted using acetone while employing a Soxhlet extractor. Throughout this paper DDGS refers to the original DDGS and STDDGS refers to the solvent treated DDGS. Extraction of polar components from PW was done using acetone employing a Soxhelt extractor to obtain solvent-treated PW (STPW). All materials were oven dried for 48 h at 100°C prior to extrusion.

13.2.2 PREPARATIONS AND PROCESSING

Table 13.1 summarizes the various treatments conducted in this research project. DDGS were modified by direct treatment with acetic anhydride and maleic anhydride solutions. Acetic anhydride and maleic anhydride were obtained from Sigma-Aldrich Chemical Company, St. Louis, MO and used as supplied without further purification. STDDGS and STPW particles were vacuum-oven dried for 24 hr at 80°C. Four hundred grams of filler were boiled in a stirred jacketed reaction vessel fitted with a distillation trap at 90°C containing 6 M acetic anhydride/acetone mixture for 24 hrs. In addition, a 5 M acetic anhydride /1 M maleic anhydride/ acetone reaction mixture was also employed. Following incubation, filler materials were filtered, washed three times with acetone, and vacuum-oven dried for 48 hr at 80°C. Hereafter, STDDGS and STPW treated with acetic anhydride (A) or acetic anhydride/maleic anhydride (AM) mixtures will be designated as STDDGS/A and STDDGS/AM, respectively. Weight percentage gains were calculated for the A and AM mixtures to be ~11 and ~12%, respectively.

TABLE 13.1 Weight Percentages in Test Formulations

Composition	HDPE	MAPE	DDGS	STDDGS	PINEW	PW	STPW	Modifiers
HDPE	100	–	–	–	–	–	–	–
HDPE-MAPE	95	5	–	–	–	–	–	–

TABLE 13.1   *(Continued)*

Composition	HDPE	MAPE	DDGS	STDDGS	PINEW	PW	STPW	Modifiers
HDPE-25DDGS	75	–	25	–	–	–	–	–
HDPE-25DDGS-MAPE	70	5	25	–	–	–	–	–
HDPE-25STDDGS	75	–	–	25	–	–	–	–
HDPE-25STDDGS-MAPE	70	5	–	25	–	–	–	–
HDPE-25STDDGS/A	75	–	–	25	–	–	–	A
HDPE-25STDDGS/A-MAPE	70	5	–	25	–	–	–	A
HDPE-25STDDGS/AM	75	–	–	25	–	–	–	AM
HDPE-25STDDGS/AM-MAPE	70	5	–	25	–	–	–	AM
HDPE-25PINEW	75	–	–	–	25	–	–	–
HDPE-25PINEW-MAPE	70	5	–	–	25	–	–	–
HDPE-12.5STDDGS/12.5PINEW	75	–	–	12.5	12.5	–	–	–
HDPE-12.5STDDGS/12.5PINEW-MAPE	70	5	–	12.5	12.5	–	–	–
HDPE-10STDDGS/30PINEW	60	–	–	10	30	–	–	–
HDPE-40PINEW	60	–	–	–	40	–	–	–
HDPE-25PW	75	–	–	–	–	25	–	–
HDPE-25PW-MAPE	70	5	–	–	–	25	–	–
HDPE-25STPW	75	–	–	–	–	–	25	–
HDPE-25STPW-MAPE	70	5	–	–	–	–	25	–
HDPE-25STPW/A	75	–	–	–	–	–	25	A
HDPE-25STPW/A-MAPE	70	5	–	–	–	–	25	A
HDPE-25STPW/AM	75	–	–	–	–	–	25	AM
HDPE-25STPW/AM-MAPE	70	5	–	–	–	–	25	AM

To investigate the influence of mixing different fillers to produce an improved composite, STDDGS was mixed with PINEW at various concentrations with and without presence of a 5% maleic anhydride coupling agent (MAPE) (Table 13.1).

The influence of the presence or absence maleic anhydride coupling agent on the physical properties of HDPE-filler blends was also investigated (Table 13.1).

Composite blends were extruded with a 27 mm corotating intermeshing twin-screw extruder, with a length/diameter ratio of 40 (Model ZSE-27 American Leistritz Extruder Corporation, Branchburg, NJ). The barrel had ten different zones, each 90 mm long, which were controlled at the following temperatures (°C): 100, 160, 170, 190, 200, 200, 210, 210, 205, and 205, respectively. The cord die temperature was set at 200°C. Premixed fillers and HDPE were dry blended in 1 gallon-resalable plastic bags. Materials were then transferred into a single drive feeder (Flex-Tuff Model 306, Schenck/AccuRate, Whitewater, WI) and fed into the extrusion feeder at the rate of 100 g/min. Extruder screw speed was set at 100 rpm. Extruded strands were cooled by immersion in a water bath and then pelletized with a strand pelletizer (Model 60E, Automatick Plastics Machinery GMbH, Grossotheim, Germany).

Molding was conducted with a 30-ton molding machine (Engel ES 30, Engel Machinery Inc., York, PA) with set point temperatures (°C) for the four zone injection molding barrel set at: feed = 160; compression = 166; metering = 177, and nozzle = 191. The mold temperature was 37 °C. An ASTM test specimen mold was used that included cavities for a ASTM D790 flexural tensile bar (12.7 mm W × 127 mm L × 3.2 mm thickness) and an ASTM D638 Type I tensile bar (19 mm W grip area × 12.7 mm neck × 165 mm L × 3.2 mm thickness × 50 mm gage L). Impact specimen bars were obtained by cutting the flexural specimens in half to 12.7 mm W × 64 mm L × 3.2 mm thickness and notched. The Type I bars were used for the tensile strength property tests. The flexural bars were used to evaluate flexural properties and also used to make impact strength measurements. The Type I bars were used to evaluate changes due to prolonged exposure to water: weight change, color change, and changes in tensile mechanical properties of the composites.

13.2.3   MECHANICAL PROPERTY MEASUREMENTS

Injection molded specimens, ASTM D638 Type I tensile bars, were tested for tensile modulus and strength using a universal testing machine (UTM), Instron Model 1122 (Instron Corporation, Norwood, MA). The speed of testing was 50 mm/min, which corresponds to a strain rate of 1 mm/mm/min at the start of the test. Specimen thickness was measured with a digital micrometer, Model 49–63 (Testing Machines Inc., Amityville, NY). Initial samples (dry) were conditioned for approximately 240 h at standard room temperature and humidity (23°C and 50% RH) prior to any test evaluations.

Three point bending flexural tests were completed according to ASTM-D790 on an Instron UTM Model 1122. The flexural tests were carried out using Procedure B with a crosshead rate of 13.5 mm/min, which corresponds to a rate of straining of the outer fiber equal to 0.1 mm/mm/min. The maximum flexural stress (flexural strength, σ_{fm}) and modulus of elasticity in bending (E_b) were calculated using the following formulas:

$$\sigma_{fm} = 3PL/2bd^2 \tag{1}$$

$$E_b = L^3 m/4bd^3 \tag{2}$$

where P is the maximum applied load, L is the length of support span, m is the slope of the tangent, and b and d are the width and thickness of the specimen bars, respectively. Five specimens of each formulation were tested. The average values and standard errors were reported.

Notched impact tests were conducted with an IZOD impact tester, Model Resil 5.5, P/N 6844.000 (CEAST, Pianezza, Italy) conformed to ASTM D256–84. Specimen bars were obtained by cutting the flexural specimens in half to 12.7 mm W × 64 mm L × 3.2 mm thickness and then notched.

13.2.4   FOURIER TRANSFORM-INFRARED SPECTROSCOPY (FT-IR)

FT-IR spectra were acquired in the attenuated total reflectance (ATR) mode at 4 cm^{-1} resolution on a Thermo Scientific Nicolet 6700 FTIR spectrometer. Samples were analyzed using the KBr pellet method. Thirty-two scans per run were averaged to capture the mid-IR absorbance between 650 cm^{-1} and 4000 cm^{-1}. The spectra were corrected for baseline shift and spectral noise.

13.2.5   WATER ABSORPTION AND THICKNESS SWELLING

The Type I tensile bars injection molded for each composite were dried in an oven for 24 h at $100 \pm 2°C$ and weighed. Tests were conducted in an incubator at $25 \pm 2°C$ under a photosynthetic photon flux density of 180 $\mu mol\,m^2\,s^{-1}$ using a photoperiod of 12 h light/12 h dark. Tensile bars were placed in distilled water at room temperature for 872 h. At predetermined time intervals the specimens were removed from the distilled water, the surface water was blotted off with paper towels, and their wet masses were determined. Water absorption, measured as weight gain percentage, was computed using the following formula,

$$\text{Weight gain (\%)} = (m_t - m_o)/m_o \times 100 \tag{3}$$

where m_o denotes the oven-dried weight and m_t denotes the weight after soak time t.

Thickness swelling (TS) was calculated using the following formula,

$$TS = (T_f - T_o)/T_o \times 100 \tag{4}$$

where *TS* is the thickness swelling (%) at time t, T_o is the initial thickness of the specimen, and T_f is the thickness at end of soak time *t* (872 h).

13.2.6 THERMAL PROPERTIES

Differential scanning calorimeter (DSC) of molded specimens was conducted with an Auto DSC-7 calorimeter with a TAC/DX controller (TA Instruments, New Castle, DE). Samples of 5–7 mg were weighed and sealed hermetically in aluminum DSC pans. First, the calorimeter was programmed to increase the temperature from 0 to 180°C at a rate of 10°C/min, kept isothermal for 3 min. Second, the samples were cooled to –50°C at a rate of 10°C/min. Finally, the samples were heated to 180°C from –50 to 180°C at the same rate. Data from the second heating cycle were used to determine the melting temperature (T_m) and enthalpy of melting (DH_m) for all composites. The heat flow rate corresponding to the crystallization of HDPE in composites was corrected for the content of the wood flour and MAPE. The value of crystallization heat was also corrected for the crystallization heat of MAPE. The crystallinity level (χ_c) of the HDPE matrix was evaluated from the following relationship:[12]

$$\chi_c = (DH_{exp}/DH_m) \times (100/W_f) \times 100 \qquad (5)$$

where DH_{exp} is the experimental heat of fusion (DH_m) or crystallization determined by DSC, DH is the assumed heat of fusion or crystallization of fully crystalline HDPE (293 J/g), and W_f is the weight fractions of HDPE in the composites.

Thermogravimetric analysis (TGA) was performed to determine the thermal characteristics of the composites. TGA was conducted using a Model 2050 TGA (TA Instruments) under nitrogen at a scan rate of 10°C/min from room temperature to 600°C. A sample of »7.5 mg was used for each run. Data was analyzed using the TA Advantage Specialty Library software (TA Instruments). The derivative TGA (wt.%/min) of each sample was obtained from the software.

13.2.7 STATISTICAL ANALYSIS

Experimental data obtained was analyzed statistically by analysis of variance for statistical significance and multiple comparisons of means were accomplished with Duncan's Multiple Range Test ($p £ 0.05$).

13.3 RESULTS AND DISCUSSION

13.3.1 FT-IR SPECTRAL INTERPRETATION

FT-IR is a common means to evaluate the chemical modification of lignocellulosic/wood fibers.[8,48,55,63] Because the effect of solvent extraction and malation are a critical aspect of this study, the FT-IR spectra are shown first, refer to Fig. 13.1.

The FT-IR of the untreated DDGS (Original DDGS) as shown in Fig. 1 indicates the following absorbance modes intensities cm^{-1}: 3377 (very strong), 3012 (weak-medium), 2926 (medium-strong), 2857 (medium), 1734 (medium), 1653 (strong), 1521(medium), 1451 (medium), 1374 (medium), 1238 (medium), and 1041 (very strong). The 3377 cm^{-1} band is a composite of the protein N-H and the O-H of the carbohydrate moieties present; whereas the small 3012 cm^{-1} band is that of the olefinic -C=C-H stretching mode of the oil content. The 2926 and 2857 cm^{-1} peaks are, respectively the alkyl groups (-CH$_2$-, -CH$_3$), whereas the 1734 cm^{-1} and 1238 cm^{-1} absorption bands are mainly those of the oil component although both the carbohydrate and protein components have a small contribution to the 2926 and 2857 cm^{-1} bands as well. The 3377 and 1041 cm^{-1} bands are mainly protein and carbohydrate contributions to the spectrum.

In the solvent (hexane/acetone) extracted-DDGS material (STDDGS) the intensity of the carbonyl band of the oil (1736 cm^{-1}) decreased from 0.25 to 0.15 absorbance units as most of the oil was removed (Fig. 13.1). Following acetylation (STDDGS/A) two bands 1743 and 1235 cm^{-1} were boosted in intensity. The ester -C=O (1743 cm^{-1}) and the ester -C-C=O stretch (1235 cm^{-1}) are observed. A competition between the malation-acetylation reaction (STDDGS/AM) resulted in an isolated product whose infrared spectrum gave a broadened 1730 cm^{-1} band accounting for the ester carbonyl.

The IR spectrum of the unmodified PW shown in Fig. 13.1 gives a characteristically strong OH band around 3419 cm^{-1} and the -CH$_2$O- band at 1047 cm$^{-1,}$ respectively. Also present is an unresolved band around 2928 cm^{-1} for the alkyls (-CH$_3$, -CH$_2$-) modes. There are also two reasonably sharp medium intensity bands at 1739 and at 1239 cm^{-1} that indicate carbonyl absorption bands betraying the presence of some oil in the wood. In addition, a pair of bands at 1622 and 1505 cm^{-1} occur that could represent the presence of some protein component in the wood.

The solvent (hexane/acetone) extracted PW (STPW) did not seem to have released all its oil components since the its spectrum displayed the same intensity in the 1739 and 1238 cm^{-1} bands as the PW spectrum shown in Fig. 13.1. A noticeable change occurred in the amide I band of the protein which was attenuated. Upon acetylation (STPW/A), two noticeable changes were evident: the 3440 cm^{-1} OH band was truncated from about 0.85 absorbance units to 0.6 absorbance units. The major evidence following acetylation was the increase in intensity of the carbonyl band at 1744 cm^{-1} from 0.24 absorbance units in the unmodified to 0.55 units in the acetylated product. Following this was the corresponding increased intensity of the 1238 cm^{-1} band (-C-C=O) from 0.4 to 0.62 absorbance units, whereas the 1047 cm^{-1} band remained unchanged. The acetylated/maleate spectrum (STPW/AM) closely mimics the STPW/A spectrum.

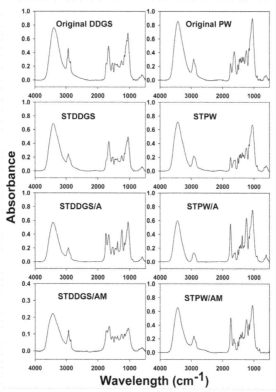

FIGURE 13.1   FT-IR spectra of DDGS and PW: original, solvent treated, and chemically modified.

13.3.2   MECHANICAL PROPERTIES OF CHEMICALLY MODIFIED DDGS AND PW FORMULATIONS

The tensile properties of tensile strength (σ_U), Young's modulus (E), and elongation strain at break (%El) of the HDPE-DDGS composites containing various composites are shown in Table 13.2. The flexural strength (σ_{fm}), the flexural modulus or modulus of elasticity in bending (E_b), and the notched IZOD impact strength for the various composites are presented in Table 13.3. The average for the five test specimens and their standard error is given for each property. Figures 13.2–13.4 graphically summarize the data in Tables 13.2 and 13.3 by normalizing the outcomes to the HDPE control material. For example, the tensile strength of HDPE-MAPE is 96% of the neat HDPE thus the bar graph of the normalized σ_U for HDPE-MAPE is 96%. This rendering is employed to clearly illustrate the effect of additives.

TABLE 13.2   Tensile Properties of HDPE and Composites*

Composition	σ_U (MPa)	E (MPa)	EI% (%)
HDPE	35.3 ± 0.8a	361 ± 24a	57.7 ± 6.1a
HDPE-MAPE	33.0 ± 0.3b	362 ± 4b	58.9 ± 0.2a
HDPE-25DDGS	25.2 ± 0.6c	432 ± 19c	43.2 ± 1.0b
HDPE-25DDGS-MAPE	29.1 ± 0.1de	524 ± 8d	28.4 ± 0.8c
HDPE-25STDDGS	25.6 ± 0.1c	435 ± 22c	34.5 ± 1.5d
HDPE-25STDDGS-MAPE	32.1 ± 0.5f	562 ± 31d	21.2 ± 0.8e
HDPE-25STDDGS/A	28.1 ± 0.2e	590 ± 5e	20.7 ± 0.4e
HDPE-25STDDGS/A-MAPE	32.2 ± 0.4f	570 ± 7e	20.8 ± 0.4e
HDPE-25STDDGS/AM	30.4 ± 0.2d	578 ± 8e	17.7 ± 0.2e
HDPE-25STDDGS/AM-MAPE	30.6 ± 0.4d	566 ± 13e	20.4 ± 0.6e
HDPE-25PINEW	32.0 ± 0.8b	871 ± 16 g	14.9 ± 0.5fg
HDPE-25PINEW-MAPE	38.1 ± 0.2h	714 ± 6h	20.0 ± 1.2h
HDPE-12.5STDDGS/12.5PINEW	26.9 ± 0.2i	719 ± 30h	17.8 ± 0.6fh
HDPE-12.5STDDGS/12.5PINEW-MAPE	29.6 ± 0.1j	503 ± 12d	25.0 ± 0.9h
HDPE-10STDDGS/30PINEW	31.0 ± 0.3 g	853 ± 16 g	11.1 ± 0.2 g
HDPE-40PINEW	25.0 ± 0.4k	962 ± 36i	6.9 ± 1.4 g
HDPE-25PW	35.2 ± 0.3a	881 ± 14j	13.4 ± 0.3b
HDPE-25PW-MAPE	40.7 ± 0.5cd	930 ± 18i	14.2 ± 1.3b
HDPE-25STPW	33.7 ± 0.7ab	911 ± 12i	11.8 ± 0.8b
HDPE-25STPW-MAPE	41.6 ± 0.6c	931 ± 40i	12.8 ± 0.4bc
HDPE-25STPW/A	35.3 ± 0.2e	853 ± 17 g	9.8 ± 0.3b
HDPE-25STPW/A-MAPE	41.2 ± 0.6ce	836 ± 27 g	14.8 ± 1.0b
HDPE-25STPW/AM	39.2 ± 0.3d	850 ± 3 g	15.8 ± 0.7b
HDPE-25STPW/AM-MAPE	38.8 ± 0.2d	817 ± 3k	16.9 ± 0.3bd

*Treatment values with different letters in the same column were significant ($p £ 0.05$). Means and standard errors derived from five different replicates are presented.

TABLE 13.3   Flexural and Impact Properties of HDPE and Composites*

Composition	E_b (MPA)	σ_{fm} (MPa)	Impact Energy (J/m)
HDPE	41.4 ± 0.2a	1169 ± 8a	921.8 ± 1.6a
HDPE-MAPE	40.0 ± 0.1b	1125 ± 5a	924.4 ± 1.3a
HDPE-25DDGS	33.7 ± 0.6c	1272 ± 31b	447.8 ± 2.4b

TABLE 13.3   *(Continued)*

Composition	E_b (MPA)	σ_{fm} (MPa)	Impact Energy (J/m)
HDPE-25DDGS-MAPE	38.7 ± 0.3b	1326 ± 8c	168.7 ± 1.2c
HDPE-25STDDGS	40.3 ± 0.2b	1609 ± 21d	272.3 ± 2.8d
HDPE-25STDDGS-MAPE	44.6 ± 0.2d	1531 ± 5e	99.2 ± 0.2e
HDPE-25STDDGS/A	39.8 ± 0.3e	1489 ± 12ef	203.6 ± 0.5f
HDPE-25STDDGS/A-MAPE	43.5 ± 0.4f	1451 ± 22fg	88.1 ± 0.3 g
HDPE-25STDDGS/AM	40.9 ± 0.3b	1440 ± 9 g	182.2 ± 0.3h
HDPE-25STDDGS/AM-MAPE	42.6 ± 0.3 g	1431 ± 11 g	102.7 ± 0.3i
HDPE-25PINEW	45.2 ± 0.3h	1966 ± 19h	109.4 ± 0.3i
HDPE-25PINEW-MAPE	48.7 ± 0.5i	1782 ± 29i	98.4 ± 0.2e
HDPE-12.5STDDGS/12.5PINEW	43.0 ±0.2f	1848 ± 12j	168.3 ± 0.3c
HDPE-12.5STDDGS/12.5PINEW-MAPE	42.7 ± 0.3 g	1818 ± 20j	154.4 ± 0.5j
HDPE-10STDDGS/30PINEW	47.8 ± 0.2h	2447 ± 29k	80.7 ± 0.1 g
HDPE-40PINEW	45.0 ± 0.6h	2994 ± 38l	69.6 ± 0.1k
HDPE-25PW	53.6 ± 0.6j	2224 ± 30 m	95.9 ± 0.1e
HDPE-25PW-MAPE	54.7 ± 0.3j	2172 ± 26n	101.9 ± 0.2i
HDPE-25STPW	54.3 ± 0.5j	2398 ± 21o	103.3 ± 0.2i
HDPE-25STPW-MAPE	57.0 ± 0.2k	2241 ± 25 m	108.3 ± 0.3i
HDPE-25STPW/A	52.9 ± 0.3l	2168 ± 19n	92.6 ± 0.3l
HDPE-25STPW/A-MAPE	55.5 ± 0.4 m	2143 ± 22n	91.4 ± 0.1l
HDPE-25STPW/AM	53.6 ± 0.5j	2135 ± 28n	91.4 ± 0.2l
HDPE-25STPW/AM-MAPE	53.2 ± 0.3j	2104 ± 19p	83.2 ± 0.1 g

*Treatment values with different letters in the same column were significant (*p* £ 0.05). Means and standard errors derived from five different replicates are presented.

All biocomposites containing DDGS exhibited much lower tensile strength but comparable modulus values compared to the neat HDPE or the HDPE-MAPE formulations. Refer to Table 13.2 and Fig. 13.2. The %El values were considerably higher in the unextracted DDGS formulation (HDPE-25DDGS) compared to the STDDGS formulation (HDPE-25STDDGS), which is attributed to the presence of residual oils in this composite (HDPE-25DDGS) which acts as a plasticizing agent interacting with the filler and the resin matrix as shown in Table 13.2 and Fig. 13.2. Similarly, Julson et al.,[25] reported the poor mechanical performance of PP- and HDPE-DDGS composites when compared to neat PP or HDPE. To improve the mechanical properties of the HDPE-DDGS composites the coupling agent MAPE was included in the formulations. Adding 5% MAPE to the DDGS composite formulation (HDPE-25DDGS-MAPE and HDPE-25STDDGS-MAPE) resulted in a slight increase in σ_U but a nominal reduction in E values compared to the corresponding DDGS composites without MAPE (HDPE-25DDGS and HDPE-25STDDGS). Refer to Table 13.2 and Fig. 13.2.

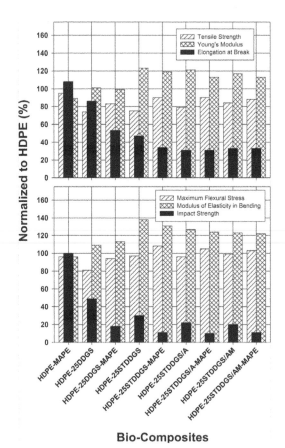

Bio-Composites

FIGURE 13.2   Effect of additives on the tensile and flexural properties of DDGS composites when compared to the control material HDPE.

Unlike in a previous study[30] where a marked increase in the tensile strength was observed when using STDDGS versus DDGS in composites (without MAPE), this study showed only a slight improvement in the tensile strength when using STD-DGS. Although, the modulus significantly increases when using the solvent treated DDGS as shown in Fig. 13.2. However, when MAPE was added to the STDDGS composite (HDPE-25STDDGS-MAPE) verses original untreated-DDGS formulations (HDPE-DDGS or HDPE-DDGS-MAPE) resulted in significantly higher σ_U values than all other DDGS formulations. In addition, the HDPE-25STDDGS-MAPE formulation compared favorably to the neat HDPE and HDPE-MAPE formulations. These results shows the importance not extrapolating and overgeneralizing the results conducted in prior studies to current studies.[30] In this study, a high melting HDPE, Paxon BA50-120 with a melting temperature of 204°C was employed; in the previous study a much lower melting HDPE, Petrothene LS 5300-00 with a melting temperature of 129°C was employed. In the previous study, the HDPE-STDDGS-MAPE formulations exhibited a significantly higher tensile strength than neat HDPE[30] while in this study the HDPE-25STDDGS-MAPE formulation exhibited a σ_U that was slightly lower than the neat HDPE. See Table 13.2 and Fig. 13.2. Nevertheless, similar trends were found between the two studies employing dissimilar HDPE resins.

Removal of extractables in order to obtain a superior filler has been previously documented.[7,30,41,49] It is notable that the HDPE-25DDGS formulation exhibited inferior mechanical (σ_U and E) and flexural (σ_{fm} and E_b) values compared to the STD-DGS formulations but had significantly higher %El and impact strength values than other formulations. Refer to Tables 2 and 3 and Fig. 13.2. DDGS contains high levels of crude protein ($\approx 26\%$), water ($\gg 5.5\%$), hexane extracted oils ($\approx 14\%$), and acetone extractables ($\approx 3\%$). The solvent extraction treatment removes oils and polar extractables to obtain the modified DDGS filler (STDDGS). Apparently, the oil and extractables in the DDGS composite formulations interacted with the resin matrix acting as plasticizing agents which allow for greater percentage of elongation at break and impact strength. Conversely, they are responsible for the lower σ_U, E, σ_{fm}, and E_b values in the composites. Adding 5% MAPE to the solvent treated DDGS formulations results in lower impact strength but higher σ_{fm} and a slight reduction in moduli compared to formulations without MAPE. PW formulations were found to exhibit similar trends in mechanical properties as the DDGS formulations previously discussed. Refer to Table 13.2 and Fig. 13.2.

Chemical modification of STDDGS particles through acetylation (A) or acetylation/malation (AM) prior to blending with HDPE produced HDPE-25STDDG/A and HDPE-25STDDGS/AM formulations. Surprisingly, these formulations exhibited lower tensile and flexural moduli, lower flexural strength, and only a modest increase in tensile strength compared to the untreated formulation (HDPE-25STD-DGS). See Tables 13.2 and 13.3 and Fig. 13.2. Further, the %El and impact strength values declined in the acetylated and acetylated/maleate formulations compared to the untreated formulation; refer to Tables 13.2 and 13.3 and Fig. 13.2. The addition of MAPE to the chemically modified formulations (HDPE-25STDDGS/A-MAPE and HDPE-25STDDGS/AM-MAPE) had little effect on changing the mechanical properties compared to the formulations without MAPE.

Although the improvements due to the chemical modifications are small, they exhibit an expected trend. The tensile strength of the HDPE-25STDDGS composite was 73% of neat HDPE and that of the HDPE-STDDGS/A was 80%. When MAPE is added to these formulations, the σ_U of HDPE-25STDDGS-MAPE composite was 91% of the neat HDPE and the HDPE-25STDDGS/A-MAPE was 91%, respectively. These results indicate the improvement is due to the degree of esterification of the hydroxyl groups by the chemical modification treatments, which were further esterified by the presence of the MAPE coupling agent. The net result is an increase mechanical properties.

Solvent treatment of the PW flour to produce the STPW composites (HDPE-STPW and HDPE-STPW-MAPE) had little effect on their tensile, flexural and impact strength properties compared to the nonsolvent treated PW composites (HDPE-PW and HDPE-PW-MAPE). See Tables 13.2 and 13.3, and Fig. 13.3. Apparently the extractables in this wood were not as critical factors affecting the mechanical properties of the composites as in the DDGS formulations.

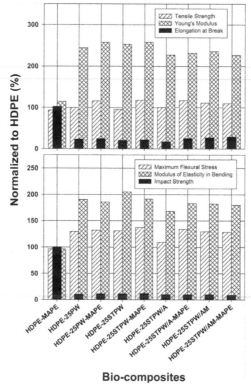

Bio-composites

FIGURE 13.3   Effect of additives on the tensile and flexural properties of PW composites when compared to the control material HDPE.

Chemically modified PW formulations (HDPE-25STPW/A, HDPE-25STPW/A-MAPE, HDPE-25STPW/AM, and HDPE-25STPW/AM-MAPE) exhibited similar trends to those seen for the DDGS formulations. Adding MAPE to the formulation had little effect on changing the mechanical properties of the chemically modified filler composites. Refer to Tables 13.2 and 13.3, and Fig. 13.3.

Despite the numerous publications dealing with the chemical modification (acetylation) of wood fiber (WF) and lignocellulosic materials in the literature, few mechanical evaluations have been conducted on the resultant biocomposites containing chemically modified wood or lignocellulosic fibers.[8,48,51,56,61,65] In addition, when the mechanical properties have been analyzed on chemically modified composites the results are rather modest or even negative.[65] For example, Ichach and Clemons[8] reported that HDPE-acetylated pine WF composites exhibited σ_{fm} and E_b values were −26 and −16%, respectively, compared to HDPE-untreated pine WF composites. However, the acetylated formulation retained its flexural properties better following weathering and fungal treatments when compared to the untreated HDPE-WF composite.[8] In another study, HDPE-acetylated-WF composites exhibited σ_U, E, %El, σ_{fm}, and E_b values of +12, –22, +8, +6, and –9%, respectively, compared to HDPE-untreated-WF composites.[51] Kaci et al.,[48] reported that maleic anhydride acetylated-low density PE (LDPE)-olive husk flour composites exhibited σ_U, E, %El valves of +9, –27 and +15%, respectively, compared to LDPE untreated- olive husk flour composites. Müller et al.,[61] reported that Polyvinyl chloride (PVC)-acetylated-WF composites exhibited σ_U, %El and impact strength values of +19, +22 and +18%, respectively, compared to PVC-untreated-WF composites. Previous reports find that acetylation of WF slightly benefits σ_U but reduces the tensile and flexural moduli in composites compared to untreated-WF composites. This study confirmed this trend with the chemically modified PW and DDGS composites (Table 13.2). The acetylation and malation of solvent treated DDGS (HDPE-STDDG/A and HDPE-STDDGS/AM) exhibited slightly higher tensile strength and flexural strength values but significantly lower tensile and flexural moduli values compared to the untreated composites (HDPE-STDDGS) as shown in Table 13.2 and Fig. 13.2. When the malation is provided by the matrix, using the MAPE coupling agent, there is a slight improvement in the mechanical strength properties but a slight reduction in the mechanical moduli values than when chemical modification is done to the fillers. This is seen by comparing HDPE-STDDGS-MAPE to HDPE-STDDGS as shown in Fig. 13.2. Impact strength of DDGS formulations were negatively affected by the malation (HDPE-STDDGS-MAPE) and chemical modification (HDPE-STDDGS/A and HDPE-STDDGS/AM) treatments compared to the untreated control (HDPE-STDDGS). See Table 13.3 and Fig. 13.2. Similar flexural results are mimicked with the maleate (HDPE-STPW-MAPE) and chemically modified (HDPE-STPW/A and HDPE-STPW/AM) PW composites compared to the untreated control PW composite (HDPE-STPW). Inclusion of the coupling agent (MAPE) with the chemically modified DDGS formulations did improve the modulus of rupture or modulus of elasticity and could significantly decreased the impact strength values compared to chemically modified formulations without MAPE. Clearly, chemical modification of the two fillers has both positive and negative effects of the mechanical properties to the resulting composites.

13.3.3   MECHANICAL PROPERTIES OF DDGS/PINEW BLENDS

Since DDGS sells for around $0.03 to $0.05/lb. and PINEW flour sells for $0.08 to $0.22/lb. there appears to be a case for combining the two ingredients to obtain a "mixed" DDGS composite and accessing the mechanical properties of the resulting PINEW/DDGS composites. Therefore, it is the contention of this study to mix these two chemically dissimilar fillers (DDGS and PINEW) together in order to determine if an enhancement of the lower-grade filler (i.e., DDGS) can be achieve by the partial mixing with higher-grade filler (PINEW). Pine wood flour was selected to be employed as the wood of choice in the DDGS mixture filler study due to its common usage in WPC.[8,11,63] PINEW formulations (HDPE-25PINEW and HDPE-25PINEW-MAPE) exhibits mechanical properties comparable or better to neat HDPE values except for %El values (Table 13.2 and 13.3; Fig. 13.4). For example, the HDPE-25PINEW formulation exhibited σ_U, E, and %El values that were −9, +141 and −74%, respectively, that of neat HDPE. Similarly, the HDPE-25PINEW-MAPE formulation exhibited σ_U, E, and %El values that were +8, +98 and −65%, respectively, that of neat HDPE (Table 13.2). Similarly, the flexural and impact strength properties of PINEW formulations were comparable or superior to neat HDPE except for impact strength values. For example, the HDPE-25PINEW formulation exhibited $\sigma_{fm,}$ E_b and impact strength values that were +9, +68 and −88%, respectively, that of neat HDPE. The HDPE-25PINEW-MAPE formulation exhibited $\sigma_{fm,}$ E_b and impact strength values that were +18, +52 and −89%, respectively, that of neat HDPE (Table 13.3). The mechanical properties of DDGS formulations have been discussed previously. The PINEW formulations were superior to the DDGS formulations in several mechanical properties.

Mixing PINEW and STDDGS fillers in equal proportions resulted in a "combination" composite that manifested somewhat different tensile, flexural and impact strength properties than that of composites composed of the individual ingredient fillers. Refer to Tables 13.2 and 13.3 and Fig. 13.4. For example, the HDPE-12.5STDDGS/12.5PINEW formulation exhibits σ_U values that were significantly less than in HDPE-25PINEW but slightly higher than in HDPE-25STDDGS. The E values of HDPE-12.5STDDGS/12.5PINEW were significantly higher than HDPE-25STDDGS but less than HDPE-25PINEW. Percent elongation values of the HDPE-12.5STDDGS/12.5PINEW formulation were significantly lower than either of the single filler composite formulations. The flexural values, σ_{fm} and E_b of HDPE-12.5STDDGS/12.5PINEW were lower than HDPE-25PINEW but higher than the HDPE-STDDGS composites. The impact strength of HDPE-

12.5STDDGS/12.5PINEW composite was significantly higher than the HDPE-25PINEW but likewise was significantly lower than the HDPE-25STDDGS composites. These trends were mimicked when the "mixed" composites contained MAPE. See Tables 13.2 and 13.3, and Fig. 13.4. Interestingly, the HDPE-12.5STDDGS/12.5PINEW-MAPE composite exhibited higher impact strength values than in the HDPE-25STDDGS-MAPE composites. Inclusion of MAPE in the "mixed" composite formulation caused a decrease in impact strength compared to composite formulations without MAPE. Obviously, some benefits and drawbacks in terms of the mechanical properties of the "mixture" composite was obtained by mixing STDDGS with PINEW over that of composites containing a single filler ingredient. To summarize, comparing the HDPE-25STDDG to HDPE-12.5STDDGS/12.5PINEW for σ_U, E, %El, $\sigma_{fm,}$ E_b, and impact strength values the following changes occurred: +5, +65, –49, +7, +15, and –40%, respectively. When comparing the HDPE-STDDG-MAPE to the HDPE-12.5STDDGS/12.5PINEW-MAPE for σ_U, E, %El, $\sigma_{fm,}$ E_b, and impact strength values the following changes occurred: +16, +15, –28, +6, +13, and –43%, respectively.

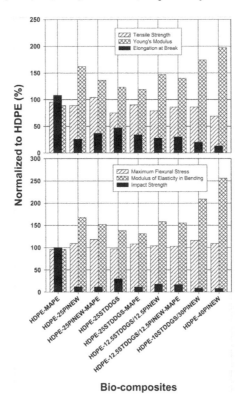

Bio-composites

FIGURE 13.4 Effect of additives on the tensile and flexural properties of PINEW-DDGS composites when compared to the control material HDPE.

When comparing the two 40% filler composites, HDPE-40PINEW and HDPE-10STDDGS/30PINEW, dissimilar mechanical and flexural properties also occurred between them. Refer to Tables 13.2 and 13.3, and Fig. 13.4. The difference between the HDPE-10STDDGS/30PINEW to the HDPE-40PINEW for σ_U, E, %El, σ_{fm}, E_b, and impact strength values were +24, –11, +60, +6, –18, and +16%, respectively. Obviously, some benefits and drawbacks were obtained comparing these "mixed" formulations. These results suggest that further work needs to be conducted to identify and maximize the merits of mixing fillers from dissimilar sources in order to obtain novel composites. This report is a preliminary evaluation employing this line of research but suggests that a useful inexpensive LPC composed of mixing DDGS and WF is feasible.

13.3.4 WATER ABSORPTION, THICKNESS CHANGES AND SPECTROPHOTOMETRIC RESPONSES

Water absorption is due to the hydrogen bonding of water molecules to the hydroxyl groups on the cells walls of the wood or lignocellulosic fibers.[48,70] The long-term water absorption as a function of time for the various LPCs at room temperature is shown in Fig. 13.5. All composites tested absorbed water during the incubation period and no distinct saturation levels were achieved after 872 h of soaking (Fig. 13.5). The HDPE and HDPE-MAPE samples exhibited inconsequential weight gains (i.e., less than a 1% increase) after the immersion incubation time (872 h) compared to the biocomposites (Table 13.4; Fig. 13.5). Absorption of water by composites is a crucial factor in evaluating the ability of biocomposite to be commercially used.[8,71,72] To improve the resistance to water absorption inclusion of MAPE into the composite formulation is routinely conducted.[71,72] However, we did not confirm this situation for the composites used in this study. In fact, in some cases inclusion of MAPE in the formulation actually resulted in greater water absorption than that from composites without MAPE. For example, HDPE-25DDGS and HDPE-25DDGS-MAPE exhibited weight gains of 1.1 and 1.8%, respectively. This trend was observed for several STDDGS, PW and STPW formulations (Table 13.4; Fig. 13.5). These results are somewhat surprising, since several other investigators have reported that inclusion of maleate olefins with the composite blend considerably reduces water absorption when using bio-fillers such as with Paulownia wood, loblolly wood, pine wood, sisal fiber, and wheat.[43,71,73] Practically no difference in water absorption rates occurred between the HDPE-PW and HDPE-PW-MAPE formulations or the HDPE-STPW and HDPE-STPW-MAPE formulations (Table 13.4; Fig. 13.5). This observation seems counterintuitive and contrary to prior observations yet it is clearly demonstrated in this study with the PW formulations employed.[43,71,73] One explanation for the discrepancy between this study and others could be the thickness of the tensile bar formulations, method of processing and/or injection molding procedures employed in various studies.

TABLE 13.4 Tensile Properties of Original and Soaked Type I Tensile Bars*

Composition	σ_U (MPa)	E (MPa)	EI% (%)	Wt. Gain (%)
HDPE	35.3, 36.4*	361, 376*	57.7, 55.1	0.0
HDPE-MAPE	33.0, 33.9*	362, 343	58.9, 58.0	0.0
HDPE-25DDGS	25.2, 26.4	432, 469	46.6, 43.4	1.1
HDPE-25DDGS-MAPE	29.1, 28.2*	524, 546*	28.4, 35.2*	1.8
HDPE-25STDDGS	25.6, 26.2*	435, 504*	34.5, 29.0*	0.9
HDPE-25STDDGS-MAPE	32.1, 31.1	562, 552	21.2, 19.8	1.3
HDPE-25STDDGS/A	28.1, 27.7*	590, 578	20.7, 17.7*	0.9
HDPE-25STDDGS/A-MAPE	32.2, 32.7	570, 622*	20.8, 17.5*	0.9
HDPE-25STDDGS/AM	30.4, 30.9	573, 581	17.7, 17.8	0.9
HDPE-25STDDGS/AM-MAPE	30.6, 31.0	566, 587	20.4, 17.8*	0.9
HDPE-25PINEW	32.0, 32.8	871, 870	14.9, 15.2	1.0
HDPE-25PINEW-MAPE	38.1, 38.8	714, 745	20.0, 18.9	0.8
HDPE-12.5STDDGS/12.5PINEW	26.9, 28.8*	719, 779*	17.8, 16.4*	1.2
HDPE-12.5STDDGS/12.5PINEW-MAPE	29.6, 31.0*	503, 549*	25.0, 18.5*	1.2
HDPE-10STDDGS/30PINEW	31.0, 30.1	853, 859*	11.1, 10.3*	1.7
HDPE-40PINEW	25.0, 26.8*	962, 987	6.9, 7.9	2.5
HDPE-25PW	35.2, 35.4	881, 913*	13.4, 10.9*	0.9
HDPE-25PW-MAPE	40.7, 41.5	930, 923	14.2, 13.6	0.8
HDPE-25STPW	33.7, 34.0	911, 926	11.8, 9.7*	0.9
HDPE-25STPW-MAPE	41.6, 41.3	931, 955	12.8, 11.6	0.9
HDPE-25STPW/A	35.3, 35.2	840, 856*	9.8, 10.0	0.8
HDPE-25STPW/A-MAPE	41.2, 40.9	836, 822*	14.8, 12.3*	0.7
HDPE-25STPW/AM	39.2, 38.8	850, 865*	15.8, 14.0	0.7
HDPE-25STPW/AM-MAPE	38.8, 39.6*	817, 804	16.9, 14.3*	0.6

Properties are given as "original" or un-soaked, "soaked" treatments and the presence of the asterisk "" after a value indicates significant difference between treatments (p £ 0.05).

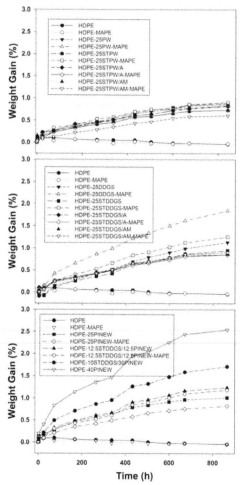

FIGURE 13.5 Comparative water absorption plots for various PW/DDGS/PINEW composites over 872 h of soaking.

The various composites exhibited distinctly different tensile bar thickness. Generally, the gate section of the tensile bar is slightly thicker than the neck, which in turn is thicker than the end section (Table 13.5). This is due to the method the plastic resin and composite blend is injected into the mold; the gate portion is subjected to more injection time than the end portion and therefore contains more plastic resin than the other portions of the tensile bar. For example, the HDPE-25DDGS composite tensile bar exhibits an initial gate, neck and end section thickness of 3.18, 3.13 and 3.11 mm, respectively (Table 13.5). No significant increase in thickness for the tensile bar sections (gate, neck or end) of the neat HDPE and HDPE-MAPE were observed. However, increases in thickness could occur in the various composite formulations tested when given a soaking treatment (Table 13.5). Significant increases in thickness for the tensile bar gate, neck and end sections occurred in the HDPE-DDGS composites (e.g., HDPE-25DDGS and HDPE-25STDDGS, HDPE-25DDGS-MAPE and HDPE-25STDDGS-MAPE (Table 13.5). In contrast, the HDPE-PW composites (HDPE-25PW, HDPE-25PW-MAPE, HDPE-25STPW, and HDPE-25STPW-MAPE) only exhibited significant increases in the end section of the tensile bar. We attribute this occurrence to the presence of more PWF and less HDPE in the end portion of the tensile bar compared to that occurring in the gate and neck sections. Bio-fillers are hydrophilic in nature due to the presence of the abundant hydroxyl groups on the cellulose, lignin and hemicellulose which readily interacts with water molecules by hydrogen bonding.[43] DDGS composites were initially thicker than PW composites and when soaked for 872 h exhibited higher percentages in thickness increases than PW composites (Table 13.5). We can conclude that DDGS composites are less dimensionally stable than PW composites. Inclusion of MAPE into the DDGS or PW composites did not notably alter the thickness measurements of the initial tensile bars compared to formulation without MAPE. Soaked composites containing MAPE exhibited slightly less thickness increases in terms of percent thickness increases than composites without MAPE (Table 13.5). Similarly this trend was also observed for "combination" composite mixtures (HDPE-12.5STDDGS/12.5PINEW and HDPE-12.5STDDGS/12.5PINEW-MAPE) where the inclusion of MAPE in the formulation reduced the thickness increase compared to the formulation without MAPE. This observation conforms to previous observations where inclusion of a maleate polyolefin in the formulation increases dimensional stability of the resulting composites through the binding of the coupling agent with the hydroxyl groups of the filler thereby preventing fillers binding to water molecules.[43]

TABLE 13.5 Influence of Soaking on the Thickness Swelling% of Tensile Bars of PW/DDGS/Pine Composites

Composition	Gate mm, mm (%)[a]	Neck mm, mm (%)	End mm, mm (%)
HDPE	3.10, 3.11 (0.24)	3.08, 3.10 (0.48)	3.09, 3.09 (-0.01)
HDPE-MAPE	3.12, 3.12 (0.07)	3.10, 3.10 (-0.09)	3.09, 3.10 (0.20)
HDPE-25DDGS	3.18, 3.21 (1.02)*	3.13, 3.17 (1.38)*	3.11, 3.016 (1.51)*
HDPE-25DDGS-MAPE	3.15, 3.19 (1.19)*	3.12, 3.15 (0.88)*	3.11, 3.15 (1.39)*
HDPE-25STDDGS	3.15, 3.18 (0.89)*	3.11, 3.14 (0.96)*	3.10, 3.13 (0.87)*

TABLE 13.5   *(Continued)*

HDPE-25STDDGS-MAPE	3.15, 3.17 (0.76)	3.08, 3.12 (1.27)*	3.09, 3.11 (0.77)*
HDPE-25STDDGS/A	3.13, 3.16 (0.83)*	3.09, 3.12 (1.09)*	3.06, 3.10 (1.25)*
HDPE-25STDDGS/A-MAPE	3.13, 3.16 (1.00)*	3.07, 3.10 (0.93)*	3.06, 3.09 (1.09)*
HDPE-25STDDGS/AM	--, 3.17 (--)	--, 3.13 (--)	--, 3.11 (--)
HDPE-25STDDGS/AM-MAPE	3.14, 3.17 (1.01)*	3.08, 3.12 (1.17)*	3.08, 3.10 (0.80)*
HDPE-25PINEW	3.11, 3.14 (1.00)	3.05, 3.07 (0.53)	3.02, 3.05 (0.89)
HDPE-25PINEW-MAPE	3.10, 3.14 (1.38)*	3.05, 3.09 (1.42)*	3.01, 3.05 (1.20)*
HDPE-12.5STDDGS/12.5PINEW	3.13, 3.15 (0.63)	3.06, 3.10 (1.15)*	3.05, 3.09 (1.34)*
HDPE-12.5STDDGS/12.5PINEW-MAPE	3.12, 3.14 (0.69)*	3.07, 3.11 (1.15)*	3.05, 3.08 (0.83)
HDPE-10STDDGS/30PINEW	–, 3.16 (–)	–, 3.10 (–)	–, 3.10 (–)
HDPE-40PINEW	–, 3.14 (–)	–, 3.08 (–)	–, 3.12 (–)
HDPE-25PW	3.12, 3.14 (0.66)	3.06, 3.07 (0.48)	3.02, 3.04 (0.71)*
HDPE-25PW-MAPE	3.11, 3.12 (0.52)	3.05, 3.06 (0.36)	3.02, 3.04 (0.77)*
HDPE-25STPW	3.11, 3.12 (0.49)	3.06, 3.05 (-0.18)	3.01, 3.04 (0.98)*
HDPE-25STPW-MAPE	3.11, 3.13 (0.48)	3.06, 3.07 (0.62)	3.03, 3.04 (0.29)
HDPE-25STPW/A	–, 3.11 (–)	–, 3.05 (–)	–, 3.02 (–)
HDPE-25STPW/A-MAPE	3.11, 3.12 (0.43)	3.04, 3.07 (1.11)*	3.02, 3.03 (0.23)
HDPE-25STPW/AM	3.11, 3.12 (0.17)	3.05, 3.07 (0.71)	3.02, 3.04 (0.56)*
HDPE-25STPW/AM-MAPE	3.11, 3.12 (0.29)	3.07, 3.07 (0.08)	3.04, 3.05 (0.47)*

[a]Thickness properties are given as "original" or un-soaked, "soaked" treatments and percent change in parenthesis. The presence of the asterisk "*" indicates significant difference between soaking treatments ($p \pounds 0.05$).

Chemical modification treatments (A and AM) performed on the HDPE-STD-DGS formulations (HDPE-25STDDGS/A, HDPE-25STDDGS/A-MAPE, HDPE-25STDDGS/AM, and HDPE-STDDGS/AM-MAPE) did not improve absorption rates (% weight gain) compared to untreated controls (HDPE-25STDDGS) (Table 13.4). Thickness of DDGS formulations that were chemically modified (A and AM) were less initially than the untreated formulation. However, both formulations increased significantly following soaking (Table 13.5). In contrast, chemical modification treatments (A and A/M) of HDPE-STPW formulations (HDPE-25STPW/A, HDPE-25STPW/A-MAPE, HDPE-25STPW/AM, and HDPE-STPW/AM-MAPE) exhibited less weight gain compared to un-modified controls (HDPE-25STPW and HDPE-25STPW-MAPE) (Table 13.4). Thickness of STPW formulations that were chemically modified (A and AM) were initially comparable to untreated controls (HDPW-25STPW). However, all soaked STPW composites whether chemically modified or not exhibited significant increases in the end section of the tensile bar only but not in the gate or neck sections, which were much less affected (Table 13.5). One explanation for the difference in responses between the two filler formulations was the abundance of protein content in the DDGS formulations which contains more hydroxyl groups that can interact with water molecules during the soaking process than present in the PW formulations.

An important attribute of WPC is their ability to retain their original color and this characteristic greatly contributes toward its commercial value.[7,19,74,75] Weathering causes color changes in WPC which is both undesirable and irreversible.[7,19,74] Water soaking is an important weathering test and is useful in determining the durable nature of a thermoplastic composite.[7,64,71,76] Weathering (e.g., water soaking) causes HDPE-composites to undergo chemical reactions such as breakdown of lignins into water soluble products which form chromophoric functional groups such as carboxylic acids, quinones, and hydroperoxy radicals.[74]

Figure 13.6 compares color values (L*, a* and b*) of the original composites to the soaked composites. Almost all of the composites exhibited lightness (L*) following the soaking treatment. This trend has been observed in other immersion studies employing WPC.[7] Coupling agents are included in the biocomposites to improve bio-filler binding to the thermoplastic resin and they may combat lightness (L* value) changes.[19,77,78] Similarly, in this study, composites containing MAPE are darker than the corresponding composites without MAPE and retained their L* values to a greater extent following the soaking process (Table 13.6; Fig. 13.6). For example, following soaking the HDPE-25DDGS formulation exhibited a 16% lightening response; while HDPE-DDGS-MAPE exhibited a 6% lightening response. Overall, DDGS formulations exhibited much higher color change values (L*, a* and b*) following soaking than PW formulations. This may be attributed to water induced chemical reactions in the DDGS; suggesting that DDGS are less chemically stable than PWF. Chemical modification of DDGS resulted in less change in color values compared to that occurring in the nonchemically modified DDGS formulations. Changes in redness (a*) and yellowness (b*) values generally followed the L* trends, but not always (Table 13.6). Changes in the C^*_{ab} (chromaticity, color quality), and H^*_{ab} (hue) values also occurred when comparing the original and soaked composites. Notable changes in color values even occurred in the neat HDPE and HDPE-MAPE polymers (Table 13.6; Fig. 13.6).

TABLE 13.6   Color Value Properties Comparison Between Original and Soaked Composites[a]

Composition	L*	a*	b*	C*$_{ab}$	H*
HDPE	62.48, 63.32*	−0.79, −1.11*	4.02, 3.85*	4.10, 4.01*	−1.38, −1.29*
HDPE-MAPE	60.53, 61.55*	−0.92, −1.02	4.65, 4.42	4.74, 4.54	−1.38, −1.34*
HDPE-25DDGS	25.62, 29.61*	0.41, 0.57*	1.56, 2.36*	1.62, 2.43*	1.31, 1.34
HDPE-25DDGS-MAPE	25.74, 27.16*	0.24, 0.35*	1.10, 1.46*	1.13, 1.50*	1.35, 1.34
HDPE-25STDDGS	30.41, 34.14*	0.71, 0.97*	2.83, 3.89*	2.92, 4.01*	1.32, 1.33
HDPE-25STDDGS-MAPE	27.50, 29.34	0.31, 0.43	1.68, 2.09	1.72, 2.14	1.37, 1.36
HDPE-25STDDGS/A	32.97, 34.95*	1.54, 1.39*	3.68, 4.41*	3.99, 4.63*	1.17, 1.27*
HDPE-25STDDGS/A-MAPE	30.52, 31.18	0.74, 0.76	2.47, 2.87*	2.58, 2.97*	1.28, 1.31
HDPE-25STDDGS/AM	32.02, 33.60	1.28, 1.20	3.33, 3.54	3.57, 3.74	1.20, 1.24
HDPE-25STDDGS/AM-MAPE	27.74, 29.79*	0.64, 0.78	1.98, 2.72*	2.08, 2.83*	1.25, 1.29
HDPE-25PINEW	43.08, 46.12	5.62, 5.11	13.19, 11.92*	14.34, 12.98*	1.17, 1.17
HDPE-25PINEW-MAPE	36.53, 36.40	4.35, 4.33	8.93, 8.14*	9.94, 9.22*	1.12, 1.08*
HDPE-12.5STDDGS/12.5PINEW	36.90, 42.84*	2.02, 2.02	7.05, 8.35	7.34, 8.59	1.28, 1.33*
HDPE-12.5STDDGS/12.5PINEW-MAPE	40.39, 42.44	2.07, 1.68*	7.02, 7.18	7.33, 7.38	1.28, 1.34
HDPE-10STDDGS/30PINEW	46.74, 50.63*	3.31, 3.30	11.39, 12.93	11.86, 13.34	1.29, 1.32*
HDPE-40PINEW	62.03, 63.02	4.36, 4.23	17.57, 17.18	18.11, 17.69	1.33, 1.33
HDPE-25PW	38.33, 38.19	3.22, 3.52	7.98, 8.07	8.61, 8.81	1.19, 1.16
HDPE-25PW-MAPE	31.44, 32.30	3.24, 3.35	6.60, 6.55	7.36, 7.36	1.11, 1.10

TABLE 13.6   *(Continued)*

HDPE-25STPW	37.82, 38.70	3.96, 4.22	9.03, 9.23	9.86, 10.15	1.16, 1.14
HDPE-25STPW-MAPE	34.55, 35.30	3.67, 3.98	7.86, 8.35	8.68, 9.25	1.13, 1.13
HDPE-25STPW/A	42.48, 42.12	2.17, 2.26	6.44, 6.52	6.80, 6.91	1.24, 1.24
HDPE-25STPW/A-MAPE	33.78, 34.49	2.29, 2.41	4.98, 5.40*	5.48, 5.91	1.14, 1.15
HDPE-25STPW/AM	38.21, 37.35	5.38, 5.46	11.15, 10.97	12.39, 12.26	1.12, 1.11
HDPE-25STPW/AM-MAPE	34.78, 35.77	5.11, 5.18	9.36, 9.17	10.67, 10.53	1.07, 1.06

[a]Color value properties are given as "original" or un-soaked, "soaked" treatments. The presence of the asterisk "*" indicates significant difference between soaking treatments (p £ 0.05).

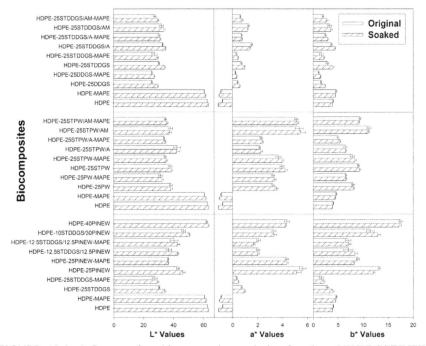

FIGURE 13.6   Influence of soaking on color analysis of various PW/DDGS/PINEW composites.

Environmental stresses such as water soaking may cause changes in the mechanical properties to occur which needs to be determined in order to assess the potential commercial value of a composite.[7,64,70,73,76,79] For example, flexural properties have been reported to decrease when LPC are weathered.[7,71,72] The response of the mechanical properties of composites as well as neat HDPE and HDPE-MAPE by water soaking are presented in Table 13.4. The neat HDPE and HDPE-MAPE blends exhibited significant changes in their σ_U and E values when given a soaking treatment (Table 13.4). Soaking caused σ_U values increased 3 and 3% for neat HDPE and HDPE-MAPE, respectively while %El values decreased 5 and 2% for neat HDPE and HDPE-MAPP, respectively. Soaking caused neat HDPE and HDPE-MAPE E values to change +4 and −5%, respectively. Changes in the mechanical properties for the composites varied considerably depending on the composition of the filler and MAPP concentration employed (Table 13.4). It is difficult to discern trends for the mechanical properties in the soaked and un-soaked formulations. Chemical modification of the PW and DDGS resulted in composites that when soaked maintained their σ_U values (Table 13.4). However, several of these chemically modified formulations exhibited significant changes occurred for the E and %El values following soaking. Absorption was found to be somewhat related to the variation of the mechanical properties of the composites. Formulations that maintained their mechanical properties after prolonged soaking generally exhibited low absorption rates (Table 13.4). Even when significant differences occurred, less than a 5% change in values occurred (Table 13.4). Further work needs to be conducted to address how water absorbance affects the long-term mechanical properties of composites.

13.3.5 THERMAL ANALYSIS

Each composite has its own unique chemical properties which is attributable to the filler type and method of preparation which in turn affects the composite's thermal properties.[26,49,80,81] For example, DDGS contains a higher concentration of protein ($\approx 26\%$) than found in most wood and lignocellulosic fiber particles (≈ 1.5–7%). The DSC thermal properties of the DDGS, PW and PINEW composites are shown in Table 13.7. Only single endothermic (melting) temperature and exothermic (crystallization) peaks were observed in the DSC curves for all formulations conducted. Little difference was found between the melting points (T_m) of neat HDPE (i.e., 130.4°C) and the original DDGS formulations (HDPE-25DDGS and HDPE-25DDGS-MAPE). Chemical modification (A or AM) treatments caused DDGS formulations to exhibit slightly higher T_ms than HDPE or the nonchemically treated-DDGS composite controls. This trend was duplicated with the PW formulations given the same chemical modification treatments. Generally, biocomposites will exhibit a slightly higher T_m compared to that the neat thermoplastic resin.[49,80] The increase in T_m in the composite is probably due to the disruption of the HDPE crystal lattice network by the presence of the filler particles. This is demonstrated by the HDPE-40PINEW formulation, which contains 40% Pine wood filler and subsequently exhibits the highest T_m value (133.5°C) recorded in this study (Table 13.7). Interestingly, the HDPE-10STDDGS/30PINEW, which also contains 40% filler had a lower T_m value (131.8°C). Although both of these formulations contained 40% filler, decidedly different mechanical and physical properties were exhibited by them (Tables 13.2–13.5). Considerably more research needs to be conducted to determine how the mixing fillers of different chemical composites interact and affect their resulting thermal properties.

TABLE 13.7   DSC Thermal Data for the HDPE Composites

Composition	T_{cc}	ΔH_{cc}	T_m	ΔH_m	X_c
	(°C)	(J/g)	(°C)	(J/g)	(%)
HDPE	118.2	190.0	130.4	165.1	56.3
HDPE-MAPE	118.3	176.6	130.7	164.5	59.1
HDPE-25DDGS	116.3	143.9	130.4	134.3	61.1
HDPE-25DDGS-MAPE	117.1	145.6	130.0	127.2	62.0
HDPE-25STDDGS	116.5	138.2	131.0	126.6	57.6
HDPE-25STDDGS-MAPE	117.5	137.1	130.8	123.3	60.1
HDPE-25STDDGS/A	116.4	140.5	131.9	127.0	57.8
HDPE-25STDDGS/A-MAPE	117.5	137.3	131.0	126.3	61.6
HDPE-25STDDGS/AM	116.6	141.6	131.1	128.6	58.5
HDPE-25STDDGS/AM-MAPE	117.6	141.4	130.9	128.7	62.7
HDPE-25PINEW	115.8	140.3	131.9	122.1	55.6
HDPE-25PINEW-MAPE	116.6	142.0	131.8	127.4	62.1
HDPE-12.5STDDGS/12.5PINEW	115.9	135.3	131.2	126.8	57.7
HDPE-12.5STDDGS/12.5PINEW-MAPE	116.4	137.0	131.9	126.9	61.9
HDPE-10STDDGS/30PINEW	115.6	121.6	131.8	105.5	60.0
HDPE-40PINEW	113.2	107.3	133.5	90.2	51.3
HDPE-25PW	116.2	140.6	132.0	127.5	58.0
HDPE-25PW-MAPE	116.6	143.6	131.2	129.1	62.9
HDPE-25STPW	116.6	140.6	131.5	127.5	58.0
HDPE-25STPW-MAPE	117.0	139.9	130.9	127.9	62.4
HDPE-25STPW/A	116.3	154.6	131.6	129.3	58.8
HDPE-25STPW/A-MAPE	116.3	137.9	131.2	122.8	59.9
HDPE-25STPW/AM	117.4	143.1	130.9	130.2	59.2
HDPE-25STPW/AM-MAPE	116.9	138.3	131.0	129.8	63.3

The addition of fillers to the HDPE results in composites with lower crystallization enthalpy (DH_c) and melting enthalpy (DH_m) values compared to neat HDPE (Table 13.7). It has been suggested that that wood fillers absorb more heat energy in the melting of composites which results in their lower DH_m values when compared to neat thermoplastic resins.[43]

The degree of crystallinity (χ_c) varied considerably depending on the filler type employed in the composite. Other investigators have also observed a variation in the degree of crystallinity values associated with various LPC.[30,49] The degree of crystallinity values for DDGS, PW and PINEW biocomposites were higher when MAPE was included in the formulation (Table 13.7). This situation may also be related to the HDPE resin employed. For example, PW blended with a different HDPE source showed distinctly lower χ_c values than neat HDPE employed in this study.[30] When the concentration of PINEW is increased to 40%, χ_c values decreased markedly below that of neat HDPE. One explanation for this phenomenon for the reduction in χ_c values is due to the amount of free volume occurring between the polymer chains capable of allowing filler to be intermixed.[81] As the volume of the filler increases less resin polymer intermolecular free volume is available for dissipating the filler material.[81]

The thermal properties of biocomposites need to be determined because the processing temperatures (extrusion and injection molding) may often exceed 200°C. Commercial HDPE products are often made with high melt temperature resins. The thermogravimetric curves for the various composites are plotted in Fig. 13.7 and these results are summarized in Table 13.8. The degradation of neat HDPE employed in this study, occurs in a single stage, beginning at 461.7°C, with a maximum decomposition peak occurring at 478.3°C. HDPE degradation was 99.7% complete at end of this stage. Similarly, the HDPE-MAPE blend mimicked these parameters, although exhibiting somewhat lower degradation and peak maximum temperatures. In contrast, there are several earlier degradation peaks for the DDGS composites. Examining the HDPE-25DDGS formulation reveals a major degradation temperature (T_d) for the DDGS flour occurring at ~ 242.2°C which subsequently results in maximum peak temperature at 322°C. Minor degradation peaks also occur and are the decomposition of low molecular weight components such as hemicellulose which degrades between 225 to 325°C.[7,82] Next a larger second higher degradation peak occurs with a maximum at 321°C, which is corresponds to the decomposition of cellulose which degrades in the 300 to 400°C.[82] Further, a third degradation peak corresponds to lignin decomposition which is reported occurring near 420°C; however it is not readily seen in this study.[82] This peak was obscured by the decomposition of the HDPE. The DDGS composite has a residual weight of 5.9% ash residue from the heterogeneous ingredients associated with the filler. Differences among the DDGS composite T_d's are due to the association of the filler material and the plastic resin. Higher T_d's and maximum peak temperatures occurred for STD-DGS composites compared to the DDGS composites; this can be attributed to the occurrence of higher levels of low-molecular-weight organic compounds in DDGS composites compared to that found in the STDDGS composites. Similarly, other investigators report that addition of extractables (clay) caused a decrease in T_d values to occur.[12] The results presented here confirm a previous study using DDGS formulations blended with a different HDPE resin.[30] The addition of the coupling agent MAPE had a somewhat complex influence on the decomposing behavior the DDGS composites. In some cases, inclusion of MAPE in the DDGS and STDDGS formulations (HDPE-25DDGS-MAPE and HDPE-25STDDGS-MAPE) resulted in occurrence of lower degradation temperatures (1st T_d) compared to formulations without MAPE (HDPE-25DDGS and HDPE-25STDDGS) (Table 13.8). Chemically modified DDGS formulations (HDPE-25STDDG/A) showed considerably higher 1st peak degradation initiation temperatures (T_d) and 1st degradation maximum peaks compared to the untreated DDGS formulations (HDPE-25DDGS). The chemical modification treatments (A and AM) improves the thermal stability of formulations. This phenomenon has been reported by other investigators where the 1st decomposition temperatures of HDPE-acetylated-WF were reported to higher when compared to HDPE-WF formulations.[51] Likewise, chemically treated PW formulations (HDPE-25STPW/A and HDPE-25STPW/AM) exhibited higher 1st peak degradation initiation temperatures (T_d) and 1st degradation maximum peaks compared to the untreated PW formulations (HDPE-25PW).

TABLE 13.8   TGA Data for DDGS Composites

Composition	1st T_d (°C)*	2nd T_d (°C)*	Peak 1 (°C)	Peak 2 (°C)	Residual (%)
HDPE	–	461.67	–	478.3	0.0
HDPE-MAPE	–	455.34	–	472.8	0.0
HDPE-25DDGS	242.17	456.74	321.9	474.2	5.9
HDPE-25DDGS-MAPE	229.98	455.24	323.5	473.0	6.7
HDPE-25STDDGS	259.03	454.70	325.4	471.5	3.8
HDPE-25STDDGS-MAPE	248.18	456.79	324.6	474.7	7.5
HDPE-25STDDGS/A	283.23	457.60	341.1	474.7	0.1
HDPE-25STDDGS/A-MAPE	272.18	455.52	335.6	473.9	6.1
HDPE-25STDDGS/AM	256.53	455.30	335.9	472.0	5.9
HDPE-25STDDGS/AM-MAPE	260.56	454.55	333.1	471.8	8.1
HDPE-25PINEW	311.77	475.34	359.8	455.7	3.7
HDPE-25PINEW-MAPE	312.68	455.23	360.5	474.2	5.6

Peak Temperatures**

TABLE 13.8 *(Continued)*

Composition	1st T_d (°C)*	2nd T_d (°C)*	Peak Temperatures **		Residual (%)
			Peak 1 (°C)	Peak 2 (°C)	
HDPE-12.5STDDGS/12.5PINEW	291.41	452.76	340.7	471.8	10.7
HDPE-12.5STDDGS/12.5PINEW-MAPE	283.01	456.63	340.4	474.5	1.6
HDPE-10STDDGS/30PINEW	307.18	454.40	347.5	473.7	2.2
HDPE-40PINEW	318.81	454.71	359.7	474.2	0.2
HDPE-25PW	300.60	455.89	346.4	474.5	6.6
HDPE-25PW-MAPE	305.21	454.00	347.1	473.4	7.2
HDPE-25STPW	306.29	454.76	348.2	473.1	6.7
HDPE-25STPW-MAPE	305.98	455.60	347.9	474.3	6.2
HDPE-25STPW/A	320.82	453.40	359.7	471.6	12.0
HDPE-25STPW/A-MAPE	321.38	453.92	359.9	473.2	4.4
HDPE-25STPW/AM	314.20	456.54	349.0	474.4	2.3
HDPE-25STPW/AM-MAPE	319.46	454.78	348.1	473.5	4.9

*Initial thermal degradation temperature (T_d).
**Maximum degradation temperature.

"Mixed" filler formulations (HDPE-12.5STDDGS/12.5PINEW) and HDPE-12.5DDGS/12.5PINEW-MAPE) exhibited much higher 1st degradation initiation temperatures (T_d) and 1st degradation maximum peaks than the nonmixed DDGS filler formulations (HDPE-25STDDGS and HDPE-25STDDGS-MAPE). This is attributed to the presence of PINEW in the mixture, which probably masks the presence of DDGS particles in these formulations. Perhaps, mixed filler composites containing PINEW flour and DDGS could be considered more "thermally stable" than the DDGS formulations tested (Table 13.8). More study is necessary to prove this contention. However, these "mixed" fillers composites were not as stable as employment of formulations containing PINEW alone. Based on the TGA analysis and since the processing temperatures did not exceed 210°C the DDGS, PW and DDGS-PINEW composites were thermally stable for the temperatures in which they were subjected to in this study.

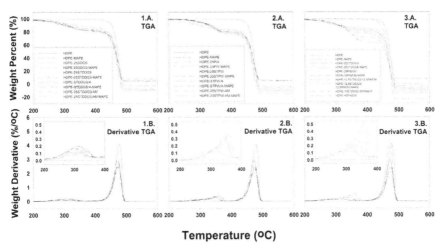

FIGURE 13.7 TGA analysis of HDPE and HDPE-DDGS/PINEW composites. (A) TGA profiles. (B) TGA derivative profiles.

13.4 CONCLUSIONS

The mechanical properties of two potential lignocellulosic material reinforcements, DDGS and PW, for use in commercial LPC was conducted in this study. Further, the benefit of "mixing" chemically dissimilar fillers, DDGS with PINEW, was assessed. The tensile, flexural, impact strength, environmental durability (soaking responses), and thermal properties of injection molded test specimens were measured. Comparison of the composites to neat HDPE that was processed with the same conditions was conducted to determine the relative merits of using a filler against a control. Using DDGS subjected to solvent extraction (STDDGS) produces a composite with superior mechanical properties (HDPE-25STDDGS) compared to the composite made with the original DDGS material (HDPE-25DDGS). Further, formulations of STDDGS with MAPE (HDPE-25STDDGS-MAPE) exhibited slightly lower tensile and flexural moduli but slightly higher ultimate stresses than similar formulations made without MAPE. The flexural properties and the tensile modulus of a solvent extracted DDGS with MAPE exceeded those of neat HDPE. The PW composites in general exhibited greater tensile and flexural properties than the DDGS composites made with similar formulations. Chemical modification by acetylation and malation of DDGS and PW fillers prior to compounding had mixed effects on improving the mechanical properties of the composites studied when compared to the untreated controls. In fact, the tensile and flexural moduli of composites containing chemically modified fillers were slightly lower than the baseline solvent treated filler with MAPE. The mixing of PINEW and STDDGS resulted in formulations (HDPE-12.5DDGS/12.5PINEW and HDPE-12.5DDGS/12.5PINEW-MAPE) with improved mechanical properties compared to the STDDGS formulations (HDPE-25STDDG or HDPE-25STDDGS-MAPE). All DDGS and PW composites soaked in water for 872 h exhibited weight gain, color changes, and some alteration in their mechanical properties, especially El%. The thermal stability of LPC formulations can be improved by mixing with wood filler and employing chemical modification.

13.5   ACKNOWLEDGEMENTS

The authors acknowledge Kimberly Pelphrey for technical assistance and Dr. N. Joshee for Paulownia wood material. Mention of a trade names or commercial products in this publication is solely for the purpose of providing specific information and does not imply recommendation or endorsement by the US Department of Agriculture. USDA is an equal opportunity provider and employer.

KEYWORDS

- **Differential Scanning Calorimetry**
- **Injection Molding**
- **Mechanical Properties**
- **Thermal Properties**

REFERENCES

1. Market Research Reports & Industry Analysis. Wood plastic composite & plastic lumber to 2015 http://www.reportsnreports.com/reports/145422 wood plastic composite plastic lumber to 2015.html (accessed Jan 14, 2012).
2. Plastics News. Plastics Resin Pricing Commodity Thermoplastics http://www.plasticsnews.com/resin pricing/commodity tps.html (accessed Feb 27, 2012).
3. Plastics Today Natural fibers growing needs. http://www.plasticstoday.com/articles/ natural-fibers-growing-needs (accessed July 22, 2013).
4. Ashori, A., Sheshmani, S., & Farhani, F. (2013). Preparation and Characterization of Bagasse/HDPE composites using multi walled carbon nanotubes *Carb Polym*, 92(1), 865–871.
5. Berglund, L., & Rowell, R. M. (2005). *Wood Composites, Handbook of Wood Chemistry and Wood Composite*, CRC Press: Boca Raton, FL, USA.
6. Carlborn, K., & Matuana, L. M. (2006). Functionalization of Wood Particles through a Reactive Extrusion Process, *J. Appl Polym Sci.,* 101(5), 3131–3142.
7. Clemons, C. M., & Stark, N. M. (2009). Feasibility of Using Salt Cedar as a Filler in Injection-Molded Polyethylene Composites. *Wood Fiber Sci.*, 41(1), 2–12.
8. Ibach, R. E., & Clemons, C. M. (2007). Effect of acetylated wood flour or coupling agent on moisture, UV, and biological resistance of extruded wood fiber-plastic composites. In *Wood Protection 2006*, Barnes, M., (Ed.), New Orleans, L. A., (March 21–23, 2006), Forest Products Society: Wisconsin, 139–147.

9. Perhac, D. G., Young, T. M., Guess, F. M., & León, R. V. (2007). Exploring Reliability of Wood Plastic Composites Stiffness and Flexural Strengths, *Intern J Reliab Appl.*, 8(2), 154–173.

10. Zahedi, M., Tabaras, T., Ashori, A., Madhoushi, M., & Shakeri, A. (2012). A Comparative Study on Some Properties of Wood Plastic Composites Using Canola Stalk, Paulownia and Nanoclay *J Appl Polym Sci.*, 129(3), 1491–1498.

11. Clemons, C. M., & Caulfield, D. F. (2010). Chapter 15 Wood Flour in *Functional Fillers for Plastics 2nd Edition*, Xanthos, M., Ed., Wiley, V. C. H. Verlag GmbH & Co. KGaA: Weinheim, Germany, 269–290.

12. Lei, Y., Wu, Q., Clemons, C. M., Yao, F., & Xu, Y. (2007). Influence of Nanoclay on Properties of HDPE/Wood Composites, *J. Appl Polym Sci.*, 106(6), 3958–3966.

13. Plastics Today. TPE resin prices, Oct. 22–26, 2012. http://www.plasticstoday.com/ articles/ tpe resin prices Oct 22–26 pe steady pp slips 0005lb spot market sees lots offers 103020123 (accessed October 28, 2012).

14. Millman, J., Sawdust shock. (March 3, 2008). A shortage looms as economy slows. *The Wall Street Journal* [Online], http://online.wsj.com/article/SB120451039119406735.html (accessed July 15, 2013).

15. Burden, D. (August 2012). Forestry Profile. Agricultural Marketing Resource Center [Online] http://www.agmrc.org/ commodities products/forestry/forestry-profile (accessed July 29, 2013).

16. Eilperin, J. (January 10, 2010). The unintended ripples from biomass subsidy program. *The Washington Post* [Online], http://www.washingtonpost.com/wp-dyn/content/article/ 2010/01/09/AR2010010902023.html (accessed July 18, 2013).

17. Le, Van, Green, S. L., & Livingston, J. (2001). Exploring the Uses for Small Diameter Trees, *J for Prod.*, 51(9), 10–21.

18. Myers, G. C., Barbor, R. J., & AbuBakr, S. M. (2003). *Small-diameter trees used for chemithermomechanical pulps*, Gen Tech Rep. FPL–GTR–141, USDA Forest Service, Madison, W. I., USA.

19. Stark, N. M., & Mueller, S. A. (2008). Improving the color stability of wood-plastic composites through fiber pretreatment *Wood Fiber Sci.* 40(2), 271–278.

20. English, G. J., & Ewing, T. W. (2002). *Vision for Bioenergy and Biobased Products in the United States*, Biomass Technical Advisory Committee http://www.usbiomassboard.gov/pdfs/ biovision03webkw.pdf.

21. Bevill, K. (January 20, 2011). Marginal land could be significant source of biofuel crops. Ethanol Producer Magazine [Online] http://www.ethanolproducer.com/articles/7428/ marginal land could be significant source of biofuel crops (accessed July 31, 2013).

22. Chinese Academy of Forestry Staff *Paulownia in China: Cultivation and Utilization*, Asian Network for Biological Sciences and International Development Research Centre, Singapore, (1986).

23. Plant Conservation Alliance's Alien Plant Working Group Princess Tree http://www.nps.gov/ plants/alien/fact/pato1.htm (accessed July 1, 2013).

24. Joshee, N. (2012). Paulownia A Multipurpose Tree for Rapid Lignocellulosic Biomass Production In *Handbook of Bioenergy Crop Plants*, Kole, C., Joshi, C. P., Shonnard, D., (Eds.), Taylor & Francis Inc.: Boca Raton, FL, USA, 671–686.

25. Julson, J. L., Subbarao, G, Stokke, D. D., Gieselman, H. H., & Muthukumarappan, K. (2004). Mechanical Properties of Biorenewable Fiber/Plastic Composites *J. Appl. Polym Sci.*, 93(5), 2484–93.

26. Onwulata, C. E., Thomas, A. E., & Cooke, P. H. (2009). Effects of Biomass in Polyethylene or Polylactic Acid Composites *J. Biobased Mat Bioener,* 3(2), 1–9.

27. Cheesbrough, V., Rosentrater, K. A., & Visser, J., (2008). Properties of distillers grains composites a preliminary investigation. *J. Polym. Environ.* 16(1), 40–50.

28. DiOrio, N. R., Tatara, R. A., Rosentrater, K. A., & Otieno, A. W. (2012). Chapter 19, Using DDGS in Industrial Materials in *Distillers Grains: Production, Properties and* Utilization, Liu, K., Rosentrater, K. A., (Eds.), CRC Press: Boca Raton, F. L., USA, 429–448.

29. Liu, K., Rosentrater, K. A. (2011). Distillers Grains: Production, Properties and Utilization, Liu, K., Rosentrater, K. A., (Eds.), CRC Press: Boca Raton, F. L., USA, 2012, 540.

30. Tisserat, B., Reifschneider, L., Harry O'Kuru, R., & Finkenstadt, V. L. (2013). Mechanical And Thermal Properties of High Density Polyethylene Dried distillers grains with soluble composites. *BioResources*, 8(1), 59–75.

31. Wisner, R. (2010). Estimated U.S. Dried Distillers Grains with Solubles (DDGS) Production & Use. *Iowa State Extension Agricultural Marketing Resource Center* [Online], http://www. extension.iastate.edu/agdm/crops/outlook/dgsbalancesheet.pdf (accessed July 17, 2013).

32. Shurson, J. (2012). Distillers grains by-products in livestock and poultry feeds. University of Minnesota, Dept. of Animal Science [Online], http://www.ddgs.umn.edu (accessed July 23, 2013).

33. Li, Y., & Sun, X. S. (2011). Mechanical and thermal properties of biocomposites from poly(lactic acid) and DDGS *J. Appl. Polym Sci.,* 121(1), 589–597.

34. Clarizio, S. C., & Tatara, R. A. (2012). Tensile strength, elongation, hardness and tensile and flexural moduli of PLA filled with glycerol-plasticized DDGS. *J. Polym Environ,* 20(3), 638–646.

35. Baboi, M, Grewell, D., & Srinivasan, G. (May 4–8, 2008). Feasibility Study of the use of DDGS plastic composites. In *66th Annual Technical Conference of the Society of Plastics Engineers*, Vol. 3, Milwaukee, WI, USA, Curran Associates, Inc: New York, 2008, 1638–1641.

36. Evangelista, R. L. (2009). Oil Extraction from Lesquerella Seeds by Dry Extrusion and Expelling *Ind. Crops Pro,* 29(1), 189–196.

37. Commodity Online. Commodity Fundamentals Global oil Seed production estimates. http://www.commodityonline.com/fundamentals/global oil seed production estimates 4577 mt in 201112-wasde/262/ (accessed Aug 2, 2013).

38. Isbell, T. A., U. S. (2009). Effort in the Development of New Crops (Lesquerella, Pennycress, Coriander and Cuphea) *Proceedings of the Journées Chevreul*, 16(4), 205–210.

39. Arvens (2011) Technology Inc. Creating a pennycress bioenergy business. USDA NIFA SBIR Grant No. 2011 33610 31157.

40. Vaughn, S. F., Isbell, T. A., Weisleder, D., & Berhow, M. A. (2005). Biofumigant Compounds Released by Field Pennycress (Thlaspi arvense) Seed Meal *J Chem Ecol,* 31(1), 167–177.

41. Reifschneider, L., Tisserat, B., & Harry O'Kuru, R. (2013). Mechanical Properties of High Density Polyethylene-Pennycress Press Cake Composites in 71st Annual Technical Conference of the Society of Plastics Engineers Cincinnati OH USA April 21–25.

42. Alibaba. (July 30, 2013). Meal Cake Product, Search http://www.alibaba.com/products/F0/meal_cake/7html.

43. Ayrilmis, N., & Kaymakci, A. (2013). Fast growing biomass as reinforcing filler in thermoplastic composites: *Paulownia elongata* wood *Ind Crops Prod,* 43, 457–464.

44. Tisserat, B., Reifschneider, L., Joshee, N., & Finkenstadt, V. L. (2013). Properties of High Density Polyethylene Paulownia Wood Flour Composites Via Injection Molding, *Bio-Resources,* 8(3), 4440–4458.

45. Finkenstadt, V. L., Liu, C. K., Evangelista, R., Liu, L. S., Cermak, S. C., Hojilla Evangelista, M., & Willett, J. L. (2007). Poly(lactic acid) green composites using oilseed coproducts as fillers, *Ind Crop Prod.,* 26(1), 36–43.

46. Hayes, M. (1997). Agricultural Residues: a promising alternative to virgin wood fiber. Resource Conversation Alliance [Online], http://www.woodconsumption.org/alts/ meghanhayes. html (accessed July 9, 2013).

47. Hill, C. (July 16, 2013). Acetylated wood: The science behind the material. Accoya Accsys Technologies [Online] 2011, http://www.accoya.com/wp-content/uploads/2011/05/Acetylated wood.pdf.

48. Kaci, M., Djidjelli, H., Boukerrou, A., & Zaidi, L. (2007). Effect of wood filler treatment and EBAGMA compatibilizer on morphology and mechanical properties of low density polyethylene/olive husk flour composites. *EXPRESS Polym Lett.*, 1(7), 467–473.

49. Kalia, S., Kaith, B. S., & Kaur, I. (2009). Pretreatment of Natural Fibers and Their Application as Reinforcing Material in Polymer Composites a Review *Polym Eng. Sci.*, 49(7), 1253–1272.

50. Lu, J. Z., Wu, Q., & Jr. McNabb, H. S. (2000). Chemical coupling in wood fiber and polymer composites: a review of coupling agents and treatments. *Soc. Wood Sci. Technol.*, 31(1), 88–104.

51. Özmen, N., Çetin, N. S., Mengeloğlu, F., Birinci, E., & Karakus, K. (2013). Effect of Wood Acetylation with Vinyl Acetate and Acetic Anhydride on the Properties of Wood Plastic Composites *BioResources,* 8(1), 753–767.

52. Rimdusit, S., Smittakorn, W., Jittarom, S., & Tiptipakorn, S. (2011). Highly filled polypropylene rubber wood flour composites *Eng. J.* 15(2), 17–30.

53. Rowell, R. M. (2006). Acetylation of wood: Journey from analytical technique to commercial reality. *For Prod J.*, 56(9), 4–12.

54. Accoya Technologies. Wood without compromise http://www.accoya.com (accessed July 17, 2013).

55. Çetin, N. S., & Özmen, N. (2011). Acetylation of wood components and Fourier transform infra red spectroscopy studies. *Afr J Biotechnol,* 10(16), 3091–3096.

56. Dobreva, D., Nenkova, S., Vasileva St., (2006). Morphology and mechanical properties of polypropylene-wood flour composites, *BioResources,* 1(2), 209–219.

57. Larsson Brelid, P., Simonson, R., & Risman, P. O. (1999). Acetylation of Solid Wood Using Microwave Heating Par 1: Studies of dielectric properties. *Holz Roh Werkst.*, 57, 259–263.

58. Li, D., Li, J., Hu, X., & Li, L. (2012). Effects of ethylene vinyl acetate content on physical and mechanical properties of wood-plastic composites. *BioResources,* 7(3), 2916–2932.

59. Lisperguer, J., Droguett, C., Ruf, B., & Nuñez, M. (2007). The Effect of Wood Acetylation on Thermal Behavior of Wood-Polystyrene Composites *J Chil Chem. Soc.*, 52(1), 1–6.

60. Matsunaga, M., Kataoka, Y., Matsunaga, H., & Matsui, H. (2010). A Novel Method of Acetylation of Wood Using Supercritical Carbon Dioxide *J. Wood Sci.*, 56, 293–298.

61. Müller, M., Radovanovic, I., Grüneberg, T., Militz, H., & Krause, A. (2012). Influence of various wood modifications on the properties of polyvinyl chloride/wood flour composites. *J. Appl Polym Sci.* 125(1), 308–312.

62. Nenkova, S., Simeonova, G., Dobrilova, T. Z. V., Vasilieva, S., & Natov, M. (2004). Modification of Wood and Wood Flour with Maleic Anhydride *Cellul Chem Technol,* 38(5–6), 375–383.

63. Özmen, N., & Çetin, N. S. (2012). A new approach for acetylation of wood: vinyl acetate. *Afr. J. Pure Appl. Chem.*, 6(6), 78–82.

64. Segerholm, B. K., Ibach, R. E., & Wålinder, E. P. (2012). Moisture sorption in artificially aged wood-plastic composites. *BioResources,* 7(1), 1283–1293.

65. Shengfei, H., Wen, C., Weihua, L., & Huaxing, L. (2007). Microwave Irradiation Treatment of Wood Flour and Its Application in PVC Wood Flour Composites *J Wuhan Univ Technol,* 22(1), 148–152.

66. Sundar, S. T. (2005). Chemical Modification of Wood Fiber to Enhance the Interface between Wood and Polymer in Wood Plastic Composites Master's Thesis, University of Idaho, Moscow, ID, USA.

67. Tullo, A. H. (2012). Making Wood Last Forever with Acetylation *C&EN*, 90(32), 22–23.

68. Westin, M., Larsson-Brelid, P. M., Sergerholm, B. K., van den Oever, M. (2008). Wood Plastic Composites From Modified Wood, Part 3. Durability of WPCs with Bioderived Matrix. In *5th Annual Meeting of International Research Group on Wood Preservation* Istanbul, Turkey, May 25–29, IRG/WP 08-40423.

69. Tisserat, B., Joshee, N., Mahapatra, A. K., Selling, G. W., & Finkenstadt, V. L. (2013). Physical and mechanical properties of extruded poly(lactic acid)-based *Paulownia elongata* biocomposites. *Ind. Crops Prod.*, 44, 88–96.

70. Thwe, M. M., & Liao, K. (2002). Effects of environmental aging on the mechanical properties of bamboo—glass fiber reinforced polymer matrix hybrid composites. *Composites, Part A.*, 33(1), 43–52.

71. Zabihzadeh, S. M. (2010). Flexural Properties and Orthotropic Swelling Behavior of Bagasse/Thermoplastic Composites *BioResources*, 5(2), 650–660.

72. Zabihzabeh, S. M. (2010). Water uptake and flexural properties of natural filler/HDPE composites *BioResources*, 5(1), 316–323.

73. Joseph, P. V., Rabello, M. S., Mattoso, L. H. C., Joseph, K., & Thomas, S. (2002). Environmental effects on the degradation behavior of sisal fiber reinforced polypropylene composites. *Comp. Sci. Technol.*, 62(10–11), 1357–1372.

74. Fabiyi, J. S., McDonald, A.G., Wolcott, M. P., & Griffths, P. R. (2008). Wood plastic composites weathering: visual appearance and chemical changes. *Polym Degrad. Stab.* 93(8), 1405–1414.

75. Zhang, Y., Zhang, S. Y., & Chui, Y. H. (2006). Impact of melt impregnation on the color of wood-plastic composites. *J. Appl. Polym. Sci.*, 102(3), 2149–2157.

76. Lopez, J. L., Sain, M., & Cooper, P. (2006). Performance of natural-fiber-plastic composites under stress for outdoor applications: effect of moisture, temperature, and ultraviolet light exposure. *J. Appl. Polym Sci*, 99(3), 2570–2577.

77. Koo, C. M., Kim, M. J., Choi, M. H., Kim, S. O., & Chung, I. J. (2003). Mechanical and Rheological properties of the maleated polypropylene layered silicate nanocomposites with different morphology. J. *Appl. Polym. Sci.*, 88(6), 1526–1535.

78. Myers, G. E., Chahyadi, I. S., Gonzalez, C., Coberly, C. A., & Ermer, D. S. (1991). Wood flour and polypropylene or high density polyethylene composites: influence of maleated polypropylene concentration and extrusion temperature on properties. *Intern. J. Polym. Matter,* 15(3–4), 171–186.

79. Kord, B. (2011). Evaluation on the effect of wood flour and coupling agent content on the hygroscopic thickness swelling rate of polypropylene composites, *BioResources*, 6(3), 3055–3065.

80. Avérous, F., & Le Digabel, F. (2006). Properties of biocomposites based on lignocelluosic fillers. *Carb Polym*, 66(4), 480–493.

81. Khalaf, M. N. (2010). Effect of alkali lignin on heat of fusion crystallinity and melting points of low density polyethylene (LDPE), medium density polyethylene (MDPE) and high density polyethylene (HDPE) *J. Thi-Qar Sci.*, 2(2), 89–95.

82. Lee, S. H., & Wang, S. (2006). Biodegradable polymers/bamboo fiber biocomposite with bio based coupling agent. *Composites, Part A.*, 37(1), 80–91.

CHAPTER 14

THE MULTIFUNCTIONAL CHEMICAL TUNABILITY OF WOOD-BASED POLYMERS FOR ADVANCED BIOMATERIALS APPLICATIONS

TERESA CRISTINA FONSECA SILVA, DEUSANILDE SILVA, and
LUCIAN A. LUCIA

ABSTRACT

This century has been witnessing the increasing development of ecofriendly materials derived from natural fibers to reinforce composites. In this chapter, wood-based polymers (i.e., cellulose, hemicellulose and lignin) have been chosen as the chief biopolymeric templates for review because together they comprise the most abundant resource on the planet, *viz.*, lignocellulosics. Moreover, although wood has long been used as a raw material for building, fuel, papermaking, and various other applications, the potential of wood-based polymers to reinforce composites has shown significant progress. One of the greatest challenges to advancing this area had been the lack of economic and abundant alternatives for natural fiber-reinforced composites. This issue is currently being addressed by the application of lignocellulosics. The chemical, thermal, and physical properties of the biopolymers and resultant composites are influenced by the molecular weight distribution and composition of the biopolymers with the miscibility of the individual components being of great significance and often being the limiting aspect to the optimization of the physical properties of the final blends. To overcome miscibility issues between many naturally occurring polymers and associated composites, chemical modification and graft polymerization of the surfaces of such biopolymers are common approaches.

This chapter will review the overall characteristics as and mechanical properties of the reinforcing biopolymers that come from wood and are used in the final biocomposites. In addition, the methods employed for polymer modification, mainly the chemical methods, will be discussed.

14.1 INTRODUCTION

Broadly defined, biocomposites tend to be composite materials formed from natural/bio-based fibers as the reinforcing elements and petroleum-derived nonbiodegradable polymers (e.g., PP, PE) or biodegradable polymers (e.g., PLA, PHA) as the matrix material. Biocomposites from plants and related biomaterials are more ecofriendly because the traditional composite structures (reinforced with epoxy, unsaturated polyester resins, polyurethanes, or phenolics) tend to have negative impacts on the environment because of their nonrenewable (nonbiodegradable) nature. Biocomposites can be classified as partially or completely biodegradable as shown in Fig. 14.1.

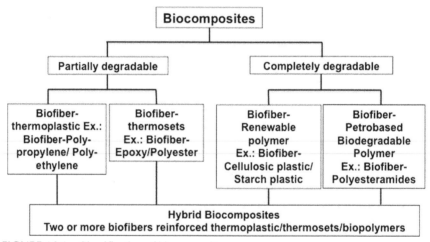

FIGURE 14.1 Classification of biocomposites.

In general, an appropriate selection of biopolymers is usually determined by the stiffness and tensile strength of the resultant composite. Other deciding factors are thermal stability, self-adhesion of fibers, and matrix, dynamic and long-term behavior, economics and processing costs.

Although bio-fibers confer some attractive properties to manufacturers such as flexibility during processing, high specific stiffness, and low cost, the formed composite still lacks the necessary thermal and mechanical properties desirable for engineering plastics compared to synthetic polymers. Also, an enhanced miscibility between natural fibers and matrix should provide better biocomposites. Intensive research on these new compositions and processes has triggered an increased level of development and application. For instance, the worldwide capacity of bio-based plastics is expected to increase from 0.36 million metric ton (2007) to 2.33 million metric ton by 2013 and to 3.45 million metric ton by 2020.

Over the last few years, a number of research efforts have focused on investigating the future development of natural fibers as load bearing constituents in composite materials. The use of such materials in composites has increased mainly due to their abundance and relative low cost, their ability to recycle, and fact that they can compete well in terms of strength per weight of material.

14.2   POTENTIAL OF WOOD TO REINFORCE COMPOSITES

Lignocellulosic fibers have been used as reinforcing composites for over 3000 years,[1] in combination with polymeric materials. Due to their moderately high specific strength and stiffness, they serve as an excellent reinforcing agent for plastics (thermosets as well as thermoplastics) besides their advantageous ecological character. The study of fibers to reinforce plastics began in 1908 with the advent of cellulose material in phenolics, extending to urea and melamine and reaching commodity status with glass fiber-reinforced plastics.[2] However, in the last several decades, much better biocomposite materials have been developed. They have higher fiber contents, better interfacial properties, improved processing technologies, and more effective additives.[1] Their biodegradability, lightness in weight, abundance and wide variety of fiber types are very important factors for their inclusion in large volume markets such as the automotive and construction industry.[3]

The use of lignocellulosic fibers derived from renewable resources as a reinforcing phase in polymeric matrix composites provides positive environmental benefits. The main advantages of lignocellulosics are: biodegradability, low costs, nonabrasive, and nonhazardous nature, low density, abundance, wide variety of fiber types, high specific strength and modulus, relatively reactive surface, which can be used for grafting specific groups. However, there are drawbacks for using lignocellulosic fibers as reinforcing materials. One of the major disadvantages is the poor compatibility exhibited between the fibers (hydrophilic nature) and the polymeric matrices (hydrophobic nature), forming flocs or aggregates during processing and thus resulting in a heterogeneous dispersion of fibers within the matrix. A low thermal stability is also a problem because the composites-based lignocellulosic fibers undergo degradation at temperatures higher than 200 °C. Another drawback is low resistance to moisture that leads to swelling and creation of pockets at the air-substrate interface leading to a compromise of mechanical properties and reduction in dimensional stability.[4]In addition, the low microbial resistance and the nonuniformity of the fiber dimensions pose further problems.

What is evident in all of these studies is that a "bandage" approach via chemical treatments is applied to solving most of the composite shortcomings. Yet, lignocellulosic fibers have distinct advantages over synthetics because they tend to deform rather than break during the manufacturing process. Also, cellulose fibers in particular have a flattened oval cross section that enhances stress transfer by a high aspect ratio.[5]

A summary of the advantages and disadvantages of lignocellulosic fibers to reinforce composites are shown in Table 14.1.[6]

TABLE 14.1 Summary of the Advantages and Disadvantages of Lignocellulosic Fibers to Reinforce Composites

Advantages	Disadvantages
Low specific weight in a higher specific strength and stiffness than glass	Lower strength especially impact strength
Renewable resources, production require little energy and low CO_2 emission	Variable quality, influenced by weather
Production with low investment at low cost	Poor moisture resistance causing swelling of the fibers
Friendly processing no wear of tools and no skin irritation	Restricted maximum processing temperature
High electrical resistant	Lower durability
Good thermal and acoustic insulating properties	Poor fire resistance
Biodegradable	Poor fiber/matrix adhesion
Thermal recycling is possible	Price fluctuation by harvest results or agricultural politics

In general, reinforcing fibers can be classified as: straw fibers, nonwood fibers and wood fibers shown in Fig. 14.2.

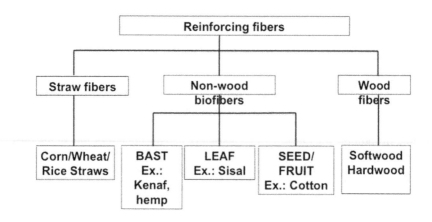

FIGURE 14.2 Classification of reinforcing fibers.

It is generally accepted that wood fibers are the most abundant biomass resource on earth. They are a class of natural composites that are principally found in trees and other vascular tissue and are composed of tubes made up of cellulose microfibrils embedded in a matrix of lignin and hemicellulose. Cellulose is a polydisperse linear polymer composed of β-D-glucopyranose monomers in which the monomers are linked together by the chemical process of dehydration condensation to form glycosidic oxygen bridges between the saccharides. In natural fibers, cellulose chains have a degree of polymerization of between 500–10,000 glucopyranose units in wood cellulose, dependent upon the type of wood examined. The respective cellulose polymer chains of saccharide units are ordered hierarchically to form nanofibrils, which are aligned along the major axis of the chain whose structural integrity is maintained by lateral hydrogen bonding forces among the hydroxyl and oxygen functionalities between chains providing wood its inherent high mechanical strength properties and high strength-to-weight ratio, in addition to rigidity. The nano-fibrils aggregate to form microfibrils that are responsible for the cell wall composition that displays a very pronounced crystalline phase interdispersed with noncrystalline (amorphous) regions. The diameters can range from 2–20 nm and possess lengths up to tens of microns thus offering very high aspect ratios as a function of biomolecular origin (e.g., valonia makes a form of cellulose that has one of the highest aspect ratios among the celluloses)[7] (Fig. 14.3).

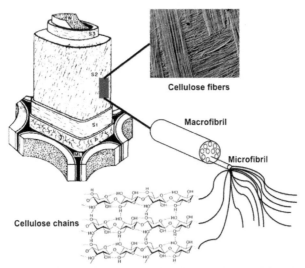

FIGURE 14.3   A representation of the top-down hierarchical nature of the cell wall from its gross compositional characteristics to its basic glucose building blocks.

The reinforcing effect imparted by cellulosic fibers is based on the nature of cellulose and its crystallinity.

Another major component of natural fibers are hemicelluloses, lignin, pectins, and waxes. Lignin is a highly cross-linked, rather amorphous polymer with a very high polydispersity consisting of substituted phenyl propane units that plays the role of the matrix. Hemicelluloses are also part of the wood biopolymer matrix and may be characterized as branched polymers of galactose, glucose, mannose, and xylose. Cellulose acts as a reinforcing material, that is, in wood fibers, cellulose fibers, microfibrils and microcrystalline cellulose, bulk materials that have varying elastic moduli as shown in Fig. 14.4. The modulus of variegated biomaterial such as wood can be up to 10GPa, from which the isolation of the cellulose component can be as high as 40GPa (upon separation by appropriate pulping/mechanical treatments). Indeed, further separation into the microfibrils allows moduli of up to 70GPa to be accessed.[8]

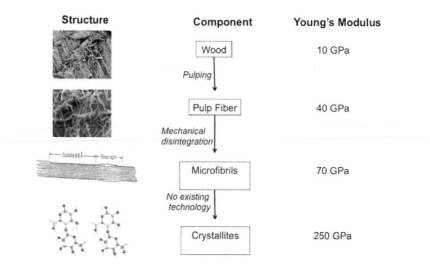

Structure	Component	Young's Modulus
	Wood	10 GPa
	Pulping ↓	
	Pulp Fiber	40 GPa
	Mechanical disintegration ↓	
	Microfibrils	70 GPa
	No existing technology ↓	
	Crystallites	250 GPa

FIGURE 14.4   Correlation between structure, process, component, and modulus (adapted from Michell[9]).

During the last several decades, a number of useful materials that use the reinforcing properties of wood cellulose have found a number of major markets. For example, wood plastic composites (WPCs) are one of the most attractive. WPCs are a type of composite that contain lignocellulosics combined with thermoset or thermoplastic polymers. Thermosets do not reversibly cure and can be represented as epoxies and phenolics, whereas thermoplastics can be repeatedly melted to allow other materials, such as wood biopolymers, to be blended with them. Polypropylene (PP), polyethylene (PE), and polyvinylchloride (PVC) are among the most widely thermoplastics applied in WPCs for building, marine, electronic, furniture, aerospace, construction, and automotive.[10] Figure 14.5 shows the uses of wood plastic composites in 2002.³ They are typically produced by blending a lignocellulosic-based polymer or composite (e.g., wood fibers) with the epoxies/phenolic resins (for example) to form a filler/polymer matrix and then pressing or molding it under high pressures and temperatures. A preeminent prerequisite for success in reinforcement of plastics is the availability of large quantities of the lignocellulosic-based fibers.¹

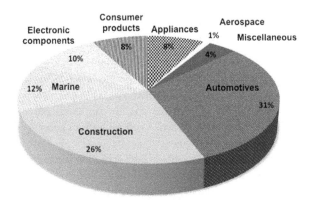

FIGURE 14.5   A breakdown of the total wood plastic composites used in 2002.³

The future growth of WPCs, cellulose-based plastics, "plastic" lumber, and analogous natural fiber composites was approximately 2.4 MM tons in 2011 with an expectation of reaching 4.6 MM tons in 2016.¹¹

PE, PVC, and PP are the predominant matrices used in WPCs although several types of WPCs with lignocellulosic matrices and conventional polymers have already received attention and subsequent development.¹² For example, composites from maple wood fibers and a bacterial polyester (poly(β-hydroxybutyrate-co-β-valerate)) have been manufactured by an extrusion-injection molding process. When the composite was reinforced with 40 wt.% of maple wood fiber, the tensile and flexural moduli of the resultant biocomposites improved by approximately 170% relative to neat bacterial polyester. Such behavior was observed to linearly depend on the regulated enrichment of the biocomposite with the wood fibers.¹³

The physical properties of lignocellulosic fibers are critically important to the successful design of biocomposites because their characteristics are highly dependent on fiber chemical and physical properties, such as the structure of fibers, cellulose content, angle of fibrils, cross-section, and the degree of polymerization. Additionally, well-defined mechanical properties are a general prerequisite for the successful use of composites and the fibers have to be specially prepared or modified with respect to the following: 1) homogenization of the fiber's properties; 2) degrees of separation and degumming; 3) degrees of polymerization and crystallization; 4) good adhesion between fiber and matrix; 5) moisture repellency; and 6) flame retardancy properties.⁸

The wood fibers, as any lignocellulosic fiber, can be processed in different ways to yield reinforcing elements having different mechanical properties. The fibrillation of pulp fiber to obtain microfibrillated cellulose is obtained through a mechanical treatment of pulp fibers consisting of refining and high pressure homogenizing processes. Also, cellulose whiskers (also known as cellulose nanocrystals) can yield individual reinforcing elements of excellent physical properties. Cellulose nanocrystals have been investigated as fillers in a number of matrix systems, including siloxanes, poly(caprolactone), glycerol-plasticized starch, styrene-butyl acrylate latex, cellulose acetate butyrate, and epoxy resins.[14]

The reinforcing ability of the cellulose whiskers lies in their high surface area and good mechanical properties. However, to obtain a significant increase in material properties, the whiskers should be well separated and evenly distributed in the matrix material.[15] Because amorphous regions are structural defects, short monocrystals can be obtained under acid hydrolysis from various sources including wood, sisal, tunicin, ramie, cotton stalks, wheat straw, bacterial cellulose, etc.

The properties of the composite materials depend on the properties of their individual components, but also on their morphology and interfacial characteristics. One of the drawbacks of cellulose whiskers with polar surfaces is poor dispersibility/compatibility with nonpolar solvents or resins. Thus, their incorporation as reinforcing materials for nanocomposites has so far been largely limited to aqueous or polar systems. To overcome this problem and broaden the type of possible polymer matrices, surface modification efforts have been made as will be discussed later.

Because the physical properties of lignocellulosic fibers are mainly determined by their composition such as structure of fibers, cellulose content, angle of fibrils, cross-section, degree of polymerization, it is necessary to give a little background on the wood types to clarify structural differences between the formed wood-based biocomposites. It is common to classify wood as softwood (gymnosperm) or hardwood (angiosperm), with basic differences in their anatomical features.

Angiosperm (hardwoods) trees present more complex and heterogeneous structures than gymnosperms (softwoods). The dominant feature separating angiosperms from gymnosperms is the presence of vessel elements for transport functions and shorter fiber cells. The vessels may show considerable variation in size, shape of perforation plates (simple, scalariform, reticulate, foraminate), and structure of cell wall, such as spiral thickenings. According to past work, the spirally layered outer secondary wall (S1 layer) restricts the flexibility of hardwood mechanical pulp fibers and thus prevents access to the subjacent inner secondary wall (S2 layer).[16]

One of the critical parameters influencing the strength properties of wood plastic composites (WPCs) is the size of the fibers. Short and tiny fibers (average particle size 0.24–0.35 mm), typically found in hardwoods, should be preferred. They provide a higher specific surface area and the fibers are distributed more homogeneously compared to composites with long fibers and so the compatibility of fiber and matrix is improved. Given this, swelling decreases and breaks during processing are reduced.[3] To support these arguments, wood polypropylene composites of different compositions (30, 40, and 50%) have been prepared using maleic anhydride–polypropylene copolymer of different percentage and from the results, it was observed that the hardwood fiber–polypropylene composites, by using maleate polypropylene (MAH-PP), show comparatively better performance to softwood fiber–polypropylene composites.[17]In another study, dissolving wood fiber pulps (Eucalyptus hardwood and conifer softwood) were used to produce composites. Surprisingly, softwood fiber biocomposites showed a tensile strength (76 MPa) significantly higher than that of hardwood.[18]

The ability of cellulose microfibrils from BSKP (bleached softwood Kraft pulp) to act as a reinforcing agent in a matrix in PVA (polyvinyl alcohol) was demonstrated by the two-fold increase in tensile and two-and-a-half-fold increase in stiffness (at 5 microfibril loading). It was further demonstrated that having a minimal aspect ratio (L/W ratio) is far more important than crystallinity in determining composite reinforcement gains when the composite was compared to MCC (microcrystalline cellulose).[19]

The overall nanometric effect exhibited by cellulose (i.e., amplification of specific macro properties by enhanced surface area or related nanoscopic parameters) is found in its nanocrystalline form. Table 14.2 illustrates several key attributes with respect to reinforcement efficiency of nanocrystals and macrofibers.[14] The dramatic enhancement in surface area, close spacing, very high stiffness and strength, and high aspect ratio allow cellulose nanocrystals to behave a high-performance reinforcement for advanced materials.[19]

TABLE 14.2 Typical Properties of Cellulose Nanofibril and Softwood Kraft Pulp Fibers

Property	Cellulose Nanofibril	Softwood Kraft Pulp
Length, nm	500	1 500 000
Diameter, nm	5	30 000
Specific surface, 1/nm	$0.048\ V_f{}^*$	$0.000008\ V_f{}^*$
Fiber spacing, nm	$5\ V_f^{-0.5}$	$30\ 000\ V_f^{-0.5}$
Aspect Ratio	100	50
Tensile strength, MPa	10 000	700
Elastic Modulus, GPa	150	20

* V_f: fiber volume fraction.

High-strength composites from softwood fibers and nanofibrillated cellulose (NFC) demonstrated increases in the tensile strength from 98 MPa to 160 MPa and the work needed to attain fracture was more than doubled with the addition of 10% NFC to wood fibers. A hierarchical structure was obtained in the composites in the form of a microscale wood fiber network and an additional NFC nanofiber network linking wood fibers and also occupying some of the microscale porosity.[20]

14.3 WOOD-BASED POLYMERS (CELLULOSE, HEMICELLULOSES, LIGNIN): CHEMICAL AND PHYSICAL CHARACTERISTICS THAT MAKE THEM AMENABLE TO NEW BIOMATERIALS APPLICATIONS

Lignocellulosic fibers are biomaterials composed of cellulose, hemicellulose, and lignin. The content of the components depends on the wood species. Table 14.3 shows a general chemical composition (cellulose, hemicelluloses and lignin content) as well as the fiber length in hardwoods and softwoods.

TABLE 14.3   General Chemical Composition of Softwoods and Hardwoods

	Softwood	Hardwood
Cellulose content	42%	45%
Xylans content	20%	5%
Mannans content	10%	20–30%
Lignin content	28%	20%
Fiber length	2–6 mm	0.2–1.5 mm

In general, hardwoods species present slightly higher cellulose content and significantly lower lignin content. Hemicellulose content between the two is variable depending on the wood species.

The hollow cellulose fibrils are heterogeneously embedded within a matrix of hemicellulose and lignin. Bonds between carbohydrates (hemicellulose and cellulose) and lignin and between hemicellulose and cellulose are present within the cementing matrix. The forming cellulose-hemicellulose network comprises the main structural component of the fiber cell. On the other hand, lignin-carbohydrate bonds (presented as benzyl esters, benzyl ethers, and phenyl glycosides) increase the stiffness of the cellulose-hemicellulose composite.

With respect to structure, the fibrils possess a thin primary wall (S1) first formed during cell wall biogenesis that is engirded by a larger, more voluminous secondary wall (S2) made up of three layers with a middle layer determining the load-bearing capacity of the wood fiber. This middle layer can be characterized as a rope-like structure of helically wound microfibrils from long chained cellulose macromolecules. An important property controlling the ultimate mechanical strength of the wood cell architecture is the angle betwixt the fiber axis and the microbrils known as the MFA (microfibril angle) that spans a gamut of values depending on the wood species.

14.3.1   CELLULOSE

Individual cellulose microfibrils have diameters ranging from 2 to 20 nm and are made up of 30–100 cellulose molecules that are the major structure elements for providing mechanical strength to the fiber.[21] The cellulose macromolecules are arranged parallel to one another with very close spacing. As indicated earlier, cellulose is a linear homo-polysaccharide composed of β-D-glucopyranose units linked together by β-1-4 glycosidic linkages.[22] Each glucose unit has three hydroxyl groups with the ability to stabilize the entire cellulosic crystalline lattice structure via later hydrogen bonds that *a priori* are responsible for the crystalline packing and controlling the overall properties of the cellulosic.

The overall properties of the cellulosic fibers are controlled by the MFA, nanocrystalline cellular dimensions, defects (number and degree), and molecular constituents. As a rule, the tensile and Young's modulus of the fibers increase with increasing cellulosic content whereas the stiffness of the fibers are controlled by the MFA. The spiral angle of the fibrils and the content of cellulose in general therefore determine the mechanical properties of the cellulose-based natural fibers. Essentially, a very high tensile strength, inflexibility, and rigidity are obtainable if the microfibrils are oriented parallel to the fibril axis.

Clearly, the overall magnitude of the final properties of the cellulosic fibers are a function of chemical composition, internal fiber structure, MFA, cell dimensions, and defects in the raw material. The mechanical properties of natural fibers also depend on the DP, crystallinity, and its degree, chain orientation, void structure (void content, size, and specific interface), and fiber diameter.

Very long cellulosic macromolecules (as observed for viscose and acetate-type fibers) give rise to very high tensile strength fibers. It has also been found that even very high DPs can give a negative correlation to strength by changing the orientation of the cellulosic chains (melting the cellulose and reforming it), by varying the crystallite dimension and crystallinity degree, by doping with contaminants or pores and by a native nonuniform fiber cross-section.[23]

Fibers that are too long would also excessively increase the viscosity and moreover would introduce rheopect behavior, which is unwanted in processing. An important parameter in fiber-reinforced materials is the strength of bonding between the fibers and the matrix material.

To repeat, the reinforcing efficiency of lignocellulosic fibers is related to the nature of cellulose and its crystallinity. Indeed, because the crystallite is composed of an aggregate of tightly bound (almost fused) parallel-aligned chains, the material exhibits very little flexibility. The overall physical and chemical properties (e.g., tensile, density, stiffness, swellability, and heat response) of the cellulosics are essentially governed by the size, shape, and organization of the crystals. A higher degree of amorphous regions will increase extensibility (less flexibility) and reduce mechanical properties.[24]

Cellulose fibril aggregate isolation from wood pulp was described previously and has been modified over the years to successfully isolate high quality NFC by using high-shear homogenization or refining.[25] Individual crystallites can be also obtained from wood pulp[26] using hydrochloric (one of several mineral acids that can be used) or sulfuric acids. The action of the acid hydrolysis is predicated on a facile deconstruction/removal of the amorphous cellulosic regions in addition to facile dissolution of the remaining polysaccharides, including the hemicelluloses and pectins. The acid acts by hydrolysis of the cellulosic macromolecules that are accessible by virtue of the noncrysallinity; what actually happens near the end of the hydrolysis is a leveling off of the DP (degree of polymerization) which corresponds to the residual (basic) highly crystalline regions of the cellulose. At this point, a rapid dilution of the acid will cause the termination of the hydrolysis. Afterwards, the nanocrystals can be isolated by the combination of centrifugation and extensive dialysis (removing acid and small molecular fragments) followed by a brief sonication to disperse the nanoparticles and provide an aqueous suspension.[14]

14.3.2   LIGNIN

Lignin displays high structural complexity characterized succinctly by a seemingly random (highly polydisperse), highly cross-linked polymeric network composed of monomeric phenylpropane units cross-linked through ether linkages and carbon–carbon bonds. It is an amorphous component of plants able to confer strength and rigidity to the cell wall. In addition to the multiplicity of lignin forms potentially available based on their origin, lignin structures and content vary widely depending on the plant. In the case of wood, the amount of lignin ranges from ca. 12% to 39%.[27]

The monomer structures in lignin consist of phenylpropane units, but differ in the degree of oxygen substitution on the phenyl ring. Basically, three structures of lignin; H-subunit (4-hydroxy phenyl), G-subunit (guaiacyl) and the S-subunit (syringyl) are conjugated to produce a three-dimensional lignin polymer that has been proposed by the majority of scientists to arise by radical (combinatorial) biosynthesis (Fig. 14.6). Thus, because enzymatic pathways for lignification have not been unequivocally identified, lignin is not expected to have a putative regular structure such as cellulose or other major macromolecules and therefore is a physically and chemically heterogeneous material with an unknown chemical structure.[27]

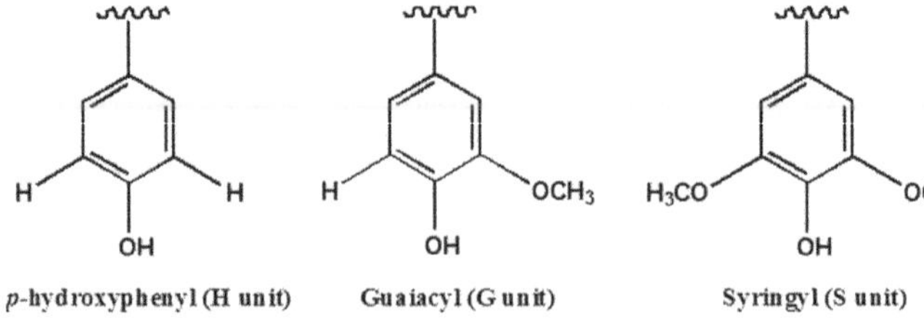

p-hydroxyphenyl (H unit)   Guaiacyl (G unit)   Syringyl (S unit)

FIGURE 14.6   Monomer structures in lignin (H, G and S-subunits).

Historically, lignin has been considered as an unwelcome by-product from the pulping process, because the main objective of pulping is to extract cellulose from wood. However, based on the biorefinery portfolio, an effective and cost-saving utilization of industrial lignin it is essential, because lignin has the possibility to even partly replace petroleum-based products because of its chemical similarity and potential cost-performance benefits.

The extraction of lignin from lignocellulosic materials always leads to fragments of low molecular weight, due to the method of extraction, thus changing its physic-chemical properties. Besides, the method of isolation, the source from which lignin is obtained also has an influence on its properties. Lignin itself possesses very attractive chemical motifs that allow it to be modified. For example, its phenolic hydroxyl is a reactive site for cross-coupling, networking, modification, and polymerization. A variety of polymers such as polyurethanes can be effectively produced from it. Indeed, by judicious control of decomposition (chemical and thermal) a panoply of chemicals can be derived for use as monomers to construct popular and valuable polymers such as polyesters, polyetheres, and polystyrenes.[27]

Manufacturing products from lignin depends on its physicochemical properties. Because of its wide varying functionality, parameters such as functional groups distribution are relevant to end-use properties of lignin. The reactivities of these lignins will impact on the attributes of the end products. Table 14.4 shows molecular weight and functional groups of lignins.[28]

TABLE 14.4   Molecular Weight and Functional Groups of Lignins

Lignin type	Mn (g mol^{-1})	COOH (%)	OH phenolic (%)	Methoxy (%)
Soda (bagasse)	2160	13.6	5.1	10.0
Organosolv (bagasse)	2000	7.7	3.4	15.1
Soda (wheat straw)	1700	7.2	2.6	16
Organosolv (hardwood)	800	3.6	3.7	19
Kraft (softwood)	3000	4.1	2.6	14

The T_g is a critical parameter to evaluate for any polymer because it is an indirect measure of crystallinity degree and crosslinking (higher crosslinking would lead to a higher temperature for phase transition to liquid state) that relate to the rubbery region of the material.[29] For example, comparing hardwood and softwood milled wood lignin, hardwoods presents a T_g lower (110–130°C) than softwoods (138–160°C). The value of the T_g will to a great degree depend on the moisture content and chemical functionalization and will be lower if the lignin has greater mobility. In addition, a greater molecular weight of the polymer typically translates to a higher T_g, but for lignin the impact of structural variation will need to be calculated into the overall rheological characteristics because it influences degree of polymerization.

The ability to introduce lignin into composite blends is greatly enhanced if its miscibility is increased; the miscibility of a hydrophobic polymer like lignin into a more hydrophilic matrix can be increased by appropriate modification of the phenolic group (hydroxypropyl, butyrate, etc.)[30] or it can be introduced into hydrophobic matrices by the formation of lignin copolymers.[28] At a value of 4 GPa, the mechanical properties of lignin are significantly lower compared to cellulose pulp (40 GPa).[31] Therefore, lignin itself is not appropriate to be used as reinforcing material as has been demonstrated for cellulose. On the other hand, many commercial applications of low-value lignin already require it to be tailored for specific end goals with high value (extrusion moldings, composites, etc.).

14.3.3   HEMICELLULOSES

Hemicelluloses (or heteropolysaccharides) can also be used as a raw feedstock for numerous polymeric biomaterial applications, but thus far they have not received as much attention as cellulose or startch polysaccharides despite their abundance and structural diversity.[32] In many lignocellulosic refining processes to obtain cellulose, hemicelluloses are partly degraded. Hemicelluloses are hydrophilic components of the cell wall and they can be extracted from plant-material either by water and/or alkaline media. Hemicelluloses are classified according to their structure, comprising monomers (D-xylose, D-mannose, D-galactose, L-arabinose and D-glucose) forming xylans, mannans, galactans, arabinans and β-glucans, respectively, in the polymer main chain. In addition, most hemicelluloses possess side groups of 1–2 monosaccharide units and also acetyl groups. The most abundant hemicelluloses are xylans and mannans. While xylans are the most common hemicelluloses and considered to be the major noncellulosic cell wall polysaccharide component of angiosperms (e.g., hardwood, grasses, and cereals), mannans (galactoglucomannans) are the predominant hemicelluloses in softwoods.

Recently, the incorporation of xylans and mannans for biodegradable films has been explored. They have the potential to be blended with other polymers or mixed with nanoparticles to achieve desirable properties.[33]

Xylan consists of β-D-xylopyranose units, linked by (1–4)-bonds, and different side groups depending on the plant source. The degree of polymerization (DP) of native cellulose is ten to one hundred times higher than that of hemicellulose. For instance, the DP of wood cellulose was reported to be 10,000 while the DP of wood arabino glucuronoxylan is about 100 and that of glucuronoxylan is about 200. In hardwood xylans, seven out of ten xylose units in the backbone contain an O-acetyl group at C-3 or C-2 (Fig. 14.7). The acetyl groups cannot be found in softwood xylans.

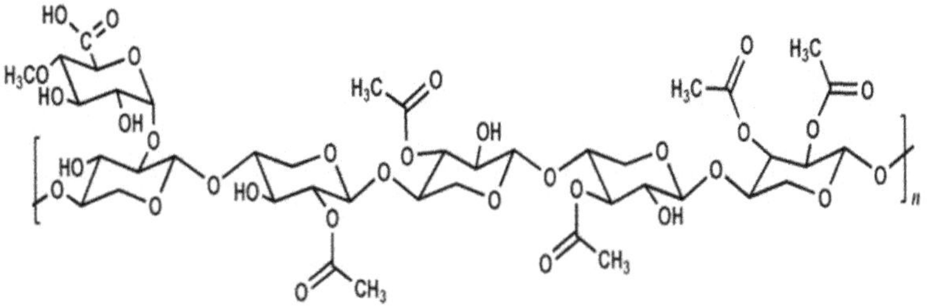

FIGURE 14.7   Structure of O-acetyl-4-O-methylglucuronoxylans structure.

Alternating D-glucopyranosyl and D-mannopyranosyl units attached by β-(1–4) bonds characterize galactoglucomannans and glucomannans. The ratio of glucose to mannose in hardwood glucomannans can vary between 1:2 and 1:1 based on the wood species. Softwood galactoglucomannans can be basically categorized into two fractions with different galactose contents that contain D-galactopyranosyl units attached by α-(1–6) linkages to the backbone mannose chain. Figure 14.8 illustrates the structure of O-acetylated galactoglucomannan.

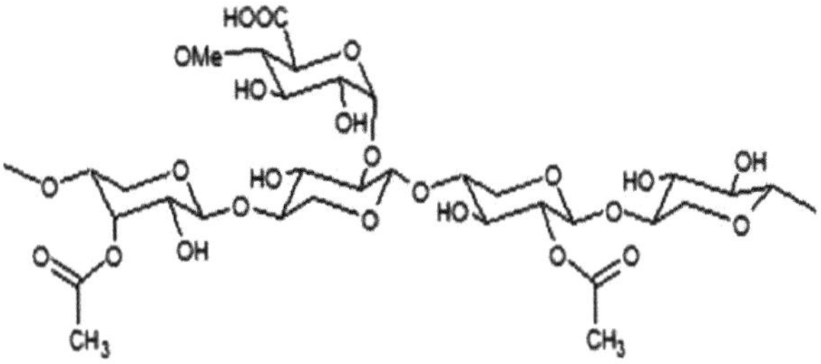

FIGURE 14.8   Structure of O-acetylated galactoglucomannan.

In terms of properties, hemicelluloses have the ability to absorb large amounts of water. They are highly hydrophilic due to the numerous free hydroxyl groups in their structure. Depending on the hemicellulose type, the solubility in water varies according to the degree of substitution: the higher the degree of substitution, the more water-soluble is the hemicellulose. The degree of polymerization also influences the solubility of hemicelluloses. In general, long chains are less water-soluble.[34]

Differently from cellulose, hemicelluloses do not have crystalline domains and they present low degrees of polymerization compared to cellulose, which probably lower their chemical and thermal stability compared to cellulose.

14.4   FACTORS INFLUENCING CHEMICAL, THERMAL AND PHYSICAL PROPERTIES OF BIOCOMPOSITES

14.4.1   MOISTURE

Moisture content of the fibers is due to its hydrophilic nature, the amorphous domains of the fiber and also the amount of interfacial area. The hydrophilic character of polysaccharides influences the overall physical properties because water on the fiber surface acts as a separating agent in the fiber-matrix interface in the fibers.[35]

Therefore, it can be considered a major problem for the use of polysaccharides as reinforcement in biocomposites because it can have a dramatic effect on the biological performance of a composite made from natural fibers, besides affecting drastically mechanical properties of the composites such as compression, flexural and tensile.[36]

Moisture can be minimized by using coupling agents such as silanes, for instance.[37] Such water repellency/inactivity can be explained because the fiber-matrix adhesion is improved via chemical silanol bonds as well as hydrogen bonds that reduce any adverse affects from moisture because it cannot penetrate the bonding system. Thus, fiber drying before processing is a key step in any successful biocomposite processing.

14.4.2   TYPE OF FIBER AND ITS CONTENT

Any successful sustainable composite must consider the natural fiber type and content. Within the type, the length, aspect ratio, and chemical composition have a great influence on the processability. Within the content, the stiffness of a composite can be increased significantly by increasing the proportion of natural fibers within the composite assembly.[38] The impact strength is also improved by higher natural fiber contents, but can also contribute to composite odor emissions and water uptake.[39] A number of other mechanical properties are also attributable to the aspect ratio which influences the quality and quantity of the networking that can be attained in a composite system.[40] For cellulose films, the mechanical properties of showed an increase in strength and stiffness with decreasing fiber size, and this stabilized after a certain number of passes in the homogenizer. Finally, in terms of reinforcement, it was found that the critical factor controlling the contribution of NFC to it was its homogeneity rather than its DP.[25]

14.4.4   MISCIBILITY BETWEEN WOOD BASED-POLYMERS AND COMPOSITES: CHEMICAL MODIFICATION OF SURFACE

The major problem in designing cellulose based-biocomposites is the lack of miscibility between components, that is, hydrophilic biopolymers from wood fibers and hydrophobic substances (matrix), leading to poor adhesion between matrix and fiber in the final composite. To overcome this problem, chemical coupling agents have been employed to improve the adhesion between fibers and matrix through a variety of approaches that include chemical linking, secondary forces, self-assembly, entanglement, and mechanical interblocking.[41] Chemical treatments, therefore, should be considered for improving any type of chemistry for the bonding/adhesion. For example, a number of compounds can promote adhesion such as sodium hydroxide, silane, acetic acid, acrylic acid, isocyanates, potassium permanganate, peroxide, etc. Later, chemical coupling agents such as amphiphilic polymers could be used in small quantities to allow the substrates of interest to bond.[15]

Cellulose based-nanocomposites (whiskers and nanofibrillated cellulose) will be explored in this section because they have been extensively investigated as reinforcing materials in recent years. Whiskers, nanocrystalline cellulose, cellulose nanoparticles, etc., all refer to the isolated crystalline regions of cellulose that possess on of the highest material mechanical strengths known. They are highly ordered and contribute greatly to reinforcing materials because of their high surface area and excellent mechanical properties.[42]

Yet, a critical parameter to good final material properties for the whiskers is that they should be well separated and homogeneously distributed in the matrix. Unfortunately, cellulose whiskers possess a very high surface energy and thus cannot be dispersed well in nonpolar media such as organic solvents or related media; their incorporation as a reinforcement filler for nanocomposites or in complex fluids has up until now be limited to aqueous or polar environments. Their flocculation or aggregation in nonpolar solvents (alkanes, olefins, etc.) can only be avoided by two routes, *viz.*, application of surfactants onto the whiskers (formation of pseudomicelles) or graft-onto or -from the whisker surface using the appropriate polymerization technique.[42]

14.4.5   SURFACE MODIFICATION METHODS FOR NATURAL FIBERS

As indicated already, the fiber-matrix interface quality is very important to promote successful reinforcement of composite materials. There are a number of physical and chemical methods to optimize the interface for varying degrees of efficiency for the adhesion.[43]

14.4.6   PHYSICAL METHODS

Physical methods for improving the adhesion of surfaces attempt to enhance or improve existing surface chemistry or functionalities. For example, surface fibrillation is one method that has been previously used; this is a type of refining or homogenization (NFC, *vide supra*) that takes a top-down approach in increase the surface area of the materials under study. What this means is that the macrofibers are "opened" up to reveal more of the micro and nano-fibrils that compose the fibers. Such an approach allows for the native high surface energy and area to be increased and promote adhesion. Another method is electric discharge that includes corona or cold plasma. Cold plasma has been used to increase bondability without significantly impacting the bulk properties of the substrate treated. Carlsson[44] was able to show via a hydrogen plasma treatment of a pure cellulose substrate that the hydroxyl content on the surface was diminished to provide lower molecular weight fragments that were more hydrophobic.

14.4.7 CHEMICAL METHODS

There are numerous chemical methods that have been used to modify the surface energy characteristics of natural fibers, but only a few prominent methods will be briefly mentioned here, including alkaline, liquid ammonia, and esterification.

In the alkaline or basic treatment, natural fibers are dissolved or mercerized so that they can be incorporated into thermoplastics or thermosets as reinforcement agents. It is one of the very best ways known to increase the amount of amorphous cellulose at the expense of the crystalline cellulose; this latter event occurs by compromising the tight spacing between the cellulose chains by disruption of the hydrogen bonding interface and loss of the crystallinity. The following reaction is a good summation of the chemistry that occurs:

$$\text{Fiber-OH} + \text{NaOH} \; à \; \text{Fiber-O}^- \, \text{Na}^+ + \text{H}_2\text{O} \tag{1}$$

The mercerization process consumes the crystalline cellulose I form as shown in the reaction above by the penetration of the alkali into the cellulose H-bonding network which forms alkali cellulose. The alkali cellulose is then washed by a water treatment to remove the unreacted alkali upon which a regenerated cellulose, cellulose II, is then formed. It has been reported that flax fibers have tremendously enhanced strength and stiffness as a result of the treatment.[45] These results are likely due to two important effects resulting from the treatment: 1-increase in surface roughness which increases the friction and resultant interlocking among the fibers, and 2-increase in the number of exposed cellulose groups on the surface thus having more hydroxyl groups available for H-bonding or reaction.

The second chemical treatment, liquid ammonia, arose as an alternative to mercerization for cotton in the 1960s. Due to its low viscosity and surface tension, it penetrates the cotton H-bonded network very effectively to form a complex compound after the rupture of the H-bonded network. The ammonia is small enough to penetrate crystalline regions very well and cause cellulose I to cellulose III which can reform I after treatment with hot water.

Esterification and the associated etherification are two of the dominant methodologies used to derivatize cellulosics. The reaction generally proceeds via the hydroxyl groups via the introduction of organic acids or anhydrides. Many esters are possible depending on the nature of the alkylating/acylating group used. Possibilities for esterification include the formation of the formate, acetate, propionate, and butyrate (1–4 carbon atoms) in addition to the longer laurate (12 carbons) and stearate (18 atoms) final products. However, the most popular esterification method is acetylation, which occurs by the use of acetic anhydride to form the natural fiber acetate and acetic acid.

With respect to nanocrystals, two routes are available to obtain nonflocculated dispersions in an appropriated organic medium:

1. Surface coating with long-chained surfactants (polar heads and long hydrophobic tails);
2. Hydrophobic chain grafting at their surface.

14.5 OVERALL CHARACTERISTICS OF REINFORCING FIBERS FROM WOOD USED IN BIOCOMPOSITES AS WELL AS THEIR MECHANICAL PROPERTIES

The interaction between fiber and matrix considerably influences the mechanical properties of reinforced composites.

The mechanical properties of biocomposites depend on a number of factors such as the quantity and type of fiber added to the material, nature and amount of polymer matrix has been pointed as the most important parameters, but also the distribution and orientation of reinforcing fillers, nature of filler-matrix interfaces, interphase region and temperature of production influences the mechanical properties of the biocomposites. The amount of coupling agents (required to make the adhesion between surface of fiber and matrix) in the composite is dependent on the fiber content and type and also influences the mechanical and other physical properties of the composites.

14.5.1   CELLULOSE

Because cellulose-based nanocomposites (whiskers and nanofibrillated cellulose) have been extensively investigated as reinforcing materials, this section will describe the properties of these nanomaterials.

Whiskers are easy to prepare because they disperse in water (if formed by sulfuric acid), can come in a variety of aspect ratios, and may be used in composites. The high surface area of the crystals because of their high aspect ratios and size offers an interphase that is very amenable to chemical and physical interactions that improve the physicochemical properties of the composite.

Nanoparticles and nanofibrils display high specific area and very high moduli that give them the highest consideration as fillers for a number of matrices. Recently, studies have focused on understanding the origin of the mechanical reinforcing effect by using predictive models that considered the following criteria: strong interactions among the whiskers and a mechanical percolation effect. [16b]

NFC has already been demonstrated to have excellent mechanical properties as a result of the homogenization process. In fact, the higher surface area of NFC tends to provide a stiffer material because of the network formation. Such a phenomenon can enhance the tensile strength when the material is dried and can lead to improved interactions with the matrix.

14.5.2 LIGNIN

Lignin blending polymers makes a greater range of potential materials properties available and, in general, results in high-performance composite materials as a result of synergistic interactions between the components. Like cellulose composites, lignin combined with immiscible polymer can lower the mechanical properties of the blend to at least of one of the individual component. When it is mixed in a blend of both natural and synthetic polymers, it can increase both the modulus and cold crystallization temperature although it decreases the melt temperature. Lignin-polymer miscibility can be improved by addition of plasticizers because they reduce the degree of self-association between lignin molecules.[46] The chemical functionalization of lignin (hydroxyl and carboxylic acid groups) allows higher compatibility with different polymer types. The opportunity to modulate the hydrophilicity of lignin is one that should be explored because it would increase its utility and potentially offer greater tensile and bulk modulus while protecting against oxidative degradation from UV radiation or elevated temperature.[28]

Lignin blending with synthetic polymers has been previously reviewed[47] within the context of protein–lignin blends, starch–lignin blends, epoxy–lignin composites, phenol-formaldehyde resins where all or part of the phenol is from lignin, polyolefin–lignin blends, lignin blends with vinyl polymers, lignin–polyester blends, lignin as a component of polyurethanes, synthetic rubber–lignin blends, graft copolymers, and lignin incorporation into other polymer systems.

Lignin-based polyurethane composites have been extensively studied. In this type of composites, depending on the type of filler used, many applications are possible. For instance, using wood powder as fillers leads to an effective and cost-saving utilization of lignin. Inorganic fillers increase thermal stability for practical use as housing materials and also residue from agriculture, food, and textile industries are described.[27]

Biocomposites based on a ternary system containing softwood Kraft lignin, poly L-lactic acid (PLLA) and polyethylene glycol (PEG) have been developed.[33,48] Binary systems containing lignin (PLLA/Lignin) show higher stiffness than PLLA/PEG system and good adhesion between the particles and the matrix. However, the ternary systems where PLLA was plasticized with 30 wt.% PEG and filled with lignin exhibited higher deformability in comparison with the unplasticized PLLA/ lignin binary system. Thus, in the ternary systems with higher amount of PEG, a good balance between flexibility and stiffness has been achieved.

14.5.3 HEMICELLULOSES

Few studies have been reported hemicelluloses as polymer comprising composites. The reason for that is probably due to insufficient quality and/or noneconomical viable materials to permit direct industrial application.

14.5.4 XYLANS-BASED COMPOSITES

Xylans cannot only form films, but they can improve film formation/properties when blended with other natural polymers such as chitosan, agar, starch, and gluten. The attraction of using xylans is because they offer the ability to provide biodegradability, increase system interactions thus enhancing mechanical properties, and decrease water vapor permeability.[33]

Several polymer blends have already demonstrated improved mechanical properties and gas/moisture barrier properties when compared to the neat materials.

Several reports of xylan-based films and composites, chemically modified or not, have been found in the literature in the recent years.[33]

The tensile data of xylan films reinforced with sulfonated nanocrystalline cellulose lead to a substantial improvement in strength properties. Addition of 7 wt.% of sulfonated nanocrystalline cellulose increased the tensile energy absorption of xylan films by 445% and the tensile strength of the film by 141%.[49] Furthermore, films to which 7% sulfonated nanocrystalline cellulose were added showed that nanocrystalline cellulose produced with sulfuric acid (sulfonated nanocrystalline cellulose) were significantly better at increasing film strength than nanocrystalline cellulose produced by hydrochloric acid hydrolysis of cellulosic fibers.

Sulfated nanocrystalline cellulose from Kraft pulp was use as a reinforce material of xylans films.[50] The results showed that xylan films reinforced by 10% sulfonated nanocrystalline cellulose exhibited a 74% reduction in specific water transmission properties with respect to xylan film and a 362% improvement with respect to xylan films reinforced with 10% softwood Kraft fibers.

Moisture barrier properties of xylan-based films were reinforced with several cellulosic sources including nanocrystalline cellulose, acacia bleached Kraft pulp fibers and softwood Kraft fibers. The films with 10% of sulfonated nanocrystalline cellulose exhibited the lowest permeability value of 174 g mil/hm^2 among the composite films studied.[51] The results showed that xylan films reinforced with 10% sulfuric nanocrystalline cellulose exhibited reductions in water transmission rates of 362%, 62% and 61% over films prepared with 10% softwood Kraft fibers, 10%

acacia fiber and 10% hydrochloric acid prepared nanocrystalline cellulose, respectively.

Mechanical properties of films made by positively and negatively xylans were studied.[52] It was found that the measured tensile strength values were the best on quaternized films (64.3 MPa) while the Young's modulus values were higher on hydroxypropyl sulfonated xylan film (3350 MPa).

14.5.6 MANNANS-BASED COMPOSITES

Few studies on mannan-based composites are emphasized on the use of glucomannan as a film-forming component. Films produced from pure glucomannan have excellent mechanical properties and the improvement of those properties has been sought by blending glucomannans with other polysaccharides, proteins (chitosan, soy protein, sodium alginate, carboxymethyl cellulose, cellulose, gelatin, starch, etc.) or synthetic polymers.[33]

14.5.7 FACTORS INFLUENCING THE CHEMICAL, THERMAL, AND PHYSICAL PROPERTIES OF BIOCOMPOSITES

In recent decades, the interest in wood-based materials has increased especially for their biodegradability and abundance in nature. The main bio-based materials as already have been targeted within this chapter are natural fibers (NF), microfibrillated cellulose (MFC), nanofibrillated cellulose (NFC), nanocrystalline cellulose (NCC), Xylan and lignin. The majority of these materials can be used alone or especially in combination with other polymeric materials. Except lignin, which has a hydrophobic nature, the combination of those materials with other polymers usually faces the issue of the hydrophilic-hydrophobic incompatibility, an issue that has generally been evident with most synthetic polymers from base oil. Therefore, the surface modification of those materials is an excellent strategy to achieve the desired properties of the final composite.

In the following sections, the effect of the hydrophilic nature of wood base materials, the type of material and its concentration, the degree of dispersibility of such materials in nonpolar solvents and their characteristics according to obtaining the targeted mechanical, thermal, and surface energy properties of the composites will be considered.

14.5.8 MOISTURE RELATED TO HYDROPHILIC NATURE OF THE WOOD-BASED MATERIALS

According to Faruk et al.[35] the main disadvantages of natural fibers in reinforcement of composites are the poor compatibility between the fiber and the matrix and their relative high moisture absorption. Therefore, natural fiber modifications are considered in modifying the fiber surface properties to improve their adhesion with different matrices. There are several treatments to modify natural fibers including physical, chemical, and enzymatic. These behaviors can be applied to other wood-based materials.

14.5.9 FIBER TYPE (OR CELLULOSE-BASED MATERIAL TYPE) AND CONTENT

The type of the cellulose-based materials such as natural fibers, nanocrystals, and nanofibrillated materials, and their content can affect the distribution of these materials on the composite structure to affect mainly the mechanical, thermal, and surface energy properties of these composites.

Coir fibers treated with water, alkali (mercerization) and bleaching were incorporated in starch/ethylene vinyl alcohol copolymers (EVOH) blends and were studied by Rosa et al. (2009).[53] Mechanical and thermal properties of starch/EVOH/coir biocomposites were evaluated. The results showed that all treatments produced surface modifications and improved the thermal stability of the fibers and consequently of the composites. The best results were obtained for mercerized fibers where the tensile strength was increased by about 53% as compared to the composites with untreated fibers, and about 33.3% as compared to the composites without fibers. The authors believe that the mercerization improved fiber–matrix adhesion, allowing an efficient stress transfer from the matrix to the fibers.

Venkateshwaran, Perumal, and Arunsundaranayagam[54] treated the surface banana fibers with alkali solution to change the fiber hydrophilic nature and the mechanical and viscoelastic behavior of the resultant composites with an epoxy matrix were evaluated. The alkali (NaOH) concentrations used were 0.5%, 1%, 2%, 5%, 10%, 15% and 20%. They found that 1% NaOH treated fiber reinforced composites behaved superiorly in terms of mechanical properties as opposed to other treated and untreated fiber composites. The authors concluded that the alkali treatment plays a significant role in improving the mechanical properties and decreasing the moisture absorption rate.

The effect of the nanocrystalline cellulose concentration of cotton fiber on the properties of starch-based nanocomposites was studied by Lu et al. (2005).[55] These authors found a positive correlation with the resistance (from 2.5 MPa to 7.8 MPa), with the modulus (from 36 MPa to 301 MPa) and with the surface energy, for 0% to 30% NCC concentration.

The study of the orientation effect of NCCs in the poly(3-hydroxybutyrate-co3-hydroxyvalerate)-(PHBV) matrix by using an electric field on the nanocomposite mechanical anisotropy was done by Ten et al.[56] These authors showed that NCC concentration strongly influenced the degree of NCC alignment under the electric field. High NCC concentration (>4 wt.%) led to high viscosity of the suspension and high restraint on CNW mobility. This caused the electric field to become ineffective in aligning the nanocrystalline cellulose particles. The aligned PHBV/NCCs nanocomposites showed substantial mechanical anisotropy. The authors suggest that the method developed in their paper can be used to prepare NCC nanocomposites with desired directional reinforcement.

The effect of the sulfuric hydrolysis time of the pea hull fibers on the isolated nanocrystalline cellulose structure and on the pea starch-based nanocomposite made by each NCC dispersion properties was reported by Chen et al. (2009).[57] The results revealed that the hydrolysis time had a great effect on the structure (including length (L), diameter (D) and aspect ratio values (L/D)) of the nanocrystalline cellulose particles, as well as on the structure and performance of the resulting nanocomposites. The authors found a negative correlation with the hydrolysis time length and the diameter of the nanocrystalline cellulose pea. The nanocomposite films exhibited higher ultraviolet absorption, transparency, tensile strength, elongation at break, and water-resistance than both the neat pea starch film and the nanocomposites with pea hull fibers without hydrolysis treatment.

Three different cellulose-based materials were used to reinforced acrylic films, acacia pulp fibers, nanocrystalline cellulose and nanocellulose balls, and their strength properties were evaluated.[58]Nanocrystalline cellulose reinforced composites had enhanced strength properties compared to the acacia pulp and nanoball composites. AFM analysis indicated that the nanocrystalline cellulose reinforced composite exhibited decreased surface roughness.

Silvério et al.[59] studied the effect of sulfated nanocrystalline cellulose from corncob using three different hydrolysis times (30, 60 and 90 min) in the mechanical and thermal properties of polyvinyl alcohol (PVA) as the polymeric matrix. They found that the NCC from 60 min. hydrolysis time resulted in nanoparticles with larger reinforcing capability. Also, the composites with these nanoparticles improved significantly the tensile strength of 140.2% when only 9% (wt.%) of these were incorporated. These particles presented a needle shaped nature, high crystallinity index (83.7%), good thermal stability (around 185 °C), an average length (L) of 210.8 ± 44.2 nm and a diameter (D) of 4.15 ± 1.08 nm, giving an aspect ratio (L/D) of around 53.4 ± 15.8. Comparing to the others NCC times, this aspect ratio of the nanocrystalline cellulose from time hydrolysis of 60 min was intermediary. Because of this, the authors concluded that the material crystallinity index was a more important parameter to consider than the aspect ratio.

Biodegradable nanocomposites prepared by casting with natural rubber and sugar cane bagasse nanocrystalline cellulose in different ratio were studied by Bras et al. (2010)[60]. The incorporation of nanocrystalline cellulose into rubber resulted in composites with enhanced thermo-mechanical properties and biodegradability. Significant improvement of Young's modulus and tensile strength was observed as a result of NCC addition to the rubber matrix especially at high whiskers' loading.

The use of NCC with hydrophobic polymer matrix such as rubber deteriorates its resistance to water vapor permeation

14.5.10 DISPERSIBILITY OF WOOD-BASED MATERIALS IN THE POLYMERIC MATRIX

Nanocomposite properties made with cellulose-based materials are strongly influenced by the dimensions and aspect ratio of these materials, as well by the dispersibility of these materials in the solvent.[61] According to Cao, Habibi and Lucia[62] the dispersion of reinforcing nanoparticles into a continuous polymer phase to form a nanocomposite has attracted a great deal of attention recently because it can provide significant improvements in thermal and mechanical properties at very low contents of the nanoreinforcement.

Ten et al.[56] observed that thermal, mechanical, dielectric, and dynamic mechanical properties exhibited abrupt transitions at 2.3–2.9 wt.% NCC concentrations due to the change in NCC dispersion state.

Dispersibility of the wood-based materials also can be influenced by the methods that originated these ones. The comparing hydrochloric acid hydrolysis with sulfuric acid one, the last leads to stable aqueous suspensions of nanocrystalline cellulose which are negatively charged and, thus, do not tend to aggregate. During the hydrolysis process, esterification of the surface hydroxyl groups from cellulose takes place and, as a consequence, sulfate groups are introduced.[59,63]

According to Peng et al.,[64] many new nanocomposite materials with attractive properties were obtained by the physical incorporation of nanocrystalline cellulose (NCC) into a natural or synthetic polymeric matrix. In addition, simple chemical modification on NCC surface can improve its dispersibility in different solvents and expand its utilization in nano-related applications, such as drug delivery, protein immobilization, and inorganic reaction template.

14.5.11 METHODS USED TO PRODUCE WOOD-BASED MATERIALS

As mentioned above, the lignocellulosic materials in the wood are arranged in the form of complex matrix following a natural organization of these constituents. The cellulose polymer is composed of glucose monomers containing three free hydroxyl groups in positions 2, 3 and 6. These hydroxyls are responsible for interactions intra and intermolecular (Fig. 14.9). These interactions result in the formation of polymers beams that are arranged in organized crystalline regions and in disorganized amorphous regions, resulting in the formation of the successive structures such as nanofibrils, microfibrils, fibrils and finally the cell wall (Fig. 14.3). Therefore, the deconstruction of the cell wall to obtain individuals materials for different applications, either by chemical, thermal or physical methods originates materials with different size and surface charge, and morphology that may influence the formation of composites.

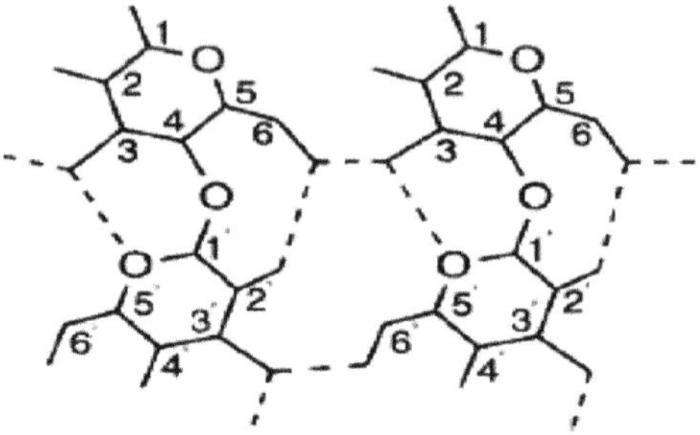

FIGURE 14.9   Hydrogen bonds to intra and intermolecular cellulose I.

According to Samir et al.,[21] nanocrystalline cellulose regions grow under controlled conditions, which enable the formation of individual nanocrystals with high purity. These nanocrystals exhibit highly ordered structure with different dimensions and morphology that can confer significant change in strength, electrical, optical, magnetic, ferromagnetic, conductive, and dielectric materials properties.

The surface characteristics of the nanocrystalline cellulose and also of the nanofibrillated cellulose are related to methods for isolating these nanomaterials. Considering the nanocrystalline cellulose the main isolation processes use strong acids such as sulfuric acid and hydrochloric acid. According Araki et al.,[65] nanocrystalline cellulose obtained with the use of sulfuric acid have a net surface negative charge due to sulfate groups present on the surface of these particles. Moreover, the use of hydrochloric acid gives the neutral net charge of the nanocrystal surface. According to these authors, the aqueous dispersion of the nanocrystals insulated with sulfuric acid provides more stable than the dispersion of the isolated nanocrystals with hydrochloric acid. Also, the morphological characteristics of these nanomaterials are related to the raw material source and to the conditions used in the isolation methods. For example, Elazzouzi-Hafraoui et al.[66] using the same conditions of temperature, type of acid, acid concentration and hydrolysis time, found close dimensions to nanocrystalline celluloses for cotton fiber and microcrystalline cellulose. For cotton, the length was between 105 and 141 nm and a width between 21 and 27 nm, and for microcrystalline cellulose, 105 nm and 12 nm for the length and the width parameters, respectively. On the other hand, cotton fibers treated with different conditions of temperature was found that there was a reduction in the size of nanocrystalline cellulose by increasing the hydrolysis temperature, and, no clear correlation was found between the effects of temperature and diameter of these nanoparticles. Also the effect of the cotton fiber hydrolysis conditions with sulfuric acid was studied by

Dong et al.[67] These authors observed a negative correlation of the nanocrystalline cellulose particles length and positive surface charge of these nanoparticles with increasing time of hydrolysis.

14.6   SURFACE MODIFICATION OF WOOD-BASED MATERIALS

Although the use of materials from biological sources is very promising as a reinforcing material in polymer matrices, some challenges must be overcome considering their dispersity and hydrophobicity. Therefore, the suspension stability of nanocrystalline and nanofibrillated celluloses in water and in other organic solvents is an important aspect for the composite preparation considering a large amount of apolar solvent available.

Nanofibrillated cellulose (NFC) refers to cellulose fibers that have been fibrillated to achieve agglomerates of cellulose microfibril units. Those materials have nanoscale (less than 100 nm) diameter and typical length of several micrometers.[68] The interest in nanofibrillated cellulose (NFC) has increased notably over recent decades mainly because its high mechanical reinforcement ability or barrier property in bionanocomposites or in paper applications, respectively. For the first application, the possibilities to interact with different polymer in matrices can be increased if different functions and/or their contents can be added in its surface. For Missoum and coauthors,[68] the two main nanofibrillated cellulose drawbacks that are associated with its mechanical properties are the high number of hydroxyl groups and the high hydrophilicity, which limits its uses for several applications. For both cases, the surface modification is recommended in order to reduce the number of hydroxyl groups and to increase their compatibility with hydrophobic polymer in matrices. Missoum and coauthors made a complete and recently review of nanofibrillated cellulose with focus on surface modification such as physical adsorption, molecular grafting or polymer grafting.

According to a recent review,[68] the surface characteristics of the NFC depend on the raw material, the process of pretreating the material, technology of production of NFC and surface modification technique itself. Thus, to obtain NFC features and predefined hydrophobicity and dispersibility, it is essential that the raw materials and production steps such as pretreatment, mechanical treatment and surface modification, are previously specified. In summary, wood pulp bleached Kraft or sulfite are most often used as a starting material for the production of NFC but also nonwood fibers have been reported. About the devices, it is common among then the use of high pressure and strong mechanical shearing to fibrillate the fibers

Dhar and coauthors,[69] changed nanocrystalline cellulose (NCC) surface from negative to positive by using surfactant, tetradecyl trimethyl ammonium bromide (TTAB). They observed that the addition of electrolyte or high amount of the surfactant the degree of phase separation in NCC suspension was reduced and the suspension became more stable. Cationically modified NCC was also studied by Zaman et al.[70] NCC, obtained from sulfuric acid hydrolysis of wood cellulose fibers, was rendered cationic by grafting with glycidyltrimethyl ammonium chloride (GTMAC). They found that the cationic surface charge density of NCC can be increased by controlling the water content of the reaction system. The optimum water content was found to be 36 wt.% for aqueous based media and 0.5 water to DMSO volume ratio for aqueous–organic solvent reaction media. As Dhar et al., Zaman and coauthors also found that the cationically modified NCC was well dispersed and stable in aqueous media due to enhanced cationic surface charge density.

Lu, Askeled and Drzal[71] studied the effect of surface modification of microfibrillated cellulose (MCC) in the mechanical properties of composites with epoxy resin matrix. Three different coupling agents were employed to modify a sample of Kraft pulp microfibrillated cellulose from a mix of wood: 3-aminopropyltriethoxysilane, 3-glycidoxypropyltrimethoxysilane, and a titanate. The surface modification changed the character of microfibrillated cellulose from hydrophilic to hydrophobic, maintaining the crystallinity of the material. Among the coupling agents, the titanate showed the most hydrophobic surface. Both treated and untreated materials were easily incorporated into the resin by using acetone as solvent. Better and stronger adhesion between the microfibrils and the epoxy polymer matrix was observed for the treated fibers, which resulted in better mechanical properties of the composite materials.

In Zaman et al. study,[72] hydrophilic surface finishing agent (glycidyl tri-methyl ammonium chloride) that contains nanocrystalline cellulose (NCC) was used to modify the quality characteristic of the polyethylene terephthalate (PET) fabric, coating durability, moisture regain, and wettability. The results showed that the surface properties of the fabric changed from hydrophobic to hydrophilic after the treatment, and the cationic NCC-containing textile surface finish showed superior adhesion onto the cationic dye able (anionic) PET surface over the unmodified NCC. Furthermore, the cationic textile surface finish was capable of withstanding multiple washing cycles.

Non-modified and modified (grafting of n-octadecyl isocyanate) sulfated nanocrystalline cellulose from *Luffa cylindrical* fibers were used to verify the effect of both NCCs on the glass transition temperature, melting point and degree of crystallinity of polycaprolactone (PCL) matrixes.[73] The nanoparticles showed an average length and diameter around 242 and 5.2 nm, respectively, with an aspect ratio around 46. The degree of crystallinity was further increased when using modified nanoparticles. Mechanical tests showed an increase of the modulus of the nanocomposites upon addition of L. cylindrica nanocrystals. This effect was more marked for modified nanoparticles and probably partly due to the increased crystallinity of the PCL matrix. Moreover, chemical grafting promotes the more homogeneous dispersion of nanocrystals within the PCL as shown by the significant improvement of the elongation at break compared to unmodified nanoparticles.

KEYWORDS

- **Bimaterials**
- **Chemical Modification**
- **Lignocellulosics**
- **Polymer**
- **Thermal, and Physical Properties**
- **Wood**

REFERENCES

1. Pritchard, G. (2004). Two technologies merge: wood plastic composites. *Reinforced Plastics*, 48(6), 26–29.
2. (a) Lubin, G. (1982). *Handbook of Composites*. van Nostrand Reinhold: New York, 786.
   (b) Piggott, M. R. (1980). *Load bearing fiber composites*. Pergamon Press: Oxford.
3. Ashori, A. (2008). Wood–plastic composites as promising green-composites for automotive industries *Bioresource Technology*, 99(11), 4661–4667.
4. (a) Bismarck, A., Mohanty, A. K., Aranberri-Askargorta, I., Czapla, S., Misra, M., Hinrichsen, G., & Springer, J. (2001). Surface characterization of natural fibers, surface properties and the water up-take behavior of modified sisal and coir fibers. *Green Chemistry*, 3(2), 100–107.
   (b) Kim, K. H., Tsao, R., Yang, R., & Cui, S. W. (2006). Phenolic acid profiles and antioxidant activities of wheat bran extracts and the effect of hydrolysis conditions. *Food Chemistry*, 95(3), 466–473.
5. John, M. J., & Thomas, S. (2008). Biofibers and biocomposites. *Carbohydrate Polymers*, 71(3), 343–364.
6. Sreekumar, P. A. (2008). Matrices for natural-fiber reinforced composites. In *Properties and performance of natural-fiber composite*, Pickering, K. L., (Ed). Woodhead Publication Limited: Brimingham, 541.
7. Nevell, T. P., & Zeronian, S. H. (1985). *Cellulose chemistry and its application*. Wiley: New York.
8. Bledzki, A. K., & Gassan, J. (1999). Composites reinforced with cellulose based fibers. *Progress in Polymer Science*, 24(2), 221–274.
9. Michell, A. J. (1989). Wood cellulose-organic polymer composites. In *Composite Asia Pacific*, Institute of Australia: Adelaide, 89, 19.
10. Panthapulakkal, S., Zereshkian, A., & Sain, M. (2006). Preparation and characterization of wheat straw fibers for reinforcing application in injection molded thermoplastic composites. *Bioresource Technology*, 97(2), 265–272.
11. Wood-Plastic Composites: Technologies and Global Markets. http://www.bccresearch.com/market-research/plastics/wood plastic composites tech markets pls034b.html (accessed October 2013).
12. (a) Selke, S. E., & Wichman, I. (2004). Wood fiber/polyolefin composites. *Composites Part A: Applied Science and Manufacturing*, 35(3), 321–326.
    (b) Li, T. Q., & Wolcott, M. P. (2004). Rheology of HDPE–wood composites. I. Steady state shear and extensional flow. *Composites Part A: Applied Science and Manufacturing*, 35(3), 303–311.

13. Singh, S., & Mohanty, A. K. (2007). Wood fiber reinforced bacterial bioplastic composites: Fabrication and performance evaluation. *Composites Science and Technology*, 67(9), 1753–1763.

14. Hamad, W. (2006). On the development and applications of cellulosic nanofibrillar and nanocrystalline materials. *Can. J. Chem. Eng.*, 84(5), 513–519.

15. Grunert, M., & Winter, W. (2002). Nanocomposites of Cellulose Acetate Butyrate Reinforced with Cellulose Nanocrystals. *Journal of Polymers and the Environment*, 10(1–2), 27–30.

16. (a) Heux, L., Chauve, G., & Bonini, C. (2000). Nonflocculating and Chiral-Nematic Self-ordering of Cellulose Microcrystals Suspensions in Nonpolar Solvents. *Langmuir*, 16(21), 8210–8212.
    (b) Favier, V., Dendievel, R., Canova, G., Cavaille, J. Y., & Gilormini, P. (1997). Simulation and modeling of three-dimensional percolating structures: Case of a latex matrix reinforced by a network of cellulose fibers. *Acta Materialia*, 45(4), 1557–1565.

17. Bledzki, A. K., Faruk, O., Huque, M. (2002). Physicomechanical studies of wood fiber reinforced composites. *Polymer-Plastics Technology and Engineering*, 41(3), 435–451.

18. Nilsson, H., Galland, S., Larsson, P. T., Gamstedt, E. K., Nishino, T., Berglund, L. A., & Iversen, T. (2010). A nonsolvent approach for high stiffness all cellulose biocomposites based on pure wood cellulose. *Composites Science and Technology*, 70(12), 1704–1712.

19. Chakraborty, A., Sain, M., & Kortschot, M. (2006). Reinforcing potential of wood pulp-derived microfibers in a PVA matrix. In *Holzforschung*, 60, 53.

20. Sehaqui, H., Allais, M., Zhou, Q., & Berglund, L. A. (2011). Wood cellulose biocomposites with fibrous structures at micro and nanoscale. *Composites Science and Technology*, 71(3), 382–387.

21. Azizi Samir, M. A. S., Alloin, F., & Dufresne, A. (2005). Review of Recent Research into Cellulosic Whiskers, Their Properties and Their Application in Nanocomposite Field. *Biomacromolecules*, 6(2), 612–626.

22. Brannvall, E. (2007). Aspect on strength delivery and higher utilization of strength potential of softwood Kraft pulp fibers. Royal Institute of Technology, Stockholm.

23. Fink, H. P., Ganster, J., & Fraatz, J. (1994). In *Challenges in cellulosic man-made fibers*, Akzo Nobel viskose chemistry seminar, Stockholm, 30 May–3 June, Stockholm.

24. (a) Tashiro, K., & Kobayashi, M. (1991). Theoretical evaluation of three-dimensional elastic constants of native and regenerated celluloses: role of hydrogen bonds. *Polymer*, 32(8), 1516–1526.
    (b) Hsieh, Y. C., Yano, H., Nogi, M., & Eichhorn, S. J. (2008). An estimation of the Young's modulus of bacterial cellulose filaments. *Cellulose*, 15(4), 507–513.

25. Zimmermann, T., Bordeanu, N., & Strub, E. (2010). Properties of nanofibrillated cellulose from different raw materials and its reinforcement potential. *Carbohydrate Polymers*, 79(4), 1086–1093.

26. Revol, J. F., Bradford, H., Giasson, J., Marchessault, R. H., & Gray, D. G. (1992). Helicoidal self-ordering of cellulose microfibrils in aqueous suspension. *International Journal of Biological Macromolecules*, 14(3), 170–172.

27. Hatakeyama, H., & Hatakeyama, T. (2010). Lignin Structure, Properties, and Applications. In *Biopolymers*, Abe, A., Dusek, K., & Kobayashi, S., (Eds.) Springer Berlin Heidelberg 232, 1–63.

28. Doherty, W. O. S., Mousavioun, P., & Fellows, C. M. (2011). Value-adding to cellulosic ethanol: Lignin polymers. *Industrial Crops and Products*, 33(2), 259–276.

29. Gargulak, J. D., & Lebo, S. E. (1999). Commercial Use of Lignin-Based Materials. In *Lignin: Historical, Biological, and Materials Perspectives*, American Chemical Society, 742, 304–320.

30. (a) Ghosh, I., Jain Rajesh, K., & Glasser Wolfgang, G. (1999). Blends of Biodegradable Thermoplastics with Lignin Esters. In *Lignin: Historical, Biological, and Materials Perspectives*,
American Chemical Society, 742, 331–350.
(b) Uraki, Y., Hashida, K., & Sano, Y. (1997). Self-Assembly of Pulp Derivatives as Amphiphilic Compounds: Preparation of Amphiphilic Compound from Acetic Acid Pulp and its
Properties as an Inclusion Compound. In *Holzforschung International Journal of the Biology,
Chemistry, Physics and Technology of Wood*, 51, 91.

31. Forss, K., Kokkonen, R., & Sagfors, P. E. (1989). Determination of molecular mass distribution studies of lignins by gel permeation chromatography. In *Lignin properties and materials*,
Glasser, W. G., Sarkanen, S., Eds. American Chemical Society: Washington, D.C. 124–133.

32. (a) Gabrielii, I., & Gatenholm, P. (1998). Preparation and properties of hydrogels based on
hemicellulose. *J. Appl. Polym. Sci.*, 69(8), 1661–1667.
(b) Silva, T. C. F., Habibi, Y., Colodette, J. L., & Lucia, L. A. (2011). The influence of the
chemical and structural features of xylan on the physical properties of its derived hydrogels.
Soft Matter, 7(3), 1090–1099.
(c) Silva, T. C. F., Ilari, F., Youssef, H., Colodette, J. L., & Lucia, L. A. (2012). A Facile Approach for the Synthesis of Xylan Derived Hydrogels. In *Functional Materials from Renewable Sources*, American Chemical Society 1107, 257–270.
(d) Kayzerilioğlu, B. Ş., Bakir, U., Yilmaz, L., & Akkaş, N. (2003). Use of xylan, an agricultural by-product, in wheat gluten based biodegradable films: mechanical, solubility and water
vapor transfer rate properties. *Bioresource Technology*, 87(3), 239–246.

33. Mikkonen, K. (2013). Recent Studies on Hemicellulose-Based Blends, Composites and Nanocomposites. In *Advances in Natural Polymers*, Thomas, S., Visakh, P. M., & Mathew, A. P.,
(Eds.) Springer Berlin Heidelberg, 18, 313–336.

34. Tenkanen, M. (2003). Enzymatic Tailoring of Hemicelluloses. In *Hemicelluloses: Science and
Technology*, American Chemical Society, 864, 292–311.

35. Faruk, O., Bledzki, A. K., Fink, H. P., & Sain, M. (2012). Biocomposites reinforced with
natural fibers: 2000–2010. *Progress in Polymer Science*, 37(11), 1552–1596.

36. Dhakal, H. N., Zhang, Z. Y., & Richardson, M. O. W. (2007). Effect of water absorption on the
mechanical properties of hemp fiber reinforced unsaturated polyester composites. *Composites
Science and Technology*, 67(7–8), 1674–1683.

37. Ismail, H., Shuhelmy, S., & Edyham, M. R. (2002). The effects of a silane coupling agent on
curing characteristics and mechanical properties of bamboo fiber filled natural rubber composites. *European Polymer Journal*, 38(1), 39–47.

38. (a) Gejo, G., Kuruvilla, J., Boudenne, A., & Sabu, T. (2010). Recent advances in green com
posites. *Key Engineering Materials*, 425, 107–166.
(b) Arbelaiz, A., Fernández, B., Ramos, J. A., Retegi, A., Llano-Ponte, R., & Mondragon, I.
(2005). Mechanical properties of short flax fiber bundle/polypropylene composites: Influence
of matrix/fiber modification, fiber content, water uptake and recycling. *Composites Science
and Technology*, 65(10), 1582–1592.

39. Bledzki, A. K., Jaszklewicz, A., Murr, M., Sperber, V. E., Lutzkendorf, R., & Reubmann,
T. (2008) Processing techniques for natural and wood-fiber composites. In *Properties and
performance of natural-fiber composites*, Pickering, K. L., (Ed.) Woodhead Publishing: Cambridge, 163–192.

40. Liu, W., Drzal, L. T., Mohanty, A. K., & Misra, M. (2007). Influence of processing methods
and fiber length on physical properties of kenaf fiber reinforced soy based biocomposites.
Composites Part B: Engineering, 38(3), 352–359.

41. Lu, J., Wu, Q., & McNabb, H. (2000). Chemical Coupling in Wood Fiber and Polymer Composites: A Review of Coupling Agents and Treatments. *Wood and Fiber Science*, 32(1), 88–
104.

42. Kamel, S. (2007). Nanotechnology and its application in lignocellulosic composites, a mini review. *Express Polymer Letters*, 1(9), 546–575.
43. Bledzki, A. K., Reihmane, S., & Gassan, J. (1996). Properties and Modification Methods for Vegetable Fibers for Natural Fiber Composites. *J. Appl. Polym. Sci.*, 59(8), 1329–1336.
44. Carlsson, C. M. G., & Stroem, G. (1991). Reduction and oxidation of cellulose surfaces by means of cold plasma. *Langmuir*, 7(11), 2492–2497.
45. Jähn, A., Schröder, M. W., Füting, M., Schenzel, K., & Diepenbrock, W. (2002). Characterization of alkali treated flax fibers by means of FT Raman spectroscopy and environmental scanning electron microscopy. *Spectrochimica Acta Part A: Molecular and Biomolecular Spectroscopy*, 58(10), 2271–2279.
46. Feldman, D., Banu, D., Campanelli, J., & Zhu, H. (2001). Blends of vinylic copolymer with plasticized lignin: Thermal and mechanical properties. *J. Appl. Polym. Sci.*, 81(4), 861–874.
47. (a) Feldman, D. (2002). Lignin and Its Polyblends A Review. In *Chemical Modification, Properties, and Usage of Lignin*, Hu, T., Ed. Springer US 81–99.
    (b) Stewart, D. (2008). Lignin as a base material for materials applications: Chemistry, application and economics. *Industrial Crops and Products*, 27(2), 202–207.
48. Rahman, M. A., De Santis, D., Spagnoli, G., Ramorino, G., Penco, M., Phuong, V. T., & Lazzeri, A. (2013). Biocomposites based on lignin and plasticized poly(L-lactic acid). *J. Appl. Polym. Sci.*, 129(1), 202–214.
49. Saxena, A., Elder, T. J., Pan, S., & Ragauskas, A. J. (2009). Novel nanocellulosic xylan composite film. *Composites Part B: Engineering*, 40(8), 727–730.
50. Saxena, A., & Ragauskas, A. J. (2009). Water transmission barrier properties of biodegradable films based on cellulosic whiskers and xylan. *Carbohydrate Polymers*, 78(2), 357–360.
51. Saxena, A., Elder, T. J., & Ragauskas, A. J. (2011). Moisture barrier properties of xylan composite films. *Carbohydrate Polymers*, 84(4), 1371–1377.
52. Šimkovic, I., Gedeon, O., Uhliariková, I., Mendichi, R., & Kirschnerová, S. (2011). Positively and negatively charged xylan films. *Carbohydrate Polymers*, 83(2), 769–775.
53. Rosa, M. F., Chiou, B. S., Medeiros, E. S., Wood, D. F., Williams, T. G., Mattoso, L. H. C., Orts, W. J., & Imam, S. H. (2009). Effect of fiber treatments on tensile and thermal properties of starch/ethylene vinyl alcohol copolymers/coir biocomposites. *Bioresource Technology*, 100(21), 5196–5202.
54. Venkateshwaran, N., Elaya Perumal, A., & Arunsundaranayagam, D. (2013). Fiber surface treatment and its effect on mechanical and viscoelastic behavior of banana/epoxy composite. *Materials & Design*, 47(0), 151–159.
55. Lu, Y., Weng, L., & Cao, X. (2005). Biocomposites of plasticized starch reinforced with cellulose crystallites from cotton seed linter. *Biomolecular bioscience*, 5(11), 1101–1107.
56. Ten, E., Jiang, L., & Wolcott, M. P. (2013). Preparation and properties of aligned poly(3-hydroxybutyrate-co-3-hydroxyvalerate)/cellulose nanowhiskers composites. *Carbohydrate Polymers*, 92(1), 206–213.
57. Chen, Y., Liu, C., Chang, P. R., Cao, X., & Anderson, D. P. (2009). Bionanocomposites based on pea starch and cellulose nanowhiskers hydrolyzed from pea hull fiber: Effect of hydrolysis time. *Carbohydrate Polymers*, 76(4), 607–615.
58. Pu, Y., Zhang, J., Elder, T., Deng, Y., Gatenholm, P., & Ragauskas, A. J. (2007). Investigation into nanocellulosics versus acacia reinforced acrylic films. *Composites Part B: Engineering*, 38(3), 360–366.
59. Silvério, H. A., Flauzino Neto, W. P., Dantas, N. O., & Pasquini, D. (2013). Extraction and characterization of cellulose nanocrystals from corncob for application as reinforcing agent in nanocomposites. *Industrial Crops and Products*, 44(0), 427–436.

60. Bras, J., Hassan, M. L., Bruzesse, C., Hassan, E. A., El-Wakil, N. A., & Dufresne, A. (2010). Mechanical, barrier, and biodegradability properties of bagasse cellulose whiskers reinforced natural rubber nanocomposites. *Industrial Crops and Products*, 32(3), 627–633.
61. Hubbe, M. A., Rojas, O. J., Lucia, L. A., & Sain, M. (2008). Cellulosic nanocomposites: a review. *BioResource*, 3(3), 929–980.
62. Cao, X., Habibi, Y., & Lucia, L. A. (2009). One-pot polymerization, surface grafting, and processing of waterborne polyurethane cellulose nanocrystal nanocomposites. *Journal of Materials Chemistry*, 19(38), 7137–7145.
63. (a) Beck-Candanedo, S., Roman, M., Gray, D. G. (2005). Effect of Reaction Conditions on the Properties and Behavior of Wood Cellulose Nanocrystal Suspensions. *Biomacromolecules*, 6(2), 1048–1054.
    (b) Lima, M. M. D., & Borsali, R. (2004). Rodlike cellulose microcrystals: Structure, properties, and applications. *Macromol. Rapid Commun.*, 25(7), 771–787.
64. Peng, B. L., Dhar, N., Liu, H. L., & Tam, K. C. (2011). Chemistry and applications of nanocrystalline cellulose and its derivatives: a nanotechnology perspective. *Can. J. Chem. Eng.*, 89(5), 1191–1206.
65. Araki, J., Wada, M., Kuga, S., & Okano, T. (1998). Flow properties of microcrystalline cellulose suspension prepared by acid treatment of native cellulose. *Colloids and Surfaces A: Physicochemical and Engineering Aspects*, 142(1), 75–82.
66. Elazzouzi-Hafraoui, S., Nishiyama, Y., Putaux, J. L., Heux, L., Dubreuil, F., & Rochas, C. (2007). The Shape and Size Distribution of Crystalline Nanoparticles Prepared by Acid Hydrolysis of Native Cellulose. *Biomacromolecules*, 9(1), 57–65.
67. Dong, X., Revol, J. F., & Gray, D. (1998). Effect of microcrystallite preparation conditions on the formation of colloid crystals of cellulose. *Cellulose*, 5(1), 19–32.
68. Missoum, K., Belgacem, M., & Bras, J. (2013). Nanofibrillated Cellulose Surface Modification: A Review. *Materials*, 6(5), 1745–1766.
69. Dhar, N., Au, D., Berry, R. C., & Tam, K. C. (2012). Interactions of nanocrystalline cellulose with an oppositely charged surfactant in aqueous medium. *Colloids and Surfaces A: Physicochemical and Engineering Aspects*, 415(0), 310–319.
70. Zaman, M., Xiao, H., Chibante, F., & Ni, Y. (2012). Synthesis and characterization of cationically modified nanocrystalline cellulose. *Carbohydrate Polymers*, 89(1), 163–170.
71. Lu, J., Askeland, P., & Drzal, L. T. (2008). Surface modification of microfibrillated cellulose for epoxy composite applications. *Polymer*, 49(5), 1285–1296.
72. Zaman, M., Liu, H., Xiao, H., Chibante, F., & Ni, Y. (2013). Hydrophilic modification of polyester fabric by applying nanocrystalline cellulose containing surface finish. *Carbohydrate Polymers*, 91(2), 560–567.
73. Siqueira, G., Bras, J., Follain, N., Belbekhouche, S., Marais, S., & Dufresne, A. (2013). Thermal and mechanical properties of bio-nanocomposites reinforced by Luffa cylindrica cellulose nanocrystals. *Carbohydrate Polymers*, 91(2), 711–717.

CHAPTER 15

LDPE/WHEAT GLUTEN HUSK BIOCOMPOSITES APPLIED TO BENZOPHENONE ABSORPTION: DETERMINATION OF PROPERTIES USING COMPUTATIONAL CHEMISTRY

NORMA-AUREA RANGEL-VAZQUEZ, ADRIAN BONILLA PETRICIOLET, and VIRGINIA HERNANDEZ MONTOYA

ABSTRACT

The development of natural fiber reinforced biodegradable polymer composites promotes the use of environmentally friendly materials. The use of green materials provides alternative way to solve the problems associated with agriculture residues. Agricultural crop residues such as oil palm, pineapple leaf, banana, and sugar palm produced in billions of tons around the world. They can be obtained in abundance, low cost, and they are also renewable sources of biomass.

Among this large amount of residues, only a small quantity of the residues was applied as household fuel or fertilizer and the rest, which is the major portion of the residues is burned in the field. As a result, it gives a negative effect on the environment due to the air pollution. The vital alternative to solve this problem is to use the agriculture residues as reinforcement in the development of polymer composites. A viable solution is to use the entire residues as natural fibers and combine them with polymer matrix derived from petroleum or renewable resources to produce a useful product for our daily applications.

Lignocellulosic materials are renewable resources that can be directly or indirectly used for the production of biomolecules and commodity chemicals. However, some of these applications are limited by the close association that exists among the three main components of the plant cell wall, cellulose, hemicellulose and lignin. Therefore, it is only through a clear understanding of this chemistry that one can identify the reasons why lignocellulosics are so resilient to biological processes such as enzymatic hydrolysis and fermentation.

Recently environmental problems caused by the conventional fuel based plastics have become public major concerns. Many countries applied various policies and managements to overcome these problems, for example, recycle reuse and reduce protocol. However, due to the enormous amount of packaging and household plastics used every day, such attempt was found to be far from succeeded. Other modern strategy is to replace the conventional plastics with biodegradable plastics such as modified starches, polylactic acids, polyhydroxyalkanoates and such. However, their prices and applications have always been considerated.

Although manufacture of a true biocomposite would demand a matrix phase sourced largely from renewable resources, the current state of biopolymer technology usually dictates that synthetic thermoplastics or thermosetting materials, such as polyethylene (PE) and polypropylene (PP), are used in commercial biocomposite production. There is still a considerable need for the development of thermosetting materials from renewable resources.

Recent examples of such developments include the use of vegetable oils to build thermosetting resins, which can then be modified to form cross-linkable molecules such as epoxides, maleates, aldehydes and isocyanates.

In recent years, there have been significant breakthroughs in the photoinitiated crosslinking of bulk PE and industrial application of photocrosslinked polyethylene (XLPE) insulated wire and cable. The mechanism and crosslink microstructures of the photocrosslinking of LDPE and its model compounds, the crystalline morphological structures, surface photo-oxidation and stabilization of the XLPE materials, and the photolytic products of benzophenone (BP) as a photoinitiator during the photocrosslinking processes.

Molecular modeling used to be restricted to a small number of scientists who had access to the necessary computer hardware and software. The reliability of the obtained results strongly improved throughout the last decades. During this period Theoretical Chemists developed new strategies to describe the reality and Computational Chemists were able to implement and test models. Nowadays many Experimental Chemists, working either in organic or physical chemistry, can easily take advantage of modern commercial software for both research and teaching purposes.

Computational chemistry is a branch of chemistry that uses principles of computer science to assist in solving chemical problems. It uses the results of theoretical chemistry, incorporated into efficient computer programs, to calculate the structures and properties of molecules and solids

The analysis techniques used were, FTIR to study this effect and an option to justify the obtained results is using theoretical calculations by means of the computational chemistry tools. Using QSAR properties, we can obtain an estimate of the activity of a chemical from its molecular structure only. QSAR have been successfully applied to predict soil sorption coefficients of nonpolar and nonionizable organic compounds including many pesticides. Sorption of organic chemicals in soils or sediments is usually described by sorption coefficients. The molecular electrostatic potential (MESP) was calculated using AMBER/AM1 method. These methods give information about the proper region by which compounds have intermolecular interactions between their units.

The electrostatic potential is the energy of interaction of a point positive charge (an electrophile) with the nuclei and electrons of a molecule. Negative electrostatic potentials indicate areas that are prone to electrophilic attack. The electrostatic potential can be mapped onto the electron density by using color to represent the value of the potential. The resulting model simultaneously displays molecular size and shape and electrostatic potential value. Colors toward red indicate negative values of the electrostatic potential, while colors toward blue indicate positive values of the potential.

15.1 INTRODUCTION

15.1.1 BIOCOMPOSITES

Biocomposites are composite materials comprising one or more phase(s) derived from a biological origin. In terms of the reinforcement, this could include plant fibers such as cotton, flax, hemp and the like, or fibers from recycled wood or waste paper, or even by-products from food crops. Regenerated cellulose fibers (viscose/rayon) are also included in this definition, since ultimately they too come from a renewable resource, as are natural 'nano fibrils' of cellulose and chitin.

Matrices may be polymers, ideally derived from renewable resources such as vegetable oils or starches. Alternatively, and more commonly at the present time, synthetic, fossil-derived polymers preponderate and may be either 'virgin' or recycled thermoplastics such as polyethylene (PE), polypropylene (PP), polystyrene (PS) and polyvinyl chloride (PVC), or virgin thermosets such as unsaturated polyesters, phenol formaldehyde, isocyanates and epoxies (Fig. 15.1). Biofibers are one of the major components of biocomposites.

Biocomposites often mimic the structures of the living materials involved in the process, in addition to the strengthening properties of the matrix that was used, but still providing bio-compatibility, for example, in creating scaffolds in bone tissue engineering. The degree of biodegradability in bio-based polymers depends on their structure and their service environment.

Natural/Biofiber composites are emerging as a viable alternative to glass fiber composites, particularly in automotive, packaging, building, and consumer product industries, and becoming one of the fastest growing additives for thermoplastics. Further, research into biological-inorganic interfaces focuses on the design, synthesis, and characterization of novel amalgams that fuse biological and inorganic materials.

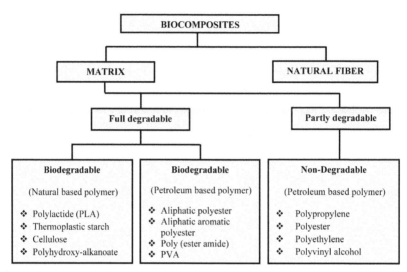

FIGURE 15.1   Classification of biocomposites.

The best known renewable resources capable of making biodegradable plastics are starch and cellulose. Starch is one of the least expensive biodegradable materials available in the world market today. It is a versatile polymer with immense potential for use in nonfood industries. Cellulose from trees and cotton plants is a substitute for petroleum feed stocks to make cellulose plastics.

Another aspect that has gained global attention is the development of biodegradable plastics from vegetable oils like soybean oil, peanut oil, walnut oil, sesame oil and sunflower oil. Green composites from soy protein based bioplastics and natural fibers show potential for rigid packing and housing and transportation applications. One of the major applications for biocomposites that recently has also gained a lot of attention in the North America is in building materials. Nowadays, biocomposites are being used to make products such as decking, fencing, siding, window, door, and so on. Use of bio-composites in building materials offers several advantages such as they are cheap, lightweight, environmental friendly, biorenewable, and more durable.

The majority of biocomposites are currently used in the automotive, construction, furniture and packaging industries, where increasing environmental awareness and the depletion of fossil fuel resources are providing the drivers for development of new more renewable products. Waste reduction is a particularly effective driver for research and experimentation. With the growing imposition of 'producer pays' policies for waste disposal across the developed world as countries realize that the ever-increasing expansion of landfill is not sustainable, more laws are being enacted to encourage the use of renewables.

15.1.2 POLYETHYLENE (PE)

Polyethylene is a thermoplastic polymer consisting of long hydrocarbon chains. Depending on the crystallinity and molecular weight, a melting point and glass transition may or may not be observable. The temperature at which these occur varies strongly with the type of PE. For common commercial grades of medium- and high-density polyethylene (MDPE, HDPE) the melting point is typically in the range 120 to 130 °C (248 to 266 °F).

Most LDPE, MDPE and HDPE grades have excellent chemical resistance, meaning that it is not attacked by strong acids or strong bases. It is also resistant to gentle oxidants and reducing agents. PE burns slowly with a blue flame having a yellow tip and gives off Adour of paraffin. The material continues burning on removal of the flame source and produces a drip. Crystalline samples do not dissolve at room temperature. PE (other than cross-linked polyethylene) usually can be dissolved at elevated temperatures in aromatic hydrocarbons such as toluene or xylene, or in chlorinated solvents such as trichloroethane or trichlorobenzene.

15.1.2.1 POLYMERIZATION

Ethylene is a rather stable molecule that polymerizes only upon contact with catalysts. The conversion is highly exothermic, that is the process releases a lot of heat. Coordination polymerization is the most pervasive technology, which means that metal chlorides or metal oxides are used. The most common catalysts consist of titanium(III) chloride, the so-called Ziegler-Natta catalysts. Ethylene can be produced through radical polymerization, but this route has only limited utility and typically requires high-pressure apparatus.

15.1.2.2. CLASSIFICATION

PE is classified into several different categories based mostly on its density and branching. Its mechanical properties depend significantly on variables such as the extent and type of branching, the crystal structure and the molecular weight. With regard to sold volumes, the most important polyethylene grades are HDPE, LLDPE and LDPE.
- Ultra-high-molecular-weight polyethylene (UHMWPE)
- High-density polyethylene (HDPE)
- Linear low-density polyethylene (LLDPE)
- Low-density polyethylene (LDPE)

(a)   Ultra-high-molecular-weight polyethylene (UHMWPE)

These include can and bottle handling machine parts, moving parts on weaving machines, bearings, gears, artificial joints, edge protection on ice rinks and butchers' chopping boards. It competes with aramid in bulletproof vests, under the trade names Spectra and Dyneema, and is commonly used for the construction of articular portions of implants used for hip and knee replacements.

(b) High-density polyethylene (HDPE)

HDPE is used in products and packaging such as milk jugs, detergent bottles, butter tubs, garbage containers and water pipes. One third of all toys are manufactured from HDPE. In 2007 the global HDPE consumption reached a volume of more than 30 million tons.

(c) Linear low-density polyethylene (LLDPE)

Lower thickness may be used compared to LDPE. Cable covering, toys, lids, buckets, containers and pipe. While other applications are available, LLDPE is used predominantly in film applications due to its toughness, flexibility and relative transparency. Product examples range from agricultural films, saran wrap, and bubble wrap, to multilayer and composite films.

(d) Low-density polyethylene (LDPE)

LDPE is created by free radical polymerization. The high degree of branching with long chains gives molten LDPE unique and desirable flow properties. LDPE is used for both rigid containers and plastic film applications such as plastic bags and film wrap.

15.1.2.3   LOW-DENSITY POLYETHYLENE (LDPE)

Low-density polyethylene (LDPE) is most commonly used as plastic and has been extensively used as a backbone for radiation grafting with different monomers (Fig. 15.2). This is essentially due to its excellent chemical resistance and high impact strength. Due to its high chemical inertness against solvents, acids and bases, polyethylene (PE) as a matrix material became a very popular membrane after grafting with various hydrophilic monomers.

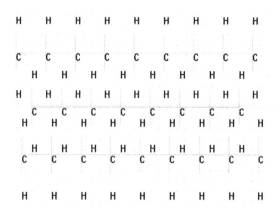

FIGURE 15.2   LDPE structure.

15.1.2.4 PROPERTIES

- Melting Point: ~115 °C
- Crystallinity: Low crystallinity (50–60% crystalline). Main chain contains many side chains of 2–4 carbon atoms leading to irregular packing and low crystallinity (amorphous).
- Strength: Not as strong as HDPE due to irregular packing of polymer chains.
- Transparency: Good transparency, since it is more amorphous (has noncrystalline regions) than HDPE.
- Density: 0.91–0.94 g/cm³, lower density than HDPE.
- Chemical properties: Chemically inert. Insolvent at room temperature in most solvents. Good resistance to acids and alkalis. Exposure to light and oxygen results in loss of strength and loss of tear resistance.
- Tensile elongation at rupture (%): 906.

15.1.2.5 APPLICATIONS

Low-density polyethylene (LDPE) is used mainly in film applications for both packaging and nonpackaging applications (Fig. 15.3). Other markets include extrusion coatings, sheathing in cables and injection molding applications. LDPE is the oldest and most mature of the polyethylenes (PEs). It is characterized by its short and long chain branching, which gives it good clarity and processability although it does not have the strength properties of the other PEs.

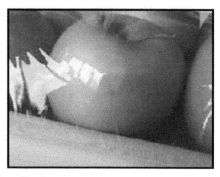

FIGURE 15.3   LDPE application.

15.1.3 LIGNOCELLULOSIC MATERIALS

Lignocellulose refers to plant dry matter, so called lignocellulosic biomass. It is the most abundantly available raw material on the Earth for the production of bio-fuels, mainly bio-ethanol. It is composed of carbohydrate polymers (cellulose, hemicellulose), and an aromatic polymer (lignin). These carbohydrate polymers contain different sugar monomers (six and five carbon sugars) and they are tightly bound to lignin.

Lignocellulosic materials include a variety of materials such as sawdust, poplar trees, sugarcane bagasse, waste paper, brewer's spent grains, switch grass, and straws, stems, stalks, leaves, husks, shells and peels from cereals like rice, wheat, corn, sorghum and barley, among others.

Lignocellulose wastes are accumulated every year in large quantities, causing environmental problems. The major constituents of lignocellulose are cellulose, hemicellulose, and lignin, polymers that are closely associated with each other constituting the cellular complex of the vegetal biomass. Basically, cellulose forms a skeleton, which is surrounded by hemicellulose and lignin. The chemical composition of plants differs considerably and is influenced by genetic and environmental factors (Table 15.1).

TABLE 15.1   Typical Chemical Composition of Various Lignocellulosic Materials

Raw material	Lignin (%)	Cellulose (%)	Hemicellulose (%)
Hardwoods	18–25	45–55	24–40
Softwoods	25–35	45–55	25–35
Grasses	10–30	25–40	25–50

On the other hand, amorphous regions within the cellulose crystalline structure have a heterogeneous composition characterized by a variety of different bonds. Ultimately, this asymmetrical arrangement, which characterizes amorphous regions, is crucial to the biodegradation of cellulose. The accessibility of cell wall polysaccharides from the plant to microbial enzymes is dictated by the degree to which they are associated with phenolic polymer.

Lignocellulose is a complex substrate and its biodegradation is not dependent on environmental conditions alone, but also the degradative capacity of the microbial population. The composition of the microbial community charged with lignocellulose biodegradation determines the rate and extent thereof.

15.1.3.1   CHEMICAL COMPOSITION

Cellulose is a high molecular weight linear homopolymer of repeated units of cellobiose (two anhydrous glucose rings joined via a β-1,4 glycosidic linkage) and is the most abundant of all naturally occurring organic compounds. The long-chain cellulose polymers are linked together by hydrogen and van der Walls bonds, which cause the cellulose to be packed into microfibrils. By forming these hydrogen bounds, the chains tend to arrange in parallel and form a crystalline structure.

Hemicelluloses contain most of the D-pentose sugars, and occasionally small amounts of L-sugars. When compared to cellulose, hemicelluloses differ thus by composition of sugar units, by presence of shorter chains, by a branching of main chain molecules, and to be amorphous, which made its structure easier to hydrolyze than cellulose.

The efficient hydrolysis of cellulose requires the concerted action of at least three enzymes: (1) endo-glucanases to randomly cleave intermonomer bonds; (2) exoglucanases to remove mono- and dimers from the end of the glucose chain; and (3) β-glucosidase to hydrolyze glucose dimers. The concerted actions of these enzymes are required for complete hydrolysis and utilization of cellulose.

The rate-limiting step is the ability of endo-glucanases to reach amorphous regions within the crystalline matrix and create new chain ends, which exo-cellobiohydrolases can attack. Although similar types of enzymes are required for hemicellulose hydrolysis, more enzymes are required for its complete degradation because of its greater complexity compared to cellulose.

Lignin is a very complex molecule constructed of phenylpropane units linked in a large three-dimensional structure. Three phenyl propionic alcohols exist as monomers of lignin: p-coumaryl alcohol, coniferyl alcohol and sinapyl alcohol. Lignin is closely bound to cellulose and hemicellulose and its function is to provide rigidity and cohesion to the material cell wall, to confer water impermeability to xylem vessels, and to form a physic–chemical barrier against microbial attack. Due to its molecular configuration, lignins are extremely resistant to chemical and enzymaticdegradation.

The amounts of carbohydrate polymers and lignin vary from one plant species to another. In addition, the ratios between various constituents in a single plant may also vary with age, stage of growth, and other conditions. However, cellulose is usually the dominant structural polysaccharide of plant cell walls (35–50%), followed by hemicellulose (20–35%) and lignin (10–25%).

15.1.3.2   WHEAT GLUTTEN HUSK

Wheat gluten (WG) is a protein composite found in foods processed from wheat and related grain species, including barley and rye. WG husk is the composite of a *gliadin* and a *glutenin* (Fig. 15.4), which is conjoined with starch in the endosperm of various grass-related grains.

The prolamin and glutelin from wheat (gliadin, which is alcohol-soluble, and glutenin, which is only soluble in dilute acids or alkalis) constitute about 80% of the protein contained in wheat fruit.

The gliadin and glutenin components contribute to dough quality either in an independent manner (additive genetic effects) or in interactive manner (epistatic effects). Commercial WG has a mean composition of 72.5% protein (77.5% on dry basis), 5.7% total fat, 6.4% moisture and 0.7% ash; carbohydrates, mainly starches, are the other major component.

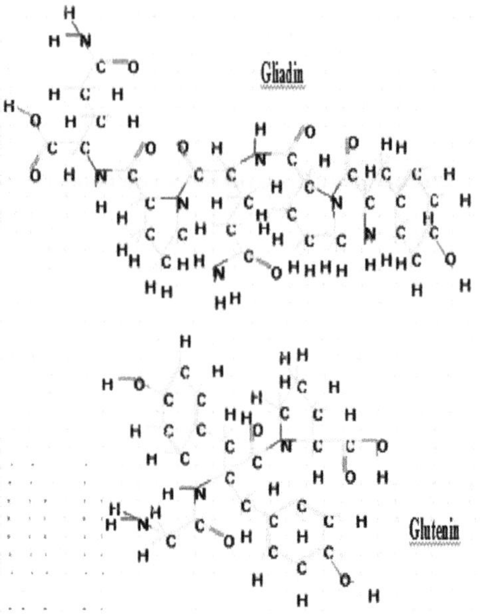

FIGURE 15.4   Structure of wheat gluten husk.

Gliadins are monomeric proteins that can be separated into four groups, alpha-, beta-, gamma- and omega-gliadins. Glutenins occur as multimeric aggregates of high molecular weight (HMW) and low-molecular-weight (LMW) subunits held together by disulfide bonds. In wheat, omega- and gamma-gliadins are encoded by genes at the Gli-1 loci located on the short arms of group 1 chromosomes, while alpha and beta-gliadin-encoding genes are located on the short arms of group 6 chromosomes. LMW glutenins are encoded by genes at the Glu-3 loci that are closely linked to the Gli-1 loci. HMW glutenins are encoded by genes at the Glu-1 loci found on the long arms of group 1 chromosomes. Each Glu-1 locus consists of two tightly linked genes encoding one 'x'-type and one 'y'-type HMW glutenin, with polymorphism giving rise to a number of different alleles at each locus.

The gliadin and glutenin components contribute to dough quality either in an independent manner (additive genetic effects) or in interactive manner (epistatic effects). It was suggested that the apparent effects of gliadins on dough quality should be attributed to the LMW glutenins due to the close linkage of the Gli-1 and Glu-3 loci.

The film forming property of hydrated wheat gluten is a direct outcome of its viscoelasticity. Whenever carbon dioxide or water vapor forms internally in a gluten mass with sufficient pressure to partially overcome the elasticity, the gluten expands to a spongy cellular structure. In such structures, pockets or voids are created which are surrounded by a continuous protein phase to entrap and contain the gas or vapor. This new shape and structure can then be rendered dimensionally stable by applying sufficient heat to cause the protein to denature or devitalize and set up irreversibly into a fixed moist gel structure or to a crisp fragile state, depending on final moisture content.

15.1.4   BENZOPHENONE

Benzophenone is the organic compound with the formula $(C_6H_5)_2CO$, generally abbreviated Ph_2CO (see Fig. 15.5). Benzophenone is a widely used building block in organic chemistry, being the parent diarylketone. Benzophenone is used as a flavor ingredient, a fragrance enhancer, a perfume fixative and an additive for plastics, coatings and adhesive formulations; it is also used in the manufacture of insecticides, agricultural chemicals, hypnotic drugs, antihistamines and other pharmaceuticals. Benzophenone an important class of organic UV filters, are widely used in sunscreen products due to their ability to absorb in the UVA and UVB ranges

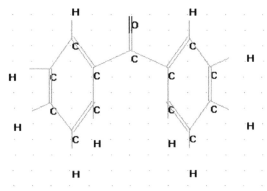

FIGURE 15.5   Benzophenone structure.

15.1.4.1   ABSORPTION PROCESS

Absorption of visible and/or ultraviolet light by a molecule introduces energy sufficient to break or reorganize most covalent bonds. From the relationship $E = hc/\lambda$, we see that longer wavelength visible light (400 to 800 nm) is less energetic (70 to 40 kcal/mole) than light in the accessible shorter wavelength (200 to 400 nm) near ultraviolet region (150 to 70 kcal/mole). Consequently, ultraviolet light is most often used to effect photochemical change. Care must also be taken to construct lamps and reaction vessels from glass that is transparent to the desired wavelength range. The low wavelength cut-off for some common glass types are given in the table on the right. The light required for a photochemical reaction may come from many sources.

Sunscreens may be defined as agents that protect the skin by absorbing damaging light rays and dissipating their energy in some harmless manner (1). When an organic compound absorbs radiation, it is either raised to a higher energy level or disassociated. An excited molecule may dissipate the absorbed energy by collision, fluorescence, or a reaction with other molecules at collision (2). The photochemistry of polyatomic molecules is quite complex and very little is known about photochemical mechanisms of organic compounds.

15.1.5   COMPUTATIONAL CHEMISTRY

Recent years have seen an increase in the number of people doing theoretical chemistry. Many of these newcomers are part time theoreticians, who work on other aspects of chemistry as well. This increase has been facilitated by the development of computer software, which is increasingly easy to use. It is now easy enough to do computational chemistry that you do not have to know what you are doing to do a computation. As a result, many people don't understand even the most basic description of how the calculation is done and are therefore successfully doing a lot of work, which is, frankly, garbage.

Many universities are now offering classes, which are an overview of various aspects of computational chemistry. Since we have had many people wanting to start doing computations before they have had even an introductory course, this document has been written as step one in understanding what computational chemistry is about. Note that this is not intended to teach the fundamentals of chemistry, quantum mechanics or mathematics, only most basic description of how chemical computations are done.

Computational chemistry is a branch of chemistry that uses principles of computer science to assist in solving chemical problems. It uses the results of theoretical chemistry, incorporated into efficient computer programs, to calculate the structures and properties of molecules and solids. Its necessity arises from the well-known fact that apart from relatively recent results concerning the hydrogen molecular ion (see references therein for more details), the quantum many-body problem cannot be solved analytically, much less in closed form.

While its results normally complement the information obtained by chemical experiments, it can in some cases predict hitherto unobserved chemical phenomena. It is widely used in the design of new drugs and materials.

Examples of such properties are structure (i.e., the expected positions of the constituent atoms), absolute and relative (interaction) energies, electronic charge distributions, dipoles and higher multipole moments, vibrational frequencies, reactivity or other spectroscopic quantities, and cross sections for collision with other particles.

Using computer simulation for analysis has many advantages over other design techniques. For example, computer simulation can allow you to see how a system might respond before you design or modify it. This avoids mistakes and one can try different ideas before the real product is produced, making it cheaper as there is no need to make different prototypes every time and testing them out.

A single molecular formula can represent a number of molecular isomers. Isomers are molecules that have the same molecular formula, but have a different arrangement of the atoms in space. That excludes any different arrangements, which are simply due to the molecule rotating as a whole, or rotating about particular bonds. Where the atoms making up the various isomers are joined up in a different order, this is known as structural isomerism. In fact, each isomer is a local minimum on the energy surface.

The determination of molecular structure by geometry optimization became routine only after efficient methods for calculating the first derivatives of the energy with respect to all atomic coordinates became available. Geometry optimization is a method to predict the three-dimensional arrangement of the atoms in a molecule by means of minimization of model energy. The phenomenon of binding, that is to say the tendency of atoms and molecules to conglomerate into stable larger structures, as well as the emergence of special structures depending on the constituting elements, can be explained, at least in principle, as a result of geometry optimization.

Geometry optimization is an essential part of quantum chemical applications. The diversity of the scaling of different methods from linear to exponential implies that there are different requirements for a chosen optimization method.

The proposed method aims to meet two requirements, good scaling with size and reliability, which would be a good match for redundant internal coordinate system-based optimization techniques with linear scaling coordinate transformation.

Evaluation of the related second derivatives allows the prediction of vibrational frequencies if harmonic motion is estimated. More importantly, it allows for the characterization of stationary points. The frequencies are related to the Eigenvalues of the Hessian matrix, which contains second derivatives. If the Eigenvalues are all positive, then the frequencies are all real and the stationary point is a local minimum. If one Eigenvalue is negative (i.e., an imaginary frequency), then the stationary point is a transition structure. If more than one Eigenvalue is negative, then the stationary point is a more complex one, and is usually of little interest.

When one of these is found, it is necessary to move the search away from it if the experimenter is looking solely for local minima and transition structures. The total energy is determined by approximate solutions of the time-dependent Schrödinger equation, usually with no relativistic terms included, and by making use of the Born–Oppenheimer approximation, which allows for the separation of electronic and nuclear motions, thereby simplifying the Schrödinger equation. It treats molecules as collections of nuclei and electrons, without any reference whatsoever to "chemical bonds."

The solution to the Schrödinger equation is in terms of the motions of electrons, which in turn leads directly to molecular structure and energy among other observables, as well as to information about bonding. However, the Schrödinger equation cannot actually be solved for any but a one-electron system (the hydrogen atom), and approximations need to be made. Quantum chemical models differ in the nature of these approximations, and span a wide range, both in terms of their capability and reliability and their "cost."

15.1.5.1   METHODS

15.1.5.1.1   AB INITIO

The term "Ab Initio" is Latin for "from the beginning." This name is given to computations, which are derived directly from theoretical principles, with no inclusion of experimental data. Most of the time this is referring to an approximate quantum mechanical calculation. The approximations made are usually mathematical approximations, such as using a simpler functional form for a function or getting an approximate solution to a differential equation.

The most common type of Ab initio calculation is called a Hartree Fock calculation (abbreviated HF), in which the primary approximation is called the central field approximation. This means that the Coulombic electron-electron repulsion is not specifically taken into account. However, it's net effect is included in the calculation. This is a variational calculation, meaning that the approximate energies calculated are all equal to or greater than the exact energy. The energies calculated are usually in units called Hartrees (1 H = 27.2114 eV). Because of the central field approximation, the energies from HF calculations are always greater than the exact energy and tend to a limiting value called the Hartree Fock limit.

15.1.5.1.2   MOLECULAR MECHANICS METHODS

Molecular Mechanics or force-field methods use classical type models to predict the energy of a molecule as a function of its conformation. This allows predictions of:
- Equilibrium geometries and transition states
- Each atom is simulated as a single particle.
- Relative energies between conformers or between different molecules

Molecular mechanics can be used to supply the potential energy for molecular dynamics computations on large molecules. Molecular mechanics expresses the total energy as a sum of Taylor series expansions for stretches for every pair of bonded atoms, and adds additional potential energy terms coming from bending, torsional energy, van der Waals energy, electrostatics, and cross terms (see Eq. (1)):

$$E = E_{str} + E_{bend} + E_{tors} + E_{vdw} + E_{el} + E_{cross} \qquad (1)$$

The term "AMBER force field" generally refers to the functional form used by the family of AMBER force fields. This form includes a number of parameters; each member of the family of AMBER force fields provides values for these parameters and has its own name.

15.1.5.1.3   SEMI-EMPIRICAL METHODS

Semi-empirical quantum chemistry methods are based on the Hartree–Fock formalism, but make many approximations and obtain some parameters from empirical data. They are very important in computational chemistry for treating large molecules where the full Hartree–Fock method without the approximations is too expensive. The use of empirical parameters appears to allow some inclusion of electron correlation effects into the methods. Within the framework of Hartree–Fock calculations, some pieces of information (such as two-electron integrals) are sometimes approximated or completely omitted. As with empirical methods, we can distinguish if: These methods exist for the calculation of electronically excited states of polyenes, both cyclic and linear.

AM1 is basically a modification to and a reparameterization of the general theoretical model found in MNDO. Its major difference is the addition of Gaussian functions to the description of core repulsion function to overcome MNDO's hydrogen bond problem. Additionally, since the computer resources were limited in 1970s, in MNDO parameterization methodology, the overlap terms, βs and βp, and Slater orbital exponent's ζs and ζp for s- and p- atomic orbitals were fixed. That means they are not parameterized separately just considered as $\beta s = \beta p$, and $\zeta s = \zeta p$ in MNDO. Due to the greatly increasing computer resources in 1985 comparing to 1970 s, these inflexible conditions were relaxed in AM1 and then likely better parameters were obtained.

Optimization of the original AM1 elements was performed manually by Dewar using chemical knowledge and intuition. He also kept the size of the reference parameterization data at a minimum by very carefully selecting necessary data to be used as reference. Over the following years many of the main-group elements have been parameterized keeping the original AM1 parameters for H, C, N and O unchanged.

Of course, a sequential parameterization scheme caused every new parameterization to depend on previous ones, which directly affects the quality of the results. AM1 represented a very considerable improvement over MNDO without any increase in the computing time needed.

AM1 has been used very widely because of its performance and robustness compared to previous methods. This method has retained its popularity for modeling organic compounds and results from AM1 calculations continue to be reported in the chemical literature for many different applications.

15.2   EXPERIMENTAL SECTION

15.2.1   GEOMETRY OPTIMIZATION

In this study the semiempirical method was used for describing the potential energy function of the system. Next a minimization algorithm is chosen to find the potential energy minimum corresponding to the lower-energy structure. Iterations number and convergence level lead optimal structure. The optimizing process of structures used in this work was started using the AM1 method, because it generates a lower-energy structure even when the initial structure is far away from the minimum structure.

The Polak-Ribiere algorithm was used for mapping the energy barriers of the conformational transitions. For each structure, 1350 iterations, a level convergence of 0.001 kcal/mol/Å and a line search of 0.1 were carried out.

15.2.2   STRUCTURAL PARAMETERS

The optimized structural parameters were used in the vibrational wavenumber calculation with AM1 method to characterize all stationary points as minima. The structural parameters were calculated select the Constrain bond and length options of Build menu for two method of analysis.

15.2.3   FTIR

The energy of each peak in an absorption spectrum corresponds to the frequency of the vibration of a molecule part, thus allowing qualitative identification of certain bond types in the sample.

The FTIR was obtained by first selecting menu Compute, vibrational, rotational option, once completed this analysis, using the option vibrational spectrum of FTIR spectrum pattern is obtained for two methods of analysis. The analysis of the structure of LDPE, benzophenone and WG husk with AM1 method was show in Tables 15.2–15.4, respectively.

TABLE 15.2   FTIR Results of LDPE

Assignment	Frequency (cm^{-1})
CH_2 scissoring stretching	4341–4318
CH stretching	3418
CH_2 symmetric stretching	3224
CH_2–CH_2 symmetric stretching	3173–3109
CH symmetric	2567
C–C	2168, 2017, 1888, 1825, 1531, 910, 894
CH_2–CH_2 balanced	1415, 1170, 1101

TABLE 15.3   FTIR Results of Benzophenone

Assignment	Frequency (cm⁻¹)
CH stretching (1° ring)	3221, 3081
CH stretching (2° ring)	3159, 2944
C=O stretching	2010
C=C (2° ring)	1760
C–C (1° ring)	1550
C–C	1436, 1401
CH (1° ring)	1339,1140,1011
CH (2° ring)	1176
CH	875–861, 819
CH (2° ring)	754

TABLE 15.4   FTIR Results of WG Husk

Assignment	Frequency (cm⁻¹)
NH	5741
CH$_2$ stretching	4279–4029
CH asymmetric stretching	4190, 3868
OH groups	3695, 3319
C=C	3588, 3291
C=O, C=C	3555, 3366, 3024
C=O	2971
NH stretching	2875, 2438
C–C, C–N, C–O	2619
C–N, C–C	2226
NH	1454
C–H, C–N, C–O	1289, 1096

15.2.4   ELECTROSTATIC POTENTIAL

After obtaining a free energy of Gibbs or optimization geometry using AMBER/ AM1 methods, we can plot two-dimensional contour diagrams of the electrostatic potential surrounding a molecule, the total electronic density, the spin density, one or more molecular orbitals, and the electron densities of individual orbitals.

HyperChem software displays the electrostatic potential as a contour plot when you select the appropriate option in the Contour Plot dialog box. Choose the values for the starting contour and the contour increment so that you can observe the minimum (typically about –0.5 for polar organic molecules) and so that the zero potential line appears.

A menu plot molecular graph, the electrostatic potential property is selected and then the 3D representation mapped isosurface for both methods of analysis. Atomic charges indicate where large negative values (sites for electrophilic attack) are likely to occur.

15.3   RESULTS AND DISCUSSIONS

15.3.1   QSAR PROPERTIES

The values of different thermodynamic parameters of different materials are described in Table 15.5. The negative value of ΔG (Gibbs free energy) reflects the spontaneity of materials. Attractive interactions between π systems are one of the principal noncovalent forces governing molecular recognition and play important roles in many chemical systems.

Attractive interaction between π systems is the interaction between two or more molecules leading to self-organization by formation of a complex structure which has lower conformation equilibrium than of the separate components and shows different geometrical arrangement with high percentage of yield (Figs. 15.6–15.9).[1]

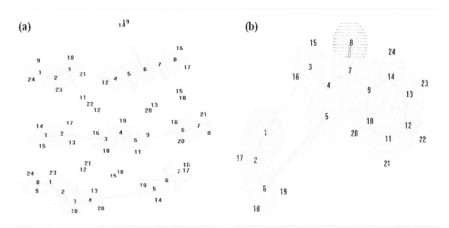

FIGURE 15.6   Geometry of optimization, (a) LDPE, (b) Benzophenone and (c) WG husk, where: red color- oxygen, white color- hydrogen, light blue- carbon and dark blue- nitrogen atom, respectively.

FIGURE 15.7   Geometry of optimization of WG husk, where: red color – oxygen, white color – hydrogen, light blue – carbon and dark blue – nitrogen atom, respectively.

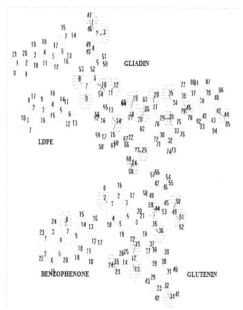

FIGURE 15.8   Geometry of optimization of Composite, where: red color – oxygen, white color – hydrogen, light blue – carbon and dark blue – nitrogen atom, respectively.

The Gibbs free energy value (–5988, –2540.75, –3514.33 and –4576.73 Kcal/mol) shows the crosslinking reaction. Gluten's attainable elasticity is proportional to its content of glutenins with low molecular weights and this content is the responsible of crosslinking.[2] The WG husk increased the compatibility with the polymer matrix and absorption of benzophenone (Fig. 15.9).[3]

Log P is the partition coefficient of solvent in an equimolar mixture of octanol and water. It is a measure of hydrophobicity, which is the rate of interaction with nonpolar molecules. Log P negative shows that these lignocellulosic materials can absorb polar solvents because of its hydrophilic character characteristic of cellulose.[1] LDPE is a hydrophobic synthetic polymer of high molecular weight.

TABLE 15.5   Thermodynamic Dates

Property	LDPE	Benzophenone	WG husk	Composite
DG (Kcal/mol)	–5988	–2540.75	–3514.33	–4576.73
Surface area (Å²)	980.56	441.68	1460.31	2131.47
Volume (Å³)	1682.78	686.91	2756.09	4039.92
Mass (amu)	322.62	182.22	1074.18	1438.75
Log P	10.96	2.68	–15.61	–20.68

15.3.2 STRUCTURAL PARAMETERS

The optimized structure parameter of LDPE/WG husk/benzophenone using AM1 is listed in Tables 1.6–1.7, respectively in accordance with the atom-numbering scheme in Figs. 15.6–15.8. According to results of bond length and bond angle, the deformations depend on the characteristic of the substituents. The carbon atoms are bonded to the hydrogen atoms with an σ bond in ring and substitution of halogen for hydrogen reduces the electron density at the ring carbon atom.

The ring carbon atoms shows a larger attraction on the valence electron cloud of the hydrogen atom resulting in an increase in the C–H force constant and a decrease in the corresponding bond length. The reverse holds well on substitution with electron donating groups. The actual change in the C–H bond length would be influenced by the combined effects of the inductive–mesmeric interaction and the electric dipole field of the polar substituent. The calculated geometric parameters can be used as foundation to calculate the other parameters for the compound.

TABLE 15.6   Length Bonds of Composite (LDPE/Benzophenone/Wheat Gluten Husk)

Bond	Length bond (A)	Bond	Length bond (A)	Bond	Length bond (A)
LDPE		C11–C13	1.6515	WG HUSK: GLUTENIN	
C1–C2	1.5840	C13–N14	1.6617	C1=C2	1.3363

TABLE 15.6 *(Continued)*

Bond	Length bond (A)	Bond	Length bond (A)	Bond	Length bond (A)
C2–C3	1.6508	N14–C15	1.6005	C2–C3	1.4660
C3–C4	1.6515	C15–C17	1.4971	C3=C6	1.3674
C4–C5	1.5514	C17–C16	1.5820	C6–C5	1.4662
C5–C6	1.6024	C16–C13	1.6730	C5=C4	1.3424
C6–C7	1.5623	N14–C18	2.3532	C4–C1	1.4508
		C18–C20	1.5022	C6–C9	1.5878
BENZOPHENONE		C20–C21	1.6740	C9–C10	1.6469
C1=C3	1.3422	C21–C22	1.9692	C10–N11	1.6086
C1–C2	1.4568	C22–C23	1.6231	N11–C12	1.4384
C3–C4	1.4865	C23–N24	1.3735	C12–C14	1.5745
C4=C5	1.3504	C20–N26	1.5369	C14–N15	1.4492
C5–C6	1.4659	N26–C27	1.3647	C10–C27	1.7473
C6=C2	1.3389	C27–C29	1.6036	C27–C28	1.6358
C4–C7	1.4988	C29–C30	1.6226	C28=C29	1.3648
C7=O8	1.2313	C30–C32	1.4520	C28–C30	1.5393
C7–C9	1.4999	C32–C33	1.4586	C30=C31	1.3408
C9=C10	1.3506	C33–N31	1.6043	C31–C32	1.4430
C10–C11	1.4659	N31–C34	1.4586	C32=C33	1.3340
C11=C12	1.3385	C34–C36	1.6437	C33–C29	1.4698
C12–C13	1.4569	C36–C37	1.6749	C10–C35	1.046
C13=C14	1.3419	C37–C38	1.6115	C35–N44	1.5608
C14–C9	1.4853	C38=C39	1.3621	N44–C45	1.6596
		C39–C40	1.4119	C45–C46	1.5439
WG HUSK: GLIADIN		C40=C42	1.3411	C46–C47	1.4405
N1–C2	1.3441	C42–C43	1.3381	C47–C48	1.4794
C2=O3	1.2302	C43=C41	1.4994	C48–N44	1.5963
C2–C4	1.5563	C41–C38	1.5513	C35=O36	1.2550
C4–C5	1.5977	C7=O9	1.2961	C45–C49	1.5807
C5–C6	1.6427	C6–N10	1.3757	C49=O50	1.2214
C6–C7	1.5592	N10–C11	1.3664	C49–O51	1.3636

TABLE 15.7   Angle of Composite (LDPE/Benzophenone/Wheat Gluten Husk)

BOND	ANGLE (Å)	BOND	ANGLE (Å)	BOND	ANGLE (Å)
LDPE		C16–C13–N14	101.928	**WG HUSK: GLUTENIN**	
C1–C2–C3	151.718	C13–N14–C18	121.911	C1=C2–C3	120.605
C2–C3–C4	158.266	N14–18=O19	127.745	C2=C3=C6	125.104
C3–C4–C5	158.488	N14–C18–C22	180.000	C3=C6–C5	110.983
C4–C5–C6	148.094	C21–C22–C23	145.466	C6–C5=C4	125.724
C5–C6–C7	132.073	C22–C23=O25	127.259	C5=C4–C1	120.382
		C22–C23–N24	121.160	C4–C1=C2	117.202
BENZOPHENONE		C14–C18–C20	143.298	C3=C6–C9	118.907
C1–C2=C6	118.959	O19=C18–C20	88.9578	C6–C9–C10	155.503
C1=C3–C4	122.224	C18–C20–C21	12.2580	C9–C10–N11	85.2091
C2=C6–C5	120.633	C18–C20–N26	120.074	C10–N11–C12	156.625
C6–C5=C4	122.151	C20–N26–C27	148.021	N11–C12=O13	126.979
C5=C4–C3	115.849	N26–C27–O28	108.072	O13=C12–C14	101.356
C4–C3=C1	122.224	N26–C27–C29	128.753	C12–C14–N15	162.194
C4–C7=O8	113.670	O28=C27–C29	123.174	C9–C10–C27	153.126
08=C7–C9	113.685	C27–C29–N31	132.242	C10–C27–C28	156.771
C9=C10–C11	122.181	C29–C30–C32	112.954	C27–C28=C29	126.854
C 1 0 – C11=C12	120.626	C30–C32–C33	115.544	C28=C29–C33	126.650
C 1 1 = C 1 2 – C13	118.925	C32–C33–N31	100.396	C29–C33=C32	121.888
C 1 2 – C13=C14	120.236	C33–N31–C29	115.704	C33=C32–C31	116.809
C13=C14–C9	122.203	C33–N31–C34	126.666	C32–C31=C30	120.404
		N31–C34=O35	120.392	C31=C30–C28	126.624
WG HUSK: GLIADIN		N31–C34–C36	137.593	C9–C10–C35	105.685
N1–C2=C3	119.523	O35=C34–C36	102.014	C10–C35=O36	118.280
O3=C2–C4	120.454	C34–C36–N45	100.618	O36=C35–N44	106.970
C2–C4–C5	124.721	N45–C36–C37	63.1441	C35–N44–C45	124.347
C4–C5–C6	134.046	C36–C37–C38	143.739	N44–C45–C46	101.903
C5–C6–C7	135.089	C37–C38=C39	116.559	C45–C46–C47	112.101
C6–C7=O8	120.661	C38=C39–C40	125.463	C46–C47–C48	113.728
C6–C7=O9	120.004	C39–C40=C42	123.560	C47–C48–N44	105.255

TABLE 15.7 *(Continued)*

BOND	ANGLE (Å)	BOND	ANGLE (Å)	BOND	ANGLE (Å)
C5–C6–N10	99.4639	C40=C42–C43	119.710	N44–C45–C49	127.539
C6–N10–C11	171.253	C42–C43=C41	118.941	C45–C49=O50	90.1501
N10–C11=O12	106.070	C43=C41–C38	123.405	C45–C49–O51	170.279
O12=C11–C13	126.586	C40=C42–O44	119.742	O50=C49–O51	80.1285
C11–C13–N14	126.586	C15–C17–C16	87.6709	C48–N44–C45	107.013
C13–N14–C15	94.4986	C17–C16–C13	121.482		
N14–C15–C17	134.420				

TABLE 15.8 FTIR Results of Composite (LDPE/WG Husk/Benzophenone)

Assignment	Frequency (cm⁻¹)
NH_2 asymmetric and symmetric stretching	5964, 3472, 3361
NH and CH stretching	5611, 3462
CH_2 asymmetric stretching (WG husk)	5411, 4876, 4319
C–H (WG husk)	4564, 784
CH aromatic	4012
NH stretching	3969, 3500
CH_2 scissoring (WG husk)	3920
C=C, C=O (benzophenone)	3715, 3677, 3234
O–H stretching	3394
C=O stretching (WG husk)	3258, 2853
C=O (benzophenone)	2743, 1750
C–C, C–N, C–O (WG husk)	2619
C–C (benzophenone)	2229
C–H (benzophenone)	1792
C–C, C–N (benzophenone)	1487
CH (LDPE)	1455, 909
C–O, C–C (WG husk)	992

15.3.3 FTIR

Table 15.8 shows the FTIR bands of composite (LDPE, wheat husk with benzophenone) where the characteristic peaks associated different components are observed. The NH stretching was assigned to 3969 and 3500 cm^{-1} [4,5] and the asymmetric and symmetric stretching vibrations of NH$_2$ grouping are observed at 5964, 3472 and 3361 cm^{-1}, respectively.[6,7] Highly intense and well defined peaks observed at 2743 and 1750 cm^{-1} are due to the C=O stretching vibration of carbonyl group correspond to strong absorption benzophenone group.[4,5] The NH and CH stretching modes arising from amino groups appear around 5611 and 3462 cm^{-1}. The CH$_2$ asymmetric stretching (wheat gluten husk) corresponds to 5411, 4876 and 4319 cm^{-1} and scissoring mode at 3920 cm^{-1}.

The bands at 4564 and 784 cm^{-1} are due to aromatic CH bends. The aromatic CH stretching vibrations appear weak just above 4012 cm^{-1}. The C=C and C=O vibrations of aromatic ring are confirmed at 3715, 1677 and 3234 cm^{-1}. At 3394 cm^{-1} was assigned at OH group of materials. The C=O stretching was attributed at 3258 and 2853 cm^{-1} (wheat gluten husk).[5] The absorbance at 1455 and 909 cm^{-1} were observed due to methylene groups which are proportional to the relative total concentration of carbonyl groups.[8] The peaks of LDPE were observed at 2914 cm^{-1} correspond at CH$_2$ asymmetric stretching, the CH$_2$ deformation were assigned at 1561 and 1439 cm^{-1} and finally the CH$_3$ symmetric stretching at 1341 cm^{-1}.[9]

15.3.4 ELECTROSTATIC POTENTIAL

The electrostatic force is a conservative force. This means that the work it does on a particle depends only on the initial and final position of the particle and not on the path followed. With each conservative force, a potential energy can be associated.

The introduction of the potential energy is useful since it allows us to apply conservation of mechanical energy, which simplifies the solution of a large number of problems. The organic compounds with electron rich (donor) and deficient (acceptor) substituents provide the asymmetric charge distribution in the π electron system and show large nonlinear optical responses.[4]

The important requirements for an efficient photoinitiator are suitable absorption coefficient and high quantum yield of initiation, together with well-adapted absorption range.[5] Pink and green areas in the MESP refer to the regions of negative and positive and correspond to the electron-rich and electron-poor regions, respectively, whereas the gray color signifies the neutral electrostatic potential.

The MESP in case of Figs. 15.9–15.11 clearly suggest that each C–OH, C–O–C bonds represent the most negative potential region of wheat gluten husk. Figures 15.9–15.11 show that NH and CH bond present neutral potential electrostatic region, glutenine structure represent the most negative potential region and finally the CH$_2$ and CH represent the most positive potential region. The values of MESP show that the electronegativity of the benzophenone produces a decrease in nucleophilic areas (green) of lignocellulosic materials.

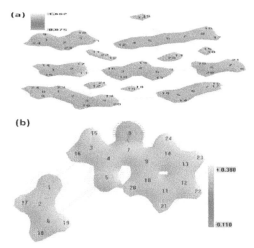

FIGURE 15.9 Electrostatic potential, (a) LDPE, (b) Benzophenone, where: red color – oxygen, white color – hydrogen, light blue –carbon and dark blue –nitrogen atom, respectively.

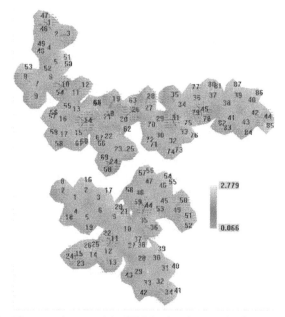

FIGURE 15.10 Electrostatic potential of WG husk, where: red color – oxygen, white color –hydrogen, light blue – carbon and dark blue – nitrogen atom, respectively.

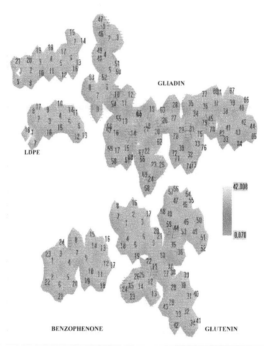

FIGURE 15.11   Electrostatic potential of Composite, where: red color – oxygen, white color – hydrogen, light blue –carbon and dark blue –nitrogen atom, respectively.

15.4   CONCLUSIONS

The use of stiff monomers able to react with the wheat gluten husk surface with only one of the functionalities, thus leaving the remaining second moiety to couple with the polymer matrix during composite processing, such a system gives rise to the formation of a covalent linkage between the matrix and the reinforcing elements and enables perfect stress transfer between the matrix and the reinforcing elements.

The LDPE and wheat gluten husk on an individual basis and such as benzophenone absorption systems were analyzed to determine the applications of lignocellulosic materials. It was determined that the negative value of the ΔG verifies that the absorption process is carried out in a way spontaneous.

The wheat gluten husk on an individual basis and absorption of benzophenone systems was analyzed to determine the applications of Lignocellulosic materials. It was determined that the negative value of the ΔG verifies that the absorption process is carried out in a way spontaneous.

The negative values of Log P show that the absorption is affected due to the hydrophilic character and additionally the absorption plays an important role in both partition and receptor binding processes of absorption (benzophenone). FTIR results show that there are shifts in the peaks of wheat gluten husk attributed to the absorption of benzophenone. The MESP values indicated the nucleophilic and electrophilic regions mainly in the NH, C-O and C=O bonds, respectively.

KEYWORDS

- **Absorption**
- **Biocomposites**
- **Biodegradable Polymer**
- **Computational Chemistry**
- **Natural Fibers**
- **Polyethylene**
- **Wheat Gluten**

REFERENCES

1. Hardinnawirda, K., & Sitirabiatull, I. (2012). Effect of rice husks as filler in polymer matrix composites *J Mech Eng Sci*, 2, 181–186.
2. Tosi, P., Ovidio, D., Napier, J. A., Bekes, F., & Shewry, P. R. (2004). Expression of epitope-tagged LMW glutenin subunits in the starchy endosperm of transgenic wheat and their incorporation into the glutenin polymers, *Theor Appl Genet*, 108, 468–476.
3. Qing, Y., & Rånby, B. (1992). Photoinitiated crosslinking of low-density polyethylene IV: Continuous extrusion application. *Polym Eng Sci.*, 32, 831–835
4. Mohamed, G., Rajarajan, K., Vimalan, M., Madhavan, J., & Sagayaraj, P. (2010). Growth and characterization of pure, benzophenone and paratoluidine doped 2A-5CB crystals. *Arch Appll Sci Res*, 2, 81–93.
5. Liang-Liang, C., Yong, Z., & Wen-Fang, S. (2011). Photoinitiating Characteristics of Benzophenone Derivatives as Type II Macromolecular Photoinitiators Used For UV Curable Resins. *Chem. Res. Chinese Universities*, 27, 145–149.
6. Bledzki, A. K., Mamun, A. A., & Volk, J. (2010). Physical, chemical and surface properties of wheat husk, rye husk and soft wood and their polypropylene composites. *Composites: Part A:* 41, 480–488.
7. Krishnakumar, V., Muthunatesan, S., Keresztury, G., & Sundius, T. (2005). Scaled quantum chemical calculations and FTIR FT-Raman spectral analysis of 3, 4-diamino benzophenone, *Spectrochim Acta A.*, 62, 1081–1088.
8. Morshedian, J., Hoseinpour, P. M., Azizi1, H., & Parvizzad, R. (2009). Effect of polymer structure and additives on silane grafting of polyethylene *eXPRESS Polym Lett*, 3, 105–115.
9. Negi, H., Gupta, S., Zaidi, M. G. H., & Goel, R. (2011). Studies on biodegradation of LDPE film in the presence of potential bacterial consortia enriched soil. *Biologija*, 57141–57147.

CHAPTER 16

NANO-CELLULOSE REINFORCED CHITOSAN NANOCOMPOSITES FOR PACKAGING AND BIOMEDICAL APPLICATIONS

PRATHEEP K. ANNAMALAI and DILIP DEPAN

ABSTRACT

Biobased nanocomposites have gained a huge attention from industrialists and academic researchers. Chitosan is N-deacetylated derivative of the most abundant nitrogen-rich polysaccharide in nature chitin. Chitosan is nontoxic, biodegradable, biocompatible and antibacterial and biologically renewable. In recent decades, by combining the benefits of chitosan and reinforcement with various nanoparticles, a wide range of materials is developed for packaging, agricultural, medical and automotive applications. In this chapter, we discuss the recent developments on chitosan-based nanocomposites using the biologically renewable nanofillers called 'nano-cellulose.' Nano-cellulose is obtained either as shorter nanocrystals or longer nanofibrils from the most abundant natural polymer cellulose, via acid hydrolysis, enzymatic treatment or mechanical shearing methods. Their reinforcement potential in the different matrices is attributed to the mechanical properties of individual nanofibers and the formation of a percolation network that connects the well dispersed cellulose nanocrystals by hydrogen bonds and provides superior performance like mechanical, barrier, controlled drug release, and antibiotic properties. Here we review the recent trends and developments on chitosan-cellulose nanocomposites including processing, properties required for packaging and biomedical engineering applications.

16.1 INTRODUCTION

16.1.1 OVERVIEW ON BIOPOLYMERS FOR PACKAGING AND BIOMEDICAL APPLICATIONS

From the view of sustainable development, the new materials associated with renewable source, low toxicity, high performance and environmental biodegradability

after disposal are enormously explored. The concerns over new materials from renewable resource have recently increased because of the economic consequences
of depleting petroleum resources, the demands from industrialists and customer
for high performance lightweight low-cost materials and the environmental regulations.[1,2] From biomass, polymers can be obtained as native biopolymers, raw
materials for monomers and bio-engineered biopolymers. Polysaccharides such as
cellulose, starch, chitosan/chitin, etc. are the abundantly available biopolymers on
the planet earth. They are replacing the materials for many industrial applications
where synthetic polymers have been materials of choice, traditionally. As the native biopolymers are not conventionally processable, research efforts have been focused on the processing and meeting the requirements of particular applications. For
packaging, the polymeric materials must exhibit flexibility, transparency, water and
gas barrier properties, biodegradability (after disposal) antimicrobial, thermal and
mechanical properties whereas surface adhesion (hydrophilicity), biocompatibility,
biodegradability and dimensional stability.

Since natural biopolymers exhibit poor mechanical and thermal properties
and processability they are very often blended with synthetic polymers synthetic
biopolymers such as, polylactic acid and polycaprolactone and reinforced with particulates.[3] With the recent breakthroughs on nanoscience and nanotechnology, which
allow tuning the materials properties at nano-scale level, the biobased nanocomposites are explored as renewable biomaterials.[4,13] Hence, this chapter focuses on the
recent developments in biobased nanocomposites based on chitosan and cellulose
nanocrystals where both matrix and filler that are biologically renewable (Fig. 16.1).

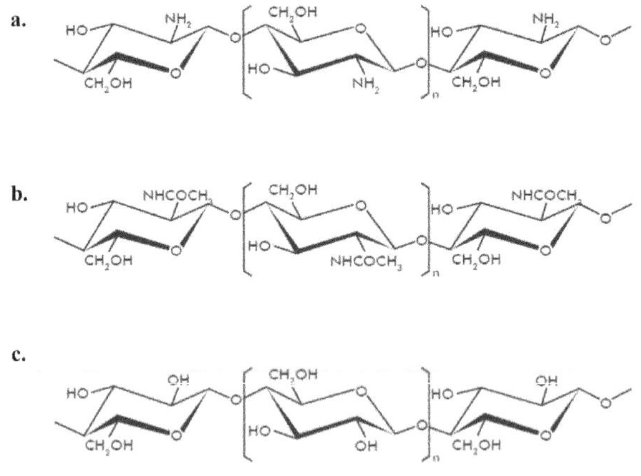

FIGURE 16.1   Chemical structure of bio-renewable polymers chitin/chitosan and cellulose.

The packaging materials based on polymer nanocomposites are recently preferred not only for extending the shelf-life of food products but also for improving

the quality of food by acting as a carrier of some active substances such as antioxidants and antimicrobials.[14] Owing to their biocompatibility, they are also explored for biomedical applications (as drug delivery system, wound dressing and bioresorbable materials and low cytotoxic scaffolds for tissue engineering).[15,18]

16.2 CHITIN/CHITOSAN: A RENEWABLE BIOMATERIAL

As discussed above, Chitosan (CH) is a unique polysaccharide derived from partial de-acetylation of chitin, which is, after cellulose, the most abundant nitrogen-rich polysaccharide. Chitosan is nontoxic, biodegradable, biocompatible and antibacterial and biologically renewable. CH has reactive amino ($-NH_2$) and hydroxyl (-OH) groups that provide many possibilities for covalent and ionic modifications (Fig. 16.1). The advantages of CH are as follows: (a) CH intrinsically possesses strong biological activity, (b) it is biocompatible, biodegradable, bioresorbable and has a hydrophilic surface, which facilitates cell adhesion, proliferation and differentiation (c) due to its cationic nature in physiological pH, CH mediates nonspecific binding interactions with various proteins.[19]

Since then, CH material has been widely investigated in a number of biomedical applications from wound dressings, drug or gene delivery systems, and nerve regeneration to space filling implant.[15,17]

16.2.1 CHITOSAN BLENDS AND NANOCOMPOSITES

CH is currently sold in the USA as a dietary supplement to aid weight loss and lower cholesterol and is approved as food additive in Japan, Italy, and Finland. However, although pure CH has very attractive properties, it lacks bioactivity and is mechanically weak.[20] These drawbacks limit its biomedical applications. For these reasons, it is highly desirable to develop a hybrid material made of CH and appropriate filler, hoping that it can combine the favorable properties of the materials, and further enhance tissue regenerative efficacy. Of particular relevance is that CH is ideally suited to complex with anionic form of cellulose (carboxymethyl cellulose), to enable a number of applications and can be conjugated with functional molecules, antibodies, biotin, and heparin.[21,25] The advantage of blending CH is not only to improve its biodegradability and its antibacterial activity, but also the hydrophilicity, which is introduced by addition of the polar groups able to form secondary interactions (–OH and $-NH_2$ groups involved in H bonds with other polymers). The most promising developments at present are in pharmaceutical and tissue engineering areas.

Clay based biocomposites using biopolymers as matrices are continuously gaining widespread attention from scientific and industrial world. Montmorillonite (MMT) clay and chitosan nanocomposites have been reported to possess excellent mechanical, thermal and bioactive properties.[11,12] Cellulose nanocrystals, have attracted a great deal of interest in the development of nanocomposites owing to their appealing intrinsic properties such as nanoscale dimensions, high surface area, unique morphology, low density, and mechanical strength. Furthermore, they are easily modified, readily available, renewable, and biodegradable and are known to improve the limited mechanical and barrier properties of biopolymers.[26,27]

16.3   NANO-SCALE CELLULOSE PARTICLES

Cellulose which is a linear homopolymer of D-anhydro glucopyranose unit (AGU) polymerized *via* β-1, 4-linkage (Fig. 16.1), can be obtained from plant biomass, bacteria and marine animals called tunicates. During its biosynthesis, cellulose is formed as linear and highly crystalline fibrils at nanoscale with outstanding mechanical properties via intra and intermolecular hydrogen bonding and organized into microscale elementary fibril-bundles separated and cemented by noncrystalline regions containing lignin, pectin, hemicelluloses. The highly crystalline nanofibrils which exhibit higher axial mechanical properties (strength and modulus), can be obtained by eliminating the other amorphous components or defibrillating them *via* acid hydrolysis, enzymatic hydrolysis, oxidation, mechanical treatment, ultrasonic treatment or combinations thereof.[27,28]

Depending on the source of cellulose and the isolation protocol used and the sources, they can be obtained as shorter cellulose nanocrystals (CNC) or whiskers (with 5–20 × 500 nm), longer cellulose nanofibers (CNF) or nanofibrillated cellulose (NFC) (3–20 nm × few microns) and microfibrillated cellulose (20–100 nm × few microns). In addition, nano-scale cellulose can also be directly by microorganisms.[27] Bacterial cellulose (BC) is an emerging and unique biopolymer characterized by very long polymer chains with a degree of polymerization in the range of 4000–10,000, high crystallinity, and high purity with an extremely large amount of water containing. The individual nanofibrils of BC exhibit a high tensile strength of equivalent to that of steel or Kevlar.[29,31]

16.4   NANO-CELLULOSE REINFORCED CHITOSAN NANOCOMPOSITES

16.4.1   PROCESSING

The nanocomposites of cellulose and chitosan has been motivated by their chemical similarities, the functional properties (antimicrobial, water transpiration) and the improvements in materials (mechanical, physical and barrier) properties achieved in parallel by blending the cellulose and chitosan by dissolution using acidic media and ionic liquids.[32] Various organic solvents and mixture of acids have been used to process the composites of chitosan and cellulose. Most reactions consume and produce toxic waste, which is hazardous to the environment. Furthermore, some components of the processing tend to release the atomic oxygen, which attacks the macromolecular chains and degrades them, and severely compromising the mechanical properties of the composites. In order to overcome the above-mentioned issues ionic liquids have been used to dissolve and process cellulose and chitosan composites. A binary system consisting of acidic ionic liquid glycine hydrochloride and neutral ionic liquid 1-butyl-3-methylimidazolium chloride is used as a cosolvent to generate electrospun fibers of chitosan and cellulose. The prepared composite fibers were shown to have excellent thermal and mechanical properties.[33] In a similar approach, environmental friendly composites of chitosan and cellulose were prepared using a mixture of sodium hydroxide and thio-urea solvent. The solvent led to chain depolymerization of both polymers by sustaining the film forming capability. The crystalline property of the prepared nanocomposites was close to that of cellulose film.[34]

Interestingly, a blend of chitosan and cellulose also showed an improvement in water vapor even when the blends were not well-miscible. The authors proposed that intermolecular hydrogen bonding of cellulose is supposed to be break down to form cellulose-chitosan hydrogen bonding, while intramolecular and intrastrand hydrogen bonds hold the network flat. These blends membranes demonstrated efficient antimicrobial activity against *E. coli* and *S. aureus*.[35]

Chitosan dissolves readily in water or acidic medium which allows one to prepare chitosan-based nanocomposites via solution casting[36,40] or spinning methods.[41] Layer-by-layer (LBL) assembly method is also adopted to develop nanocomposites, by alternatively dipping in to chitosan solution and nano-cellulose suspension.[42,43]

16.4.2   OPTICAL PROPERTIES

In general, by the reinforcement with inorganic particulates, the transparency of films is significantly affected which can limit the potential applications in packaging. In case of cellulose-chitosan nanocomposites, the optical transparency of the films is often considered as auxiliary criterion to judge the miscibility of the components. In an earlier investigation, the transmittance of UV/Vis light (200–1000 nm) of the casted films of cellulose nanocrystals reinforced nanocomposites was shown to decrease slightly from 90 to 85% with 0 to 10 wt.% of CNCs /CNWs. Above 10 wt.% of CNCs loading, the transmittance is significantly decreased.

Fernandes et al.[39] have demonstrated an improved transparency with nanocomposites prepared using longer NFCs nanocomposites in two different types of chitosan with different molecular weight. Both low and high molecular weight chitosan (LCH and HCH) were modified to dissolve in acetic acid and water. The four types of nanocomposites with 0–20 wt.% NFC in HCH and water-soluble HCH (WSHCH) and 0–60 wt.% NFC in LCH and water-soluble LCH (WSLCH) prepared via solution casting exhibited a transparency up to 20-90% depending NFC content. Up to 5 wt.% of NFC, the optical transmittance was unaffected. They have also exhibited a significant increase in thermal stability, for example, the initial-degradation temperature (T_{di}) from 227°C to 271°C.[39] In a similar approach, highly transparent (up to 90% optical transmittance) and flexible films were also produced using bacterial cellulose (BC) (Fig. 16.2).[44]

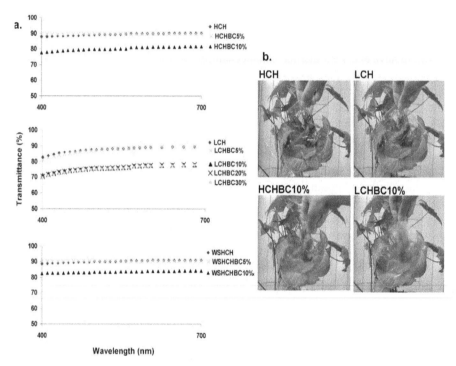

FIGURE 16.2 (a) Transmittance of unfilled chitosan films and corresponding nanocomposites films with different bacterial cellulose contents, and (b) images of chitosan and chitosan-bacterial cellulose nanocomposites films placed in front of a green plant. Reprinted with permission from Royal Society of Chemistry, London, (Green Chemistry, 2009, 11, 2023-2029).

These materials have the great potential to be exploited in food and antimicrobial packaging applications where the transparency, mechanical properties, thermal stability are required with inherent antimicrobial activity and biodegradability (after disposal).

16.4.3 MECHANICAL PROPERTIES

Mechanical properties (strength, stiffness and elongation) play a vital role in determining the capability for using in structural and semistructural packaging materials. The remarkable reinforcement of polymers by cellulose nanocrystals or nanofibrils are attributed to the high axial mechanical properties (strength and modulus) of individual particles and their ability to form percolating network structure via hydrogen bonding between fillers within the matrix. In case of chitosan, the miscibility of filler and matrix due to the chemical similarities is an added advantage. Reinforcement of chitosan was demonstrated to significantly improve the tensile properties depending on the aspect ratio of nanofillers. In an earlier report tensile strength of CNC reinforced chitosan nanocomposites, was shown to increase up to 41% (85 to 120 MPa) with 20 wt. CNC loading in comparison of neat chitosan matrix.[40] A similar trend was observed by Khan et al.,[36] and the optimum filler content for improving tensile strength was found to be 5% (w/w) with 25% improvement. The improvement in stiffness, that is, was also shown to increase upto 87% from 1.6 GPa to 3.0 GPa with 5 wt.% of loading of CNC in the chitosan matrix.[36] However, the elongation at break which relates to the resilience of materials was found to decrease with increasing content of nanofillers, for example, it decreases from 20% to 14%[40] and 9% to 4%[36] depending on chosen chitosan matrix.

16.4.3.1 EFFECT OF COVALENT COMPATIBILIZATION

Surface modified cellulose nanocrystals are also gaining a significant attention to improve the compatibility between the filler and matrix.[45] In a recent study, nanocomposites of chitosan were prepared using methyl adipoyl chloride functionalized cellulose nanocrystals in which the primary amine groups of chitosan were shown to conjugate with functionalized cellulose, as shown in Fig. 16.3.

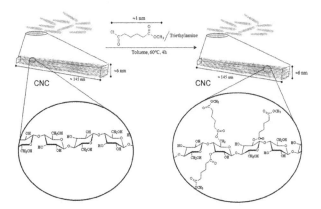

FIGURE 16.3 Schematic illustration of the functionalization of cellulose nanocrystals with methyl ester end group. Reprinted with permission from Elsevier Ltd., (Carbohydrate Polymers, 2012, 90, 210–217).

This phenomenon of conjugation allowed to prepare nanocomposites with high amount of nanofillers, that is, up to 60% (MA-CNC 0.8). These nanocomposites exhibited a linear improvement in the tensile strength and modulus from 45 MPa to 108 MPa and 1.2 GPa to 3.7 GPa, respectively.[37]

16.4.3.2   EFFECT OF HIGHER ASPECT RATIO NANO-CELLULOSE

The highly transparent and flexible films prepared using NFCs exhibited a drastic improvement mechanical properties, with a maximum enhancement in Young's modulus of 78% and 150% for high molecular weight chitosan and water-soluble high molecular weight chitosan (WSHCH), respectively with 20 wt.% of NFC loading; and of 200% and 320% for low molecular weight chitosan and water-soluble low molecular weight chitosan, respectively with 60 wt.% of NFC loading. The tensile strength values were in closer agreement with the trend of modulus values. As it can be assumed from a classical reinforcement effect, the values for elongation at break were reduced.[39] Similar trends was also observed using bacterial cellulose.[44]

16.4.4   WATER ABSORPTION AND GAS BARRIER PROPERTIES

The main poor attributes of natural polymers like chitosan that limits them in packaging applications are water / moisture absorption and gas barrier (high permeability of gas) properties. As the mechanical properties are highly influenced by water absorption (through plasticization), it is very important to improve the gas barrier properties and moisture absorption or water uptake for chitosan based materials.

16.4.4.1   REDUCTION IN WATER UPTAKE OR MOISTURE ABSORPTION

The water absorption behavior of the cellulose reinforced chitosan films was followed by gravimetrically exposing the films to the moisture at relative humidity (RH) from 55 to 100% or by immersing in deionized water and sometimes the mechanical properties dependence up on moisture absorption. The water uptake at equilibrium of nanocomposite films in water was reduced from 71 to 40% linearly with CNC loading up to 30 wt.%.[40] A similar trend in the reduction in the water uptake was also observed with even 4 times shorter nanocrystals and the longer NFCs are used to prepare nanocomposites when the films were incubated in water and at RH 75%, respectively.[36] The reduction in water uptake can be attributed to the increase in highly crystalline cellulose which lesser hydrophilic than chitosan matrix and a strong interfacial interaction between them. Due to the reduced water uptake, the moisture-dependent mechanical properties chitosan-cellulose nanocomposite films were less affected in comparison of neat chitosan. However, the water uptake can be still reduced by covalently linking the cellulose and chitosan as represented in Fig. 16.4 that compares the unmodified CNC and surface modified CNCs.[37]

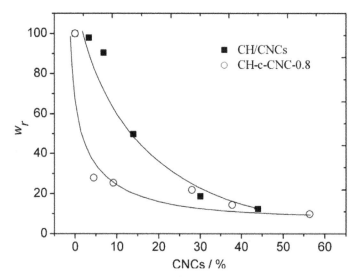

FIGURE 16.4 Comparison of water vapor permeability chitosan-CNC nanocomposites using unmodified and surface modified CNC. Reprinted with permission from Elsevier Ltd., (Carbohydrate Polymers, 2012, 90, 210–217).

Contrarily, the cellulose-chitosan nanocomposites systems can be tuned to absorb higher amount of water for biological and sanitary applications, when required. The in-situ grafting of acrylic acid onto chitosan in presence of CNCs (10% w/w) has led to a super absorbent hydrogel materials. The superabsorbent materials exhibit an improved the swelling capacity in ca.100 units, from 381 to 486 with faster equilibrium. In this case, the averaged dimension of pores was increased by the incorporation of CNCs to increase the porous capacity.[46] These hydrogels can be used as functional pH/ionic materials.

16.4.4.2   IMPROVEMENT IN BARRIER PROPERTIES

The incorporation of cellulose nanocrystals in the chitosan matrix was observed to decrease water vapor permeability values, for example, from 3.31 to 2.23 gm/m^2 day kPa, with 10 wt.% loading of CNC.[36] The protection against oxidation of food or meat products can be reduced by packaging films with reduced oxygen permeability. The significant reduction in oxygen permeability was demonstrated with nanocomposite films prepared via layer by layer assembly by alternatively dipping in chitosan solution and CNC dispersion.[43] The reduction in oxygen permeability of 180 μm thick A-PET substrate was about 94%, after 30 numbers of bilayers coating of chitosan and CNC via LBL assembly (Fig. 16.5).

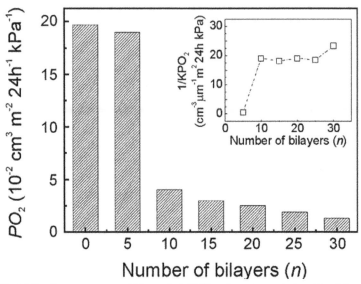

FIGURE 16.5 Oxygen permeability of A-PET substrate coated with chitosan/CNC nanocomposites via layer-by-layer assembly and inset: reciprocal value of the oxygen permeability coefficient (KPO$_2$). Reprinted with permission from Elsevier Ltd., (Carbohydrate Polymers, 2013, 92, 2128–2134).

16.4.5 BIOLOGICAL PROPERTIES

As mentioned above, chitosan is being explored for wound healing, antibacterial, antifungal, drug delivery, wound healing and absorption of organic and inorganic pollutants.[17,19,47–51]

Antimicrobial properties: Since bacterial infections (such as from *Listeria monocytogenes*) are the leading cause for foodborne human infection and diseases, the development of antimicrobial packaging material is still growing. Antimicrobial packaging materials can be classified into three following categories (a) first category antimicrobial packaging materials involves the direct use of antimicrobial additive in the packaging film, (b) in the second category, a carrier for antimicrobial additive can be used as a coating, while (c) the more advanced option is the use of cationic polysaccharide, which can make membranes and possess antimicrobial activity.[14,52,59]

In this regards, chitosan, notably possess a polycationic structure that interacts with anionic bacterial cell membrane, thereby disrupting the integrity of bacterial cells. Similar antimicrobial activity can be expected from chitin, which is the raw material of chitosan, but never been explored, despite the fact that chitin nanocrystals accelerate wound healing.[17,60,61]

Multicomponent nanocomposite films of polysaccharides (starch, chitosan and cellulose nanofibers) were shown to have bactericidal activity against *S. aureus*, depending on chitosan content. Since the chitosan (CH) and water soluble CH are the only components responsible for antimicrobial activity, the bactericide activity strongly depend on the CH and WCH content.[62] In comparison of NFC, BC nanofibers did not significantly affect the bactericidal activity. Recently, the nanocomposites of bacterial cellulose nanofibers and chitin nanocrystals prepared via both in-situ biosynthesis of BC and solution casting were demonstrated to possess superior antibacterial properties against *E. Coli.*[63] The antibacterial activity of nanocomposites increased with increasing deacetylated chitin content, by forming strong interaction with bacterial cellulose network which indicates them suitable for antimicrobial applications.

A novel polyelectrolyte-macroion complex composed of chitosan and cellulose nanocrystals have been examined for its potential drug delivery applications.[64] As shown in Fig. 16.6, the particles were primarily composed of cellulose nanocrystals only and are governed by the strong mismatch in the densities of the ionizable groups of the two components. Further, higher cellulose nanocrystals concentration in the composite leads to the formation of larger and highly aggregated particles.

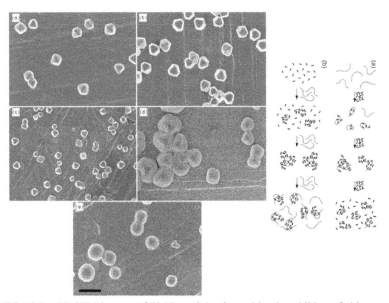

FIGURE 16.6 FE-SEM images of PMC particles formed by the addition of chitosan with different ratios. Proposed mechanism of chitosan-cellulose nanocrystals complexation. Reprinted with permission from American Chemical Society, USA (Biomacromolecules, 2011, 12, 1585–1593).

16.5   MULTI-COMPONENT SYSTEMS AND THEIR PROPERTIES

Chitosan is very often used as matrix or compatibilizer for preparing multicomponent nanocomposites with high modulus, significant strength and toughness as well as fire retardancy and low oxygen transmission rate (OTR). The three phase nacre-like structured nanopaper containing chitosan montmorillonite (MMT) and nanofibrillated cellulose was prepared.[38] These nanocomposites exhibited a linear improvement in tensile strength and modulus with increasing NFC content with improved the thermal stability, indicating the enhanced stress transfer primarily the strong interaction between clay nanolayers, NFC through chitosan organic matrix (Fig. 16.7).

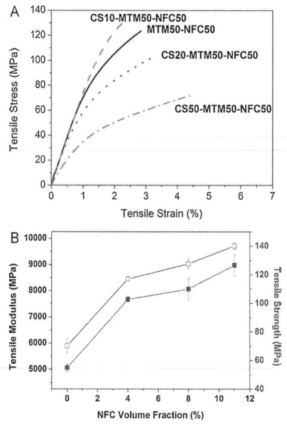

FIGURE 16.7   Mechanical properties. (A) The stress-strain curves for CS-MTM50-NFC50 with different contents of chitosan. (B) Tensile modulus and strength of CS50-MTM50 as a function of NFC volume fraction. Reprinted with permission from Elsevier Ltd., (Carbohydrate Polymers, 2012, 87, 53–60).

In a very interesting and innovative method, chitosan was used to coat the cellulose microfibers by a layer-by-layer self-assembly process with silver nanoparticles. In this case, the strong metal binding capability of chitosan was harnessed to bind silver ions along with titania.[65] These hybrid films exhibited a cable-like core-shell structure, displaying a cellulose microfibers bundle core as well as nano-titania/chitosan/silver nanoparticles shell consisting of uniformly dispersed 4–20 nm diameter silver nanoparticles, as shown in Fig. 16.8a.

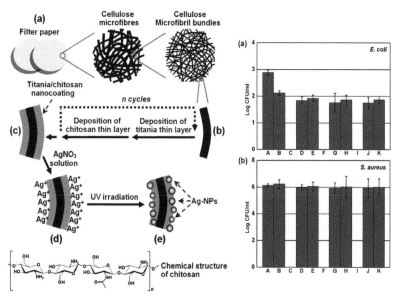

FIGURE 16.8 Schematic illustration of nanocoating of natural cellulose microfibril bundles of filter paper with titania/chitosan/Ag-NP composite films. Histograms illustrating the antibacterial activity of the cellulose/titania/chitosan/Ag-NP composite sheets against E. coli (a) and S. aureus (b). Reprinted with permission from Royal Society of Chemistry, London, (J. Mater. Chem. B, 2013, 1, 3477–3485).

This approach is unique in the sense that various functional molecules can be incorporated and their inherent properties can be harnessed providing a pathway to accomplish the combination of the internal excellent properties of cellulose (large surface area, flexibility, biodegradability), and antimicrobial properties imparted due to the presence of silver nanoparticles, chitosan and nano-titania. The composite film disinfected almost all the incubated bacteria, hence illustrating excellent material in antibiotic wound dressing and antibacterial packaging material (Fig. 16.8b).

16.6 CONCLUDING REMARKS AND FUTURE PERSPECTIVES

Owing to its abundance, renewability, biodegradability, biocompatibility (low cytotoxicity), antimicrobial activity and chemical and physical functionalities, chitin/chitosan is being explored as potential candidate for packaging and biomedical applications. Due to their poor thermal processability and low solubility in neutral media, the applications of chitin/chitosan alone are still underexplored. Through chemical modification (esterification, etherification, N-acylation, and grafting), blending with other biopolymers, reinforcement by nanoscale particles, chitosan is being explored as potential candidates for these applications. With aid of nanoscience of nanotechnology, chitosan based nanocomposites are being explored for packaging including smart active and antimicrobial packaging for food, meat and dairy products, antimicrobial sanitary products and biomedical applications including wound dressing, drug-delivery and tissue engineering. This chapter reviewed the recent developments on the nanocomposites of chitosan using nano-cellulose, which is also bio-renewable. They are usually processed using acidic media or in water after specific chemical modification. The nanocomposites of chitosan/cellulose exhibited remarkable improvement mechanical properties, reduction in water absorption/uptake, decrease in permeability coefficient for water vapor and oxygen and retained transparency and antimicrobial or microbicidal activity, making them suitable candidates for packaging and applications. These improvements have been attributed to the property of highly crystalline individual nanocrystals and their strong interactions with the chitosan matrix. This hetero-phasic domains phenomenon was also observed when the chitosan and cellulose blended by dissolution. The approach of covalent conjugation has been demonstrated to increase the higher loading of nanocellulose, to improve compatibility and to decrease the water uptake/permeability. The approach of layer-by-layer assembly for preparing nanocomposite films or coating can be explored for high barrier coatings, and packaging without compromising the transparency. Superabsorbent materials for sanitary applications or antimicrobial applications can also be prepared via in-situ grafting of chitosan in presence of cellulose nanocrystals. Biological applications such as drug delivery, antimicrobial nonwoven materials for wound dressing are also accessible via multicomponent nanomaterials based on chitosan and cellulose. It is observed that the technological advancements on the chitosan-based nanocomposites are still being explored. For larger scale commercialization, the industrial scale production of these nanocomposites is still to be demonstrated. By portraying the recent developments in this area, this chapter may stimulate further research activities on the innovative nanomaterials based on chitosan and nanocellulose at favorable production cost/performance ratio.

16.7 ACKNOWLEDGEMENT

Authors are acknowledging the financial support from Queensland Government under Smart Futures Research Partnerships Program (2012–2014).

KEYWORDS

- **Antimicrobial**
- **Biomedical**
- **Cellulose**
- **Chitosan**
- **Nanocomposites**
- **Nanocrystals**
- **Nanofibers**
- **Packaging Applications**

REFERENCES

1. Mohanty, A. K., Misra, M., & Hinrichsen, G. (2000). Biofibers, biodegradable polymers and biocomposites: An overview. *Macromol Mater Eng*, 276(3–4), 1–24.
2. Kumar, A. P., & Singh, R. P. (2008). Biocomposites of cellulose reinforced starch: Improvement of properties by photo-induced crosslinking. *Bioresource Technol*, 99(18), 8803–8809.
3. Ray, S. S., & Bousmina, M. (2005). Biodegradable polymers and their layered silicate nano composites: In greening the twenty-first century materials world. *Prog Mater Sci*, 50(8), 962–1079.
4. Depan, D., & Misra, R. D. K. (2013). The interplay between nanostructured carbon-grafted chitosan scaffolds and protein adsorption on the cellular response of osteoblasts: Structure-function property relationship. *Acta Biomater*, 9(4), 6084–6094.
5. Depan, D., & Misra, R. D. K. (2012). Processing-structure-functional property relationship in organic-inorganic nanostructured scaffolds for bone-tissue engineering: The response of preosteoblasts. *J Biomed Mater Res A*, 100A(11), 3080–3091.
6. Depan, D., Girase, B., Shah, J. S., & Misra, R. D. K. (2012). Structure-process-property relationship of the polar graphene oxide-mediated cellular response and stimulated growth of osteoblasts on hybrid chitosan network structure nanocomposite scaffolds (vol. 7, p. 3362, 2011). *Acta Biomater*, 8(3), 1395–1395.
7. Depan, D., Surya, P. K. C. V., Girase, B., & Misra, R. D. K. (2012). Organic/inorganic hybrid network structure nanocomposite scaffolds based on grafted chitosan for tissue engineering (vol 7, pg 2163, 2011). *Acta Biomater*, 8(3), 1394–1394.
8. Depan, D., Shah, J., & Misra, R. D. K. (2011). Controlled release of drug from folate-decorated and graphene mediated drug delivery system: Synthesis, loading efficiency, and drug release response. *Mat Sci Eng C-Mater*, 31(7), 1305–1312.
9. Depan, D., Girase, B., Shah, J. S., & Misra, R. D. K. (2011). Structure-process-property relationship of the polar graphene oxide-mediated cellular response and stimulated growth of

osteoblasts on hybrid chitosan network structure nanocomposite scaffolds. *Acta Biomater*, 7(9), 3432–3445.

10. Depan, D., Surya, P. K. C. V., Girase, B., & Misra, R. D. K. (2011). Organic/inorganic hybrid network structure nanocomposite scaffolds based on grafted chitosan for tissue engineering. *Acta Biomater*, 7(5), 2163–2175.

11. Depan, D., Kumar, B., & Singh, R. P. (2008). Preparation and characterization of novel hybrid of Chitosan-g-PDMS and sodium montmorrilonite. *J Biomed Mater Res B*, 84B(1), 184–190.

12. Depan, D., Kumar, A. P., & Singh, R. P. (2009). Cell proliferation and controlled drug release studies of nanohybrids based on chitosan-g-lactic acid and montmorillonite. *Acta Biomater*, 5(1), 93–100.

13. Kumar, A. P., Depan, D., Tomer, N. S., & Singh, R. P. (2009). Nanoscale particles for polymer degradation and stabilization-Trends and future perspectives. *Prog Polym Sci*, 34(6), 479–515.

14. Sorrentino, A., Gorrasi, G., & Vittoria, V. (2007). Potential perspectives of bio-nanocomposites for food packaging applications. *Trends Food Sci Tech*, 18(2), 84–95.

15. Tharanathan, R. N., Kittur, F. S. (2003). Chitin–The undisputed biomolecule of great potential. *Crit Rev Food Sci*, 43(1), 61–87.

16. Liu, W. G., Sun, S. J., Zhang, X., & De Yao, K. (2003). Self-aggregation behavior of alkylated chitosan and its effect on the release of a hydrophobic drug. *J Biomat Sci-Polym E* 14(8), 851–859.

17. Rinaudo, M. (2008). Main properties and current applications of some polysaccharides as biomaterials. *Polym Int*, 57(3), 397–430.

18. Li, Z. S., Ramay, H. R., Hauch, K. D., Xiao, D. M., & Zhang, M. Q. (2005). Chitosan-alginate hybrid scaffolds for bone tissue engineering. *Biomaterials*, 26(18), 3919–3928.

19. Suh, J. K. F., & Matthew, H. W. T. (2000). Application of chitosan-based polysaccharide biomaterials in cartilage tissue engineering: a review. *Biomaterials*, 21(24), 2589–2598.

20. Metcalfe, A. D., & Ferguson, M. W. J. (2007). Tissue engineering of replacement skin: the crossroads of biomaterials, wound healing, embryonic development, stem cells and regeneration. *J R Soc Interface*, 4(14), 413–437.

21. Zhang, L., Jin, Y., Liu, H. Q., & Du, Y. M. (2001). Structure and control release of chitosan/carboxymethyl cellulose microcapsules. *J Appl Polym Sci*, 82(3), 584–592.

22. Rusmini, F., Zhong, Z. Y., & Feijen, J. (2007). Protein immobilization strategies for protein biochips. *Biomacromolecules*, 8(6), 1775–1789.

23. Slutter, B., Soema, P. C., Ding, Z., Verheul, R., Hennink, W., & Jiskoot, W. (2010). Conjugation of ovalbumin to trimethyl chitosan improves immunogenicity of the antigen. *J Control Release*, 143(2), 207–214.

24. Fernandez-Megia, E., Novoa-Carballal, R., Quinoa, E., & Riguera, R. (2007). Conjugation of bioactive ligands to PEG-grafted chitosan at the distal end of PEG. *Biomacromolecules*, 8(3), 833–842.

25. Gondim, D. R., Lima, L. P., de Souza, M. C. M., Bresolin, I. T. L., Adriano, W. S., Azevedo, D. C. S., & Silva, I. J. (2012). Dye Ligand Epoxide Chitosan/Alginate: A Potential New Stationary Phase for Human IgG Purification. *Adsorpt Sci Technol*, 30(8–9), 701–711.

26. Wu, X. W., Moon, R. J., & Martini, A. (2013). Crystalline cellulose elastic modulus predicted by atomistic models of uniform deformation and nanoscale indentation. *Cellulose*, 20(1), 43–55.

27. Moon, R. J., Martini, A., Nairn, J., Simonsen, J., & Youngblood, J. (2011). Cellulose nanomaterials review: structure, properties and nanocomposites. *Chem Soc Rev*, 40(7), 3941–3994.

28. Mendez, J., Annamalai, P. K., Eichhorn, S. J., Rusli, R., Rowan, S. J., Foster, E. J., & Weder, C. (2011). Bioinspired Mechanically Adaptive Polymer Nanocomposites with Water-Activated Shape-Memory Effect. *Macromolecules*, 44(17), 6827–6835.

29. Klemm, D., Kramer, F., Moritz, S., Lindstrom, T., Ankerfors, M., Gray, D., & Dorris, A. (2011). Nanocelluloses: A New Family of Nature-Based Materials. *Angew Chem Int Edit*, 50(24), 5438–5466.

30. Yano, H., Sugiyama, J., Nakagaito, A. N., Nogi, M., Matsuura, T., Hikita, M., & Handa, K. (2005). Optically transparent composites reinforced with networks of bacterial nanofibers. *Adv Mater*, 17(2), 153-+.

31. Guhados, G., Wan, W. K., & Hutter, J. L. (2005). Measurement of the elastic modulus of single bacterial cellulose fibers using atomic force microscopy. *Langmuir*, 21(14), 6642–6646.

32. Stefanescu, C., Daly, W. H., & Negulescu, I. I. (2012). Biocomposite films prepared from ionic liquid solutions of chitosan and cellulose. *Carbohyd Polym*, 87(1), 435–443.

33. Ma, B. M., Zhang, M., He, C. J., & Sun, J. F. (2012). New binary ionic liquid system for the preparation of chitosan/cellulose composite fibers. *Carbohyd Polym*, 88(1), 347–351.

34. Almeida, E. V. R., Frollini, E., Castellan, A., & Coma, V. (2010). Chitosan, sisal cellulose, and biocomposite chitosan/sisal cellulose films prepared from thiourea/NaOH aqueous solution. *Carbohyd Polym*, 80(3), 655–664.

35. Wu, Y. B., Yu, S. H., Mi, F. L., Wu, C. W., Shyu, S. S., Peng, C. K., & Chao, A. C. (2004). Preparation and characterization on mechanical and antibacterial properties of chitsoan/cellulose blends. *Carbohyd Polym*, 57(4), 435–440.

36. Khan, A., Khan, R. A., Salmieri, S., Le Tien, C., Riedl, B., Bouchard, J., Chauve, G., Tan, V., Kamal, M. R., & Lacroix, M. (2012). Mechanical and barrier properties of nanocrystalline cellulose reinforced chitosan based nanocomposite films. *Carbohyd Polym*, 90(4), 1601–1608.

37. de Mesquita, J. P., Donnici, C. L., Teixeira, I. F., & Pereira, F. V. (2012). Bio-based nanocomposites obtained through covalent linkage between chitosan and cellulose nanocrystals. *Carbohyd Polym*, 90(1), 210–217.

38. Liu, A. D., & Berglund, L. A. (2012). Clay nanopaper composites of nacre-like structure based on montmorrilonite and cellulose nanofibers Improvements due to chitosan addition. *Carbohyd Polym*, 87(1), 53–60.

39. Fernandes, S. C. M., Freire, C. S. R., Silvestre, A. J. D., Neto, C. P., Gandini, A., Berglund, L. A., & Salmen, L. (2010). Transparent chitosan films reinforced with a high content of nanofibrillated cellulose. *Carbohyd Polym*, 81(2), 394–401.

40. Li, Q., Zhou, J. P., & Zhang, L. N. (2009). Structure and Properties of the Nanocomposite Films of Chitosan Reinforced with Cellulose Whiskers. *J Polym Sci Pol Phys*, 47(11), 1069–1077.

41. Devarayan, K., Hanaoka, H., Hachisu, M., Araki, J., Ohguchi, M., Behera, B. K., & Ohkawa, K. (2013). Direct Electrospinning of Cellulose-Chitosan Composite Nanofiber. *Macromol Mater Eng*, 298(10), 1059–1064.

42. de Mesquita, J. P., Donnici, C. L., & Pereira, F. V. (2010). Biobased Nanocomposites from Layer-by-Layer Assembly of Cellulose Nanowhiskers with Chitosan. *Biomacromolecules*, 11(2), 473–480.

43. Li, F., Biagioni, P., Finazzi, M., Tavazzi, S., & Piergiovanni, L. (2013). Tunable green oxygen barrier through layer-by-layer self-assembly of chitosan and cellulose nanocrystals. *Carbohyd Polym*, 92(2), 2128–2134.

44. Fernandes, S. C. M., Oliveira, L., Freire, C. S. R., Silvestre, A. J. D., Neto, C. P., Gandini, A., & Desbrieres, J. (2009). Novel transparent nanocomposite films based on chitosan and bacterial cellulose. *Green Chem*, 11(12), 2023–2029.

45. Lin, N., Huang, J., & Dufresne, A. (2012). Preparation, properties and applications of polysaccharide nanocrystals in advanced functional nanomaterials: a review. *Nanoscale*, 4(11), 3274–3294.

46. Spagnol, C., Rodrigues, F. H. A., Pereira, A. G. B., Fajardo, A. R., Rubira, A. F., & Muniz, E. C. (2012). Superabsorbent hydrogel composite made of cellulose nanofibrils and chitosan graft poly(acrylic acid). *Carbohyd Polym*, 87(3), 2038–2045.
47. Bordenave, N., Grelier, S., & Coma, V. (2010). Hydrophobization and Antimicrobial Activity of Chitosan and Paper-Based Packaging Material. *Biomacromolecules*, 11(1), 88–96.
48. Ngah, W. S. W., & Isa, I. M. (1998). Comparison study of copper ion adsorption on chitosan, Dowex A-1, and Zerolit 225. *J Appl Polym Sci*, 67(6), 1067–1070.
49. Rhim, J. W. (2007). Potential use of biopolymer-based nanocomposite films in food packaging applications. *Food Sci Biotechnol*, 16(5), 691–709.
50. Sanchez-Garcia, M. D., Lopez-Rubio, A., & Lagaron, J. M. (2010). Natural micro and nano-biocomposites with enhanced barrier properties and novel functionalities for food biopackaging applications. *Trends Food Sci Tech*, 21(11), 528–536.
51. Fernandes, S. C. M., Freire, C. S. R., Silvestre, A. J. D., Neto, C. P., & Gandini, A. (2011). Novel materials based on chitosan and cellulose. *Polym Int*, 60(6), 875–882.
52. Weng, Y. M., Chen, M. J., & Chen, W. (1999). Antimicrobial food packaging materials from poly(ethylene-comethacrylic acid). *Food Sci Technol Leb*, 32(4), 191–195.
53. Cooksey, K. (2005). Effectiveness of antimicrobial food packaging materials. *Food Addit Contam*, 22(10), 980–987.
54. Akbari, Z., Ghomashchi, T., & Moghadam, S. (2007). Improvement in Food Packaging Industry with Biobased Nanocomposites. *Int J Food Eng*, 3(4).
55. Siracusa, V., Rocculi, P., Romani, S., & Dalla Rosa, M. (2008). Biodegradable polymers for food packaging: a review. *Trends Food Sci Tech*, 19(12), 634–643.
56. Fernandez-Saiz, P. (2011). Chitosan polysaccharide in food packaging applications. *Woodhead Publ Mater*, 571–593.
57. Silvestre, C., Duraccio, D., & Cimmino, S. (2011). Food packaging based on polymer nano-materials. *Prog Polym Sci*, 36(12), 1766–1782.
58. Lagaron, J. M., & Nunez, E. (2012). Nanocomposites of moisture-sensitive polymers and biopolymers with enhanced performance for flexible packaging applications. *J Plast Film Sheet*, 28(1), 79–89.
59. Youssef, A. M. (2013). Polymer Nanocomposites as a New Trend for Packaging Applications. *Polym-Plast Technol*, 52(7), 635–660.
60. Watthanaphanit, A., Supaphol, P., Tamura, H., Tokura, S., & Rujiravanit, R. (2008). Fabrication, structure, and properties of chitin whisker-reinforced alginate nanocomposite fibers. *J Appl Polym Sci*, 110(2), 890–899.
61. Mututuvari, T. M., Harkins, A. L., & Tran, C. D. (2013). Facile synthesis, characterization, and antimicrobial activity of cellulose chitosan hydroxyapatite composite material A potential material for bone tissue engineering. *J Biomed Mater Res A*, 101(11), 3266–3277.
62. Tome, L. C., Fernandes, S. C. M., Perez, D. S., Sadocco, P., Silvestre, A. J. D., Neto, C. P., Marrucho, I. M., & Freire, C. S. R. (2013). The role of nanocellulose fibers, starch and chitosan on multipolysaccharide based films. *Cellulose*, 20(4), 1807–1818.
63. Butchosa, N., Brown, C., Larsson, P. T., Berglund, L. A., Bulone, V., & Zhou, Q. (2013). Nanocomposites of bacterial cellulose nanofibers and chitin nanocrystals: fabrication, characterization and bactericidal activity. *Green Chem*, 15(12), 3404–3413.
64. Wang, H. Z., & Roman, M. (2011). Formation and Properties of Chitosan-Cellulose Nanocrystal Polyelectrolyte-Macroion Complexes for Drug Delivery Applications. *Biomacromolecules*, 12(5), 1585–1593.
65. Xiao, W., Xu, J. B., Liu, X. Y., Hu, Q. L., & Huang, J. G. (2013). Antibacterial hybrid materials fabricated by nanocoating of microfibril bundles of cellulose substance with titania/chitosan/silver-nanoparticle composite films. *J Mater Chem B*, 1(28), 3477–3485.

CHAPTER 17

BIONANOCOMPOSITES: A GREENER ALTERNATIVE FOR FUTURE GENERATION

MURSHID IMAN and TARUN K. MAJI

ABSTRACT

Modern scenario about the conservation of natural resources and recycling has led to the renewed interest concerning biomaterials with the focus on renewable raw materials. The use and removal of traditional composite materials usually made of glass, carbon, aramid fibers being reinforced with unsaturated polyester, epoxy or phenolics are considered crucially due to increasing global awareness and demands of legislative authorities. Therefore, the growing awareness of the pressing need for greener and more sustainable technologies has focused attention on use of bio-based polymers instead of conventional petroleum based polymers to fabricate biodegradable materials with high performance. Another aspect, which is receiving more attention, is the use of alternate resource prior to the use of the conventional materials. The important aspect of composite materials is that they can be designed and tailored to meet different desires. Natural fibers such as hemp, ramie, jute, etc. are cheap, biodegradable and most importantly easily available worldwide. The bio composites prepared by using natural fibers and varieties of natural polymers such as soy flour, starch, gluten, poly(lactic acid), etc. have evoked considerable interest in recent years due to their ecofriendly nature. These natural polymers have some negative aspects also. Thus modification by cross-linking, grafting, blending and inclusion of nanotechnology provide desired properties and widen the spectrum of applications of bio composites. Biocomposites offer modern world an alternative solution to waste-disposal problems associated with conventional petroleum based plastics. Therefore, the development of commercially viable "green products" based on natural fibers and polymers for a wide range of application is on the rise. Moreover, using nanotechnology for the synthesis of biocomposites provide better mechanical properties and thermal stability. In short, the use of bionanocomposites may provide us a healthier environment owing to its multifaceted advantages over conventional polymers. This chapter discusses on the potential efficacy of natural polymers and its various derivatives for preparation of biocomposites to be used for varieties of applications.

17.1   INTRODUCTION

With the advancement of science and technology, the people of the current civilized world are becoming more dependent on the advanced materials. In this regard, the chemist from all over the world has contributed a lot for the modernization of our society. One of such major gift-that chemist has ever bestowed to the human society is the "polymer" or "polymeric materials" without which the world have been in a totally different situations. However, as the environmental and health effects of a chemical or chemical process have begun to be considered, therefore, there has been an expanding search for new materials with high performance at affordable costs in recent years. With growing environmental and health awareness, there has been a significant focus within the scientific, industrial, and environmental communities on the use of ecofriendly materials, with terms such as "renewable," "recyclable," "sustainable," and "triggered bio-degradable" becoming buzzwords. The development or selection of a material to meet the desired structural and design requirements calls for a compromise between conflicting objectives. This can be overcome by resorting to multiobjective optimization in material design and selection. Composite materials, which are prepared using natural reinforcements and a variety of renewable matrix, are included in this chapter.

Since, most of the renewable materials are associated with bio-logical and plant based products as a source of raw materials, particularly to plastic industries, and these could generate a non–food source of economic development for farming and rural areas in developing country. The development of such materials has not only been a great motivating factor for materials scientists, but also an important provider of opportunities to improve the living standard of people around the world. This can also provide a potential for economic improvement based on these materials even though major thrust for their use has been driven by the needs in industrialized countries. For example, natural fibers such as jute, sisal, hemp, pineapple, etc., whose extraction is an important process that determines the properties of fibers, can generate rural jobs since those fibers have established their potential as reinforcing fillers in many polymers, and products based on these have found increasing use on a commercial scale in recent years.[1,31] Another example for the generation of jobs by agro-based materials is provided by the use of rice husk, which constitutes more than 10% of a world rice production. These examples underline not only the development of new materials, but also the possible generation of additional employment through the collection, transportation and development of new materials. It is reported that increasing use of renewable materials would create or secure employment in rural areas, the distribution of which would be agriculture, forestry, industry, etc.

The use of natural polymers was superseded in the twentieth century as a wide-range of synthetic polymers was developed based on raw materials from low cost petroleum. However, since 1990s, there is a simultaneous and growing interest in developing bio-based products and innovative process technologies that can reduce the dependence of fossil fuel and move to a sustainable materials basis. The main reasons for development of such material are stated below:[32]

1. Growing interest in reducing the environmental impact of polymers or composites due to increased awareness to ecofriendliness;

2. Finite petroleum resources, decreasing pressures for the dependence of petroleum products with increasing interest in maximizing the use of renewable materials; and

3. The availability of improved data on the properties and morphologies of natural materials such as lignocellulosic fibers, through modern instruments at different levels, and hence better understanding of their structure-property correlations.

These factors have greatly increased the understanding and development of new materials such as biocomposites.

Commodity polymer-based composite materials are now well established all over the world. Because of their high specific strength, modulus and long durability compared to conventional materials such as metals and alloys, these materials have found wide applications. However, the use of large volumes of polymer-based synthetic fiber composites in different sectors has led to disposal problems. Therefore, scientists have been looking for the reduction of such environmentally abusive materials, and triggering greater efforts to find materials based on natural resources in view of the letter's ecofriendly attributes. Such natural resources are organic in nature and also a source for carbon and a host of other useful materials and chemicals, particularly for the production of "green" materials.[1,5,6,22,33]

In parallel, researchers have focused their works on the processing of nanocomposites (materials with nanosized reinforcement) to enhance mechanical properties. Similar to traditional microcomposites, nanocomposites use a matrix where the nanosized reinforcement elements are dispersed. The reinforcement is currently considered as a nanoparticle when at least one of its dimensions is lower than 100 nm. This particular feature provides nanocomposites unique and outstanding properties never found in conventional composites. Bio-based nanocomposites are the next generation of materials for the future.

17.2 BIO-BASED POLYMERS

Now a days, 80% of the polymer market is occupied by synthetic polymers and most of these polymers are non degradable. The nondegradable nature of polymer causes disturbance in the earth ecosystem. Besides this, the earth has finite resources in terms of fossil based fuel. The escalating increase of price of petroleum based products and alternative disposal method are also a great concern. Hence, the use of fossil-based products is not sustainable. So, there is an urgent need to overcome the dependence on such conventional polymers by using bio-degradable polymers and composites. In order to produce fully renewable and biodegradable composites, both the polymeric matrix and the reinforcement must be derived from renewable natural resources such as agricultural and biological origin. Also, the use of natural polymers, which are normally biodegradable, can pave new direction in designing of newer greener composites and could widen the spectrum of applications in different sectors such as automobiles, furniture, packing and construction industrial parts.

17.2.1   BIO-BASED MATRIX

"Bio" is a Greek word that means "life." Bio-based materials, therefore, refer to products that consist mainly of a substance, or substances, derived from living matter (biomass) and either occurs naturally or is synthesized. The term "bio-based materials" should not be confused with "biomaterials," which has another meaning and relates to biocompatible materials used in and adapted to medical applications, which include implantable medical devices, tissue engineering, and drug delivery systems.[34]

The range of bio-based materials, from natural fibers to biopolymers, is making significant advances in petroleum-based materials industries.[35] Renewable resource-based chemicals and bio-based polymers, such as 1,3-propanediol, soy, polyol, polylactic acids, and so on, are gaining momentum in commercialization as supplements and possible replacements for petroleum based products. For decades, cellulosic polymers have played a key role in a wide range of applications, such as apparel, food, and varnishes. Since the 1980s, an increasing number of starch polymers have been introduced, which have made them one of the most important groups of commercially available bio-based materials.

Bio-based materials, are commonly thought to be greener alternatives than their petroleum-based counterparts, which are nonbiodegradable, have potentially devastating effects on animal and ocean life, and for the most part, have an inherently toxic life cycle from their production through their final disposal. Bio-based materials frequently are labeled as produced from "renewable" resources, although this term is used loosely because biomass production requires nonrenewable inputs, which include fossil fuels, and ties up other finite resources such as land and water. The claim that bio-based materials are friendlier to the environment than their petroleum-based counterparts is being scrutinized closely.[36,39]

17.2.1.1   STARCH POLYMERS

Starch is one of the most exciting and promising raw materials for the production of biodegradable products. It is the major polysaccharide reserve material of photosynthetic tissues and of many types of plant storage organs such as seeds and swollen stems. The primary crops used for its production consist of potatoes, corn, wheat and rice. In all of these sources, starch is produced in the form of granules, which vary in size and somewhat in composition based on the resources. Starch granule is composed of two main polysaccharides, amylose and amylopectin with some minor components such as lipids and proteins. Amylose is linear polymer of (1→4)-linked α-D-glucopyranosyl units with some slight branches by (1→6)-α-linkages (Fig. 17.1). Amylose can have a molecular weight between 10^4 and 10^6 g mol^{-1}, but it is soluble in boiling water.[40,41] Amylopectin is a highly branched molecule composed of chains of α-D-glucopyranosyl residues linked together mainly by (1→4)-linkages but with (1→6)-linkages at the branched points. Amylopectin consists of hundreds of short chains of (1→4)-linked α-D-glucopyranosyl interlinked by (1→6)-α-linkages (Fig. 17.2). It is an extremely large and highly branched molecule with a molecular weights ranging from 10^6 to 10^8 g mol^{-1}. Therefore, it is insoluble in boiling water, but in their use in foods, both fractions are readily hydrolyzed at the acetal link by enzymes. Amylases attack the α-(1-4)-link of starch while the α-(1–6)-link in amylopectin is by glucosidases. The crystallinity of the starch granules is attributed mainly to the amylopectin and not to amylose, which although linear, presents a conformation that hinders its regular association with other chains.[42,43]

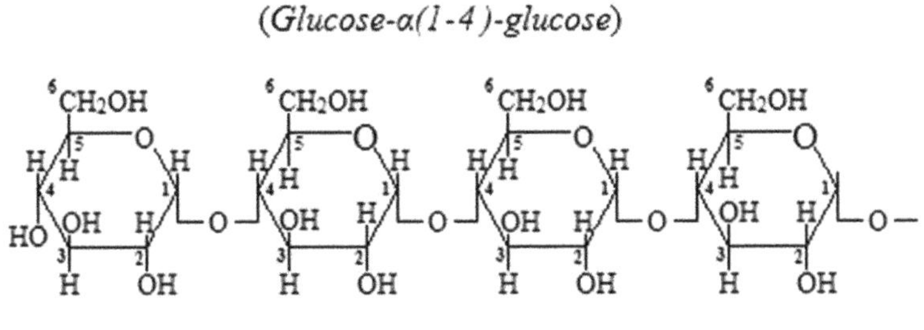

(Glucose-α(1-4)-glucose)

FIGURE 17.1 Structure of amylose.

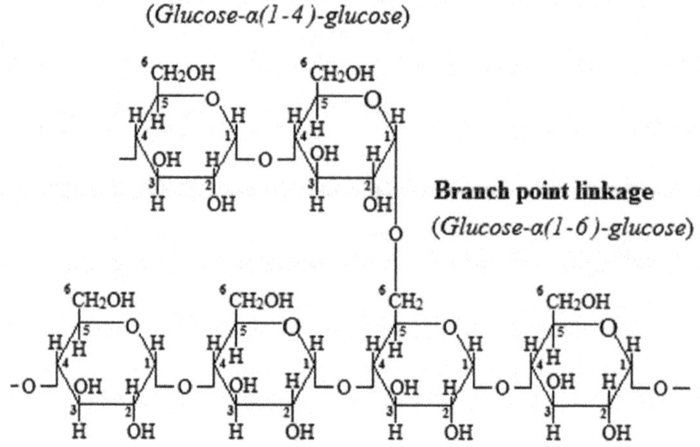

(Glucose-α(1-4)-glucose)

FIGURE 17.2   Structure of amylopectin.

Starch has received significant interest during the past two decades as a biodegradable thermoplastic polymer. Starch offers an attractive and cheap alternative in developing degradable materials. Starch is not truly thermoplastic as most synthetic polymers. However, it can be melted and made to flow at high temperatures under pressure and shear. It has been widely used as a raw material in film production because of increasing prices and decreasing availability of conventional film-forming resins based on petroleum resources. Starch films possess low permeability and are thus attractive materials for food packaging. Starch is also useful for making agricultural mulch films because it degrades into harmless products when placed in contact with soil microorganisms.[44,45]

By itself, starch is a poor alternative for any commodity plastic because, it is mostly water soluble, difficult to process, and brittle. Therefore, research on starch includes exploration of its water adsorptive capacity, the chemical modification of the molecule, its behavior under agitation and high temperature, and its resistance to thermo mechanical shear. Although starch is a polymer, its stability under stress is not high. At temperatures higher than $150°C$, the glucoside links start to break, and above $250°C$ the starch grain endothermally collapses. At low temperatures, a phenomenon known as retrogradation is observed. This is a reorganization of the hydrogen bonds and an aligning of the molecular chains during cooling. In extreme cases under $10°C$, precipitation is observed. Thus, though starch can be dispersed into hot water and cast as films, the above phenomenon causes brittleness in the film.[46]

Plasticized starch is essentially starch that has been modified by the addition of plasticizers to enable processing. Thermoplastic starch is plasticized to completely destroy the crystalline structure of starch to form an amorphous thermoplastic starch. Thermoplastic starch processing involves an irreversible order-disorder transition termed gelatinization. Starch gelatinization is the disruption of molecular organization within the starch macromolecules and this process is affected by starch-water interactions. Most starch processing involves heating in the presence of water and some other additives like sugar and salt to control the gelatinization in the food industry, or glycerol as a plasticizer for biodegradable plastics applications. Most of the commercial research on thermoplastic starches has involved modified starches and or blends with additives and other appropriate polymers for its application as biodegradable plastics.[47] The starch molecule has two important functional groups, the –OH group that is susceptible to substitution reactions and the C–O–C bond that is susceptible to chain breakage. The hydroxyl group of glucose has a nucleophilic character. To obtain various properties starch can be modified through its –OH group. One example is the reaction with silane to improve its dispersion in polyethylene.[48] Crosslinking or bridging of the –OH groups changes the structure into a network while increasing the viscosity, reducing water retention and increasing its resistance to thermo mechanical shear.

One of the approaches to modify this starch is by acetylation to from starch acetate. Acetylated starch does have several advantages as a structural fiber or film-forming polymer as compared to native starch. The acetylation of starch is a well-known reaction and is a relatively easy to synthesize. Starch acetate is considerably more hydrophobic than starch and has been shown to have better retention of tensile properties in aqueous environments. Another advantage is that starch acetate has an improved solubility compared to starch and is easily cast into films from simple solvents. The degree of acetylation is easily controlled by trans esterification, allowing polymers to be produced with a range of hydrophobicities. Starch has been acetylated [with a high content (70%) of linear amylose] and its enzymatic degradation has been studied. Apart from acetylation and esterification, some other modification of starch such as carbonilation of starch with phenyl isocyanates, addition of inorganic esters to starch to produce phosphate or nitrate starch esters, production of starch ethers, and hydroxypropylation of starches via propylene oxide modification has been performed. Generally all these modifications involve hydroxyl group substitution on the starch that will lower gelatinization temperatures, reduce retro-degradation and improve flexibility of final product.[42]

Starch has been used for many years as an additive to plastic for various purposes. Starch was added as a filler[49] to various resin systems to make films that were impermeable to water but permeable to water vapor. The use of starch as a biodegradable filler in LDPE was reported.[50] A starch-filled polyethylene film was prepared which became porous after the extraction of the starch. This porous film could be readily invaded by microorganisms and rapidly saturated with oxygen, thereby increasing polymer degradation by biological and oxidative pathways.[51] Otey et al. in a study on starch-based films, found that a starch– polyvinyl alcohol film could be coated with a thin layer of water-resistant polymer to form a degradable agricultural mulching film.[47] Starch-based polyethylene films were formulated and containing up to 40% starch, urea, ammonia and various portions of low density polyethylene (LDPE) and poly(ethylene-co-acrylic acid) (EAA). The EAA acted as a compatibilizer, forming a complex between the starch and the PE in the presence of ammonia. The resulting blend could be cast or blown into films, and had physical properties approaching to those of LDPE.[52,53]

Additionally, crosslinked starch may be induced by the addition of organic/inorganic esters, hydroxyethers, aldehydes and irradiation. Kulicke et al. examined solution phase crosslinking of starch with epichlorohydrin and trisodium trimetaphosphate.[54] Jane et al. examined the crosslinking of starch/zein cast films for improving water resistance.[55] Iman et al. studied the crosslinking of starch/jute composite with glutaraldehyde to improve its performance characteristics such as mechanical properties, thermal properties, flame retardancy, etc.[56] The possibility of chemically combining starch or starch-derived products with commercial resins in such a manner that the starch would serve as both filler and a crosslinking agent may provide a possible approach for incorporating starch into plastics.

Commercial starch polymer based products are provided in Table 17.1 given below:

TABLE 17.1 Starch Polymer Based Products and Suppliers.[42,57]

Base Polymer	Source Type	Advantages	Disadvantages	Potential Applications	Manufacturer (Product name)
Starch	Renewable	Low cost, Fast biodegradation	Poor mechanical properties, Hydrophilicity	Foams, Films and bags, Molded items, Starch-based composite	Novament (Mater-bi™), Biotec (Bioplast®, Bioflex®, Biopur®), National Starch (ECO-FOAM), Buna Sow Leuna (Sconacell), Starch Tech (ST1, ST2, ST3), Novon (Poly NOVON®)

One of the first starch-based products was developed probably by the National Starch in the brand name ECO-FOAM™ and used as packaging material. ECO-FOAM™ materials are derived from maize or tapioca starch and include modified starches. This relatively short-term, protected-environment packaging use is ideal for thermoplastic starch polymers. National Starch now has additional thermoplastic starch materials, blends and specialty hydrophobic thermoplastic starches for a range of applications including injection molded toys, extruded sheet and blown film applications [http://www.ecofoam.com/loosefill.asp]. Novament has been developing thermoplastic starch based polymers since 1990. Mater-Bi™ polymers are based on starch-blend technologies and product applications include biodegradable mulch films and bags, thermoformed packaging products, injection molded items, personal hygiene products and packaging foam [http://www.novament.com]. Similarly, Biotech GmbH produces Bioplast™ based on starch for a wide range of applications including accessories for flower arrangements, bags, boxes, cups, cutlery, edge protectors, golf tees, horticultural films, mantling for candles, nets, packaging films, packaging materials for mailing, planters, planting pots, sacks, shopping bags, straws, strings, tableware, tapes, technical films, trays and wrap films [http://www.biotech.de/engl/index_engl.htm]. Recently, Plantic Technologies Ltd. produced soluble Plantic™ thermoformed trays for confectionery packaging.[42]

17.2.1.2   CELLULOSE AND CELLULOSE DERIVATIVES

Cellulose is one of the most fascinating organic resources, an almost inexhaustible raw material, and a key source of sustainable materials on an industrial scale in the biosphere. Natural cellulose based materials (cotton, wood, linen, hemp, etc.) have been used by our society as engineering materials for millennia and their use continues today as verified by the extent of the world wide industries in building materials, paper, textiles, etc. Generally, cellulose is a fibrous, tough, water-insoluble natural polymer that plays a vital role in maintaining the structure of plant cell walls. It was first discovered and isolated by Anselme Payen[58] in 1838, and since then, numerous physical and chemical prospects of cellulose have been extensively studied. As a chemical raw material, cellulose has been used for about 150 years for wide spectrum of products and materials in daily life. Many polymer researchers are of the opinion that polymer chemistry had its origins with the characterization of cellulose. Cellulose differs in some respects from other polysaccharides produced by plants, the molecular chain being very long and consisting of one repeating unit. Cellulose can be characterized as a high molecular weight homopolymer of β-1,4-linked anhydro-D-glucose units in which every unit is corkscrewed 180° with respect to its neighbors, and the repeat segment is frequently taken to be a dimer of glucose, known as cellobiose (Fig. 17.3).

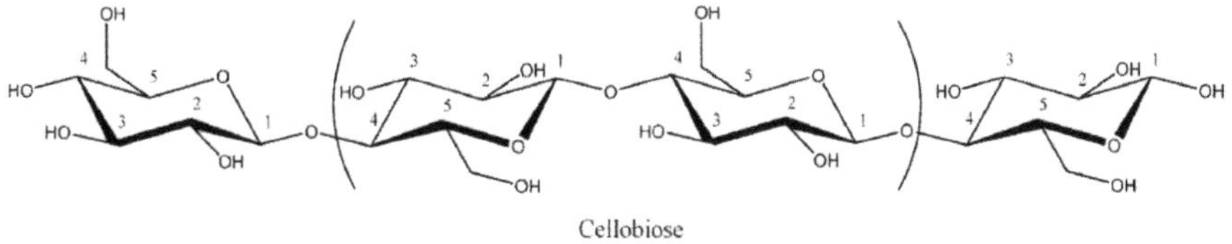

Cellobiose

FIGURE 17.3   Structure of cellobiose (Chem. Rev., 2010, 110 (6), 3479–3500).

Naturally, it occurs in a crystalline state. From the cell walls, cellulose is isolated in microfibrils by chemical extraction. In all forms, cellulose is a very highly crystalline, high molecular weight polymer. Because of its infusibility and insolubility, cellulose has driven the step-by-step creation of novel types of materials. Highlights were the development of cellulose esters and cellulose ethers as well as of cellulose regenerates and the discovery of the polymeric state of molecules. The very first thermoplastic polymeric material of cellulose was manufactured by Hyatt Manufacturing Company in 1870 to make celluloid in which they had reacted cellulose with nitric acid to form cellulose nitrate. The chemical modification of cellulose on an industrial scale led to a broad range of products based on cellulose from wood. The first example was the fabrication of regenerated cellulose filaments by spinning a solution of cellulose in a mixture of copper hydroxide and aqueous ammonia.[59]

Natural cellulose has earned in the materials society a tremendous level of awareness that does not emerge to be yielding. The cellulose biopolymer imprimatur such interest not only because of their unsurpassed quintessential physical and chemical properties but also because of their inherent renewability and sustainability in addition to their abundance. They have been the subject of a wide array of research efforts as reinforcing agents in nanocomposites due to their availability, low cost, renewability, light weight, nanoscale dimension, unique morphology and most importantly they have low environmental, animal/human health and safety risks. Currently, the isolation, characterization, and search for applications of novel forms of cellulose, variously termed crystallites, nanocrystals, whiskers, nanofibrils, and nanofibers, is generating much activity. Novel methods for their production range that begins at the highest conceptual level and works down to the details methods involving enzymatic/chemical/physical methodologies for their isolation from wood and forest/agricultural residues to the bottom-up production of cellulose nanofibrils from glucose by bacteria.[60,61] Some fungi can secrete enzymes that catalyze oxidation reactions of either cellulose itself or the lower molecular weight oligomers produced from the enzymatic hydrolysis of cellulose. Of these, the peroxidases can provide hydrogen peroxide for free radical attack on the C_2–C_3 positions of cellulose to form 'aldehyde' cellulose, which is very reactive and can hydrolyze to form lower molecular weight fragments while other oxidative enzymes can oxidize glucose and related oligomers to glucuronic acids. Such isolated cellulosic materials with one dimension in the nanometer range are referred to generically as nanocelluloses.[46] These nanocelluloses provide important cellulose properties—such as hydrophilicity, wide spectrum of chemical-modification capacity, and the formation of versatile semicrystalline fiber with very large aspect ratio which is the specific features of nanoscale materials. On the basis of their dimensions, functions, and preparation methods, which in turn depend mainly on the cellulosic source and on the processing conditions, nanocelluloses may be classified in three main subcategories.

17.2.1.2.1 MICROFIBRILLATED CELLULOSE (MFC)

MFC is normally produced from highly purified wood fiber (WF) and plant fiber (PF) pulps by high pressure homogenization according to the procedures developed at ITT Rayonnier.[62,63] Pulp is produced by using a mixture of sodium hydroxide and sodium sulfide and thus so-called Kraft pulp (almost pure cellulose fibers) is obtained. Pulping with salts of sulfurous acid leads to cellulose named sulphite pulp (which contains more by-products in the cellulose fibers). MFC particles are considered to comprise of several elementary fibrils. Each one of them consisting of 36 cellulose chains has a high aspect ratio or ~10–100 nm wide and 0.5–10 μm in length. MFCs are ~100% cellulose, and contain both amorphous and crystalline regions. In food and cosmetic industries, MFCs have been used as a thickening agent.[64]

17.2.1.2.2 NANOCRYSTALLINE CELLULOSE (NCC)

NCC is the term frequently used for the cellulose nanocrystals or cellulose whiskers prepared from natural cellulose by acid hydrolysis. NCC is the enlightened crystalline segments of elementary nanofibrils after the amorphous segments have been removed via the treatment with strong acids at eminent temperature. The nanocrystals formed from wood pulp are shorter and thinner than the MFC. NCCs have a high aspect ratio (3–5 nm wide, 50–500 nm in length), are ~100% cellulose, and are highly crystalline (54–88%). Most likely, from the result of acid hydrolysis process, the end of the cellulose nanocrystals are narrowed due to which they are look like whiskers. This hierarchical structure of natural fibers, based on their elementary nanofibrilar components, leads to the unique strength and high-performance properties of different species of plants. The mechanical properties of cellulose can be characterized by its properties in both the ordered (so-called crystalline) and disordered (so-called amorphous) regions of the molecule. The chain molecules in the disordered regions contribute to the flexibility and the plasticity of the bulk material, while those in the ordered regions contribute to the elasticity of the material. As they are almost defects free, the modulus of cellulosic nanocrystals is close to the theoretical limit for cellulose. It is potentially stronger than steel and similar to Kevlar.[65,69]

17.2.1.2.3 BACTERIAL NANOCELLULOSE (BNC)

BNCs are also known as bacterial cellulose, microbial cellulose, or biocellulose. BNCs are microfibrils concealed by aerobic bacteria, such as acetic acid bacteria of the genus Gluconacetobacter, as a pure component of their biofilms. The resulting microfibrils are microns in length, have a large aspect ratio with morphology depending on the specific bacteria and culturing conditions. These bacteria are widespread in nature where the fermentation of sugars and plant carbohydrates takes place. In contrast to other forms of cellulose, that is, MFC and NCC, materials isolated from cellulose sources, BNC is formed as a polymer and nano material by biotechnological assembly processes from low-molecular weight carbon sources, such as d-glucose. The bacteria are cultivated in common aqueous nutrient media, and the BNC is excreted as exopolysaccharide at the interface to the air. The resulting form-stable BNC hydrogel is composed of a nanofiber network (fiber diameter: 20–100 nm) enclosing up to 99% water. This BNC is proved to be very pure cellulose with a high weight-average molecular weight (MW), high crystallinity, and good mechanical stability. The bio-fabrication approach opens up the exciting option to produce cellulose by fermentation in the sense of white biotechnology and to control the shape of the formed cellulose bodies as well as the structure of the nanofiber network during biosynthesis. The resulting unique features of BNC lead to new properties, functionalities, and applications of cellulose materials.[70,78]

17.2.1.3   CHITIN AND CHITOSAN

Chitin, a natural polymer, is the second most abundant organic resource on the earth next to cellulose. It is an exoskeleton of crustacean, cuticle of insects, cell wall of fungi and micro organisms. It consists of 2-acetamido-2-deoxy-β-1, 4-D-glucan through the β-(1–4)-glycoside linkage.[79] Chitin can be degraded by chitinase. Chitin fibers have been used for making artificial skin and absorbable sutures. Although chitin is structurally similar to cellulose, much less attention has been paid to chitin than cellulose, primarily due to its inertness. Therefore, it has remained an almost unused resource. Deacetylation of chitin yields chitosan, which is relatively reactive and can be produced in numerous forms, such as powder, paste, film, fiber, and more. The materials are biocompatible and have antimicrobial activities as well as the ability to absorb heavy metal ions. They also find applications in the cosmetic industry because of their water-retaining and moisturizing properties. Using chitin and chitosan as carriers, a water-soluble prodrug has been synthesized.[80]

Chitosan is polysaccharides comprising copolymers of glucosamine and N-acetylglucosamine. Chitosan is usually prepared from chitin and chitin has been found in a wide range of natural resources (cruslaceans, fungi, insects, annelids, molluscs, coelenterate, etc.). Chitosan has interesting biopharmaceutical characteristics such as pH sensitivity, biocompatibility and low toxicity.[81] Moreover, chitosan is metabolized by certain human enzymes, especially lysozyme, and is considered as biodegradable.[82] Due to these favorable properties, the interest in chitosan and its derivatives as excipients in drug delivery has been increased in recent years. Modified chitosans have been prepared with various chemical and biological properties.[83] N-Carboxymethylchitosan and N-carboxybutylchitosan have been prepared for use in cosmetics and in wound treatment.[84] Chitin derivatives can also be used as drug carriers[85], and a report on the use of chitin in absorbable sutures shows that chitins have the lowest elongation among suture materials consisting of chitin, poly(glycolic acid) (PGA), plain catgut and chromic catgut.[86] The tissue reaction of chitin is similar to that of PGA.

17.2.1.4 SOY PROTEIN

Proteins are abundant in nature and widely available in various forms from plants (soy, corn, whey protein, and wheat gluten) and animals (collagen, gelatin) and have been used to develop bioplastics for various applications.[87] Among all those proteins, soy protein is one of the less expensive bio-polymers abundant worldwide. It was first introduced by Henry Ford in automobile manufacturing as alternating source for plastics and fibers.[88] Its purification process is benign and environment friendly. This protein can also be used as resin due to its ability to form ductile and viscous polymers. Soy protein is generally available in four different forms as soy protein isolate (SPI), soy protein concentrate (SPC), soy flour (SF), and soy meal. Chemically, SPI contains 90% protein and 4% carbohydrates, SPC contains 70% protein and 18% carbohydrates, SF, which requires less purification, contains about 55% protein and 32% carbohydrate and finally soy meal has 40% protein. Soy protein is globular, reactive and often water soluble, as compared to helical or planar, nonreactive and water resistant synthetic polymers.[89,90]

Generally soy proteins are classified on the basis of their sedimentation rate in fractional ultracentrifugation. Approximately 90% of the proteins in soybeans are globulins, and exist as dehydrated storage proteins. Based on Svedberg numbers (S), soy protein has mainly four fractions that include 2S, 7S, 11S, and 15S.[91] The main constituents of soy protein, the 7S fraction, is also called conglycinin and it comprises of many important enzymes and storage proteins. The 7S fraction is about 30% of the total soy protein by weight. The 11S fraction comprises about 52% of the total soy protein and is usually called glycinin.[92] The other two minor fractions are present as 2S (8%) and 15S (5%). It can be seen that storage proteins, 7S (conglycinin) and 11S (glycinin) are the principal components of soy protein. 7S has a quaternary structure and is highly heterogeneous according to Kineslla. Its principal component is beta-conglycinin, a sugar containing globulin. The fraction also comprises of enzymes (beta-amylase and lipoxygenase) and hemagglutinins. The 11S fraction consists of glycinin, the principal protein of soybeans. 11S has also a quaternary structure and is composed of three acidic and three basic subunits with isoelectric points between pH 4.7–5.4 and 8.0–8.5, respectively. The polypeptides in native glycinin are tightly folded and stabilized via intermolecular disulfide bonds. The ability of soy proteins to undergo association–dissociation reactions under known conditions is related to their functional properties and particularly to their texturization. The presence of various polar groups and reactive amino acids such as cystine, arginine, lysine, and hystidine in the structure of soy protein enable them to convenience for chemical and physical modifications, thereby improving the tensile and thermal properties of the biopolymer.[95]

17.2.1.5   POLY (LACTIC) ACID (PLA)

PLA is a renewably derived thermoplastic polyester and is completely biodegradable and bioabsorbable.[96] PLA, one of the oldest and most promising biodegradable polymers (aliphatic polyester) which is obtained from agricultural products such as corn, sugarcane, etc., is at the forefront of emerging biodegradable polymer used in industries through improved manufacturing practices that lower its production cost.[97] Poly(lactic acid) and polylactide are the same chemical products and both are abbreviated as PLA. The only difference between them is how they are produced. Lactic acid is a chiral molecule existing as two stereoisomers, l- and d-lactic acid which can be produced in different ways, that is, biologically or chemically synthesized.[98] In the first case, lactic acid is obtained by fermentation of carbohydrates from lactic bacteria, belonging mainly to the genus *Lactobacillus*, or fungi.[99] This fermentative process requires a bacterial strain and is a sources of carbon (carbohydrates), nitrogen (yeast extract, peptides, etc.) and mineral elements to allow the growth of bacteria and the production of lactic acid. The lactic acid as-formed exists almost exclusively as l-lactic acid and leads to poly(l-lactic acid) (PLLA) with low molecular weight by polycondensation reaction. However, Moon et al.[100] have proposed an alternative solution to obtain higher molecular weight PLLA by the polycondensation route. In contrast, the chemical process could lead to various ratio of l- and d-lactic acid. Indeed, the chemical reactions leading to the formation of the cyclic dimer, the lactide, as an intermediate step to the production of PLA, could form macromolecular chains with l- and d-lactic acid monomers. This mechanism of ring-opening polymerization ROP from the lactide explains the formation of two enantiomers. This ROP route has the advantage of reaching high molecular weight polymers[101] and allows control of the final properties of PLA by adjusting the proportions and the sequencing of l- and d-lactic acid units. At present, due to its availability on the market, PLA has one of the highest potentials among biopolyesters, particularly for packaging and medical applications.[102]

17.2.2   BIO-BASED FIBERS AS POLYMER REINFORCEMENTS

In biocomposites, biofibers become a promising option for improving the mechanical strength and stiffness of bio-based polymeric materials. The conventional fibers like carbon, aramid, glass, etc., can be produced with a definite range of properties. The characteristic properties of natural fibers[103] vary significantly depends on whether the fibers are taken from plant stem or leafs[104], the quality of the plants locations[105], the age of the plant[106] and the preconditioning step[107]. Relying on the origin of the natural fibers, they may be rounded up into: leaf, seed, bast, and fruit origin. Some of the well known examples are: (i) Leaf: pineapple leaf fiber, sisal, and henequen; (ii) Seed: Cotton; (iii) Bast: jute, ramie, kenaf, hemp and flax; (iv) Fruit: coir. The natural fibers are lignocellulosic (LC) in nature which is generally the most ample renewable biomaterial of photosynthesis on earth. In terms of mass units, the net primary production of natural fiber per year is estimated to be 2×10^{11} tons[108] as compared to synthetic polymers by 1.5×10^{8} tons. Lignocellulosic materials are widely distributed in the biosphere in the form of trees (wood), plants and crops. Cellulose, in its several forms, constitutes approximately half of all polymers used in the industry worldwide.[57]

17.2.3 SURFACE TREATMENTS

The production of polymer composites comprising of lignocelluloses (LC) fibers will often result in fibers physically dispersed in the polymeric matrix. But in most of the cases, poor adhesion and consequently inadequate mechanical properties result. Hence, surface treatment of the fibers will play a vital role. Generally surface treatment of LC fibers is not required to develop the bonding for the synthesis of biopolymer based composites, in view of the comparable chemical scenery of both the biofiber and biopolymer matrix, which have a hydrophilic nature, unlike the situation with commodity polymers, which have a tendency to be hydrophobic. To improve many specific aspects, such as providing greater adhesion and reduced moisture sensitivity, surface treatment can be useful even in the case of biodegradable composites. Although better adhesion between the biopolymers and fibers is contributed by the similar polarities of the two materials yet these results in an increase in water absorption of the composite. Hence, these fibers require suitable surface treatments.

Surface treatment normally involves one of four methods, namely chemical, physical, physical–chemical and physical–mechanical. Chemical methods involve treatment with silanes or other chemicals through chemical functionalization reactions and leaching of the surface through alkali or bleaching.[109] Physical methods involve treatment by plasma, corona, laser or γ-ray and subjected to steam explosion.[110] Steam explosion process, a high pressure steaming, involves heating of LC materials at high temperatures and pressures followed by mechanical disruption of the pretreated material by violent discharge (explosion) into a collecting tank.[111] Mechanical methods involve rolling or swaging and those may damage the fibers. Finally, physical–chemical methods involve solvent extraction of surface gums and other soluble components of the fibers. As an alternative to the methods described above, drying of LC fibers may be an effective process for surface modification, both in terms of cost and improvement in properties.[112] It should be noted that all the above mentioned treatments of LC fibers have helped to improve their interaction with the matrix materials, increase adhesion of fibers with the matrix through surface roughness of fiber, leading to increased strength or other properties of composites through higher fiber incorporation and possibly providing greater durability of the composites.

17.2.4   NANOMATERIALS

Nanomaterials are such a stuff, which has at least one dimension in nanometer scale, that is, 1 to 100 nm. Nanomaterials can be classified into two categories viz. nanostructured material and nanophase/nanoparticle materials. Nanostructured materials usually refer to condensed bulk materials that are made of grains (agglomerates), with nanometric size range. The latter are generally the dispersive nanoparticles. Nanotechnology is the study and control of nanomaterial which also deals with the design, fabrication and application of nanostructures. Nanomaterials, a new branch of materials research, are attracting a great deal of attraction because of their potential applications in areas such as optics, electronics, magnetic data storage, catalysis and polymer nanocomposites (PNs).

Incorporation of inorganic/organic nanoparticles as additives into polymer systems has resulted in PNs displaying multifunctional, high performance polymer characteristics beyond what conventional filled polymeric materials acquire. Multifunctional features attributable to PNs consist of improved mechanical properties, thermal properties and/or flame retardancy, moisture resistance, chemical resistance, decreased permeability, and charge dissipation. Through control/alteration of the nanoscale additives, one can maximize the property enhancements of selected polymers to meet or exceed the needs. Uniform dispersion of these nanoscale materials produces super interfacial area per volume between the nanoparticle and the polymer. There are different types of commercially available nanoparticles such as montmorillonite organoclays, carbon nanofibers, carbon nanotubes, nanosilica, nanotitanium dioxide, nano ZnO and others that can be incorporated into the polymer matrix to form PNs.[113]

17.3   BIONANOCOMPOSITES: PROCESSING ASPECTS AND PRODUCTS

Bionanocomposites are a novel class of nanosized materials in the modern day world. The terminology "bionanocomposites" is introduced several years ago to classify an emerging class of biohybrid materials. Bionanocomposites are the combination of biopolymers such as proteins, polysaccharides, nucleic acids, etc. with a reinforcing agent having at least one dimension in the nanometer range. The reinforcing agent may include plant fibers and by products from lignocellulosic renewable resources or synthetic inorganic fraction of finely divided solids, spanning from clays to phosphates or carbonates, whose origin can be either natural or synthetic. The most important challenge in bionanocomposites is to achieve materials with improved performance characteristics, by the elusive management of the individual properties of the incorporated components. There are some similarities of bionanocomposites with nanocomposites prepared by using commodity polymers but also have fundamental differences in the methods of preparation, functionalities, properties, biodegradability, and applications. Processes and structure of bionanocomposites are regulated by water that is added in an amount to only hydrate functional groups in the carbohydrate macromolecule. In particular, the biodegradability and biocompatibility nature of biopolymers, along with the thermal and mechanical properties of the reinforcing counterpart, bridge the gap between functional and structural materials.[114,117] Engineered biopolymer-layered silicate nanocomposites are reported to have markedly improved physical properties including higher gas barrier properties, tensile strength, and thermal stability.[118,120] Chemically treated nanoscale silicate plates incorporated with appropriate polymers can provide effective barrier performance against water, gases, and grease.[121] These hyper-platy, nanodimensional thickness crystals create a tortuous path structure that inherently resists penetration.

17.4 APPLICATION AND MARKET

Unlike many biopolymer products being developed and marketed, very few biodegradable composites have been developed, with most of their technologies still in the research and development stages. This is despite the fact that the environmentally friendly composites, where biodegradability is important, provide designers new alternatives to meet challenging requirements. These include aquatic and terrestrial environments, municipal solid waste management and compostable packaging, while those for automobiles include parcel shelves, door panels, instrument panels, armrests, headrests and seat shells. Accordingly, a wide range of biodegradable products have been produced using LC fibers and biopolymers for different applications, ranging from automotive vehicles including trucks, construction (hurricane resistant housing and structures, especially in the USA) and insulation panels, to special textiles (geotextiles and nonwoven textiles).[122] The hurricane resistant housing, structures and a variety of products developed using soy oil with LC fibers could be the predecessor for diverse range of applications for the biodegradable composites. Other identified uses for these materials include bathtubs, archery bows, golf clubs and boat hulls. This is further underlined with the estimated global market of about 900,000 metric tons of wood plastics and natural fiber composites as per Steven Van Kourteren, Consultant, Principia Partners.[123] Hence the market for biodegradable composites can be expected to grow in the future. This is based on continued technical innovations, identification of new applications, persistent political and environmental pressures, and investments mostly by governments in new methods for fiber harvesting and processing of natural fibers.[124,125]

17.5 CONCLUDING REMARKS

Renewable resources based products finding privilege particularly because of environmental friendliness and dwindling petroleum resources. Biopolymers reinforced with natural fibers have developed significantly over the past two decades because of their significant processing advantages, biodegradability, low cost, low relative density, high specific strength and renewable nature. These composites are preordained to find more and more application in the near future. Interfacial adhesion between natural fibers and matrix will remain the key issue in terms of overall performance, since it dictates the final properties of the biocomposites. Research on biodegradable polymer and its composites has been very impressive due to their environmental friendliness, carbon dioxide sequestration, sustainability, nontoxicity and varieties of other reasons. The potential areas of applications for these composites are packaging, structural, transportation and automotive, agriculture and various consumer products. The market scenario has been changing continuously due to the development of newer biodegradable polymers, processing techniques and imposition of stringent environmental laws. Raw materials, processing techniques and application of biocomposites have been studied and well documented in recent years. Still there are lot of issues need to be addressed for further improvement pertaining to those above areas.

One of such issue is the nonavailability of quality fiber used as reinforcing agent in the composites. The production of quality fiber may be obtained through better cultivation, which includes the use of generic engineering. Exploration of nontraditional fiber as a source of reinforcing agent is another important area. In order to achieve proper reinforcement, the introduction of hybrid nanocomposite may be attempted. The processibility and development of new biodegradable polymers with much improved properties in terms of moisture resistance, mechanical strength, thermal stability and biodegradability are some of the areas which require much attention. The variation of properties along with the high cost of the bio-composites prevents their uses in various application sectors. The possibility of using high percentage of reinforcing fiber may be tried in order to achieve a reduction in cost. Therefore, the requirement of improving interfacial interaction between reinforcing agent, filler and matrix is another critical area to be looked upon. The development of newer processing tools at lower temperature is another important aspect that needs to address.

The introduction of nanomaterials in the biocomposites is one of the effective ways to enhance the properties. Research effort should be directed towards development of nanowhiskers and nanofibers from different lignocellulosic materials and their inclusion in biocomposites for improving various properties. Efforts may also be required to derive resin, reinforcing agent and coupling agent from renewable resources. Efforts may be directed towards searching for new and improved bioresin, fiber with better properties or new composite manufacturing technology to meet with the future environmental goals. The concept of biodegradability should be directed to 'triggered' biodegradability.

The price of biodegradable polymers for making composites is expected to reduce further in the coming years due to development of raw material, manufacturing techniques and hence it may be considered as a valid alternative to conventional composites. It is also envisaged that further research and development on biodegradable composite may lead to open up new avenues to meet the local as well as global challenges and thus may expand the horizon of applications.

KEYWORDS

- **Green Materials**
- **Nanocomposites**
- **Natural Fibers**
- **Natural Resources**
- **Polymer Matrix**
- **Surface Treatments**

REFERENCES

1. Satyanarayana, K. G., Ramos, L. P., & Wypych, F. (2005). Development of new materials based on agro and industrial wastes towards ecofriendly society. In *Biotechnology in energy management*, Ghosh, T. N., Chakrabarti, T., & Tripathi, G., (Eds.), APH Publishing Corporation: New Delhi, 583.
2. Uddin, M. K., Khan, M. A., & Ali, K. M. I. (1997). Degradable jute plastic composites, *Polym Degrad. Stab*, 55, 1–7.
3. Wollerdorfer, M., & Bader, H. (1998). Influence of natural fibers on the mechanical properties of biodegradable polymers. *Ind. Crops Prod.*, 8, 105–112.
4. Luo, S., & Netravali, A. N. (1999). Effect of Co-60 gamma-radiation on the properties of Poly(hydroxybutyrate-cohydroxyvalerate). *J. Appl. Polym. Sci.*, 73, 1059–1067.
5. *Advancing Sustainability through Green Chemistry and Engineering,* Lankey, R. L., & Anastas, P. T. Eds., (2002). ACS Symposium Series 823, American Chemical Society: Washington, DC.
6. Mohanty, A. K., Misra, M., & Drzal, L. T. (2002). Sustainable bio-composites from renewable resources: Opportunities and challenges in the green materials world. *J. Polym. Environ*, 10, 19–26.
7. Mohanty, A. K., Drzal, L. T., & Misra, M. (2002). Engineered natural fiber reinforced polypropylene composites: Influence of surface modifications and novel powder impregnation processing. *J. Adhes. Sci. Technol.*, 16, 999–1015.
8. Coates, G. W., & Hillmyer, M. A. (2009). A virtual issue of Macromolecules: "Polymers from renewable resources." *Macromolecules*, 42, 7987–7989.
9. Ryberg, Y. Z. Z., Edlund, U., & Albertsson, A. C. (2011). Conceptual Approach To Renewable Barrier Film Design Based On Wood Hydrolysate. *Biomacromolecules*, 12, 1355–1362.
10. DeWit, M. A., & Gillies, E. R. (2009). A cascade biodegradable polymer based on alternating cyclization and elimination reactions. *J. Am. Chem. Soc.*, 131, 18327–18334.
11. Hartman, J., Albertsson, A. C., Lindblad, M. S., & Sjöberg, J. (2006). Oxygen barrier materials from renewable sources: Material properties of softwood hemicellulose-based films. *J. Appl. Polym. Sci.*, 100, 2985–2991.

12. Tian, H., Tang, Z., Zhuang, X., Chen, X., & Jing, X. (2012). Biodegradable synthetic poly-mers: Preparation, functionalization and biomedical application. *Prog Polym. Sci.*, 37, 237–280.
13. Odelius, K., Ohlson, M., Höglund, A., & Albertsson, A. C. (2013). Polyesters with small struc-tural variations improve the mechanical properties of polylactide. *J. Appl. Polym. Sci.*, 127, 27–33.
14. Muggli, D. S., Burkoth, A. K., Keyzer, S. A., Lee, H. R., & Anseth, K. S. (1998). Reac-tion behavior of biodegradable, photo-cross-linkable polyanhydrides, *Macromolecules*, 31, 4120–4125.
15. Metters, A. T., Anseth, K. S., & Bowman, C. N. (2000). Fundamental studies of a novel bio-degradable PEG-b-PLA hydrogel *Polymer*, 41, 3993–4004.
16. Ray, S. S., Yamada, K., Okamoto, M., & Ueda, K. (2002). Polylactide-layered silicate nano-composite: A novel biodegradable material. *Nano Lett.*, 2, 1093–1096.
17. Nam, J. Y., Ray, S. S., & Okamoto, M. (2003). Crystallization Behavior and Morphology of Biodegradable Polylactide/Layered Silicate Nanocomposite *Macromolecules*, 36, 7126–7131.
18. Tang, X., & Alavi, S. (2012). Structure and physical properties of starch/poly vinyl alcohol/laponite RD nanocomposite films *J. Agric. Food Chem.*, 60, 1954–1962.
19. Lim, S. T., Hyun, Y. H., & Choi, H. J. (2002). Synthetic Biodegradable Aliphatic Polyester/Montmorillonite Nanocomposites *Chem. Mater.*, 14, 1839–1844.
20. Dash, B. N., Sarkar, M., Rana, A. K., Mishra, M., Mohanty, A. K., & Tripathy, S. S. (2002). A study on biodegradable composite prepared from jute felt and polyesteramide (BAK). *J. Reinf. Plast Compos* 21, 1493–1503.
21. Mishra, S., Tripathy, S. S., Misra, M., Mohanty, A. K., & Nayak, S. K. (2002). Novel eco-friendly biocomposites: Biofiber reinforced biodegradable polyester amide composites-Fabri-cation and properties evaluation. *J. Reinf. Plast Compos.*, 21, 55–70.
22. Netravali, A. N., & Chabba, S. (2003). Composites get greener. *Mater Today*, 6, 22–29.
23. Wambua, P., Ivens, J., & Verpoest, I. (2003). Natural fibers: can they replace glass fiber in reinforced plastics. *Compos Sci. Technol.*, 63, 1259–1264.
24. Plackett, D., Andersen, T. L., Pedersen, W. B., & Nielsen, L. (2003). Biodegradable compos-ites based on L-polylactide and jute fibers. *Compos Sci. Technol.*, 63, 1287–1296.
25. Mohanty, A. K., Wibowo, A., Misra, M., & Drzal, L. T. (2004). Effect of process engineering on the performance of natural fiber reinforced cellulose acetate biocomposites. *Composites A*, 35, 363–370.
26. Geeta, M., Mohanty, A. K., Thayer, K., Misra, M., & Drzal, L. T. (2005). Novel biocopmpos-ites sheet molding compounds for low cost housing panel applications. *J. Polym. Environ.* 13, 169–175.
27. Liu, W., Misra, M., Askeland, P., Drzal, L. T., & Mohanty, A. K. (2005). "Green" composites from soy based plastic and pineapple leaf fiber: fabrication and properties evaluation. *Poly-mer*, 46, 2710–2721.
27. Liu, W., Mohanty, A. K., Drzal, L. T., & Misra, M. (2005). Novel biocomposite from native grass and soy based bioplastic: processing and properties evaluation. *Ind. Eng. Chem. Res.*, 44, 7105–7112.
28. Bhardwaj, R., Mohanty, A. K., Drzal, L. T., Pourboghrat, F., & Misra, M. (2006). Renewable resource-based green composites from recycled cellulose fiber and poly(3-hydroxybutyrate-co3-hydroxyvalerate) bioplastic. *Biomacromolecules*, 7, 2044–2051.
29. Maya, J. J., & Thomas, S. (2008). Biofibers Biocomposites *Carbohydr Polym.*, 71, 343–364.
30. Williams, C. K., & Hillmayer, M. A. (2008). Polymers from renewable resources: A perspec-tive for a special issue of polymer reviews, *Polym. Rev.*, 48, 1–10.

31. Cheng, M., Attygalle, A. B., Lobkovsky, E. B., & Coates, G. W. (1999). Single–site catalysts for ring opening polymerization Synthesis of heterotactic poly(lactic acid) from rac-lactide. *J. Am. Chem. Soc.*, 121, 11583–11584.

32. Satyanarayana, K. G., Arizaga, G. G. C., & Wypych, F. (2009). Biodegradable Composites Based on Lignocellulosic Fibers an overview *Prog Polym Sci.*, 34, 982–1021.

33. Bledzki, A. K., & Gassan, J. (1999). Composites reinforced with cellulose based fibers. *Prog Polym. Sci.*, 24, 221–274.

34. Medical Device and Diagnostic Industry. Biomaterials: Developments in medical polymers for biomaterials applications from http://www.devicelink.com/mddi/archive/01/01/contents. html (accessed July 7, 2013).

35. Carole, T. M., Pellegrino, J., & Paster, M. D. (2004). Opportunities in the industrial biobased products industry. *Appl. Biochem. Biotechnol* 115, 871–885.

36. Clean Production Action. Biobased Materials, from http://www.cleanproduction.org/Steps. BioSociety.Biobased.php (accessed July 7, 2013).

37. National Research Council, Biobased Industrial Products: Research and Commercialization Priorities, National Academies Press, Washington, D.C., 2000, 162.

38. Doornbosch, R., &. Steenblik, R. (2007). Biofuels: Is the Cure Worse than the Disease? Organization for Economic Co-Operation and Development, New York, 57.

39. Kanzig, J., Anex, R., & Jolliet, O. (2003). International Workshop on Assessing the Sustainability of Bio-Based Products *Int. J. Life Cycle Assess.* 8, 313–314.

40. Buleon, A., Colonna, P., Plachot, V., & Ball, S. (1998). Starch granuels: Structure and biosynthesis. *Int. J. Biol. Macromol.*, 23, 85–112.

41. Roger, P., Tran, V., Lesec, J., & Colonna, P. (1996). Isolation and characterization of single chain amylose. *J. Cereal Sci.* 24, 247–262.

42. Halley, P. J. (2005). Thermoplastic starch biodegradbale polymers, In *Biodegradable Polymers for Industrial Applications*, Smith, R. Ed., CRC Press: Washington, DC, 1, 140.

43. Thomson, D. B. (2000). On the nonrandom nature of amylopectin branching. *Carbohydr Polym.*, 43, 223–239.

44. Dufresne, A., & Vignon, M. R. (1998). Improvement of Starch Film Performances Using Cellulose Microfibrils *Macromolecules,* 31, 2693–2696.

45. Otey, F. H., Westhoff, R. P., & Russell, C. R. (1977). Biodegradable Films from Starch and Ethylene-Acrylic Acid Copolymer *Ind. Eng. Chem. Prod. Res. Dev.*, 16, 305–308.

46. Chandra, R., & Rustogi, R. (1998). Biodegradable Polymes *Prog Polym Sci.*, 23, 1273–1335.

47. Otey, F., Mark, A., Mehtretter, C., & Russel, C. (1974). Starch-Based Film for Degradable Agricultural Mulch *Ind. Eng Chem Prod Res. Dev.*, 13, 90–95

48. Huang, J. C., Shetty, A. S., Wang, M. S. (1990). Biodegradable plastics: A review. *Adv. Polym. Technol.*, 10, 23–30.

49. Shulman, J. & Howarth, J. T. (June 16, 1964) Waterproof Plastic Films of Increased Water Vapor Permeability and Method of Making Them. US Patent 3(137), 664.

50. Griffin, G. J. L. (1973). Biodegradable Fillers in Thermoplastics *Am. Chem Soc. Div. Org Coat Plast Chem.*, 33, 88–96.

51. *Biodegradable fillers in thermoplastic*, Deanin, R. D., Schott, N. R., (Eds.), (1974). Advances in Chemistry Series, 134, American Chemical Society Washington, DC.

52. Otey, F. H., & Doane, W. M. (June 1987). A starch based degradable plastic film, in Proc. Society of the Plastic Industry. *Symp Degrad. Plast* Washington, DC, 39–40.

53. Otey, F. H., Westhoff, R. P., & Doane, W. M. (1980). Starch based blown films. *Ind. Eng. Chem. Prod. Res. Dev.*, 19, 592–595.

54. Kulicke, W. M., Aggour, Y. A., Nottelmann, H., & Elsabee, M. Z. (1989). Starch Sodium Trimetaphosphate Hydrogels *Starch*, 41, 140–146.

55. Jane, J. L., Lim, S., Paetau, I., Spence, K., & Wang, S. (1993). Biodegradable plastics made from agricultural biopolymers. In *Polymmers from agricultural coproducts*, Fishman, M., Friedman, R., Huang, S. J., Eds., ACS Symposium Series.

56. Iman, M., & Maji, T. K. (2012). Effect of crosslinker and nanoclay on starch and jute fabric based green nanocomposites. *Carbohydr Polym.*, 89, 290–297

57. Mohanty, A. K., Misra, M., & Hinrichsen, G. (2004). Biofibers, biodegradable polymers and biocomposites: An overview. *Macromol Mater Eng.*, 276/277, 1–24.

58. Payen, A. (1838). Mémoire sur la composition du tissue proper des plantes et du ligneux. *Compt. Rend.*, 7, 1052–1056.

59. Habibi, Y., Lucia, L. A., & Rojas, O. J. (2010). Cellulose nanocrystals: Chemistry, self-assembly, and applications. *Chem. Rev.*, 110, 3479–3500.

60. Klemm, D., Kramer, F., Moritz, S., Lindstrm, T., Ankerfors, M., Gray, D., & Dorris, A. (2011). Nanocelluloses: A new family of nature-based materials. *Angew Chem. Int. Ed.*, 50, 5438–5466.

61. Dufresne, A. (2012). Nanocellulose: Potential reinforcement in Composites. In *Natural Polymers: Volume 2: Nanocomposites*, John, M. J., Thomas, S. Eds., RSC Publication, London, 1–32.

62. Herrick, F. W., Casebier, R. L., Hamilton, J. K., & Sandberg, K. R. (1983). Microfibrillated cellulose: morphology and accessibility. *J. Appl. Polym. Sci. Appl. Polym. Symp* 37, 797–813.

63. Henriksson, M., Berglund, L. A., Isaksson, P., Lindstr□m, T., & Nishino, T. (2008). Cellulose nanopaper structures of high toughness. *Biomacromolecules*, 9, 1579–1585.

64. Beck-Candanedo, S., Roman, M., & Gray, D. G. (2005). Biomacromolecules, 6, 1048–1054

65. Turbak, A. F., Snyder, F. W., & Sandberg, K. R. (1983). Microfibrillated cellulose, a new cellulose product: properties, uses, and commercial potential. *J. Appl. Polym. Sci. Appl. Polym. Symp* 37, 815–827.

66. De. Rodriguez, N. L., Thielemans, G., & Dufresne, W. (2006). A Sisal Cellulose Whisker Reinforced Polyvinyl Acetate Nanocomposites. *Cellulose*, 13, 261–270.

67. Beck-Candanedo, S., Roman, M., & Gray, D. G. (2005). Effect of Reaction Conditions on the Properties and Behavior of Wood Cellulose Nanocrystal Suspensions *Biomacromolecules*, 6, 1048–1054.

68. Bai, W., Holbery, J., & Li, K. C. (2009). A technique for production of nanocrystalline cellulose with a narrow size distribution. *Cellulose*, 16, 455–465.

69. Moon, R. J., Martini, A., Nairn, J., Simonsen, J., & Youngblood, J. (2011). Cellulose nanomaterials review: Structure, properties and nanocomposites. *Chem. Soc. Rev.*, 40, 3941–3994.

70. Tokoh, C., Takabe, K., Fujita, M., & Saiki, H. (1998). Cellulose synthesized by acetobacter xylinum in the presence of acetyl glucomannan. *Cellulose*, 5, 249–261.

71. Klemm, D., Schumann, D., Udhardt, U., & Marsch, S. (2001). Bacterial synthesized cellulose artificial blood vessels for microsurgery. *Prog Polym. Sci.*, 26, 1561–1603.

72. Yamanaka, S., Watanabe, K., Kitamura, N., Iguchi, M., Mitsuhashi, S., Nishi, Y., & Uryu, M. (1989). The structure and mechanical properties of sheets prepared from bacterial cellulose. *J. Mater. Sci.*, 24, 3141–3145.

73. Cannon, R. E., & Anderson, S. M. (1991). Biogenesis of Bacterial Cellulose *Crit. Rev. Microbiol.*, 17, 435–447.

74. Vandamme, E. J., Baets, S. De, Vanbaelen, A., Joris, K., & De, Wulf, P. (1998). Improved Production of Bacterial Cellulose and its Application Potential *Polym Degrad. Stab.*, 59, 93–99.

75. Salmon, S., & Hudson, S. M. (1997). Crystal Morphology Biosynthesis and Physical Assembly of Cellulose Chitin and Chitosan *J Macromol Sci. Rev. Macromol Chem. Phys.*, C, 37, 199–276.

76. Jonas, R., Farah, L. F. (1998). Production and application of microbial cellulose, *Polym. Degrad. Stab.*, 59, 101–106.

77. Sakairi, N., Asano, H., Ogawa, M., Nishi, N., & Tokura, S. (1998). A Method for Direct Harvest of Bacterial Cellulose Filaments during Continuous Cultivation of Acetobacter Xylinum *Carbohydr Polym.*, 35, 233–237.

78. Guhados, G., Wan, W., & Hutter, J. L. (2005). Measurement of the Elastic Modulus of Single Bacterial Cellulose Fibers using Atomic Force Microscopy *Langmuir,* 21, 6642–6646.

79. Chandy, T., & Sharma, C. P. (1990). Chitosan as a Biomaterial *Biomat Art Cells Art Org.*, 18, 1–24.

80. Hosokawa, J., Nishiyama, M., Yoshihara, K., & Kubo, T. (1990). Biodegradable film derived from chitosan and homogenized cellulose. *Ind. Eng. Chem. Res.*, 29, 800–805.

81. Prabaharan, M., & Mano, J. F. (2006). Chitosan Derivatives Bearing Cyclodextrin Cavities as Novel Adsorbent Matrices *Carbohydr Polym.* 63, 153–166.

82. Prabaharan, M., Rodriguez-Perez, M. A., de Saja, J. A., & Mano, J. F. (2007). Preparation and characterization of Poly(L-lactic acid)-chitosan hybrid scaffolds with drug release capability. *J Biomed Mater Res. Part B: Appl. Biomater.* 81B, 427–434.

83. Muzzarelli, R. A. A. (1986). Chitin in Nature and Technology Plenum Press New York.

84. Muzzarelli, R. A. A. (1990). Modified Chitosans and Their Chemical Behavior *Am. Chem. Soc. Polym. Prep Div Polym Chem.,* 31, 626.

85. Tokura, S., Miura, Y., Uraki, Y., Watanabe, K., Saiki, I., & Azuma, I. (1990). Biodegradable Chitin Derivative as Various Types of Drug Carriers, *Am Chem Soc Polym Prep. Div. Polym. Chem.,* 31, 627.

86. Tachibana, M., Yaita, A., Taniura, H., Fukasawa, K., Nagasue, N., & Nakamura, T. (1988). The Use of Chitin as a New Absorbable Suture Material an Experimental Study *Jpn. J. Surg.*, 18, 533–539.

87. Pommet, M., Redl, A., Morel, M. H., & Guilbert, S. (2003). Polymer Study of Wheat Gluten Plasticization with Fatty Acids *Polymer*, 44, 115–122.

88. Johnson, L. A., & Myers, D. J. (1995). In Industrial Uses for Soybeans in Practical Handbook of Soybean Processing and Utilization, David R. Erickson, Ed., AOCS Press: Champaign, IL 565–584.

89. Chabba, S., Matthews, G. F., & Netravali, A. N. (2005). 'Green' Composites Using Cross Linked Soy Flour and Flax Yarns *Green Chem* 7, 576–581.

90. Iman, M., & Maji, T. K. (2013). Effect of crosslinker and nanoclay on jute fabric reinforced soy flour green composite *J. Appl. Polym. Sci.*, 127, 3987–3996.

91. Khorshid, N., Hossain, M. M., & Farid, M. M. (2007). Precipitation of Food Protein Using High Pressure Carbon Dioxide *J. Food Eng.* 79, 1214–1220.

92. Kumar, R., Liu, D., & Zhang, L. (2008). Advances in Proteinous Biomaterials *J. Biobased Mater Bioenergy*, 2, 1–24.

93. Kinsella, J. E. (1979). Functional Properties of Soy Proteins *J. Am. Oil Chem. Soc.*, 56, 242–258.

94. Reddy, M., Mohanty, A. K., & Misra, M. (2010). Thermoplastics from Soy Protein: A Review on processing, blends and composites. *J. Biobased Mater. Bioenergy*, 4, 298–316.

95. Iman, M., Bania, K. K., & Maji, T. K. (2013). Green jute-based cross-linked soy flour nanocomposites reinforced with cellulose whiskers and nanoclay *Ind. Eng. Chem. Res.,* 52, 6969–6983.

96. Ray, S. S. (2012). Polylactide-based bionanocomposites: A promising class of hybrid materials. *Acc. Chem. Res.*, 45, 1710–1720.

97. Yu, L., Dean, K., & Li, L. (2006). Polymer blends and composites from renewable resources *Prog. Polym Sci.*, 31, 576–602.

98. Averous, L. (2008). *Polylactic acid: synthesis, properties and applications*. In: Belgacem N, Gandini A, editors. Monomers, Oligomers, Polymers and Composites Fromrenewable Resources Oxford: Elsevier Limited Publication, 433–450.

99. Garlotta, D. (2001). A Literature Review of Poly (lactic acid) *J. Polym Environ* 9, 63–84.
100. Moon, S. I., Lee, C. W., Taniguchi, I., Miyamoto, M., & Kimura, Y. (2001). Melt/solid poly-condensation of l-lactic acid: an alternative route to Poly (l-lactic acid) with high molecular weight. *Polymer*, 42, 5059–5062.
101. Okada, M. (2002). Chemical Syntheses of Biodegradable Polymers *Prog Polym Sci.*, 27, 87–133.
102. Bordes, P., Pollet, E., & Avérous, L. (2009). Nano-biocomposites: Biodegradable polyester/nanoclay systems *Prog. Polym Sci.*, 34, 125–155.
103. Bledzki, A. K., Reihmane, S., & Gassan, J. (1996). Properties and Modification Methods for Vegetable Fibers for Natural Fiber Composites *J. Appl Polym Sci.*, 59, 1329–1336.
104. Bisanda, E. T. N., & Ansell, M. P. (1992). Properties of Sisal-CNSL Composites *J. Mater Sci.* 27, 1690–1700.
105. Barkakaty, B. C. (1976). Some Structural Aspects of Sisal Fibers *J. Appl Polym Sci.*, 20, 2921–2940.
106. Chand, N., & Hashmi, S. A. R. (1993). Mechanical properties of sisal fiber at elevated temperatures. *J. Mater Sci.*, 28, 6724–6728.
107. Ray, P. K., Chakravarty, A. C., & Bandyopadhyay, S. B. (1976). Fine structure and mechanical properties of jute differently dried after retting. *J. Appl. Polym. Sci.*, 20, 1765–1767.
108. Hon, D. N. S. (1988). Cellulose: A wonder material with promising future. *Polym News* 13, 34–140.
109. Mohanty, A. K., Khan, M. A., & Hinrichsen, G. (2000). Surface Modification of Jute and Its Influence on Performance of Biodegradable Jute-Fabric/Biopol Composites *Compos Sci. Technol.*, 60, 1115–1124.
110. Singh, R. P., Pandey, J. K., Rutot, D., Degée, Ph., & Dubois, P. (2003). Biodegradation of poly(ε -caprolactone)/starch blends and composites in composting and culture environments: The effect of compatibilization on the inherent biodegradability of the host polymer *Carbohyd. Res.*, 338, 1759–1769.
111. Keller, A., Zerlik, H., & Wintermantel, E. (1999). Hemp fiber reinforced biodegradable polyesters effect of fiber refinement by steam explosion. *Angew Makromol Chemie.*, 272, 46–50.
112. Moraes, G. S., Alsina, O. L. D., & Carvalho, L. H. (2003). In *7th Annual Brazilian Congress of Polymers*, Organized by the Brazilian polymers Association, 1062–1064.
113. *Polymer Nanocomposites: Processing, Characterization, and Applications*, Koo, J. H., (Eds.), McGraw-Hill (2006). Nanoscience and Technology Series, McGraw-Hill: New York.
114. *Introduction to cellulose nanocomposites*, Oksman, K. S. M., (Ed.) (2006). ACS Symposium Series 938, American Chemical Society: Washington, DC.
115. Kamel, S. (2007). Nanotechnology and its applications in lignocellulosic composites: A mini review. *Express Polym. Lett*, 1, 546–575.
116. Kalia, S., Dufresne, A., Cherian, B. M., Kaith, B. S., Avérous, L., Njuguna, J., & Nassiopoulos, E. (2011). Cellulose-based bio and nanocomposites: a review. *Int. J. Polym. Sci.* 2011, 1–35.
117. Darder, M., Aranda, P., & Ruiz-Hitzky, E. (2007). Bionanocomposites: A new concept of ecological, bio inspired, and functional hybrid materials. *Adv. Mater.*, 19, 1309–1319.
118. Alexandre, M., & Dubois, P. (2000). Polymer-layered silicate nanocomposites: Preparation, properties and uses of a new class of materials. *Mater Sci. Eng.*, 28, 1–63.
119. Ray, S. S., & Okamoto, M. (2003). Polymer/layered silicate nanocomposites: A review from preparation to processing *Prog Polym. Sci.*, 28, 1539–1641.
120. Zhao, R. X., Torley, P., & Halley, P. J. (2008). Emerging biodegradable materials: Starch-and protein-based bionanocomposites. *J. Mater. Sci.*, 43, 3058–3071.
121. Cabedo, L., Gimenez, E., Lagaron, J. M., Gavara, R., & Saura, J. J. (2004). Development of EVOH/Kaolinite Nanocomposites *Polymer*, 45, 5233–5238.

122. Craig, M. C., & Daniel, F. C. (2005). *Natural fibers* In: Xanthos M, Ed. Functional fillers for plastics. Weinheim: Wiley VCH Verlag, 195–206.
123. Marikarian, J. (2008). Outdoor living space drives growth in wood plastic composites. *Plast Addit. Compd.*, 10, 20–25.
124. O'Donnell, A., Dweib, M. A., Shenton, H. W., & Wool, R. P. (2004). Natural fiber composites with plant oil based resin. *Compos Sci. Technol.*, 64, 1135–1145.
125. Dweib, M. A., Hu, B., O'Donnell, A., Shenton, H. W., & Wool, R. P. (2004). All natural composite sandwich beams for structural applications. *Compos Struct,* 63, 147–157.

INDEX

I

Milton Keynes UK
Ingram Content Group UK Ltd.
UKHW030901141024
449569UK00025B/1276